AA006294

3rd IORA International Conference on Operations Research (IORA-ICOR 2018)

Optimal Decisions for Marine Tourism

IOP Conference Series: Materials Science and Engineering
Volume 567

Manado, Indonesia
20 - 21 September 2018

Editors:

Altien J. Rindengan
Chriestie E.J.C. Montolalu
Nelson Nainggolan
Rinancy Tumilaar
Toni Bakhtiar

ISBN: 978-1-5108-9393-1
ISSN: 1757-8981

Printed from e-media with permission by:

Curran Associates, Inc.
57 Morehouse Lane
Red Hook, NY 12571

Some format issues inherent in the e-media version may also appear in this print version.

Institute of Physics
Dirac House, Temple Back
Bristol BS1 6BE UK

Phone: 44 1 17 929 7481
Fax: 44 1 17 920 0979

techtracking@iop.org

Additional copies of this publication are available from:

Curran Associates, Inc.
57 Morehouse Lane
Red Hook, NY 12571 USA
Phone: 845-758-0400
Fax: 845-758-2633
Email: curran@proceedings.com
Web: www.proceedings.com

IOP Conf. Series: Materials Science and Engineering **567** (2019) 011001 doi:10.1088/1757-899X/567/1/011001

INDONESIAN OPERATIONS RESEARCH ASSOCIATION

The 3rd International Conference on Operations Research

Sam Ratulangi University, Manado, Indonesia, 20 – 21 September 2018

Optimal Decisions for Marine Tourism

IOP Conference Series:
Materials Science and Engineering

Editors:

Altien J. Rindengan, Chriestie E. J. C. Montolalu, Nelson Nainggolan, Rinancy Tumilaar (*Sam Ratulangi University*), Toni Bakhtiar (*IPB University*)

Organized by: Faculty of Mathematics and Natural Sciences, Sam Ratulangi University
Supported by: Indonesian Operations Research Association, Monash University, IPB University, Padjadjaran University, University of Waterloo, Universitas Terbuka, Universiti Brunei Darussalam, University of Sumatra Utara, Atmajaya Catholic University, Pakuan University, Universiti Sultan Zainal Abidin

IOP Conf. Series: Materials Science and Engineering **567** (2019) 011001 doi:10.1088/1757-899X/567/1/011001

INDONESIAN OPERATIONS RESEARCH ASSOCIATION

The 3rd International Conference on Operations Research 2018

Abstract

The Indonesia Operations Research Association International Conference on Operations Research 2018, in short IORA-ICOR2018, is the third conference in the field held by Indonesian Operations Research Association (IORA). This year conference took place at Manado and Minahasa, North Sulawesi, 20-21 September 2019, hosted by Faculty of Mathematics and Natural Sciences, Universitas Sam Ratulangi. The main objectives of the conference are to provide OR/MS community an ideal platform for sharing experiences, communication and networking and to promote and disseminate the scientific field of operations research in Indonesia. The theme of this conference is Optimal Decisions for Marine Tourism.

Date and Venue

Date: 20-21 September 2018, 08.00 – 17.00

Venue: Auditorium of Universitas Sam Ratulangi
Kampus Kleak UNSRAT
Manado 95115, Indonesia

Tetempangan Hills
West Koha Village, Mandolang
Minahasa 95661, Indonesia

In the spirit to promote decisions based analytics through OR/MS, the theme of the conference is

Optimal Decisions for Marine Tourism

The primary objectives of the conference are:

1. to provide OR/MS researchers, academicians and practitioners an appropriate platform for sharing experiences, communication and networking with other experts within the nation and from around the world, and
2. to promote and disseminate the scientific field of operations research in Indonesia through the Indonesian Operations Research Association (IORA).

IOP Conf. Series: Materials Science and Engineering **567** (2019) 011001 doi:10.1088/1757-899X/567/1/011001

INDONESIAN OPERATIONS RESEARCH ASSOCIATION

The 3rd International Conference on Operations Research 2018

From the Conference Chair

Dr. Nelson Nainggolan, *Sam Ratulangi University*

Operation Research as well as Management Science (OR/MS) are viewed as a body of established mathematical models and techniques to solve complex management problems. It encompasses a growing number of technical areas arising in the direction and management of large systems of men, machines, materials and money in industry, business, government, defence, and other decision-making fields.

This conference can be held in collaboration of Indonesian Operations Research Association (IORA) and Faculty of Mathematics and Natural Sciences, Sam Ratulangi University. The same conference had previously been held twice, initially hosted by Universitas Pakuan Bogor, and followed by Universitas Terbuka Tangerang as second host. This conference is now held at Sam Ratulangi University for the third, that why it is known as the 3rd International Conference on Operations Research (ICOR) under a main theme entitled Optimal Decisions for Marine Tourism. The point of this conference is to provide a perfect event for researchers, academics and practicioners of Operations Research, to share experience, build communication and network with experts from all over world.

Thank you!

IOP Conf. Series: Materials Science and Engineering **567** (2019) 011001 doi:10.1088/1757-899X/567/1/011001

INDONESIAN OPERATIONS RESEARCH ASSOCIATION
The 3rd International Conference on Operations Research 2018

From the President
Indonesian Operations Research Association (IORA)
Prof. Sudradjat Supian

Distinguished guests, all invited speakers, participants, ladies and gentlemen.

It is great pleasure for me on behalf of the Indonesian Operations Research Association, I would like to welcome you all at this special event of International Conference on Operations Research at the Universitas Sam Ratulangi, Manado. This event is the third event for IORA-ICOR and congratulations to the Universitas Sam Ratulangi be able to host and also thanks to the Minister of Maritime Affairs and Fisheries and Manado Provincial Governor.

The theme of the conference, *Optimal Decisions for Marine Tourism,* reflects our belief that many future challenges in our life need involvement of operations research and typical analytic operations research. Our future and our capacity to reach sustainable development goal such as ensure availability and sustainable management of water and sanitation for all; ensure access to affordable, reliable, sustainable and modern energy for all; take urgent action to combat climate change and its impacts; end poverty in all its forms everywhere; conserve and sustainably use the oceans, seas and marine resources for sustainable development; end hunger, achieve food security and improved nutrient and promote sustainable agriculture, ensure healthy lives and promote well-being for all at all ages, and other challenges require the advances the roles of operations research in collaboration with other disciplines. Operations Research is the application of scientific & mathematical methods to the study & analysis of problems involving complex systems. Analytics is defined as the scientific process of transforming data into insights for making better decisions.

Operations research is a multidiscipline therefore interaction with other fields of science is indispensable and proven to have given rise to new areas that improve the ability in decision making and techniques used are modeling.

Typically, applications of operations research in these and other areas deal with decisions involved in planning the efficient allocation of scarce resources - such as material, skilled workers, machines, money and time to achieve stated goals and objectives under conditions of uncertainty and over a span of time. Efficient allocation of resources may entail establishing policies, designing processes, or relocating assets. OR analysts solve such management decision problems with an array of mathematical

IOP Conf. Series: Materials Science and Engineering **567** (2019) 011001 doi:10.1088/1757-899X/567/1/011001

methodologies. Completely of the operations research field can be seen in AMS 2000 or MCS 2010.

And now, I need to clarify that IORA is new association in Indonesia he is beginning piloted in workshops in Operations Research and Optimization modelling on June 4, 2011 in the Department of Mathematics Faculty of Mathematics and Natural Science, Universitas Padjadjaran, then be disseminated to several universities, government and industry.

IORA members came from a variety of fields, education, Researcher, government, industry, practitioners etc., in 2017 members numbered 130 members and until now IORA members numbered 240 members, and we wait for those who have become members through www.iora.or.id.

Ladies and gentlemen, we need to inform that at the current conference it will also be held the IORA board meeting.

Finally, have a nice International Conference on Operations Research. I hope there are plenty of benefits we can share and empower through this and hopefully your participations and contributions will make this conference a productive and successful one

IOP Conf. Series: Materials Science and Engineering **567** (2019) 011001 doi:10.1088/1757-899X/567/1/011001

INDONESIAN OPERATIONS RESEARCH ASSOCIATION
The 3rd International Conference on Operations Research 2018

From the Dean
Faculty of Mathematics and Natural Sciences, Sam Ratulangi University
Prof. Benny Pinontoan

Welcome to Universitas Sam Ratulangi, welcome to The 3rd International Conference on Operations Research (ICOR) organized by the Indonesian Operations Research Association (IORA) and the Faculty of Mathematics and Natural Sciences, Universitas Sam Ratulangi.

The theme this year is *Optimal Decisions for Marine Tourism*. This theme is based on the fact that Indonesia is a maritime country as well as tourism is targeted to be the prime mover of the country development and that Operations Research is relevant and essential knowledge to apply to these areas.

We have more than 200 participants registered, which shows a wide range of interests on this topic. Hopefully all papers can be published in the IOP Proceeding indexed in Scopus.

It is a great honor for us to organize this event. We thank IORA for appointed us and for the collaboration which I believe will be a door that open to more collaborations with more partners in the future. We get full support from and therefore we thank the Rector of Universitas Sam Ratulangi and the Governor of North Sulawesi Province.

Finally, we welcome you to Manado, for many of the participants, it's maybe still long destination, but we hope that during your stay you will enjoy and take advantage of the many sights to see in the city and also experience wonderful moments in and on Bunaken Island as well as the many natural wonders in the surrounding areas.

Pakatuan wo Pakalawiden. God bless you.

INDONESIAN OPERATIONS RESEARCH ASSOCIATION

The 3rd International Conference on Operations Research 2018

IOP Conf. Series: Materials Science and Engineering **567** (2019) 011001 doi:10.1088/1757-899X/567/1/011001

INDONESIAN OPERATIONS RESEARCH ASSOCIATION

The 3rd International Conference on Operations Research 2018

Table of Contents

No	Authors	Paper Title
15	Sukono, D Susanti, E S Hasbullah, Y Hidayat, Subiyanto	*Expansion of the investment portfolio performance assessment model based on Value-at-Risk using a time series approach*
16	B Pinontoan, J Titaley, C E J C Montolalu	*Book Embedding of 3-crossing-critical Graphs with Rational Average Degree between 3.5 and 4.*
17	D P W Putra	*On Model of Supply Chain Using Max-Plus Algebra*
18	I Aisah, P R E Jayanti, A K Supriatna	*2-Cut Splicing and 4-Cut Splicing on DNA Molecule*
19	I Aisah, N U Sayyidatunnisa, B Subartini, A Kartiwa	*Vector Space of Codons Sequence over Galois Field GF (7^3)*
20	M I A Prayitno, S Utama, S Aminah	*Properties of Anti-Adjacency Matrix of Directed Cyclic Sun Graph*
21	D A Sumilat, T Oda, D S Wewengkang, M Namikoshi, H Yamazaki	*Inhibition of Interleukin-8 Production in Interleukin-1 Stimulated Human Monocytic THP-1 Cells by N,N Didesmethylgrossularine-1 Obtained from an Ascidian Polycarpa Aurata Collected in North Sulawesi*
22	E O Ogello, S Wullur, A Hagiwara	*Blending Fishwastes and Chicken Manure Extract as Low-Cost and Stable Diet For Mass Culture of Freshwater Zooplankton, Optimized for Aquaculture*
23	I F M Rumengan, L Mandey, G Citraningtiyas, A H Luntungan	*Antihyperglicemic capacity of basil (Ocimum basilicum L.) leaves extracts coated with the marine fish scales derived nanochitosan*
24	N Nurfadillah, S Tarina, E Miswar, I Dewiyanti, S Agustina	*Relationship of Primary Productivity and Phytoplankton Abundance in Muara Kuala Raja, Bireuen District, Aceh*
25	I Dewiyanti, N Nurfadillah, T Setiawati, F Yanti, S A Elrahimi	*Litter Production and Decomposition of Mangrove in The Northern Coast of Aceh Besar District, Aceh Province*
26	K W A Masengi, I F Mandagi, L Manu, F Silooy, I L Labaro, A W R Masengi, N Sebua, E I K G Masengi, B Pinontoan, Y Hutabarat, F Hukom, M Iwata, Y Abe, Y Sato, R Kimura, K Yamahira	*Study on Existence of the Fisheries Resources Abundance By Using Environmental Deoxyribo Nucleic Acid (e-DNA) Approach at Fishing Grounds in the Sulawesi Sea, Indonesia*
27	N Auliyah	*Lenght-Weight Relationship and Condition Factor of Huluu Fish (Gurius margitacea) in Limboto Lake*
28	P A Angmalisang, L Manu, H F Sangian	*Analysis of Chemical Elements of the Rocks on the Coastal Areas in the Eastern Bolaang-Mongondow District, North Sulawesi Indonesia*
29	S Y Noor	*Trophic Status of Limboto Lake in Gorontalo Province*
30	S Wullur, E L Ginting, V Waraow, I F M Rumengan, E Ogello, A Hagiwara	*Growth Responses of Rotifer on a Bacterial-Based Diet Made from Fishwastes*
31	Y Salih, M A Hamoda, Sukono, M Mamat	*The Convergence Properties of New Hybrid Conjugate Gradient Method*

IOP Conf. Series: Materials Science and Engineering **567** (2019) 011001 doi:10.1088/1757-899X/567/1/011001

No	Authors	Paper Title
32	K Priyadi, Chin-Tu Lu, H Sutanto	*Optimization of Impeller Design for Stirred Tank using Computational Fluid Dynamics*
33	A R Nasution, Arviansyah	*Container Terminal Landside Operation Analysis and Discrete Event Simulation in Container Terminal in Port. A Case Study of Terminal 3 Ocean-going PT Pelabuhan Tanjung Priok*
34	J W Jufiardi, F A Pinagara	*Analysis of Supply Chain Management in ABC Poultry using System Dynamics Approach*
35	Farizal, R M Dewi, D S Gabriel	*Supplier Evaluation and Order Allocation using Fuzzy Analytical Hierarchy Process and Augmented Epsilon Constraint Methods*
36	E Hertini, C Harisbaya, J Nahar, Sudradjat	*Queuing model using Sojourn Time Distribution with Single Working Vacation and Vacation Interruption*
37	H N Ahmad, T Bakhtiar, P T Supriyo	*Heuristic approach for solving capacitated location-routing problems*
38	Sukono, E Lesmana, E S Hasbullah, V Rosario, Y Hidayat, Subiyanto	*Determination of optimal insurance company reimbursement by using exponential utility functions*
39	S Maulana, F Hanum, T Bakhtiar	*Mixed integer formulation for multiproduct maritime inventory routing problems*
40	F Fatimah, J J Pelealu, S Gugule, A R H Korompot, V W Monintja, T E Tallei, J A Rorong	*Chemical Characteristics of Tuna Fish Bakasang*
41	G Citraningtyas, W I Wiyono, W A Lolo	*The Potency of Drug Reconciliation Need on Elderly Diabetic Patient at Kandou Hospital Manado*
42	G H Tamuntuan, A Tanauma, G Pasau, H S Sangian	*Grain Size Distribution, Morphology, and Elemental Composition of Iron Sand from North Sulawesi*
43	H I R Mosey, D P Pandara, M D Bobanto, H S Sangian	*A Simple Low-Cost Video-Based Surveillance System for a Flash Flood Warning System*
44	H L Rampe, S D Umboh, R Siahaan, P V Maabuat	*Anatomical Characters of Stomata, Mesophyll and Petiole of Six Varieties Sweet Potatoes (Ipomoea batatas L.) After Organic Fertilizer Induction*
45	J M E Mamahit, J Manueke, V B Montong	*Host Plant and Intensity Pest Attack Paracoccus marginatus on Papaya Plant in North Minahasa Regency*
46	R Setiadi, Franky	*The Profile of Heart Attack Symptoms of Cardiovascular Patients at XYZ Hospital*
47	A Supriatna, E Lesmana, L Aridin, Sukono, H Napitupulu	*Comparison Between Multiplicative Holt Winter and Decomposition Method in Predicting The Number of Incoming International Tourists to Indonesia*
48	C E Mongi, Y A R Langi, C E J C Montolalu, N Nainggolan	*Comparison of hierarchical clustering methods (case study: data on poverty influence in north sulawesi)*
49	E Lesmana, A Supriatna, Riaman	*Prediction of Foreign Tourist Visits to Indonesia using Moving Average and Exponential Smoothing Methods and Acceptance of Admission Analysis*

IOP Conf. Series: Materials Science and Engineering **567** (2019) 011002 doi:10.1088/1757-899X/567/1/011002

Peer review statement

All papers published in this volume of *IOP Conference Series: Materials Science and Engineering* have been peer reviewed through processes administered by the proceedings Editors. Reviews were conducted by expert referees to the professional and scientific standards expected of a proceedings journal published by IOP Publishing.

IOP Conf. Series: Materials Science and Engineering **567** (2019) 012001 doi:10.1088/1757-899X/567/1/012001

A hybrid data mining model for Indonesian telematics SMEs empowerment

E T Tosida[1], F Andria[2], I Wahyudin[1], R Widianto[1], M Ganda[1], R R Lathif[1]

[1]Computer Science Department, Faculty of Mathematic & Natural Science, Universitas Pakuan, Bogor, Indonesia

[2]Management Department, Faculty of Economy, Universitas Pakuan, Bogor, Indonesia

Corresponding author: enengtitatosida@unpak.ac.id

Abstract. The power of information technology and communication (telematics) is one of the vital forces for every country. In the Industrial Revolution 4.0 era, the development of telematics was one of the priorities of the Indonesian government nawacitas. The development of the field of telematics in Indonesia for a decade is inseparable from the role of SMEs. The role of telematics SMEs in the strength of national development can be mapped through the optimization of National Economic Census data (Susenas). The detailed 2016 Susenas data has not been released by BPS. Therefore, this research still uses 2006 Susenas data. The 2016 Susenas recapitalization shows that Indonesian telematics has a very large power, consisting of 2.6 million players. This great strength needs to be optimized to have high competitiveness so as to be able to support Indonesia's development. The purpose of this study was to conduct hybrid data mining modeling to be used as a decision model in mapping the classification of Indonesian telematics SMEs. The classification map includes the feasibility of assistance for the empowerment of Indonesian telematics SMEs, business prospects and development plans for Indonesian telematics SMEs. The hybrid data mining model with K-Medoids & C4.5 technique shows better performance compared to other models, , with an average accuracy rate of 71.87%. This model validation test also involves K-fold cross validations.

1. Introduction

Indonesia has economic power which is supported by SMEs. One of the rapidly developing SMEs is the SMEs in the field of telematics. In the joy of the Indonesian government, the field of telematics is one of the development priorities. This is in line with global economic conditions which are increasingly competing in the era of the 4.0Industrial Revolution (RI). Moreover, this competition was strengthened by the birth of the Asian Economic Community. Since 2015 competition in the Asian Economic Community has become a benchmark for the economic success of countries in Asia.

Data from Indonesian telematics SMEs show considerable strength. The results of the 2006 National Economic Census alone show 12 thousand SMEs strengths in the field of telematics services [1] and have been mapped through web-based applications [2]. The scope of SMEs in Indonesian telematics services is stated in the Indonesian Minister of Industry Regulation No. 16 / M-Ind / PER / 7/2011. The Efforts to empower SMEs Indonesian telematics services have been modeled through data mining clustering and classification approaches [3][4][5].This effort was carried out to identify the strengths and weaknesses of Indonesian SME telematics services. The results of the identification can be used as a reference to focus the target of empowering SMEs, so as to increase competitiveness [6]. The most influential aspect of empowering SMEs in Indonesian telematics services is remuneration [5]. This will be closely related to the level of education of managers and human resources owned by SMEs as a whole [7]. This can be explored through the LVQ approach, and shows the performance of a very good model, with a highaccuracy. This aspect is very relevant to the rapidly changing conditions of telematics technology. Empowerment of human resources is needed to answer the challenges of the competitiveness of Indonesian telematics SMEs. Moreover, it is associated with the challenges of MEA and RI 4.0 which are currently being faced by Indonesia and globally by all countries [8] [9] [10] [11]. Therefore research on the condition of SME telematics is very important, because SMEs telematics is one of the sources of development power in Indonesia. The results of the 2016 National Economic Census (Susenas) for telematics SMEs in Indonesia showed a very significant development. However, the results of the Susenas have not been fully accessible, due to constraints on the prevailing policies. The results of the 2016 Susenas in West Java, the number of telematics SMEs reached 62 thousand units. This condition needs to be studied more deeply so that this enormous strength can be optimized to support the nation's development in Indonesia.

In a previous research on the empowerment of Indonesian telematics SMEs, the data mining approach used was limited to single classifier techniques. This technique is able to be used to describe the important aspects and recommendations for empowering Indonesian telematics SMEs through the scheme of eligibility for assistance [3] [4] [5]. Efforts to build a model of feasibility of assistance for the empowerment of more accurate telematics SMEs are no longer significant [12]. Even though there are still many aspects that can be studied more deeply related to efforts to empower SME telematics, including aspects of development plans and business prospects [13]. Therefore in this study the development of a hybrid mining model is proposed for SMEs in Indonesian telematics services. The proposed hybrid mining model is carried out through two stages, namely clustering and classification. The first stage is clusterin, which is done to classify telematics SMEs that have homogeneous characteristics related to the level of eligibility for assistance. The second stage, namely classification, is intended to identify the plans and prospects of Indonesian telematics SMEs. Thus the purpose of this study is to build a hybrid mining model for Indonesian telematics SMEs in order to empower them to face the competition of AEC and 4.0 RI.

2. Hybrid Mining Model for Management and Empowerment Strategies, and Data Mining Techniques Used in This Part

Empowerment of SMEs can be done through providing assistance. Assistance is provided in a variety of ways, and is tailored to the main needs of the SMES. The process of providing assistance to telematics SMES has a mechanism similar to the process of providing credit assistance at banks. Therefore the basic literature used in this study refers to [12] [14]. Research on the empowerment of SMEs that implement data mining models is still relatively small (only 1%) [14] [15]. The study of the application of hybrid mining models for empowering SMEs is still a potential and novelty for this paper. Special study on the empowerment of Indonesian telematics SMEs on an ongoing basis [1] [2] [3] [4] [5] [13].

The proposed hybrid mining model [12] uses two stages of the mining process. The first stage of the clustering model is used to classify accepted credit proposers and new proposers. The second stage of the classification model (SVM) is used to build a credit scoring model. With this hybrid mining model, rules for determining proposers can be identified that have the potential to be in arrears in credit, so that they can assist management in the process of risk management.

Hybrid mining models also applied into credit scoring [16]. In the research carried out through three mining stages, the first stage performs feature selection and is tested using SVM classification technique, the best feature selection performance is shown by PCA (compared to Genetic Algorithm, information gain ratio and relief. The second stage processes the credit scoring model using artificial neural network (ANN) Adaptive Boosting (There is a boost) with the best FS technique (PCA). The performance of this hybrid mining model shows a model that is performing strongly especially for the credit scoring model.The application of hybrid mining models to the survey data on job career dynamics has produced a potential model to be implemented [17]. The proposed method is a model-based combination and heuristic clustering. The model is capable of forming cluster sequences without losing information about their dynamics. There are two the research phase, first transformed the categorical data sequences using hidden markov models, so that it formed into probabilistic data.The second stage carried out the clustering process using hierarchical clustering.The identification strategy of factors that influence the management of innovation was carried out through the application of hybrid mining models Search Algorithm (HGA) combined with K-NN can eliminate factors that are irrelevant in determining innovation performance decisions of manufacturing companies [18].

The application of hybrid mining models has been carried out [19] by combining M5 decision tree and evolutionary tree for categorical data and regression algorithms for numeric data, showing better model performance. The model has been validated using 12 data set decision problems. Ada also [20] proposed a hybrid mining model using artificial neural network (ANN) and self-organizing map (SOM) to predict turnover rate technology professionals. This can affect the performance of human resources to improve organizational performance.In this study proposed a hybrid mining model for empowerment of Indonesian telematics SMEs services that have mixed data in the form of numeric and categorical. The study was conducted in two stages, and used a combination of clustering techniques (first stage) and classification (second stage). There are 3 hybrid mining scenarios, namely k-means - interative dychotomizer three (ID3), k-medoids - C45 and k-medoids - C5. Based on the three scenarios selected which have the highest level of accuracy and tested its relevance to the real conditions and regulations that occur in Indonesia.

Clustering is one of the unsupervised learning techniques and aims to group data or objects into clusters, so that each cluster will be filled with data as closely as possible. A good clustering method will produce high cluster quality with a high degree of similarity in one class and a low level of similarity between classes. The quality of clustering results is very dependent on the size of the similarity used by the method and its implementation. The quality of the clustering method is also measured by its ability to find some or all hidden patterns [21].

k-means is one of the non-hierarchical data clustering methods that partition data into one or more clusters / groups, so that data that has the same characteristics are grouped in the same cluster and data that has different characteristics are grouped into other groups [21]. K-means algorithm is :

1. Determine the number of clusters
2. Calculating the centroid
3. Calculating distance (distance between data and centroid). Use the euclidean distance formula as follows.

$$D_{ij} = \sqrt{\sum_{k=1}^{p} (X_{ik} - X_{jk})^2} \qquad (1)$$

Information :

D_{ij} = Distance between object i and j
p = Data Dimension
X_{ik} = coordinates of object i in dimension k
X_{jk} = coordinates of object j in dimension k
k = amount of cluster

4. Group by minimum distance
5. If there is a moving object, then do the next iteration again by repeating steps 2 to 4, if it is not there then the calculation is complete.

IOP Conf. Series: Materials Science and Engineering **567** (2019) 012001 doi:10.1088/1757-899X/567/1/012001

The k-medoids algorithm, also known as Partitioning Around Medoids (PAM), is a variant of the k-means algorithm. This is based on the use of medoids rather than the use of means possessed from each cluster, with the aim of reducing the sensitivity of the partition produced in relation to the extreme values present in the dataset [21] [22]. The formula for calculating the distance k-medoids is referred to in equation (1) with Euclidean Distance. PAM algorithms include the method of partitioning clustering to group a set of n objects into a cluster of k clusters. Cluster representation in PAM is an object of a set of objects representing clusters called medoids. Clusters are built by calculating the proximity between the medoid and non-medoid objects using a distance measure. So the method of partitioning can still be done based on the principle of minimizing the number of dissimilarities between each object and medoid that are appropriate and this which forms the basis of the PAM method [21] [22].Whereas to calculate the value of difference (S) is stated by equation 2 as follows:

$$S = new\ total\ cost - long\ total\ cost \qquad (2)$$

Where :

$new\ total\ cost$: Amount of cost / distance of non-medoids
long total cost : total cost / distance of medoids

The process of validating the clustering model can be done through an internal approach, and the Davies Bouldin Index (IDB) is one of them [23]. This validation technique is based on maximizing the inter-cluster distance and minimizing intra-cluster distance. It is denoted by the cluster average (μ_i), the dispersion of points around the cluster average (σ_μ), and the total variant of var(C_i), then the following equations (3) and (4) are obtained:

$$\mu_i = \frac{1}{n_i} \sum_{x_j \epsilon C_i} x_j \qquad (3)$$

$$\sigma_{\mu_i} = \sqrt{\frac{\sum_{x_j \epsilon C_i} \delta(x_j, \mu_i)^2}{n_i}} = \sqrt{var(C_i)} \qquad (4)$$

DB in equation (5) measures the ratio of clusters C_i and $C_j$with the equation:

$$DB_{ij} = \frac{\sigma_{\mu_i} + \sigma_{\mu_j}}{\delta\left(\sigma_{\mu_i}, \sigma_{\mu_j}\right)} \qquad (5)$$

DB_{ij} measure the compactness of clusters compared to the distance between cluster averages, so IDB is defined by the equation (6):

$$DB = \frac{1}{k} \sum_{i=1}^{k} max\{DB_{ij}\} \qquad (6)$$

The smallest DB value is the best clustering result. Between clusters are well separated (high cluster average distance), and each cluster is well represented based on its average value (low distribution value).

Clasification is one method that is used as a method of supporting decision making by looking at the behavior and attributes of defined groups. This technique can provide classification of new data by manipulating existing data that has been classified and by using the results to provide a number of rules [21].

ID3 algorithm uses the concept of information entropy. This algorithm conducts greedy searches on all possible decision trees. Algorithms in this method use the concept of entropy. The Entropy concept is used to measure "how informative" a node (which is usually called how good). Entropy (S) = 0, if all instances in S are in the same class. Anthropy (S) = 1, if the number of samples is positive and the number of negative samples in S is equal. 0 <Entropy (S) <1, if the number of examples positive and negative in S are not the same.

C4.5 algorithm is a group of Decision Tree algorithms. This algorithm has input in the form of Samples and Samples Training. Samples Training is sample data that will be used to build a tree that has been tested for its truth. Whereas Samples is data fields that will later be used as parameters in classifying data [22]. The calculation formula of the C4.5 algorithm is referenced in equations 7 and 8.

$$Entropy(S) = \sum_{i=1}^{n} -p_i \log_2 p_i \ldots\ldots (7)$$

Information :
S = set of case
n = amount of S partition
p_i = proportion from the S_i to S

$$Gain\ (S,A) = Entropy(S) - \sum_{i=1}^{n} - \frac{|S_i|}{|S|} Entropy\ (S_i) \ldots\ldots (8)$$

Information :
A = attribute
n = number of partition attributes of A
$|S_i|$ = number of cases on the i-partition
$|S|$ = number of cases in S.

C5.0 algorithm is one of the data mining algorithms that are specifically applied to the decision tree technique. C5.0 is an improvement of the previous algorithm formed by Ross Quinlan in 1987, namely ID3 and C4.5. In this algorithm the selection of attributes will be processed using information gain. In selecting attributes for object breakers in several classes attributes must be chosen which produce the greatest information gain. Attributes with the highest information gain value will be selected as parent for the next node [21].

3. Methods

This research is based on the initial research in the form of a map-based Information System of SME telematics services [2] and factor analysis that influence the provision of assistance through the Multi Attribute Decision Making (MADM) [1][13]. Pre-process data also refers to previous research using single classifier [4] [5] and single clustering [3]. The data used is Susenas 2006 based, because until now the 2016 Susenas data has not been accessed in detail.

The hybrid mining model is proposed to delve deeper into the conditions of Indonesian telematics SMEs, through the determination of the factors presented in the Susenas results. The Susenas results provide data on the condition of SMEs, including HR conditions, basic technology, remuneration, training, business difficulties, partnerships, prospects, and business development plans [24][25][26] [27]. The focus of this study is aimed at predicting prospects and business development plans that are currently not widely discussed by other studies. This factor can be used as a reference for the second stage of the empowerment process after SMEs telematics services are grouped through the clustering process. The establishment of hybrid mining models is carried out using Rapidminer. The overall stages of research are presented in Figure 1.

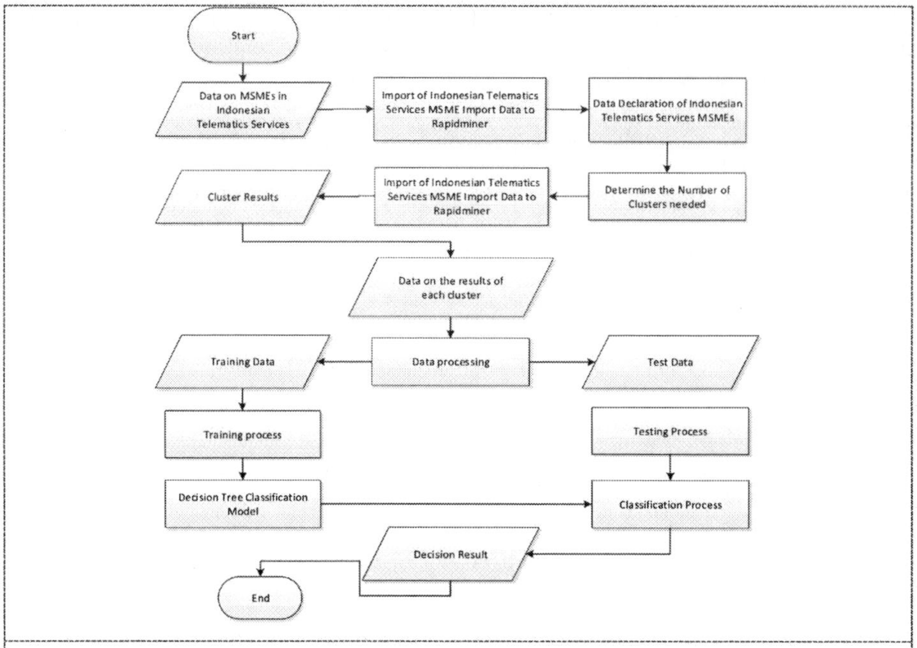

Figure 1. Research stage

The stages of the research were carried out for all scenarios. The scenario that produces the best hybrid mining model performance is then implemented in HTML, PHP, and My SQL programming as the database.

4. Result and Discussion

The characteristics of SMEs in Indonesian telematics services in 2016 did show significant changes but were still dominated by internet cafe businesses. Because the detailed data of the 2016 Susenas results are not yet accessible, this paper is still based on the 2006 Susenas results. The basic characteristics of the 2006 and 2016 Susenas data patterns still have similarities, only the dominance of SMEs in Indonesian telematics services shifts from kiosks to internet cafes. This condition shows that the growth of SMEs in Indonesian telematics services does not yet have a strong foundation for the basic framework of RI 4.0 [7] [8]. However, from the results of the mining hybrid model this time it can be shown the condition of telematics service SMEs that have business prospects and development plans. That description became the novelty of this research.

The results of the clustering process as the first stage of the formation of hybrid mining models indicate that the number of clusters that are more optimal for testing this model are 5 clusters. This is also in line with the results of the study [3]. The use of a number of 5 clusters is also proven by the results of the validation test using IDB [23] for all three scenarios. The IDB level trial was conducted on three alternative number of clusters, shown in Table 1.

Table 1. Test the IDB validity of the clustering process

Amount of cluster	IDB value		
	k-means – ID3	k-medoids – C45	k-medoids – C5
3	3.397	2.084	2.074
5	3.262	2.060	2.050
7	3.448	2.075	2.083

This clustering process generates groups of SME telematics services that are clustered based on the similarity of their characters. The intended character shows the level of feasibility to be given assistance [25][26]. The clustering results are input to be processed in the second stage, namely the classification process. The classification process is used to see the prospects and plans for the development of Indonesian telematics services SMEs that have not been widely studied so far in previous studies.

The process of forming the classification learning machine this time was tested using the k-fold cross validation scenario and without k-fold cross validation. The scenario of the distribution of training data and test data is very important to see the opportunities for even distribution of data in each scenario performed [19]. Moreover, the data set used is data inbalance, which is indicated to affect the performance of the hybrid mining model. The k-fold cross validation scheme used is 5-fold cross validation, this is adjusted also by the number of clusters used. The performance of hybrid mining models is shown in Table 2.

Table 2. Comparison of the accuracy of hybrid mining models

Cluster	Accuration level (%)					
	k-means – ID3		k-medoids – C45		k-medoids – C5	
	a	b	A	B	a	b
1	36.45	91.90	58.36	68.42	44.87	49.21
2	35.03	90.99	59.41	68.55	48.18	51.87
3	33.72	97.58	59.78	80.59	0.08	94.24
4	35.12	93.42	59.20	73.68	46.42	48.99
5	35.53	94.96	59.00	68.11	59.36	70.48

[a]k-fold cross validation
[b]no k-fold cross validation

Based on Table 2. It shows that the 5-fold cross validation scheme has a significant effect on the performance of the hybrid mining model. All scenarios show that the accuracy of the hybrid mining model has a lower level of accuracy when the 5-fold cross validation is applied. K-means - ID3 hybrid mining model shows a very significant difference in performance after being tested 5-fold cross validation. The decline in model performance in the first scenario averaged 58.6%. This condition shows the model's instability, and has a high sensitivity to changes in the distribution of training data and test data. In the second scenario (k-medoids - C45) showed almost the same condition, but the decline in performance was relatively lower (12.72%). The second scenario provides better model performance, and this is also in line with research [12] [19]. The third scenario (k-medoids - C5) gives an unexpected picture, because the 5-fold cross validation test results provide better performance. However, there are indications that the distribution of clustering results results in an imbalance resulting in the performance of the third scenario being more unstable. This is also indicated by the performance of the model which only achieved an average accuracy rate of 39.78%. The decline in the performance of the hybrid mining model for the third scenario also showed a significant difference to reach 23.17%.

Analysis of the hybrid mining model is then carried out for the selected model, the model formed from the second scenario. Based on the results of the implementation of the hybrid mining model, the conditions for the prospects of SMEs in Indonesian telematics services were obtained as shown in Figure 2.

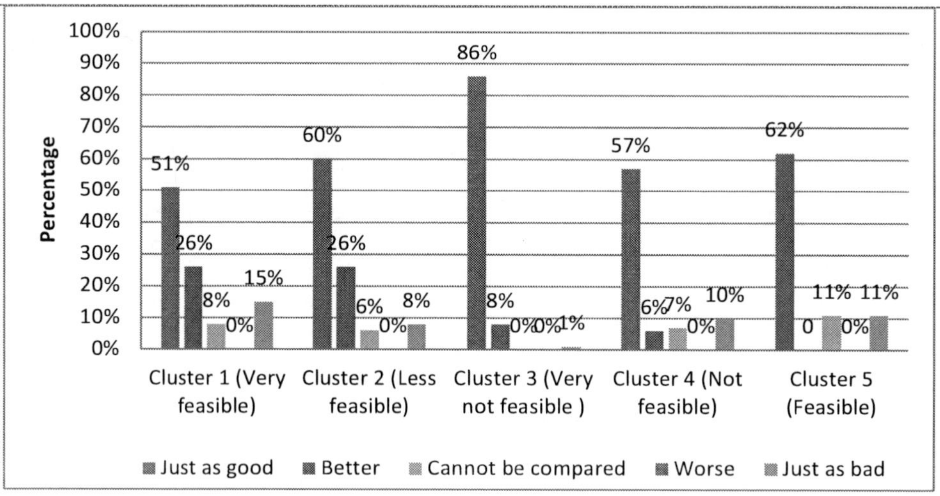

Figure 2. Prospect prediction of Indonesian telematic SMEs

Prediction of the prospect of SMEs in Indonesian telematics services based on the results of the ybrid mining model capable of describing conditions relevant to government policies related to the provision of assistance and empowerment of SMEs [24][25][27]. In cluster 1, it is a group of SMEs that are very feasible to be given assistance, there are 26% of SMEs that have better business prospects. This can be used as a reference for the government or even those who will provide assistance so that this group is the priority of the beneficiaries. In this study, the types of assistance that are really needed by SMEs have not been identified. But the latest findings related to the focus and priority groups that will be given assistance are seen from the business prospect factors, which have not been able to be discussed in previous studies [3][4][5].

The description of other conditions that can be extracted from the results of implementing this hybrid mining model is being able to predict business development plans for SMEs in Indonesian telematics services. This condition is detailed in Figure 3. The business development plan is an important factor for evaluating the feasibility of providing assistance to SMEs, so that the assistance provided is truly a generator of empowerment to increase the competitiveness of SMEs. The assistance provided is not only transactional and grants, but must be able to provoke and generate SMEs to improve the quality of their business, so they can be highly competitive. Moreover, the field of telematics as one of the priority areas for MEA competition and facing IR 4.0 is very relevant if the strength of its SMEs needs to be improved. The ability to develop a business development plan is certainly very influenced by the creativity of human resources owned by SMEs. This cycle plays an important role in empowering SMEs, so it is expected that Indonesian telematics SMEs are able to make a real contribution in the development of human resources in particular, even the economic development of Indonesia in general [6][7][13][24][27].

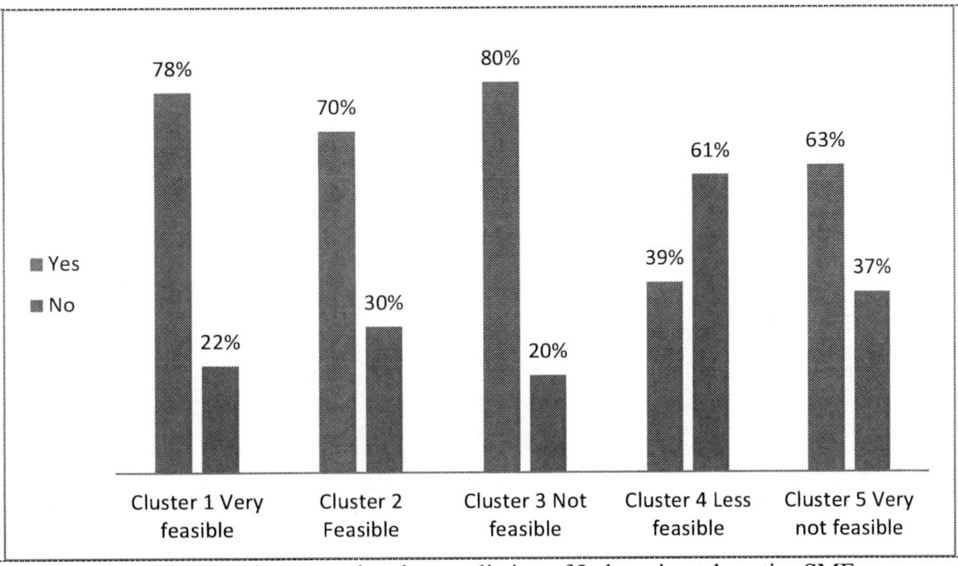

Figure 3. Development planning prediction of Indonesian telematics SMEs

In Figure 3 shows that the focus of providing assistance can be assigned to cluster 1 and cluster 2 which are predicted to have business development plans. The types of business development plans of each SMES are certainly different. This finding is also one of the novelty, because in previous studies [3] [4] [5] have not been able to dig up information to this information. The shortcomings of this system have not been able to show in detail the types of business development plans of the SMES in question.

This hybrid mining model is implemented on a user friendly interface system. This is intended to facilitate stakeholders who will use this system. The process of providing assistance and empowerment of Indonesian telematics services SMEs can be done easily, accurately and is able to provide more detailed recommendations regarding the priorities of beneficiaries (especially those related to prospects and business development plans of each assessed SMES). This system interface is shown in Figure 4.

Figure 4. User Interface of Decision Support System of Indonesian Telematics SMEs

The user interface for loading data is shown in Figure 5. The process of loading data can be input independently directly using this feature, (specifically for small amounts of data, suppose data only amounts to under 10 data). If the data to be loaded is grouped data, the user can use the "import data" facility so that the loading process is more effective and efficient. This feature has also not been developed in the previous system [5].

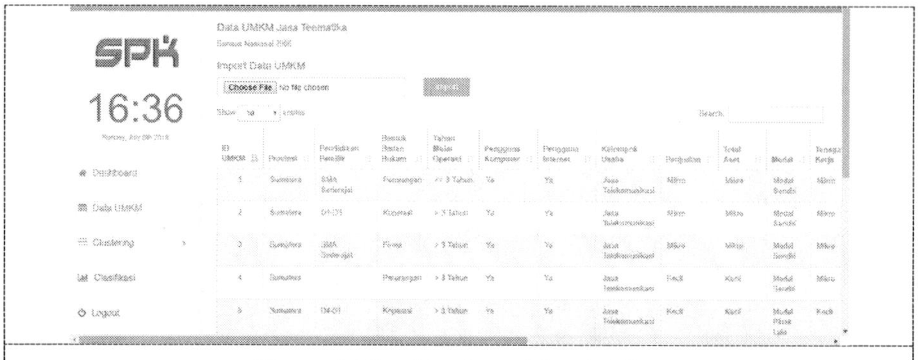

Figure 5. User interface of data loading

In this system the process of supporting decisions using hybrid mining models is carried out simultaneously. The clustering process (using the k-medoids technique) in this system produces SME clusters grouped based on the level of eligibility for assistance. The next process of the system will immediately display the prioritized clusters to predict prospects and business development plans using the C45 classification technique. The use of simple clustering and classification techniques is able to provide an accuracy value of 71.87%.

One disadvantage of this system is the process of predicting prospects and business development plans for SMEs that are assessed, still carried out in parallel (each one data inputted). The input data will be read by the learning machine, then stored by the system and will improve the next hybrid mining model that is formed. This developed system has not been able to generate predictive processes in groups, because the hybrid mining model implemented is still carried out separately using rapidminer. The lack of this system can be used as input for the construction of subsequent research.

5. Conclusion

The empowerment of Indonesian telematics SMEs will have a positive impact on increasing the nation's competitiveness, especially in the face of MEA and IR 4.0 competition. Indonesia has enormous strength because it is supported by a large number of telematics SMEs. The strength of this SMEs has proven to be a pillar of the Indonesian economy in the face of the monetary crisis. Therefore the government and other stakeholders need to have a good scheme in the process of empowering SMEs, especially SME telematics. The results of this study provide an alternative process for assessing the feasibility of assistance for SMEs in Indonesian telematics services using hybrid mining models.

The hybrid mining model that is implemented into decision support systems is the best model of the three scenarios analyzed. The use of k-medoids and C45 techniques which are simple data mining techniques are still able to produce model performance with an accuracy rate of 71.87%. The system is built with a user friendly interface, making it easier for stakeholders to use it. The lack of this system has not been able to show more specific business development (which refers to the development of certain fields). Another disadvantage of this study is that the model formation process is still separate using rapidminer, so the prospect prediction process and business development plan are still carried out in stages. The system has not been able to generate predictive processes for large groups of data, because the process of synchronizing new data into the renewal of hybrid mining models is still simultaneous. The results of this study have been able to provide novelty which is able to provide priority information for SME groups of telematics services that deserve to be given assistance by showing predictions of prospects and business development plans. Thus the focus of empowerment will be more targeted and the nature of the provision of assistance will be mutually reinforcing in line with the prediction of prospects and business development plans owned by the SMEs.

Acknowledgments

1. DRPM RistekDikti, as the main sponsor, which gives us Competitive Grants Scheme
2. Computer Science Department, Mathematics and Natural Science Faculty, Pakuan University,and Research Institute Pakuan University, for supporting, coordinating and facilitating to achieve this grants.
3. Agro Industrial Engineering IPB and Computer Science IPB who have became the companion research team in this research grant scheme
4. Vocational Schools (200 schools) in Jabodetabek, BPS, ICT Associations and any stakeholdersfor active participation in the activities of interviews and user requirement

References

[1] Tosida, ET, Seminar KB, Herdiyeni Y. 2016. Atribut selection of Indonesian Telematic Services MSMEs Assisstance Feasibility, using AHP. Kursor 8 (2), 2016. DOI: 10.21107/kursor.v8i2. 1299.

[2] Tosida ET, Maryana S, Thaheer H, Damin FA. 2015. Visualization model of Small Medium Enterprises (SMEs) telematics services potentiality map in Indonesia. International Conference on Information & Communication Technology and Systems (ICTS), 2015. DOI: 10.1109/ICTS.2015.7379890.

[3] Tosida ET, Maryana S, Thaheer H. 2017a. Implementation of Self Organizing Map (SOM) as decision support: Indonesian telematics services MSMEs empowerment. IOP Conference Series: Materials Science and Engineering 166 (1), 012017.

[4] Tosida ET, Hairlangga O, Amirudin F, Ridwanah M. 2018a. Appication of Decision Rules for Empowering of Indonesian Telematics Services SMEs. IOP Conference Series: Materials Science and Engineering 332 (1), 012018.

[5] Tosida ET, FD Wihartiko, I Lumessa. 2018. Learning Vector Quantization Implementation to Predict The Provision of Assisstance for Indonesian Telematics Services SMEs. International Journal of Engineering & Technology, Vol. 7 No 3. 20

[6] Jaswadi, Iqbal M, Sumiadji. 2015. SME Governance in Indonesia – A Survey and Insight from Private. Procedia Economics and Finance 31 (2015) 387 – 398.

[7] Andra Benesova, Jiri Tupa. 2017. Requirements for Education and Qualification of People in Industry4.0. Procedia Manufacturing 11 (2017) 2; 27th International Conference on Flexible Automation and Intelligent Manufacturing, FAIM2017, 27-30 June 2017, Modena, Italy. doi: 10.1016/j.promfg.2017.07.366

[8] Jian Qin, Ying Liu, Roger Grosvenor. 2016. A Categorical Framework of Manufacturing for Industry 4.0 and Beyond. Procedia CIRP 52 (2016) 173-178. Elsevier. doi: 10.1016/j.procir.2016.08.005

[9] Sommer L. 2015. Industrial Revolution - Industry 4.0: Are German Manufacturing SMEs the First Victims of this Revolution?. Journal of Industrial Engineering and Management JIEM, 2015 – 8(5): 1512-1532. DOI: 10.3926/jiem.1470.

[10] Jones C, Pimdee P. 2017. Innovative Ideas: Thailand 4.0 and the fourth industrial revolution. Asian International Journal of Social Sciences, 17(1), 4–35. DOI: 10.29139/aijss.20170101.

[11] Pereira T, L Barreto, A Amaral. 2017. Network and Information Security Challenges within Industry 4.0. Paradigm. 2017. Procedia Manufacturing 13 (2017) 1253-1260.

[12] Chen W, Xiang G, Liu Y, Wang K. 2012. Credit risk Evaluation by hybrid data mining technique. System Engineering Proceedia 3 (2012), 194-200. DOI: 10.1016/j.sepro.2011.10.029.

[13] Tosida ET, H Thaheer, S Maryana. 2017. Strategi Peningkatan Daya Saing melalui Framework Rantai Nilai untuk Kompetensi Usaha JasaTelematika Indonesia. Jurnal Penelitian Pos dan Informatika, 5 (1), 1-18.

[14] Sadatrasoul, SM, GholamianMR, Siami M, HajimaohammadiZ. 2013. Credit scoring in banks and financial institutions via data mining techniques: A literatur review. *Journal of AI and Data Mining*. Vo. 1., No. 2, 2013, 119-129.

[15] Zhu, Y., C. Xie, Gang-Jin Wang, Xin-Guo Yan. 2016. Comparison of individual, ensemble and integrated machine learning methods to predict China's SME credit risk in supply chain finance. Neural Comput & Applie. The Natural Computing Applications Forum 2016. CorssMark, Springer. DOI 10.1007/s00521-016-2304-x.

[16] Koutanaei FN, H Sajedi, M Khanbabaei. 2015. A hybrid data mining model of feature selection algorithms and ensemble learning classifiers for credit scoring. Journal of Retailing and Consumer Services, Vol. 27, November 2015, 11-23. DOI: 10.1016/j.jretconser. 2015.07.003.

[17] LucaDe Angelis,José G.Dias 2014. Stochastics and Statistics Mining categorical sequences from data using a hybrid clustering method. European Journal of Operational Research, Vol. 234, Issue 3, 1 May 2014, 720-730. DOI: 10.1016/j.ejor.2013.11.002.

[18] Eroglu DY, Kilic K. 2017. A novel Hybrid Genetic Local Search Algorithm for feature selection and weighting with an application in strategic decision making in innovation management. Information Sciences, Vol. 405, 18-32. DOI: 10.1016/j.ins.2017.04.009.

[19] Kim K, Hong J. 2017. A hybrid decision tree algorithm for mixed numeric and categorical data in regression analysis. Pattern Recognition Letters, Vol. 98, 39-45. DOI: 10.1016/j.patrec. 2017.08.011.

[20] Fan CY, Fan PS, Chan TY, SH Chang . 2012. Using hybrid data mining and machine learning clustering analysis to predict the turnover rate for technology professionals. Expert Systems with Applications, Vol. 39, Issue 10, 8844-8851. DOI: 10. 1016/j.eswa.2012.02.005.

[21] Han, J., Kamber M, Pei J. 2012. Data mining: Concepts and techniques. Third Edition. Morgan Kaufmann, Waltham, MA 02451, USA.

[22] Mishra, D., Hiranwal, S. 2014. Analysis & implementation of item basedcollaboration filtering using K-Medoid. *Paper presented at 2014 International Conference on the Advances in Engineering and Technology Research (ICAETR).*

[23] Zaki MJ, Wagner M Jr. 2013. Data Mining and Analysis: Fundamental Concepts and Algorithms. Troy, Belo Horizonte.

[24] Palacios-Marqués, D., Soto-Acosta, P., Merigó, J.M. 2015. Analyzing the effects of technological, organizational and competition factors on Web knowledge exchange in SMEs. J. Telematics and Informatics 32(1): 23–32. DOI: 10.1016/j.tele.2014.08.003.0736-5853/@2014.

[25] Hadighi SA, Sahebjamnia N, Mahdavi I, Shirazi MA. 2013. A framework for strategy formulation based on clustering approach: A case study in a corporate organization. *Knowledge-Base Systems*, 49 (2013), 37-49. DOI: 10.1016/j.knosys.2013.04.008.

[26] Lee H, Lee S, Byungun Y. 2011. Technology clustering based on evolutionary patterns : The case of information and communications technologies. *J. Technological Forecasting & Social Change*, 78 (2011) 953-967. DOI : 10.1016/j.techfore.2011.02.002.

[27] McGuirk H, Lenihan H, Hart M. 2015. Measuring the impact of innovation human capital on small firms propensity to innovative. *J. Research Policy,* Vol. 44 (4): 965-976. DOI: 10.1016/j.respol.2014.11.008.

IOP Conf. Series: Materials Science and Engineering **567** (2019) 012002 doi:10.1088/1757-899X/567/1/012002

Development of data reconciliation system for billing process at PT Perusahaan Gas Negara

R Hadianti[1], A Ihsan[2], A Mujiyanto[3], S. Uttunggadewa[4]

[1]Industrial and Financial Mathematics Research Group, Faculty of Mathematics and Natural Sciences, Institut Teknologi Bandung, Jalan Ganesha no. 10 Bandung 40132, Indonesia
[2]PT Scada Prima Cipta, Bandung, Indonesia
[3]Energy Management Division, PT. Perusahaan Gas Negara, Jakarta, Indonesia
[4]Combinatorial Mathematics Research Group, Faculty of Mathematics and Natural Sciences, Institut Teknologi Bandung, Jalan Ganesha no. 10 Bandung 40132, Indonesia

Corresponding author: hadianti@math.itb.ac.id

Abstract. We study a problem faced by PT. Perusahaan Gas Negara (abbreviated as PGN), the Indonesian company in the natural gas transportation and distribution segment, in developing data reconciliation system for billing process. PGN serves its residential, commercial and industrial customers' natural gas need through transmission pipelines. We focus our study on the billing process for industrial customers, where gas consumptions are measured by combining measurement results from turbine meter, temperature measurement device and pressure measurement device. Normally, customers' bills can be obtained directly from volume metered by volume measurement device. But in every month there are a number of anomaly behaviour in customers' gas consumption data or reports on meter failure, although statistically there are less than 1% of meter failure or anomaly behaviour from all customers' data. Due to the large number of on-site meters which located in very wide area in Sumatera and Java islands, sometimes defective meter problems are already resolved after several days since the problem was reported/indicated, and this can generate loss for PGN. In this paper, we perform analysis to answer: 1. how to determine the beginning of the period where a volume meter is fail, 2. how to estimate gas volume delivered to the customers during the meter failure. The first problem is solved by analysing historical volume, temperature, and pressure data of each customer. The analysis results, in line with Spitzglass (Energy) Equation, that gas flow is linear to the pressure drop between the pressure at distribution pipe and the pressure at the customer's pipe. We derive a procedure for detecting anomaly in gas volume data based on this phenomenon, where we reconcile the volume data and the pressure data and categorize the reconciled data into normal type or anomaly type data. The period during the meter failure is then determined as the period where the data in this period are of anomaly type. As soon as the period during the meter failure is determined, we then need to estimate the gas volume delivered during this period. We apply some forecasting methods for time series data such as AR(p) and Holt Winters models, and we choose the estimation from the forecasting model with the smallest error. The solution of those problems will be used for calculating bills for those customers which experience meter defective in the billing month. We develop a graphical user interface based on the anomaly detection method and the forecasting method. This system has been implemented, tested, and used by Energy Management Division of PGN, and will be developed into an energy management system.

1. Introduction

PT. Perusahaan Gas Negara (abbreviated as PGN) is the largest national company in the natural gas transportation and distribution segment. This company has pipeline networks in North Sumatera, Southern part of Sumatera, Riau Islands, and Java Sea to connect the natural gas sources location with the natural gas end user location through the transmission pipelines mode. PGN plays a significant role in meeting domestic natural gas needs, serves natural gas to 196,221 residential, commercial and industrial customers in Indonesia's main population centers (in the end of 2017). For the past 50 years, PGN has strengthened its energy business by investing in upstream oil and gas, mid-stream transmission and downstream distribution. Today, PGN has transformed from a natural gas transmission and distribution company into a provider of integrated energy solutions. By integrating the assets and expertise bound within its subsidiaries, PGN is unlocking new synergies across its enterprise, enabling it to develop effective solutions for customers' specific energy needs.

In the competitive market situation, PGN needs to focus on value-added services and service options that can differentiate PGN from competitors. The customer measurement system is now fully digitized, ensuring greater efficiency and transparency. Customers can also access their accounts, support services, and other products through PGN mobile applications [1]. All innovations are strived in order to enhance current PGN level of service.

An effort in ensuring greater efficiency, accuracy and transparency is performed by developing data reconciliation system for billing process. Industrial customers' gas consumptions are measured by combining measurement results from turbine meter, temperature measurement device and pressure measurement device. All measurement results are converted to base condition with Electronic Volume Converter. Normally, customers' bills can be obtained directly from volume metered by volume measurement device. But this device, also the other two devices as well, can be defect although happen rarely. Statistically, in every month there are only less than 1% anomaly behaviour in customers' gas consumption data or reports on meter failure from all customers' data. But if it happens to the volume meters of industrial customers which consume large volume of gas, the unmetered consumption during the defective meter problem unresolved may lead to a massive revenue loss for PGN. Due to the large number of on-site meters which located in very wide area in Sumatera and Java islands, sometimes defective meter problems are already resolved after several days since the problem was reported/indicated. No one knew when the meter start to failure so that it is difficult to estimate gas volume delivered to the customers during the meter failure. Briefly, there are two data validation problems to be solved due to meter failure: 1. how to determine the beginning of the period of meter failure, 2. how to estimate gas volume delivered to the customers during the meter failure. These problems are common problems in pipeline network, so that we may find many approachs to solve the problems [2][3][4].

The solutions of the two problems will be used for calculating bills for those customers which experience meter defective in the billing month. Without any scientific methods for solving the problems, PGN is always difficult to convince customers the gas volume receipt by customers. A collaborative research[1] is then performed to solve the problems, and to develop an application based on these data reconciliation.

The first problem is solved by analysing historical volume, temperature, and pressure data of each customer. The analysis results, in line with the Spitzglass (Energy) Equation, that gas flow is linear to the pressure drop between the pressure at distribution pipe and the pressure at the customer's pipe. We derive a procedure for detecting anomaly in gas volume data based on this phenomenon, where we reconcile the volume data and the pressure data and categorize the reconciled data into normal type or

[1] Among PGN, PT. Scada Prima Cipta and Research Consortium Optimization on Pipeline Network (OPPINET) P2MS ITB

anomaly type data. The period during the meter failure is then determined as the period where the data in this period are of anomaly type. We perform some validations to historical data, and we can say that our procedure work effectively in determining the period of anomaly data which indicate the period during the meter failure. As soon as the period during the meter failure is determined, we then need to estimate the gas volume delivered during this period. The estimation is performed by first analysing the trend, seasonal and cyclical behaviour of the customers' historical consumption data. Most of the data show strong seasonal behaviour so that we try to estimate the gas volume by applying Exponential Smoothing Forecasting Methods such as Holt Winters model [5]. Estimation on the data without non-seasonal behaviour is performed by applying Auto Regressive Forecasting Model (p) for various values of p. We estimate the gas volume based on the forecasting model with the smallest forecast error.

For practical purpose, we develop a graphical user interface based on the anomaly detection method and the forecasting method described above. This system has been implemented, tested, and used by Energy Management Division of PGN. Up to now, this system has been helping PGN in monthly billing process. In the future, this system will be developed into an energy management system. PGN needs to know consumption behaviour of each PGN customers' sector such as manufacture sector, energy sector, chemical sector, etc. By knowing the consumption behaviour and the economic development prediction for each sector, PGN can decide the optimal plan for future natural gas supply for each sector. Moreover, by the system PGN can offer dynamic pricing to the customers. Dynamic pricing is reported can reduce unutilized supply and can give an enormous potential cost saving both for PGN and the customers [6][7][8].

This paper is written as follows. After this introduction, in section 2 we discuss the historical data and its analysis. The result of the data analysis lead us to a derivation of anomaly detection procedure and the forecast method we have to choose for forecasting gas consumption. In section 3 we then discuss how it results when we implement the anomaly detection procedure and the forecast method in real case billing process. This paper is ended by a summary and discussions that we present in section 4.

2. Data analysis, anomaly detection, and forecasting

The PGN data reconciliation system is constructed by first performing three mathematical steps, i.e. data analysis, anomaly detection and estimation/forecasting. This section gives brief explanations for those three steps.

2.1. Data analysis

We analysed the trend, seasonal and cyclical behaviour of the industrial customers' historical consumption data. Most of the data have strong cyclical or intermittent behaviour as shown in the following example.

Figure 1 and 2, in the following, show an example of hourly consumption volume and pressure of one customer of PGN. In longer period of time the data show that this customer consumption has a cyclical or intermittent behaviour with weekly period, as can be seen on figure 1 from the 968th data to 1135th data. We can say that every cycle is started by a period of time with zero volume and followed by a period of time with non-zero volume. For brevity, we call the period of time with zero volume by **Zero Volume Period**, and the period of time with non-zero volume by **Non-Zero Volume Period.**

Due to a customer's production system shut down, the gas flow was blocked as can be seen on figure 1 from the 285th data to 529th data. If we refer to the pressure data, we can see that in the same period the pressure are as high as the pressure in other periods of time with zero volume. This bring us to a conclusion that there is a strong correlation between the behaviour of volume data and the pressure data. This conclusion scientifically can be explained by the Spitzglass energy equation, which described in the following.

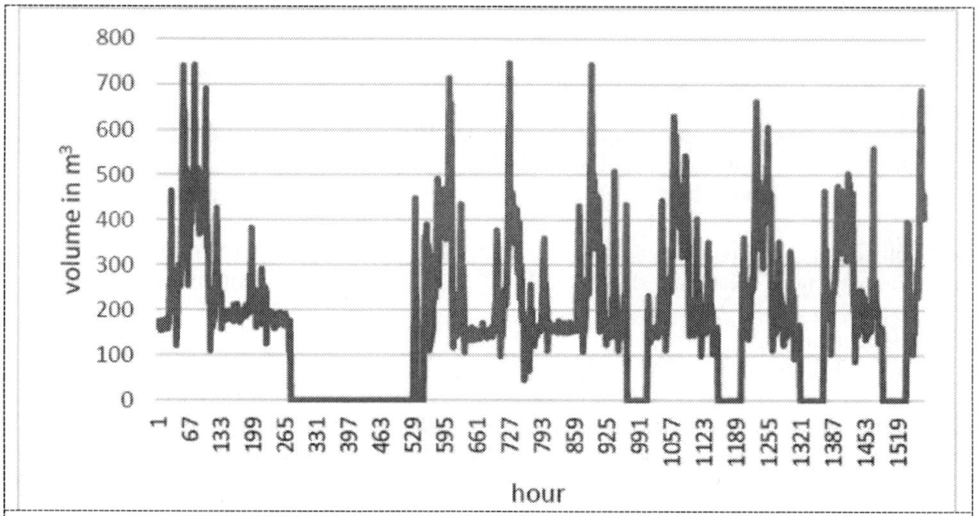

Figure 1. Time series data of natural gas hourly consumption volume of customer X

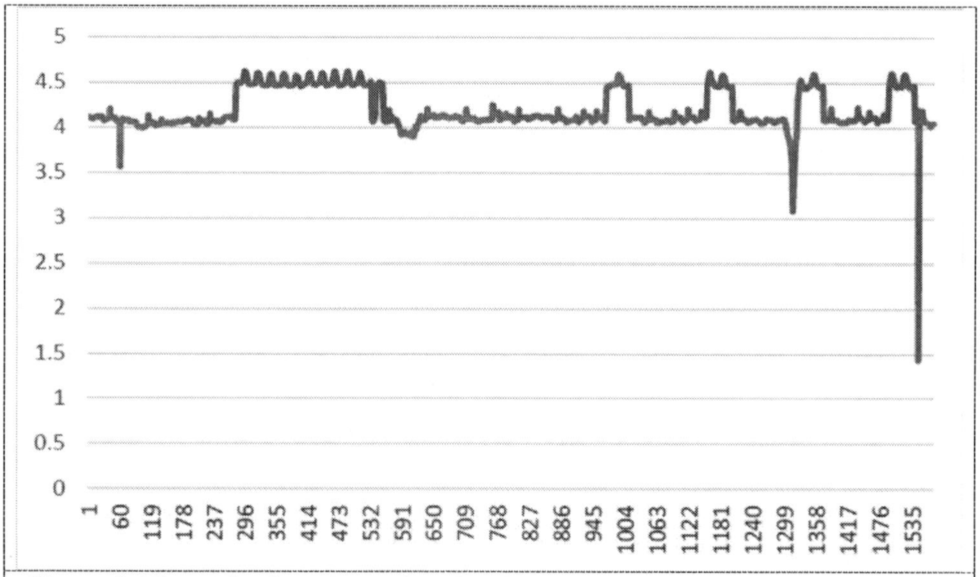

Figure 2. Time series data of natural gas hourly consumption volume of customer X

For gas flowing in a pipeline between two points (A and B), there are a number of energy equations which relate the properties of the gas to the flow rate, pipe diameter and length, the temperature and the pressure in the pipe. One of the energy equation is the Spitzglass equation, i.e.

$$Q = c_1 c_2 E \left(\frac{T_b}{P_b}\right)\left(\frac{P_A - P_B}{G T_f L_e Z\left(1+\frac{c_3}{D}+c_4 D\right)}\right)^{0.5} D^{2.5},$$

where

Q	: is the gas flow rate	T_f	: is the average gas flowing temperature
E	: is the pipeline efficiency,	L_e	: is the pipe segment length between point A and point B
T_b	: is the base temperature,	Z	: is the gas compressibility factor at the flowing temperature,
P_b	: is the base pressure,	D	: is the pipeline inside diameter,
P_A	: is the gas pressure at point A,	c_1, c_2, c_3, c_4	: are some constants (see (**Error! Reference source not found.**).
P_B	: is the gas pressure at point B,		
G	: is the gas gravity,		

In the condition where Q goes to zero, from the Spitzglass equation we will have $P_A - P_B$ goes to zero since the remaining parameter in the right-hand side of the equation are constants. If A and B are the starting point and the end point of the segment from the distribution pipeline to the customer's location, respectively, then when there is no flow to the customer the pressure at a customer's meter is almost the same as the pressure at the distribution pipe. So, normally during Zero Volume Period the volume is around zero and the pressure is high (around the pressure at the distribution pipeline). During Non-Zero Volume Period, if the volume is far from zero then the pressure is also far from the high-pressure band.

The problem we study in this paper is motivated by our finding on the other customers' data where we found certain long periods of time, even longer than one cycle, where the hourly volumes are zero but the pressure is far from high-pressure band. This phenomenon breaks the Spitzglass equation, and we have to categorize it as an anomaly. This anomaly behaviour indicates volume meter failure in anomaly period. Our discussion on the anomaly behaviour can be summarized in the following table.

Table 1. Normal and anomaly conditions

Data behavior	Volume and pressure conditions
Normal	1. Volumes tends to zero & pressures are in the high-pressure band, or
	2. Volumes are far from zero & pressures are far from the high-pressure band.
Anomaly	Volumes tends to zero & pressures are far from the high-pressure band

2.2. Anomaly detection

The anomaly detection procedure we derive in this sub section is based on Table 1, and here we articulate it in mathematical model.

Let T be a set of time where historical data are observed, P be a set of gas pressure data in T, and V be a set of gas volume data in T. For each $t \in T$, let $v_t \in V$ be the volume at time t and $p_t \in P$ be the pressure at time t. We model the statement 'volume tends to zero' as $v_t \leq 1$ and the statement 'pressures are in the high-pressure band' as $p_t \geq \rho$, where ρ is the 95th percentile of P. Then, we have mathematical expression of Table 1 in the following.

Table 2. Normal and anomaly conditions in mathematical expressions

Data behavior	Volume and pressure conditions
Normal	1. $v_t \leq 1$ and $p_t \geq \rho$, or
	2. $v_t > 1$ and $p_t < \rho$
Anomaly	$v_t \leq 1$ & $p_t < \rho$

The range of data in P depends on the diameter of customer's pipe, so that we set the value of ρ as 95th percentile of P in order to accommodate the variant of P among customers.

2.3. Forecasting and estimation
If the anomaly period has been determined by the anomaly detection procedure, we then need to determine a number of forecasting methods that can capture the behaviour of customers' gas consumption. We mean by the behaviour of customers' gas consumption is the behaviour in a certain period of time close to the anomaly period, which we call the training period. If the training period is close to the anomaly period, we can assume that the behaviour of customers' gas consumption in both period are the same. Mathematically, the behaviour are characterized by the trend, seasonal and cyclical behavior of the time-series data in the training period. For the data that has strong seasonal behavior we can choose Exponential Smoothing Forecasting Methods such as Holt Winters model [5]. For the data without non-seasonal behavior is performed by applying Auto Regressive Forecasting Model (p) for various values of p. Since we do not know the behavior of customers' gas consumption data beforehand, we perform a number of forecasting with various forecast methods to accommodate any behaviour.

In order to measure forecast accuracy, we tried some forecast accuracy measures such as Mean Squared Error (MSE), Mean Absolute Percentage Error (MAPE), Relative Geometric Root Mean Square Error (RGRMSE), Mean Absolute Deviation (MAD) and Mean Absolute Percentage Error (MAPE). But after analyze the errors between the training data and the fitting data resulted by any forecasting method we use, we think that most suitable forecast accuracy measure for our need to estimate the total consumption value in the anomaly period is Percentage Cumulated Forecast Error (PCFE), which we create based on the Cumulated Forecast Error (CFE).

If we note the training data as $X_1, X_2, \dots, X_T$ and its forecast data as $\widetilde{X_1}, \widetilde{X_2}, \dots, \widetilde{X_T}$, the CFT is given by

$$CFE = \sum_{i=1}^{T} (X_i - \widetilde{X_\imath}),$$

and our PCFT is given by

$$PCFE = \frac{\sum_{i=1}^{T} (X_i - \widetilde{X_i})}{\sum_{i=1}^{T} X_i}.$$

We come up to consider the CFE after we realize that MSE and MAD values are quite difficult to be understood by the PGN customers. It is though that the customers will easy to understand a measure in term of percentage, so that we then consider MAPE as an alternative for the measure. But the MAPE value depends on the training data scale [10][11], and furthermore the existence of zero volume data make the really high value of MAPE even for the case where the forecast results is very close to the training data. Since all we need is to estimate the total consumption value in the anomaly period, we then consider the CFE. The 'percentage' version of CFE is given by the PCFE. We have test that there is a strong linear correlation between MSE and PCFE values, and between MAPE and PCFE as well.

We estimate the volume during the anomaly period from the forecast result with the smallest PCFE.

3. The implementation of the anomaly detection procedure and the forecast method inreal case billing process

We tested the anomaly detection procedure and the forecast method in real case billing process to a large number of historical PGN customers' data. The test showed that:

1. The anomaly detection procedure work effectively in determining the period of anomaly data which indicate the period during the meter failure. We test the procedure to some customers' historical data in which these customers reported meter failure in certain periods of time. By implementing our procedure we obtained periods of meter failure that cover such a period reported by the customer.

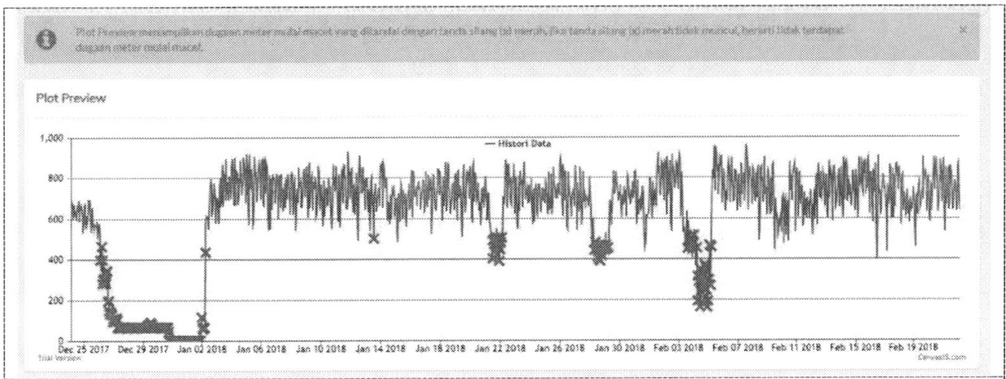

Figure 3. Anomaly data, indicated by red-cross signs.

2. For cases in which the customers' historical data have strong seasonal behavior, we obtain the best forecast model that give quite small error (under 20%, for some cases even less than 5%). Our experience show that (p) model oftenly occurs as the best model compare to the other models (Holt Winters model and the general Exponential Smoothing Models).

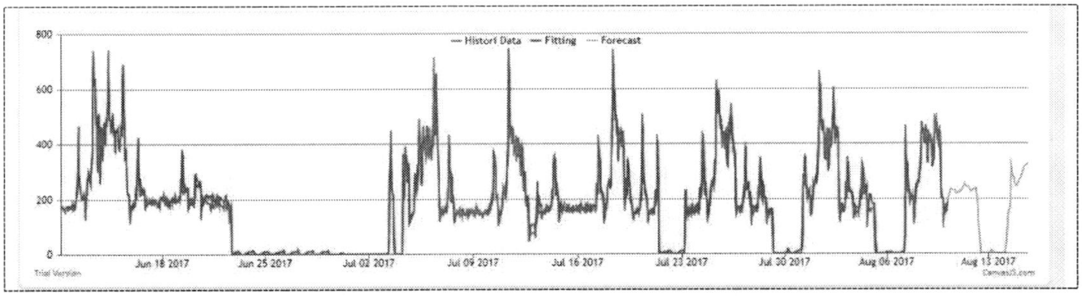

IOP Conf. Series: Materials Science and Engineering **567** (2019) 012002 doi:10.1088/1757-899X/567/1/012002

Figure 4. Example of forecast result, given by the green curve.

For practical purpose, we develop a web-based graphical user interface based on the anomaly detection method and the forecasting method described above. The following figure is the screen shoots of the interface which show the results on the anomaly detection and forecasting. In this example, we need to forecast gas consumption during period 4 am on 1 October 2017 to 3 am on 4 October 2017 (92 hours long). We use three month historical data as training data. In this example, the best fitting model is the Holt-Winter Multiplicative with Damped model that give us 18% error in training data period. The interface shows the historical data and its fitting data, and the forecasting result shown in green colour while the hourly quantities are given on the right bottom box.

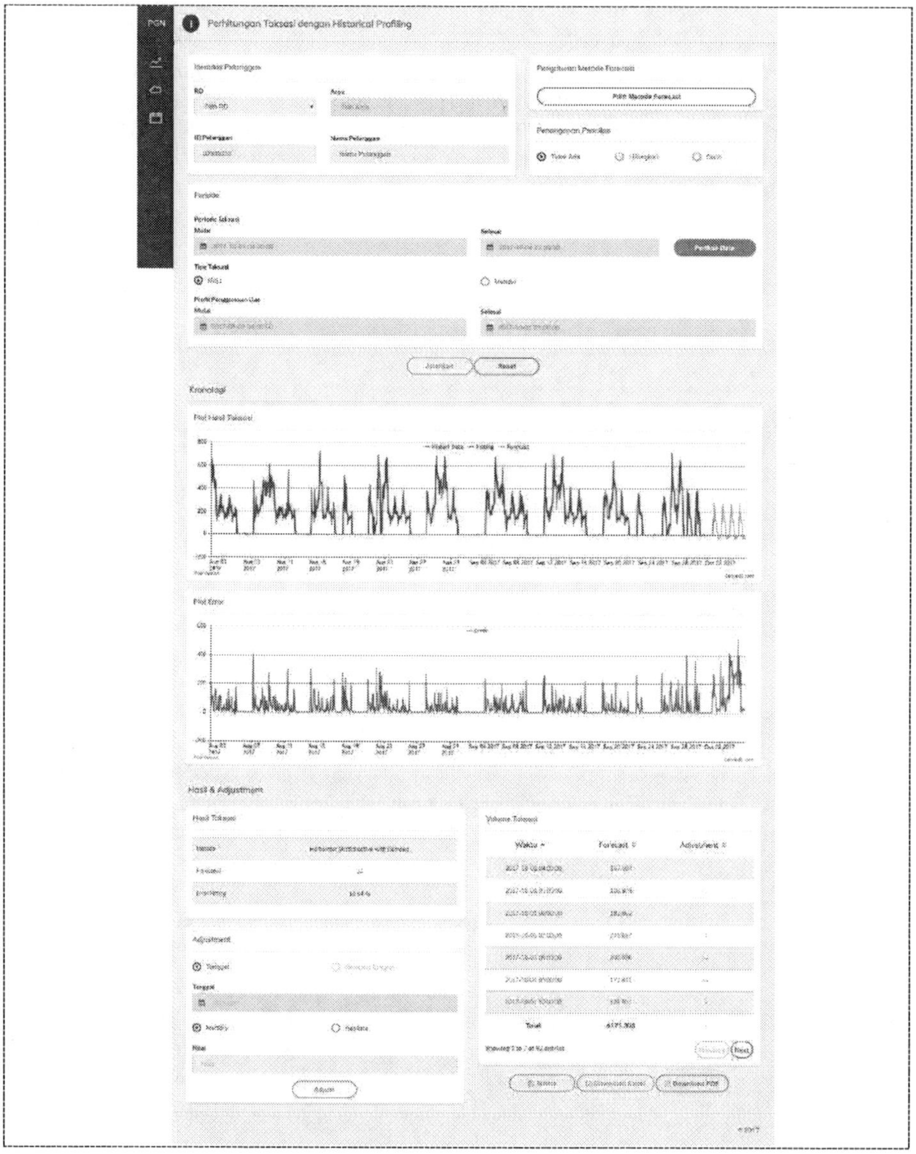

Figure 5. Graphical User Interface showing farecast result and its error.

We also found other case where it is impossible to pick training data from historical data before the anomaly period. This case may occur to a new customer of PGN where there is no enough historical data and the anomaly period happen in the beginning of service. In this case, we implement backward estimation after fitting data on the period after the anomaly period. Our procedure still work effectively as shown in the following figure.

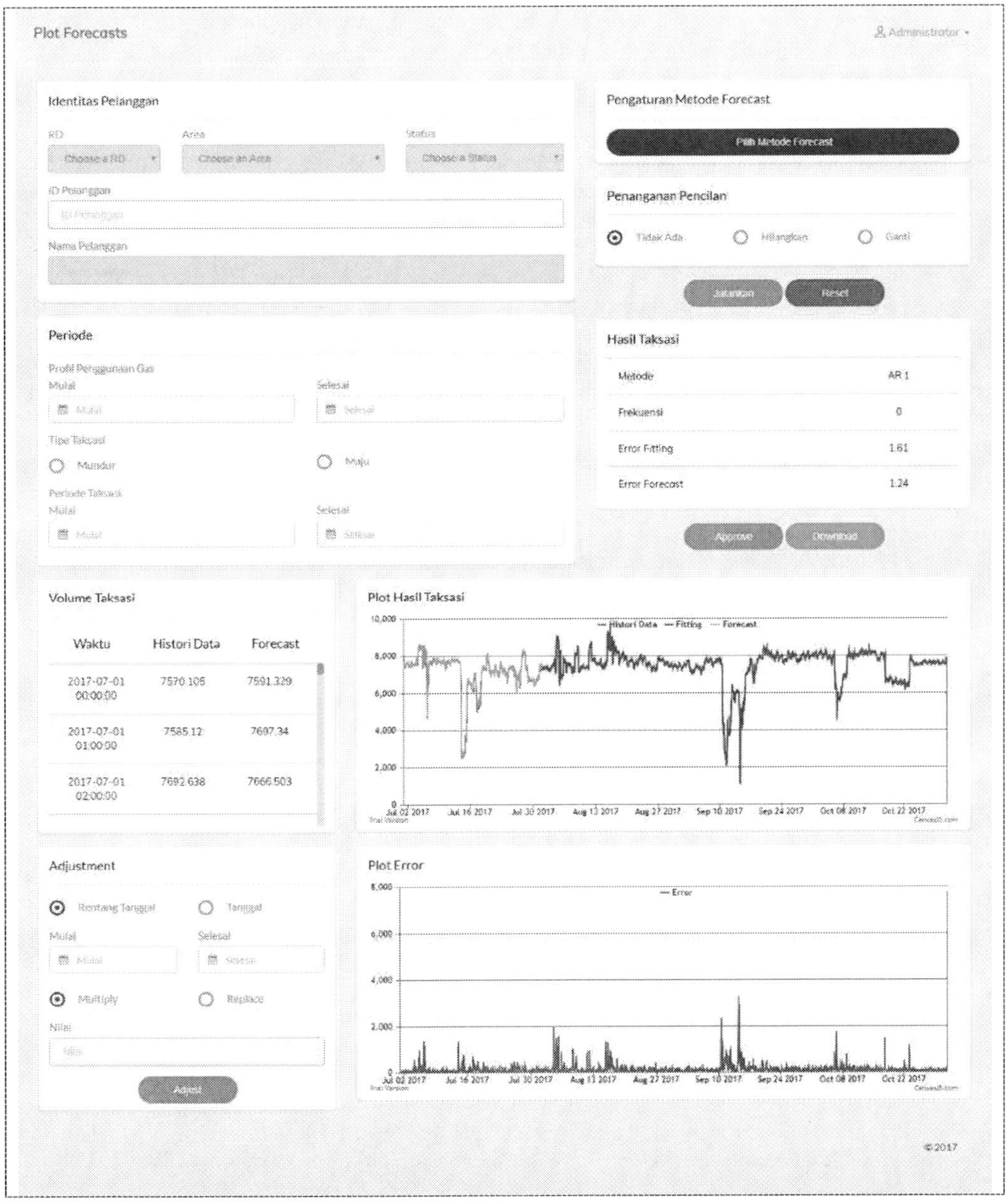

Figure 6. Example of backward estimation

4. Summary and discussion

A collaborative research has been performed for solving the estimation for total natural gas consumption due to volume meter failure, which is needed for billing process. By analysing historical data on gas consumption, temperature and pressure, the problem is then formulized as anomaly detection and estimation in the anomaly period. We derive a procedure for anomaly detection, and choose some appropriate forecasting methods for estimation of gas volume in the anomaly period. After some test or validation for those procedures, we developed a graphical user interface for practical purposes.

The system we derive has been implementing in monthly billing process. Although it still needs minor refinements, the end users of this new system have seen some possibilities to develop it to new energy management system. The energy management system not only content the anomaly detection procedure and the forecast procedure for billing process, but is expected to be able to analyse overall customers' consumption behaviour. Since PGN supply natural gas to the pipeline network in bulk volume, we guess there are some period of time where the gas supply is excessive. For developing the energy management system, we first need to perform analysis for determining the periods where the gas supply are excessive, and making an inference whether such periods are periodic or not. If the analysis results show that the excessive supply periods are periodic, we will try to reduce unutilized gas supply by offering dynamic pricing to the customers that their production can be shifted to the excessive supply periods or to the customers that can increase their productivity. The energy management system will monitor the impact of the dynamic pricing implementation to the unutilized supply reduction, so that can give an enormous cost saving both for PGN and the customers as reported in some scientific articles.

Acknowledgement

We are grateful to our research assistant Muhammad Ridwan Reza Nugraha and Nita Ratnawati who have implemented the anomaly detection procedure and the forecast method in R. We also thank to Luthfi Pratama for developing the graphical user interface.

References

[1] PT PERUSAHAAN GAS NEGARA (Persero) Tbk 2017 *Annual Report 2017*
[2] Akpinar S and Yumusak N (F. Martinez-Alvarez, Ed.,) 2016) *Energies 727*
[3] Bagajewicz J and Cabrera E 2008 Data Reconciliation in Gas Pipeline Systems *Industrial & Engineering Chemistry Research* **42** pp. 5596-5606
[4] Gschneidner Jr K A and Erwin L 2015 *Gas Reconciliation Requirements and Procedures* Gas Industry Company Limited
[5] De Gooijer G and Hyndman R J 2006 25 Years of Time Series Forecasting *International Journal of Forecasting* p. 443-473
[6] Coşgun Ö and Kaya G O 2015 Stochastic Dynamic Programming for natural gas pricing in the Turkish Energy Market *J. of Mult.-Valued Logic & Soft Computing* **26** pp. 353–361
[7] Flaim T 2018 *Fundamentals of Dynamic Pricing* (Presentation file https://emp.lbl.gov/sites/default/files/oh120809.pdf) dowloaded on 16 July 2018
[8] Harvard's Electricity Policy Group 2018 *Dynamic Pricing* (presentation file: https://sites.hks.harvard.edu/hepg/Papers/2016/June%202016/P2.%20Kelly..pdf,) dowloaded on 16 July 2018
[9] Carroll L and Hudkins R W 2018 Advance Pipeline Design *The University Oklahoma Report, www.ou.edu/class/che-design/a.../Pipeline%20Design.pdf* dowloaded on 16 July 2018
[10] Makridakis S, Wheelwright S C and Hyndman R J 1998 *Forecasting,Methods and application 3rd edition*

IOP Conf. Series: Materials Science and Engineering **567** (2019) 012002 doi:10.1088/1757-899X/567/1/012002

[11] Wallström P 2009 *Evaluation of Forecasting Techniques and Forecast Errors with Focus on Intermittent Demand* Licentiate Thesis, Department of Business Administration and Social Sciences Division of Industrial Marketing, e-commerce and Logistics Industrial

IOP Conf. Series: Materials Science and Engineering **567** (2019) 012003 doi:10.1088/1757-899X/567/1/012003

Niche theory within video platform competition: traditional vs modern

R Sukwadi[1], S C Lasiman[1], M M W Inderawati[1], F Suprata[1]

[1]Innovation Systems Laboratory, Department of Industrial Engineering Atma Jaya Catholic University of Indonesia, Jl. Raya Cisauk, BSD City, Tangerang Selatan 15345, Indonesia

Corresponding author: ronald.sukwadi@atmajaya.ac.id

Abstract. Rapid proliferation of the usage of internet-based modern television (IPTV and OTT) in Indonesia as well as limited and similar resources threatens the existence of old/traditional video platforms (satellite television and cable television). Adapted from ecological studies, niche theory is used to measure the level of dynamic competition among media. Through three concepts of measurement of niche theory, niche breadth, nice overlap, and competitive superiority, the competition relationship between these four video platforms can be analyzed. The questionnaire was spread out online and offline to 502 respondents of each platform. Focus group discussions were conducted in two groups with 6 respondents each group. The results of this study were also analyzed based on the media life cycle. The conclusions of this study are: 1) OTT is a generalist platform, IPTV and cable television tend to be generalist, and satellite television is classified as a specialist; 2) The highest competition occurs between cable television and IPTV; 3) OTT is the most competitive platform whereas satellite television is the least competitive; 4) Satellite television is in defensive resistance life cycle, cable television in defensive resistance and adaptation cycles, IPTV in growth cycle, and OTT in maturation cycle.

1. Introduction

Service providers including video platform providers need to pay attention to customer feedback to better provide service [1][2][3] and must rely more on customers' experiences in service processes [4]. The providers have to organize around the customer in order to be a successful firm. They are forced to change their activities from a provider's point of view towards a customer's point of view [5]. Customer delight has to be created during the process of service delivery in order to really outperform competitors and win the hearts of customers [6][7][8][9]. While a good quality of service has been considered a competitive advantage, many researchers have now pointed out that a quality service aiming to satisfy expectations can no longer win the hearts of customers. How to provide high quality experiences for customers is more critical now. A memorable service experience could improve customer satisfaction, loyalty and reputation of service providers leading to greater profit. Service innovations have contributed to the growth in the number of service firms and to the scale on their operations, which in turn has increased their economic impact [1].

Penetration of internet users in Indonesia in 2016 increased to 51.8% or reaches 132.7 million people. Users often use the internet as entertainment (96.8%). Of the various types of entertainment that can be accessed by the internet, entertainment content that is most frequently visited by users is watching online movies with a percentage of 41% or about 54.4 million people [6]. Nevertheless, television is still at the top of the peak as the main medium. The result of the Nielsen Consumer Media View survey of more than 17,000 people over the age of 10 in 11 major cities in Indonesia shows that

IOP Conf. Series: Materials Science and Engineering **567** (2019) 012003 doi:10.1088/1757-899X/567/1/012003

television is still ranked first (96%), followed by outdoor media (53%), and internet (44%)[7]. IPTV (Internet Protocol Television) is an interactive (two-way) television, subscribers can rewind or pause the broadcast, where data is transmitted over an IP network [8]. While, OTT (Over The Top) according to U. S. Federal Communication Commission, is defined as an online video distributor that sends video content to customers via the internet [9].

Nielsen Cross-Platform 2017 survey found that watching via traditional television (cable television and satellite television) is still the top choice (77%) in Indonesia. Satellite television transmits television signals from satellites which will then be received by a television antenna. While, cable television delivers its programming through radio waves transmitted into coaxial cables. On the other hand, the modern platform for accessing video through the internet also has a high penetration, streaming sites such as YouTube, Vimeo, and so on (51%), online television portals (44%), and subscription internet television such as Netflix, Iflix , HOOQ, VIU, and so on (28%). If user's penetration of those modern platform are combined, they can threaten traditional platforms. Nielsen Cross-Platform is an annual study of digital content in the Asia Pacific region using an online survey methodology [10]. Based on research conducted by Ericsson ConsumerLab on future technology trends conducted in 2011 to 2014 in nine countries (Brazil, China, Germany, South Korea, Spain, Sweden, Taiwan, United Kingdom, United States), it was found that from year to year the use of television has decreased. Based on the research, President of Ericsson Indonesia, Thomas Jul, predicts that the same will happen to other countries including Indonesia [11].

Theory of the niche is originated from an ecological study of competing living populations and it is adapted to explain competition in the mass media with the same consumer resources as in a video platform between traditional platforms and modern platforms that live together. Analysis of niche theory is done by applying the three measurement concepts namely niche breath, niche overlap, and competitive superiority. Li's study [12] also adapted from Dimmick's niche theory by using gratification obtained and gratification opportunities to test the competitive relationship between three television media -OTT, IPTV, and digital cable. Among the three media, OTT is the most competitive medium [12]. Therefore, in this study, niche theory will be applied to test dynamic competition relationships among video platforms - satellite television, cable television, IPTV, and OTT in Indonesia.

2. Literature review

2.1. Niche theory

Three important concepts in understanding niche theory are the niche breath, niche overlap, and competitive superiority. Niche breadth shows the amount of resources consumed by a population of media or the dependence relation of media population with the resources. Niche breadth measures the level of ability of a medium to give gratification. Based on the niche breadth, media can be categorized as generalists and specialists. Niche overlap is an index that measures similarity or deference between two media regarding on their resource utilization. High overlap shows strong competition among two media. Competitive superiority is used to find out which platforms are among a pair of platforms that have the capability of better resource utilization, meaning that it provides a wider gratification to the user [13]. Major resource dimensions that are the object of competition – gratification utilities obtained and gratification opportunities, media content, consumer spending, time spent by consumers on the media, and advertising expenditures [14].

2.2. Media life cycle

Media life cycle stages are divided into: (1) birth; (2) penetration; (3) growth; (4) maturation; (5) defensive resistance; (6) adaptation, convergence, or obsolescence [15]. Birth, often refers to a previous medium developed as a continuous innovation from the previous one [16]. Penetration, new medium enter the market, develop new usage and attract users. Growth, inventors, developers, and users learn to utilize, apply, and extend the capabilities of the medium. In this phase, the market share penetration of a medium has increased by 16 to 50 percent. Maturation, the new medium has found its place in a dynamic communication environment. Medium has grown, it is using and applying its capability maximally. The penetration of market share obtained ranges from 50 to 90 percent.

Defensive resistance, competition between old media and new media pushes old media to seek new directions in order to retain their old users. Older media users will experience a decrease from 90 percent to 50 percent.

The last stage is divided into three possibilities: adaptation, convergence, or obsolescence. If the old medium is able to adapt to the new situation by developing different functions that retain its users or find new users, the medium goes into the adaptation stage. If the old medium cannot survive on its own but survive by joining or cooperating with the new medium, the medium goes into the convergence stage. If the traditional medium fails to adapt to change, the medium goes into the obsolescence stage, decreases and then disappears.

3. Research methods

3.1. Research design

This study used gratification dimension of the niche resource dimension. The dimension of gratification is described by a factor or so-called macro dimension. In this research, there are six factors that are broken down into micro dimension which can be seen in Table 1.

Table 1. Identification of gratification factors.

Gratification Factor (Macrodimensions)	Microdimensions	Sources
Ease of Use	Program information is easy to understand	[12]
	Pleasing format programs	
	Easy to operate	
Financial Benefit	Affordable price	[17]
	Discounts for multiple purchases (packages)	
	Reasonable price	
Easy Interactivity	Minimal advertisement disturbance	[12]
	View programs not found in other platform	
	Easily find the desired program	
Convenience	Portable (can be watched anytime, anywhere)	[9]
	The program runs smoothly	
	Easy to save favorite programs	
Content	The program is more update than on other media	[12]
	Wide variety of programs	
System Quality	The picture quality is good	
	The signal quality is good	
	Have copyrights	

3.2. Data Collection

Non-probability sampling technique (purposive sampling) is applied for questionnaire with certain specific criterion. The criteria are traditional or modern television users or viewers aged between 18 and 58 years old. The required sample size ranges from 340 to 680 respondents. Focus group discussions were conducted in two groups. Both questionnaire distribution and focus group discussions were done in February 2018 and March 2018.

3.3. Methods and testing

3.3.1. Validity and reliability test

The questionnaire was tested its accuracy in measuring what it wanted to measure and its consistency as a reliable measuring tool using SPSS 16 software. The test results showed that the questionnaire was valid and reliable.

IOP Conf. Series: Materials Science and Engineering **567** (2019) 012003 doi:10.1088/1757-899X/567/1/012003

3.3.2. Niche breadth

Niche breadth of each video platform on each macro dimension is calculated by using the formula in equation (1) [13].

$$B=\frac{\sum_{n=1}^{N}\frac{\left[\sum_{k=1}^{K}GO_n\right]-Kl}{K(u-l)}}{N} \tag{1}$$

where, u and l denote upper and lower bounds of a scale. GO indicates gratification obtained rating on scale. N is the number of respondents using a medium, n is the first respondent. K denotes the number of scales on a dimension, k is the first gratification scale. The maximum value of the niche breadth is 1 and the minimum is 0.

3.3.3. Niche overlap

Niche overlap each pair of video platforms on each macrodimension is calculated by using the formula in equation (2) [13].

$$O_{i,j}=\frac{\sum_{n=1}^{N}\sqrt{\sum_{k=1}^{K}\frac{(GO_i-GO_j)^2}{K}}}{N} \tag{2}$$

where, i and j denote media i and media j. GO is gratification obtained rating on a scale for media i and media j. N is the number of respondents using both i and j, n is the first respondent. K denotes the number of scales on a dimension, k is the first gratification scale.

3.3.4. Competitive superiority

Competitive superiority between media pairs or video platforms in a macrodimension is measured using the equations (3) and (4)[13].

$$S_{i>j}=\frac{\sum_{n=1}^{N}\sum_{k=1}^{K}(m=_{i>j})}{N} \tag{3}$$

$$S_{j>i}=\frac{\sum_{n=1}^{N}\sum_{k=1}^{K}(m=_{j>i})}{N} \tag{4}$$

where, i and j indicate media i and j. $m_{i>j}$ means the value of a respondent's rating for those scale items on which i is rated greater than j (the sum of the actual values); $m_{j>i}$ means the value of a respondent's rating for those scale items on which j is rated greater than i (the sum of the actual values). N is the number of respondents using both i and j, n is the first respondent. K denotes the number of scales on a dimension, k is the first gratification scale. There are 2 things that should happen if platform i is superior compared to platform j, the value of $S_{i>j}$ should be higher than $S_{j>i}$ and the average gratification obtained rating of the two platforms differ significantly on the t-test.

3.3.5. Verbatim transcripts, coding, categorization, and filtering focus group discussion data

Discussions conducted in focus group discussions are transcribed into verbatim. The coding is done by marking the part of the discussion that is considered to be an important topic. Categorization is done by sorting topics that have been coded into major themes. There are eight major themes in this study. The seven major themes consist of 4 platform types (satellite television, cable television, IPTV, and OTT) and 4 other themes, video platform purchases, video platform consumptions, media comparisons, and content.

The first four themes are grouped in accordance with the advantages of each platform and the reasons participants tend to choose the platform. The video platform puchases discusses how participants can enjoy the video, viewed from subscribers' subscribing behaviour on video media. Then, the consumption of the video platform outlines the reason participants watch television or video, its relation to social aspects, as well as the time and place of the watch participants. Meanwhile, the comparison of media related to other entertainment media outside the video media, such as books, magazines, comics, radio, and so forth. How participant preferences in choosing entertainment media and more time spent. This section also compares satellite television, cable television, IPTV, and OTT. From grouping into 8 major themes, it was found that there were at least 3 themes that were not primary in relation to interpreting the media life cycle of each platform. The three not primary themes are video platform purchases, video platform consumptions, and content.

4. Result and discussion

4.1. Profile of respondents

The questionnaire was distributed to 582 people, there were 502 respondents matching the criteria (405 online and 97 offline). For ages, 16.53% of respondents were under 18 years old, 73.90% were between 18-38 years old, 8.77% were between 39-58 years old, and 0.80% were older than 58 years. The percentage of men from 502 respondents was 45.82%. Focus group discussions were conducted in two groups. The first group consisted of 6 participants, 4 aged 21 and 2 aged 22, all women. The second group consisted of 6 participants, where 1 was 20 years old, 3 were 21 years old, 1 was 28 years old, and 1 was 29 years old, of which 3 were male.

4.2. Niche breadth

Niche breadth of each platform is shown in Table 2. OTT has the ability to provide the highest gratification among other platforms on all macrodimension. This shows that OTT has a dependency relationship with various resources. Meanwhile, the relationship of satellite television depends only on a few resources. Satellite television gratuities are the lowest on any macrodimension except in financial benefit. IPTV provides the lowest gratification on that macrodimension.

Table 2. Niche breadth on each video platform

Platform	Ease of Use	Financial Benefit	Easy Interactivity	Convenience	Content	System Quality
Televisi Satelit	0.635	0.698	0.485	0.451	0.485	0.583
Televisi Kabel	0.741	0.629	0.683	0.584	0.693	0.721
IPTV	0.702	0.623	0.694	0.666	0.705	0.71
OTT	0.758	0.723	0.745	0.81	0.831	0.733

Niche breadth moves from 0 to 1. If divided into 4 categories, specialists, tend to be specialists, tend to generalists, and generalists, then the range of each interval is 0.25. Niche breadth categorization can be seen in Table 3.

4.3. Niche overlap

Niche overlap of each pair of video platforms can be seen in Table 4. The niche overlap moves from 0 to 4 where, the nearer to the zero the higher overlap between two platforms, the stronger the competition or substitution relationship the two platforms have. Cable television and IPTV have the most intense competition relationships among other platform pairs in the six macro dimensions. The most likely explanations are 1) The operation of the television is similar from the remote as well as the appearance and arrangement of the television channel, 2) Both require a subscription fee with similar price range, 3) The ease of finding the desired program is not much different, although the advertising interference on IPTV is less, 4) Both of them are less portable than OTTs, 5) Both have national television channels and foreign television channels, and 6) Providers of both platforms are almost certainly have copyrights of their program and provide good image quality, although how to transmit data is different, cable television via coaxial cable and IPTV over IP network.

While, the weakest overlap indicated by the large niche overlap value occurs in satellite television pairings and OTT, except for the financial benefit macrodimension. The most likely causes are 1) The arrangement of the OTT program is much more modern than the satellite television; 2) The satellite television contains a national television channel that has a lot of advertising disruption compared to the foreign television channel (cable television and IPTV), moreover with OTT, there are paid/subcription OTT which is free from advertisement; 3) OTT is portable and easy to save favorite programs; satellite television is not; 4) There are types of programs that only exist in OTT, OTT even includes programs that exist in other platforms; variations of satellite television programs are very narrow; and 5) Sometimes it is difficult to find the position of satellite television antenna where all channels have good picture quality. Meanwhile, OTT system quality is better overall, many OTT have HD resolution picture quality even 4K resolution. In the financial benefit macrodimension, the weakest overlap occurs in satellite television and IPTV (0.999), followed by satellite television and cable television (0.986). Satellite television does not require a subscription fee, whereas IPTV and cable television have similar and relatively high subscription fees, while the OTT has two services, paid OTT services and free OTT services.

4.4. Competitve superiority

Calculated competitive superiority and t-test results of each pair of video platforms can be seen in Table 5. Satellite television is inferior in all macrodimensions with each of other platforms. Satellite television is only superior in financial benefit macrodimension to cable television and IPTV. It means satellitetelevision provides a greater gratification to consumers related to finance compared to cable television and IPTV.

The ability of IPTV and cable television to provide gratification is no different from financial benefit, easy interactivity, content, and system quality macrodimensions. Cable television is superior to IPTV in the macrodimension of ease of use; IPTV is superior to cable television in the convenience macrodimension. The ability of OTT and cable television to provide gratification is no different in the macrodimensions of ease of use and system quality. However, OTT is able to provide greater gratification on the macrodimensions of financial benefits, easy interactivity, convenience, and content. OTT is also superior to IPTV in all macro dimensions.

4.5. Media life cycle

Satellite television is on a defensive resistance lifecycle. Sooner or later, old media will be threatened by new media. New media (cable television, IPTV, and OTT) offer facilities that satellite television does not have, for example international television programs and better video resolution. Speed of old media response is economical: the greater the investment in infrastructure, the higher the reluctance of the old media to change or adapt – "network lock-in" [18], as published by Lehman-Wilzig and Cohen-Avigdor [15]. Infrastructure investment is carried out in the form of satellite television service launches such as Telkom 2 in November 2005, Indostar 11 in May 2009 and Telkom 3S in February 2017. Therefore, satellite television is still reluctant to change from the current video delivery system.

Table 3. Niche breadth on each video platform

Platform	Ease of Use	Financial	Interactivity	Convenience	Content	System Quality	Notes
Satellite Television	Tend to be Generalist	Tend to be Generalist	Tend to be Specialist	Tend to be Specialist	Tend to be Specialist	Tend to be Generalist	Many ads, limited program variations, limited image quality, cheap, easy to use
Cable Television	Tend to be Generalist	Tend to be Generalist	Tend to be Generalist	Tend to be Generalist	Tend to be Generalist	Tend to be Generalist	Subscription required, wide program variations, smooth program
IPTV	Tend to be Generalist	Tend to be Generalist	Tend to be Generalist	Tend to be Generalist	Tend to be Generalist	Tend to be Generalist	Subscription required, broad program variations, minimal ads, free enough to select videos
OTT	Generalist	Tend to be Generalist	Tend to be Specialist	Generalist	Generalist	Tend to be Generalist	Subscriptions based on preferences, exclusive content, a wide variety of programs, portable, free to choose videos

Table 4. Niche breadth on each video platform

Macrodimension	Satellite TV-Cable TV	Satellite TV-IPTV	Satellite TV-OTT	Cable TV-IPTV	Cable TV-OTT	IPTV-OTT
Ease of Use	0.83	0.892	0.99	0.557	0.67	0.654
Financial Benefit	0.986	0.999	0.808	0.457	0.89	0.768
Easy Interactivity	1.101	1.172	1.412	0.546	0.852	0.737
Convenience	0.867	1.246	1.76	0.737	1.315	0.919
Content	1.017	1.081	1.546	0.464	0.779	0.687
System Quality	0.799	0.821	1.065	0.343	0.613	0.525

Table 5. Niche breadth on each video platform

		Satellite TV-Cable TV		Satellite TV-IPTV		Satellite TV-OTT		Cable TV-IPTV		Cable TV-OTT		IPTV-OTT	
		S	t-Value	S	t-Value	S	t-Value	S	t-Value	S	t-Value	S	t-Value
Ease of Use	$S_{i>j}$	1.434	-9.815*	2.321	-5.966*	1.622	-10.837*	2.996	3.730*	2.169	-1.572	1.41	-5.066*
	$S_{j>i}$	5.131		4.45		5.797		1.528		3.297		3.751	
Financial Benefit	$S_{i>j}$	5.151	6.073*	5.291	6.691*	2.651	-2.215*	1.894	0.558	1.578	-8.532*	0.98	-9.203*
	$S_{j>i}$	2.317		2.375		3.263		1.504		5.018		4.725	
Easy Interactivity	$S_{i>j}$	0.735	-16.352*	0.845	-17.545*	0.811	-22.419*	1.966	-0.978	1.841	-5.743*	4.002	-4.837*
	$S_{j>i}$	6.47		6.791		8.199		2.185		4.426		4.725	
Convenience	$S_{i>j}$	0.741	-10.226*	0.932	-17.178*	0.552	-29.837*	1.223	-7.037*	0.972	-20.263*	0.769	-13.564*
	$S_{j>i}$	4.639		6.488		9.327		3.464		7.4		5.837	
Content	$S_{i>j}$	0.291	-16.412*	0.388	-17.536*	0.277	-27.911*	1.155	-1.068	0.446	-12.284*	0.339	-11.333*
	$S_{j>i}$	4.604		4.791		6.809		1.4		4.141		3.833	
System Quality	$S_{i>j}$	0.518	-12.246*	0.863	-11.414*	1.375	-13.239*	1.54	1.022	2.084	-1.071	1.472	-2.104*
	$S_{j>i}$	5.221		5.09		6.058		1.096		2.643		2.57	

Cable television is ongoing two cycles at once, defensive resistance and the sixth stage, adaptation. The declination of cable television can be seen in the performance of the two largest cable television providers in Indonesia, First Media and MNC Sky Vision, each experienced a loss of 279.05 billion rupiahs and 293.80 billion rupiahs in the first half of 2015 [19]. The decline was caused by the threat of IPTV and OTT. Supported by FGD participants' statement that IPTV is better because the ads can be accelerated, OTT is more up to date and cheaper than cable television. The adaptation made by cable television is they begin to migrate to IPTV technology, based on news articles from industri.bisnis.com [20]. This method does not guarantee the survival of the old media, but allows the survival of media organizations when the old media (cable television) disappears.

The birth of IPTV in 2011 with a similar display to cable television gives IPTV an advantage because users do not need to take long to get to know and understand the use of IPTV. IPTV can offer facilities that did not exist in cable television, it can rewind and pause the broadcast, also watch the programs that have passed. IPTV is in the third phase, growth, IPTV has not entered the stage of maturation because until now IPTV has not passed more than half of the video media market.

OTT is a video services provider through the internet. The Nielsen Cross-Platform 2017 survey in Indonesia shows the penetration users of streaming site such as YouTube, Vimeo (51%), online television portals (44%), subscription platforms such as Netflix, Iflix, HOOQ (28%). It is not known to the number of users who use more than one OTT platform such as one user using YouTube and Netflix in the same time. However, it can be ascertained that overall OTT penetration is more than 50%. This clearly indicates that OTT is already in the fourth stage, maturation. Corporate Secretary Division General Manager of PT Indosat Mega Media, Syachrial Syarif, suggests that from the research he knows only 40% of consumers who want to subscribe to internet television, the rest are not. Paid television is solely the home internet addition [21]. Internet-television package is one of the reasons users subscribe to paid television. This is supported by FGD participants' statement, "Cable TV is usually bundling with home internet. So, it feels more profitable." and on one of the reasons participants chose IPTV," the price is already one package with my home internet."

5. Conclusion and Suggestion

The conclusions of this research are: 1) Based on the obtained niche breadth, satellite television is a specialist platform, cable television and IPTV are classified as tend to be generalist, and OTT is a generalist platform; 2) Based on the niche overlap acquired, the strongest competition occurs between cable television and IPTV compared to other platform pairs; 3) Among the four video platforms, OTT is the most competitive platform and satellite television is the least competitive in all macrodimension. Cable television is superior to IPTV in ease of use macro dimension, whereas IPTV is more competitive in convenience macrodimension; the other macro dimensions show no difference in providing gratification to users between cable television and IPTV; 4) Satellite television is in a defensive resistance cycle (fifth cycle), while cable television is in two cycles at once - defensive resistance and adaptation (sixth cycle). IPTV is in the growth cycle (third cycle), whereas OTT is already in the maturation cycle (fourth cycle).

Suggestions for each platform in order to survive or extend the period of its success include: 1) Satellite television focuses its target consumers on areas with minimal communications services in Indonesia (countryside or remote area), 2) Cable television switches technology using IP networks and privileges by products (internet and OTT) to attract and retain users, 3) IPTV providers extend fiber networks to a wider range of users than before, collaborate with local authorities in its realization, and 4) Each OTT markets itself massively and continuously through internet networks such as social media and pop-up ad, in order to last longer in the maturation cycle.

Suggestions for future research based on the limitations of this study are to distinguish the OTT platform into paid OTT and free OTT services.

Acknowledgments

The authors express thanks to Faculty of Engineering, Atma Jaya Catholic University of Indonesia for providing material and also financial support to this research.

IOP Conf. Series: Materials Science and Engineering **567** (2019) 012003 doi:10.1088/1757-899X/567/1/012003

References

[1] Sukwadi R 2017 The moderating role of service innovation on the relationship between customer satisfaction and customer value: A case of 3-star hotels in Jakarta *International Journal of Services, Economics and Management* **8**(1/2) 18

[2] Sukwadi R, Suef M and Widawati E 2016 Translating company internal data into customer needs: Text mining analysis approach *ARPN Journal of Engineering and Applied Sciences* **11**(9) 5773

[3] Sukwadi R 2015 Utilizing Customer Experience Management framework to create a delightful service experience *International Journal of Industrial Engineering and Management* **6**(1) 29

[4] Suef M, Suparno, Singgih M L, Sukwadi R and Widawati E 2014 Utilizing claims, complaints, and company initiatives as VOC in a product development using QFD-Kano approach *International Journal of Applied Engineering Research* **9**(22) 18013

[5] Sukwadi R, Yang C C and Fan L 2012 Capturing customer value creation based on service experience - A case study on News Café *Journal of the Chinese Institute of Industrial Engineers* **29**(6) 383

[6] Asosiasi Penyelenggara Jasa Internet Indonesia 2017 *Asosiasi Penyelenggara Jasa Internet Indonesia* Available: https://www.apjii.or.id/survei2016

[7] Lubis M 26 Juli 2017 Available: http://www.nielsen.com/id/en/press-room/2017/TREN-BARU-DI-KALANGAN-PENGGUNA-INTERNET-DI-INDONESIA.html

[8] Maisonneuve J, Deschanel M, Heiles J, Li W, Liu H, Sharpe R and Wu Y 2009 An overview of IPTV standards development *IEEE Transactions on* Broadcasting **55**(2) 315

[9] Kim J, Kim S and Nam C 2015 Competitive dynamics in the Korean video platform market: Traditional pay TV platforms vs. OTT platforms *Telematics and Informatics* **33**(2) 711

[10] Micom RO 26 Juli 2017 Available: http://www.mediaindonesia.com/index.php/news/read/114722/survei-nielsen-masyarakat-indonesia-makin-gemar-internetan/2017-07-26

[11] Yusuf O 23 Januari 2015 Available: http://www.kompas.com

[12] Li S C S 2017 Television media old and new: A niche analysis of OTT, IPTV, and digital cable in Taiwan *Telematics and Informatics* **4**(7) 1024

[13] Dimmick J 2003 *Media Competition and Coexistence: The Theory of the Niche* (New Jersey: Laurence Erlbaum Associates Inc.)

[14] Dimmick J 2006 *Media Competition and Levels of Analysis* in *Handbook of Media Management and Economics* (New Jersey: Lawrence Erlbaum Associates) p 345

[15] Lehman-Wilzig S and Cohen-Avigdor N 2004 The natural life cycle of new media evolution *New Media & Society* **6**(6) 707

[16] Haniff Z 2012 Niche theory in new media: Is digital overtaking the print magazine industry? *UNLV Theses, Dissertations, Professional Papers, and Capstones* 1571

[17] Seol H, Park G, Lee H and Yoon B 2012 Demand forecasting for new media services with consideration of competitive relationships using the competitive bass model and the theory of the niche *Technological Forecasting and Social Change* **79**(7) 1217

[18] Antonelli C ed 1991 *The Economics of Information Networks* (Amsterdam: North-Holland)

[19] Rayana U 23 September 2015 Available: https://selular.id/2015/09/ditinggal-ribuan-pelanggan-bisnis-tv-berbayar-terancam-buyar/

[20] Maskur F 31 Oktober 2017 Available: http://industri.bisnis.com/read/20171031/105/704564/pengusaha-tv-kabel-migrasi-televisi-analog-ke-teknologi-iptv

[21] Jamaludin F 3 Agustus 2017 Available: https://www.merdeka.com/teknologi/tv-berbayar-sekadar-pemanis-internet-rumah.html

IOP Conf. Series: Materials Science and Engineering **567** (2019) 012004 doi:10.1088/1757-899X/567/1/012004

Coral reef's healthy level measurement system design using digital image processing on Bunaken National Sea Park

A J Rindengan[1*], B Pinontoan [1], L A Latumakulita[1], C E Mongi[1], C E J C Montolalu[1], Y A R Langi[1]

[1]Department of Mathematics, Faculty of Mathematics and Natural Sciences, Sam Ratulangi University, INDONESIA

Corresponding author: altien@unsrat.ac.id

Abstract.The existence of coral reefs have been beneficial for life of living things in the sea. However, their existence todays is threaten by some factors which can be seen from its color bleaching. This means that coral reefs' level of healthy is decreasing. This is happening in Bunaken National Sea Park North Sulawesi, which is one of tourism destinations in Indonesia. Data used on this research is 40 reef images collected randomly from 10 diving spots around Bunaken. This research will design a system to analyse coral reefs' healthy level using digital image analysis. Reef healthy observation is accomplished based on Coral Health Chart issued by coralwatch.org. Based on the chart, RGB average is calculated using digital image processing and applying least square method to obtain formula in determining coral reefs' healthy level. Furthermore, the 40 coral reef images collected are classified based on regression equation obtained. This system is built in computer application system. Results show that application system using digital image to determine reef healthy level can be used. Healthy level of coral reefs in Bunaken National Sea Park is 72.8184%.

1. Introduction

Sea is a massive area which makes it interesting to be observed. Moreover, sea is a big source of economic profit of a nation because of its sectors in fisheries and tourism, as well as its coral reefs ecosystem. Coral reefs' existence todays has been in a warning level, this is because of increasing earth water temperature, which leads to coral destruction marked by coral's color bleaching. Coral bleaching is a coral's color fading because of the diseapearance of symbiotic algae which acts as energy and color suppliers in coral tissues [1]. Less was known in coral bleaching process, however, increasing temperature can interrupt algae's power in photosynthesis and trigger the production of dangerous chemical substance which can harm their cells [2].

Vision is important in human's point of view, this also acts in an image. In this case, unhealthy coral can be observed on its color changing pattern, more brighten the color is, the more it is unhealthy. In 2006 did research about Monitoring Coral Bleaching Using a Colour Reference Card [3]. In that research, images of some reefs were analysed on *Adobe Photosop V6* using histogram function to determine *hue*, *saturation* and *brightness*. Reef bleaching pattern was then treated to obtain four color classes and generate a *Coral Health Chart* to observe reef's healthy level. The chart was used in fieldwork by comparing reef's color to the colors in the chart, which then continued by listing the corresponding code. This research use *Coral Health Chart* and digital image processing in obtaining information of colors *RGB* (*red, green, blue*) points as well as least square method to form formula. From that formula, calculation of coral reefs' healthy level can be completed.

2. Literature Review

2.1 *Coral* and *Coral Health Chart*

Coral reef has important role for sea organisms. In a healthy reef, algae (*dinoflagelata simbiotik*) lives in reef's tissues. Algae supplies energy to the reef in form of sugar and amino acid, and gives color characteristics. Severe environment causes algae to be detached from the tissues, so that the reef's color turn into white or well known as reef bleaching [1]. Major cause of massive reef bleaching on these decades corresponds to increasing of sea surface temperature [4]. Increasing average temperature because of climate change causes reefs to be a frequent and extreme subject of bleaching [5]. Mass bleaching in 1998 was the worst in history, because one sixth of reef colonies die every year [1].

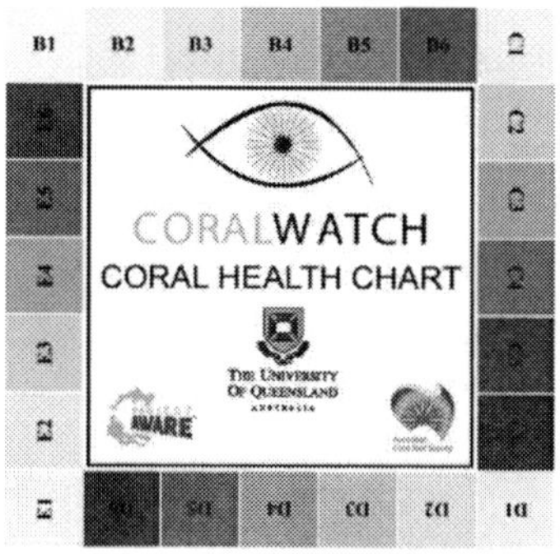

Figure 1. Coral Health Chart

CoralWatch is an organisation started from a research project in Queensland, Brsibane, Australia. *CoralWatch* uses *Coral Health Chart*, an undesrtuctible simple method to observe bleaching and reef's healthy. There are four classification colors in the chart, each represents coral reef, i.e. green, red, brown and brownish [1].

2.2 Digital Image Processing

Digital image processing is a rapidly growing field with improving application in science and technology. Digital image processing term generally refers to two dimensional image processing using digital computer. Digital image processing has important role in application such as remote detecting via satelite and spaceship, image transmission and storage for applications in business, medical process, radar, sonar, acoustic image processing, robotics, etc [6]. Digital *image* is function $f(x,y)$ with size M rows and N columns, where x and y are spatial coordinates and f on coordinates (x,y) is called grey level or intensity of image on that coordinates.

In matrix notation:

$$f(x,y) \approx \begin{bmatrix} f(0,0) & f(0,1) & \dots & f(0,M-1) \\ f(1,0) & f(1,1) & \dots & f(1,M-1) \\ \vdots & \vdots & \vdots & \vdots \\ f(N-1,0) & f(N-1,1) & \dots & f(N-1,M-1) \end{bmatrix}$$

A digital image can be considered as large array of discrete points. These points are called image elements or *pixel*. In *true color*, each *pixel* has specific color which is represented by amount of red, green and blue in it [7]. Each color is produced by mixing of three primary colors or RGB [8].

For *RGB* or red, green, blue , the color has range between 0-255, this gives total $255^3 = 16.777.261$ colors which can be different in the image [7]. Color model *RGB* based on color responded by human eyes which sensitive to wavelength around 400 nm (blue) to 700 nm (red) [9].

Random colors can be produced from the mix of primary colors *R*, *G*, *B* with color specific percentage (*W*):

$$W = R + G + B \tag{1}$$

2.3 Last Square Method

In science and technology, often appear case in trial which gives sets of data points $(x_i, y_i), \ldots, (x_N, y_N)$ where abscissa $\{x_i\}_{i=1}^{N}$ finite. One of the numerical method goals is determining formula $y = f(x)$ which relates all variables [10]. In this case, linear function is used in the form:

$$y = f(x) = ax + b \tag{2}$$

Sometimes there are errors in measurement, so that the real value of $f(x_i)$ holds:

$$f(x_i) = y_i + e_i \tag{3}$$

Where e_i is error in measurement.

In determining best linear estimation of form (2), it has to be discussed the error

$$e_i = f(x_i) - y_i \tag{4}$$

To measure the length of line $y = f(x)$ on the data, *root-mean-square error*

$$E(f) = \left(\frac{1}{N}\sum_{i=1}^{N}(f(x_i) - y_i)^2\right)^{1/2} \tag{5}$$

is used.

Theorem *(Least-squares Line)*

If $\{(x_i, y_i)\}_{i=1}^{N}$ are points *N*, where abscissa $\{x_i\}_{i=1}^{N}$ finite. Coefficient of *least-squares* line

$$y = ax + b$$

is the solution of linear system known as *normal equations*:

$$\begin{gathered}\left(\sum_{i=1}^{N} x_i^2\right)a + \left(\sum_{i=1}^{N} x_i\right)b = \sum_{i=1}^{N} x_i y_i \\ \left(\sum_{i=1}^{N} x_i\right)a + Nb = \sum_{i=1}^{N} y_i\end{gathered} \tag{6}$$

with,

$$a = \frac{N\sum_{i=1}^{N} x_i y_i - \sum_{i=1}^{N} x_i \sum_{i=1}^{N} y_i}{N\sum_{i=1}^{N} x_i^2 - (\sum_{i=1}^{N} x_i)^2}$$

$$b = \frac{\sum_{i=1}^{N} y_i \sum_{i=1}^{N} x_i^2 - \sum_{i=1}^{N} x_i \sum_{i=1}^{N} x_i y_i}{N\sum_{i=1}^{N} x_i^2 - (\sum_{i=1}^{N} x_i)^2} \tag{7}$$

where,

a= gradient.
b= intercept to vertical axis.

2.4. Bunaken National Park

Bunaken National Park is a sea park located in North Sulawesi Province and part of Manado city territory. This park located on Coral Triangle Reef and is home of 390 coral reef species (http://id.wikipedia.org/wiki/Taman_Nasional_Bunaken) as well as various species of fish, molluscs, sea reptiles and mammals. Bunaken National Park is symbol of Indonesia's sea ecosystem, including seagrass meadows, coral reefs and beach ecosystems.

This national park was established in 1991 with SK Menteri Kelautan RI No 730/Kpts-II/1991 as its legal document and covers areas of 890.65 km². (http://www.dephut.go.id/ INFORMASI/TN%

20INDO-ENGLISH/tn_bunaken.htm). Geographic location of National Sea Park is on 1^0 35' – $1^0$49'LU dan $124^0$39' – $124^0$35' BT. Of all habitat in this national park, there are 97% of sea habitat and 3% of land habitat. Bunaken National park covers five island, i.e Bunaken, Manado Tua, Mantehage, Naen and Siladen. For Bunaken Island with its diving point, given in Figure 1.

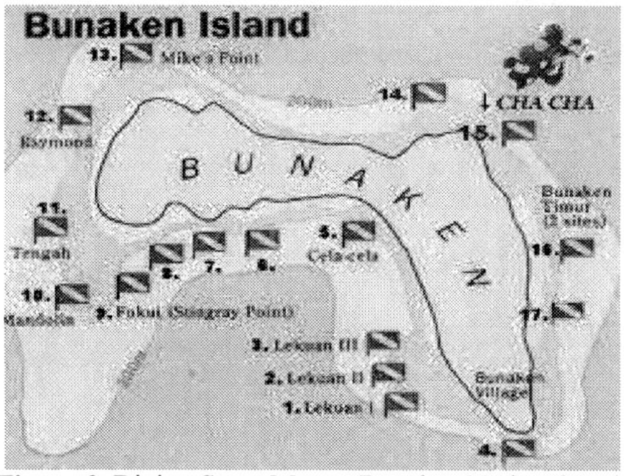

Figure 2. Diving Spots Map at Bunaken National Park

3. Research Method

Steps:

1) Taking digital image Coral Health Chart from CoralWatch.
2) Applying digital image processing on Coral Health Chart to calculate RGB average points from the four color classes.
3) Determining regression equations from Coral Health Chart with least square method based on RGB average points from the four color classes obtained before with computer application system.
4) Taking coral reef image sampel in Bunaken National Park.
5) Determining classification of reef's color class.
6) Applying digital image processing on coral reefs' image to find RGB average points.
7) Determining and concluding reef's healthy level using formula from Coral Health Chart.

In this research, the image was taken directly by diving into the place where the reef was located. The camera used is a camera with a resolution of 10 mega pixels. Also do not consider to the level of turbidity of sea water, lighting and others. This results in the possibility of distortion in taking pictures of the reef and this research has not considered the distortion.

4. Result and Discussion

4.1 Research Data

Coral reef image data is randomly collected from 10 diving spots in Bunaken National Sea Park, i.e. spots 2, 3, 5, 6, 9, 11, 13, 14, 16, 17, with 4 images from each spot. Therefore, data used is 40 images.

4.2. Regression Equations

Regression equations for class B, C, D, and E (from Coral Health Chart) are :

1. Class B : $y = -0.032x + 9.22$
2. Class C : $y = -0.026x + 7.45$
3. Class D : $y = -0.028x + 8.01$
4. Class E : $y = -0.030x + 8.63$

These regression equations become a model for determining the level of coral health in the application system that is built. The value of x is the RGB average of each reef in its class and the y value is the level of freshness of the reef expressed in the interval of values 1 - 6, ranging from unhealthy to healthy.

IOP Conf. Series: Materials Science and Engineering **567** (2019) 012004 doi:10.1088/1757-899X/567/1/012004

4.3. Application System

Application system is constructed with *Guide User Interface* (GUI) created is as in Figure 3.

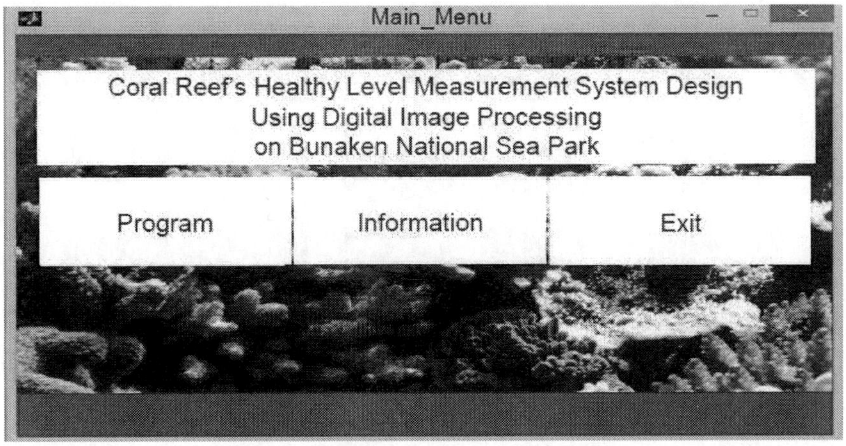

Figure 3. Application System Main Menu

In the Figure 4, as application system main menu, there are **program**, **information** and **exit** buttons. **Program** contains application to input coral reef image and determine its healthy level. **Information** contains information about coral health chart as reference to determine reef healthy level. **Exit** is the exit button from main menu.

If the program button is pressed, then the window will appear as in Figure 4.

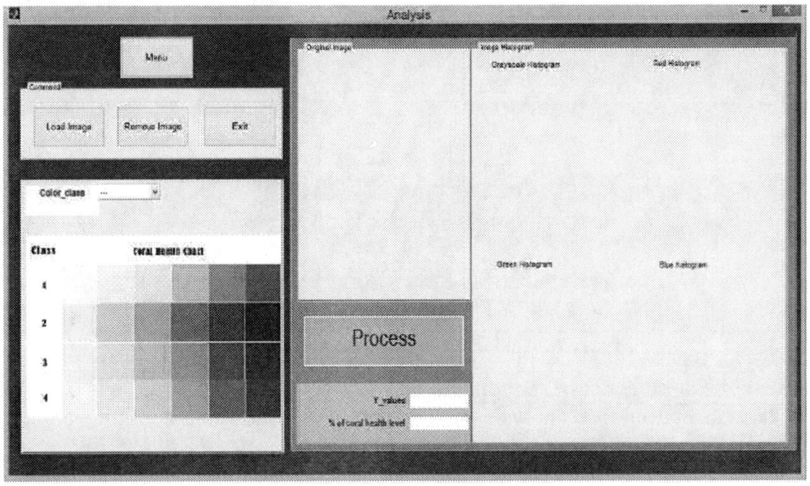

Figure 4. Analysis window

Steps to determine reef healthy level on this program are:

1. Press ***Load Image*** button then there will appear a window as in Figure 5..
2. Then choose coral reef image file which will be analysed from the image storage folder and press ***Open.***

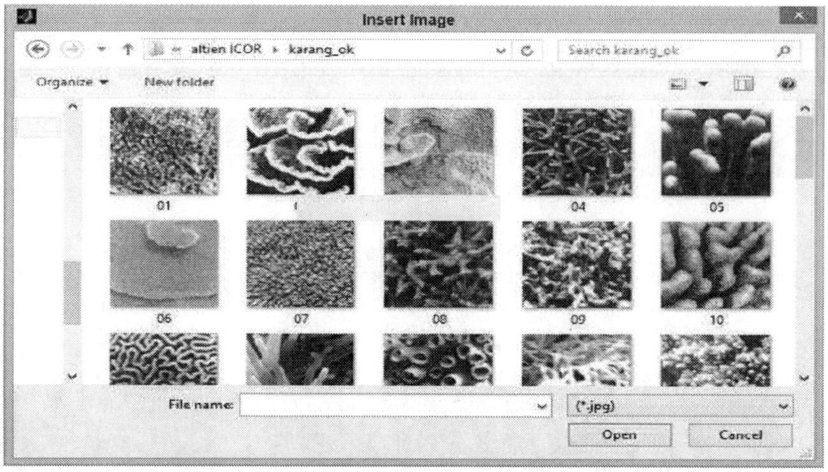

Figure 5. Window to choose coral reef image file

On the analysis window, there will rise appearance as in Figure 6.

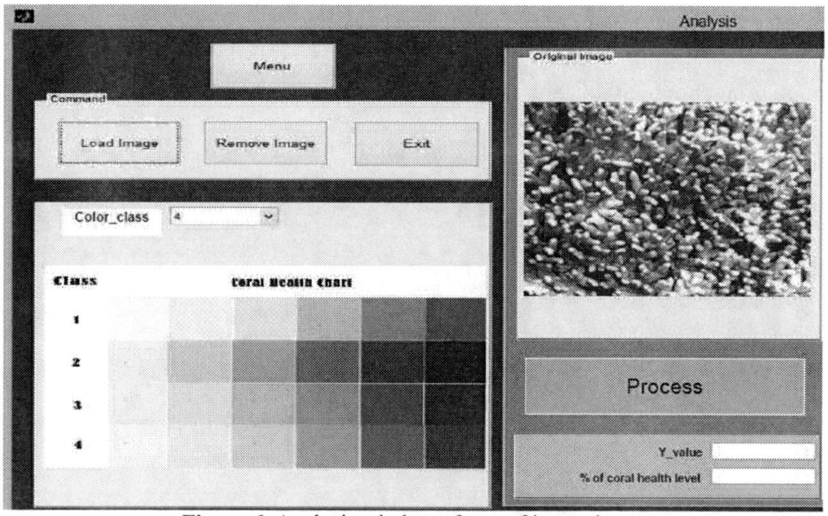

Figure 6. Analysis window after reef image input

3. Choose reef color class corresponding to *coral health cart* (On this case is color class 4)
4. Press **Process** button, then analysis window will give conclusion about the coral reef's healthy level, as in Figure 7.
5. Do the same steps for other reef's images.

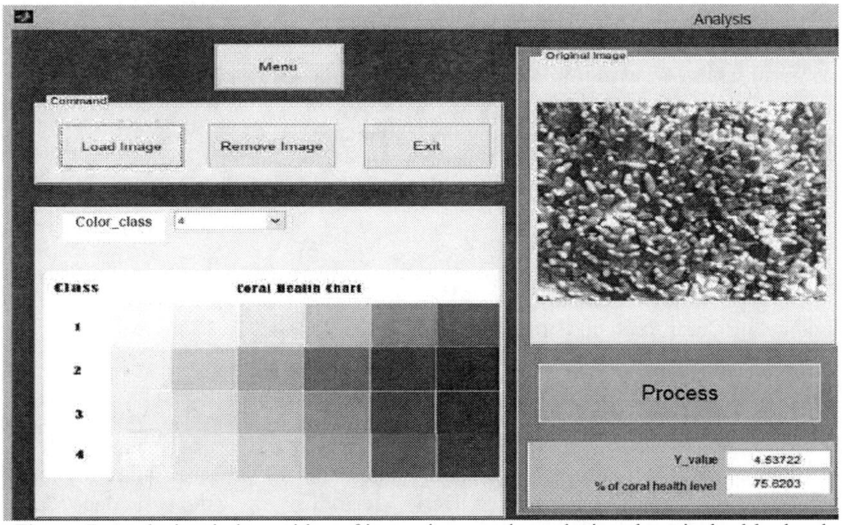

Figure 7. Analysis window with reef image input and conclusion about its healthy level.

4.4. Healthy Level of Reef of Bunaken National Sea Park

Based on system analysis for 40 images of coral reefs in Bunaken National Sea Park, results obtained can be shown on table 1.

Table 1. Healthy Level of 40 Coral Reefs Sample

Sample Number	Healthy Level Points	Healthy Level Percentage	Sample Number	Healthy Level Points	Healthy Level Percentage
01	4.53722	75.6203	21	2.07688	34.6147
02	4.01406	66.9011	22	3.12380	52.0634
03	4.77539	79.5899	23	3.24342	54.0750
04	5.69843	94.9738	24	4.37487	72.9145
05	5.62807	93.8012	25	5.30097	88.3496
06	5.12022	85.3371	26	2.31989	38.6648
07	3.81096	63.5160	27	3.17864	52.9774
08	5.87616	97.9361	28	4.99504	83.2507
09	4.11918	68.6529	29	4.29341	71.5568
10	3.92341	65.3902	30	4.3026	71.7101
11	5.71350	95.2251	31	4.74353	79.0588
12	4.74016	79.0027	32	3.69888	61.6479
13	4.48465	74.7441	33	5.68900	94.8167
14	3.91990	65.3317	34	4.03312	67.2187
15	4.00790	66.7984	35	5.61147	93.5244
16	5.36629	89.4382	36	4.92388	82.0646
17	4.56291	76.0485	37	5.29551	88.2586
18	3.26236	54.3727	38	4.14463	69.0771
19	3.80768	63.4613	39	4.36936	72.8226
20	2.36645	39.4409	40	5.30931	88.4885
			Average	**4.36908**	**72.8184**

Based on table 1, it can be seen that the average of healthy level of reefs in Bunaken National Sea Park is 4.36908 or its percentage is 72.8184%. This is fairly at good condition, nevertheless there must be serious concern to preserve or even improve the healthy level.

Descriptively, there are some snorkeling and diving spots where the coral reefs in it have died because of unaware tramples of some beginner snorkelers and divers. The spot is spot number 6

IOP Conf. Series: Materials Science and Engineering **567** (2019) 012004 doi:10.1088/1757-899X/567/1/012004

(Tawara). Whereas almost all spots in Eastern Bunaken (number 16 & 17) are still in good and conserved conditions.

Because in taking coral pictures does not consider the effect of camera resolution, the level of turbidity of the sea, light, etc., then there is a possibility of distortion of coral images. This is an opportunity for further research by considering these matters.

5. Conclusion

Based on this research, it can be concluded that:

1. Application system using digital image processing can be used to determine healthy level of coral reefs in Bunaken National Sea Park.
2. Healthy level of Coral Reefs in Bunaken National Sea Park is 72.8184%.

6. References

[1] Reid C, Marshall J, Logan D and Kleine D 2009 *Coral Reefs and Climate Change. The Guide for Education and Awareness* CoralWatch The University of Queensland

[2] Jones R, Guldberg O H, Larkum A W L, and Schreiber U 1998 Temperature Induced Bleaching of Corals begins with Impairment of Dark Metabolism in Zooxanthellae *Plant Cell and Environment* 21(12):1219–1230

[3] Siebeck U E, Marshall N J, Kluter A, and Guldberg O H 2006 Monitoring Coral Bleaching using Colour Reference Card *Coral Reefs* 25:453-460

[4] Guldberg O H 1999 Climate Change, Coral Bleaching and the Future of the World's Coral Reefs *Marine and Freshwater Research* 50(8): 839–866

[5] Westmacott S, Teleki K, Wells S, and West J 2000 *Pengelolaan Terumbu Karang Yang Telah Memutih Dan Rusak Kritis. Terjemahan Jan Henning Steffen* The Nature Conservation Bureau Ltd Newbury England

[6] Jain A K 1989 *Fundamental of Digital Image Processing* Prentice-Hall Internasional Inc Singapore

[7] McAndrew A 2004 *An Introduction to Digital Image Processing with MATLAB*. Victoria University of Technology. Australia

[8] Young T 1992 On the Theory of Light and Colors *Philosophical Transactions of the Royal Society of London* 18(02):20-71

[9] Presetyo E 2011 *Pengolahan Citra Digital dan Aplikasinya Menggunakan MATLAB.* Penerbit Andi Yogyakarta

[10] Mathews J and Fink K 1999 *Numerical Methods using MATLAB 3^ed^* Prentice Hall Upper Saddle River USA

IOP Conf. Series: Materials Science and Engineering **567** (2019) 012005 doi:10.1088/1757-899X/567/1/012005

Numerical modelling for tsunami wave propagation (case study: Manado bays)

G Pasau[1], G H Tamuntuan[1], A Tanauma[1]

[1]Department of Physics, Faculty of Mathematics and Natural Sciences, Sam Ratulangi University, 9500-Manado, Indonesia

Corresponding author: guntur_pasau@unsrat.ac.id

Abstract. The north arm of Sulawesi is one of the most vulnerable areas of the earthquake and tsunami. This region is a meeting place for several tectonic plates such as the Indo-Australian Plate, the Pacific Plate, the Philippine Sea Plate and the Eurasian Plate. As a result, collisions of the plates, subduction zones are formed in this region. The North of Sulawesi subduction zone and Molluca double subduction zone are sources of earthquakes that can generate tsunamis in the region. This study aims to examine the potential tsunami hazard on the north coast of Sulawesi's northern arms specifically in the Manado bays. Data were collected from various tsunami catalogues and literature review to determine the input parameters. The modelling is done by using TUNAMI N2 software which is a tsunami modelling using shallow water wave theory. The simulation results show that the tsunami wave height that can hit the Manado beaches varies from about 1.5 to 3 meters, while the tsunami propagation time from the source to the beach is about 24 minutes after earthquake.
Keywords: Earthquake, Tsunami, potential, TUNAMI

1. Introduction

Indonesia is the third largest country in the world within 500 years, with a population of around 5.5 million people killed by the tsunami waves. Even within a hundred years Indonesia has been hit by at least 24 tsunamis with the lives of no less than 231,900 people [1]. As a country located in the seismic zone of Southeast Asia, Indonesia is one of the countries with the highest level of seismicity in the world [1][2]. This is because the territory of Indonesia is a meeting place for some of the world's large plates such as the Indo-Australian plate, the Pacific plate and the Philippine plate which are under the Eurasian plate.

One of Indonesia's with high seismicity regions is the northern arm of Sulawesi. The North Arm of Sulawesi Island is geologically a meeting point of several tectonics plates. North Arm of Sulawesi located between three plates: Eurasian Plate, the Pacific Plate, and the Indo-Australian Plate and a micro plate, the Philippine Sea Plate. As a result of the complexity of the movement of these plates resulted in the formation of Molluca sea double subduction and North Sulawesi Trench which could potentially cause medium to large earthquakes in megathrust zones which could trigger a tsunami. Historical tsunami data show that the area of the north arm of Sulawesi has been hit several times by the tsunami wave causing loss of life, devastating property and environmental damage [3].

Tsunami is an ocean waves that occur due to an impulsive undersea disturbance or activity near the coast or in the ocean, when a sudden displacement of a large volume of water occurs, or if the sea floor is suddenly raised or dropped by an earthquake, big tsunami waves can be formed. The waves travel out of the area of origin and can be extremely dangerous and damaging when they reach the shore. This exclusive disruption occurs due to sudden sea deformation, especially in the vertical direction. Deformation can be caused by three main sources, namely earthquakes, volcanic eruptions, and landslides on the seabed. In general, tsunamis are caused by earthquakes, where the vertical tectonic shift of the seabed along the fracture zones on the earth's crust causes vertical disturbance of the water body[1][4].

Mitigation efforts need to be done as early as possible. Tsunami disaster mitigation is one of them through tsunami modelling or simulation. Tsunami modelling is an effort to determine the spread of tsunami waves from a particular source mechanism, the time needed for a tsunami wave (travel time) from the trigger source to one of the coastal areas and tsunami height (run up) in the area affected by the tsunami wave. Predictions of the time of propagation and height of tsunamis on the coast are very important as input in making evacuation plans and maps, construction of sea walls on the beach, plans for development and development of coastal areas

2. Tectonics Setting of North Arm Sulawesi

The northern arm of Sulawesi is one of the regions that has a very high level of seismicity (Figure 1) when compared to other regions in Sulawesi Island. The last largest earthquake in the northern arm was occurred in 1996 with a magnitude of M7.9. The source of the earthquake in this region came from several sub-districts such as North Sulawesi subduction (Minahasa Trench), Moluccan sea double collisions, subduction of the Philippine sea plate, and several active faults on the mainland north of Sulawesi [3]

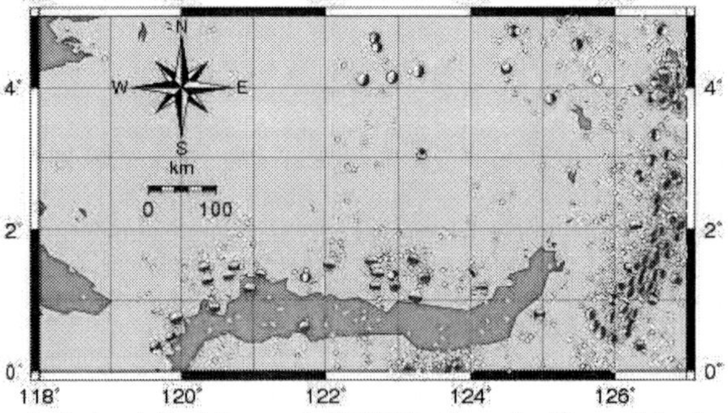

Figure 1. Seismicity Map and CMT Earthquake Focus Mechanism

Tsunami

Most tsunamis are caused by disturbances in the earth's crust on the seabed such as seabed earthquakes, landslides, or volcanic eruptions on the seabed causing rising water levels in large areas [5][6]. Tsunamis caused by earthquakes, where tectonic shifts in the vertical direction of the seabed along the fracture zones of the earth's skin cause vertical disturbances of the water body.

Shallow Water Equation

The tsunami wave propagation equation is approached by using a non-linear equation which is a shallow water equation [7]. Wave propagation model in shallow water is expressed by mass conservation / conservation equation and momentum equation in 3-dimensional cases as shown in equation (1).

$$\frac{\partial \eta}{\partial t}+\frac{\partial u}{\partial x}+\frac{\partial v}{\partial y}+\frac{\partial w}{\partial z}=0$$

$$\frac{\partial u}{\partial t}+u\frac{\partial u}{\partial x}+v\frac{\partial u}{\partial y}+w\frac{\partial u}{\partial z}+\frac{1}{\rho}\frac{\partial p}{\partial x}+\frac{1}{\rho}\left(\frac{\partial \tau_{xx}}{\partial x}+\frac{\partial \tau_{xy}}{\partial y}+\frac{\partial \tau_{xz}}{\partial z}\right)=0$$

$$\frac{\partial v}{\partial t}+u\frac{\partial v}{\partial x}+\upsilon\frac{\partial v}{\partial y}+w\frac{\partial v}{\partial z}+\frac{1}{\rho}\frac{\partial p}{\partial x}+\frac{1}{\rho}\left(\frac{\partial \tau_{xx}}{\partial x}+\frac{\partial \tau_{xy}}{\partial y}+\frac{\partial \tau_{xz}}{\partial z}\right)=0$$

$$g+\frac{1}{\rho}\frac{\partial \rho}{\partial z}=0 \tag{1}$$

with x, y denotes the horizontal axis, z represents the vertical axis, t represents time, η states the vertical change in sea water immediately after the disturbance, h is the depth of sea water, u, v and w represent the velocity of the water particle in the direction of the x, y and z axes, g is the gravitational acceleration of the earth[7][8].

Tsunami wave propagation modeling is divided into two, namely near field tsunami and far field tsunami. In far field the tsunami uses a spherical coordinate system while near field tsunami is approached using a Cartesian coordinates system. This study used near field tsunamis so that in tsunami wave propagation modeling uses Cartesian cooperatives. The use of equations in Cartesian coordinates is based on the assumption of the shallow-water theory which states that the vertical acceleration of water particles is negligible when compared to gravitational acceleration except for the distribution of tsunamis in the ocean. We can solve the tsunami wave propagation problem by using equation (1) on boundary conditions. Kinetic and dynamic boundary conditions on the surface and at the base are given with the terms:

$$p=0 \qquad \text{at } z=\eta \tag{2}$$

$$w=\frac{\partial \eta}{\partial t}+u\frac{\partial \eta}{\partial x}+v\frac{\partial \eta}{\partial y} \qquad \text{at } z=\eta \tag{3}$$

$$w=-u\frac{\partial h}{\partial x}-v\frac{\partial h}{\partial y} \qquad \text{at } z=-h \tag{4}$$

The dynamic and kinetic conditions of the equation (2)-(4) above finally obtained the nonlinear equation of movement (conservation of momentum) and the continuity equation (conservation of mass) in 2 dimensions (also called by shallow water wave theory):

$$\frac{\partial \eta}{\partial t}+\frac{\partial M}{\partial x}+\frac{\partial N}{\partial y}=0$$

$$\frac{\partial M}{\partial t}+\frac{\partial}{\partial x}\left(\frac{M^2}{D}\right)+\frac{\partial}{\partial y}\left(\frac{MN}{D}\right)+gD\frac{\partial \eta}{\partial x}+\frac{\tau_x}{\rho}=A\left(\frac{\partial^2 M}{\partial x^2}+\frac{\partial^2 M}{\partial y^2}\right)$$

$$\frac{\partial N}{\partial t}+\frac{\partial}{\partial x}\left(\frac{MN}{D}\right)+\frac{\partial}{\partial y}\left(\frac{N^2}{D}\right)+gD\frac{\partial \eta}{\partial y}+\frac{\tau_y}{\rho}=A\left(\frac{\partial^2 N}{\partial x^2}+\frac{\partial^2 N}{\partial y^2}\right) \tag{5}$$

where D is the total water depth of h, $h+\eta$, η_x and η_y are the basic friction in the direction of the x and y axes, A is the eddy viscosity in the horizontal direction which is assumed to be constant and the shear stress (movement due to wind) on the surface is ignored. M and N are components for releasing water mass flux in the direction of the x and y axes stated in the equation:

$$M=\int_{-h}^{\eta} u\,dz=u(h+\eta)=uD$$

$$N=\int_{-h}^{\eta} v\,dz=v(h+\eta)=vD \tag{6}$$

A tsunami moves at the same speed as the square root of the product of the multiplication between the gravitational acceleration of the earth and the depth of the sea [8][9]:

$$v = \sqrt{gh} \qquad (7)$$

where:
$\boldsymbol{v}$ = tsunami propagation speed (m/s)
g = gravity acceleration (m/s^2)
h = sea depth (m)
Each tsunami produces inundation and run up that vary according to energy and several factors such as coastline, bathymetry and coastal slopes

3. Methodology

The simulation uses a shallow water wave theory model in shallow sea. The simulation uses the help of the TUNAMI N2 (Tohoku University's Numerical Analysis Model for Investigation of near Field Tsunami software, number 2). TUNAMI N2 is a tsunami modeling that uses a leap-frog scheme [7]. This application uses tsunami generator source parameters based on the analysis of potential earthquake energy. The next step is to determine the area to be modeled using the nested grid method. The modeled area is then divided into several spatial resolutions (layers), each layer will be divided based on the tsunami generation area and the tsunami affected area. The widest area of work is layer 1 and is narrowing down to the extent of the layer located within it. In this study used four layers of each scenario.

The next stage of the research is to create a scenario using the source of the tsunami generator, namely the subduction of North Sulawesi (North Sulawesi subduction). The earthquake hypocenter is assumed to occur in the megathrust area or at a depth of less than 50 km. The North Sulawesi subduction consists of two segments namely the west and east segments. In this scenario using two segments at once namely the east and west segments as a tsunami generator.

Tsunami generation and propagation due to earthquakes with different fault parameters were studied. This scenario requires data such as strike, dip, slip, length and width of the fault with earthquake mechanisms interpreted in source parameters as presented in Table 1 below:

Table 1. Source Parameter

Parameter	Scenario 1	Scenario 2
Longitude	120,735	121,558
Latitude	1,879	2,29
L (m)	100 km	200 km
W (m)	60 km	60 km
Strike	60	100
Dip	20	20
Slip angle	85	85
Depth	10 km	10 km

The next step is to collect depth or bathymetry data from GEBCO (30 arcseconds) and 1 arcsecond SRTM Digital Elevation Model (DEM) topography data from USGS. The simulation results in the form of snapshots are displayed for 1 minute intervals, with a simulation time of 60 minutes (3600 seconds) after the earthquake. A minute wave that describes the time of arrival of a tsunami wave (travel time) and the tsunami wave height (run up) is obtained based on a predetermined parameter.

4. Discussion

Modeling results in the form of snapshots per minute which describe the time of propagation of tsunami waves (travel time) and tsunami height (run up) on each layer are explained as follows.

IOP Conf. Series: Materials Science and Engineering **567** (2019) 012005 doi:10.1088/1757-899X/567/1/012005

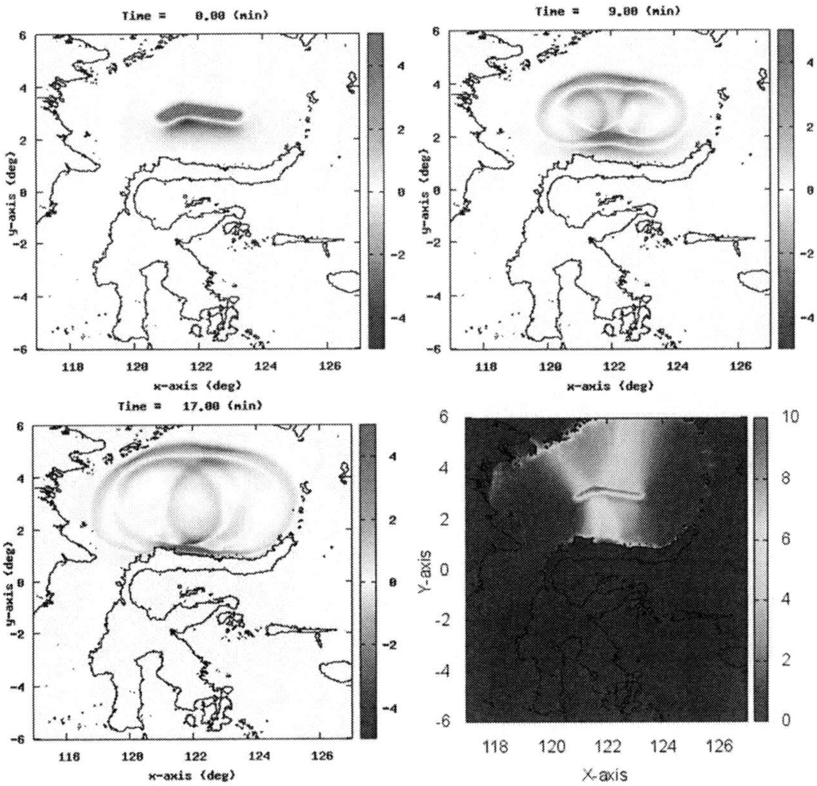

Figure 2. Visualization of tsunami wave propagation in the North Arm Sulawesi at 0 minute, 9 minute, 17 minute and maximum run up

From Figure 2 above it shows that shortly after the earthquake when t = 0 minutes most of the coastal area especially Tolitoli Regency, Gorontalo North Coast and North Bolaang Mongondow experienced a decrease in water level / low tide (blue). Likewise at t = 4 minutes the water still recedes to the large part of the north coast but once entering the 16th minute there is an increase in water level (red). From the simulation large maximum water levels ranging from 6-10m found along the northern coastline Buol Regency, Central Sulawesi.

The along north coast of Sulawesi's northern arm show the tsunami wave is expected to hit at 16-24 minutes after earthquakes. Especially the Manado Bay area is predicted that a tsunami can occur in the 24th minutes (Figure 3) after earthquake with the water level maximum is 1.5-3m. The water level maximum can greater than 3m due to amplification of tsunami waves that occur in the bay region.

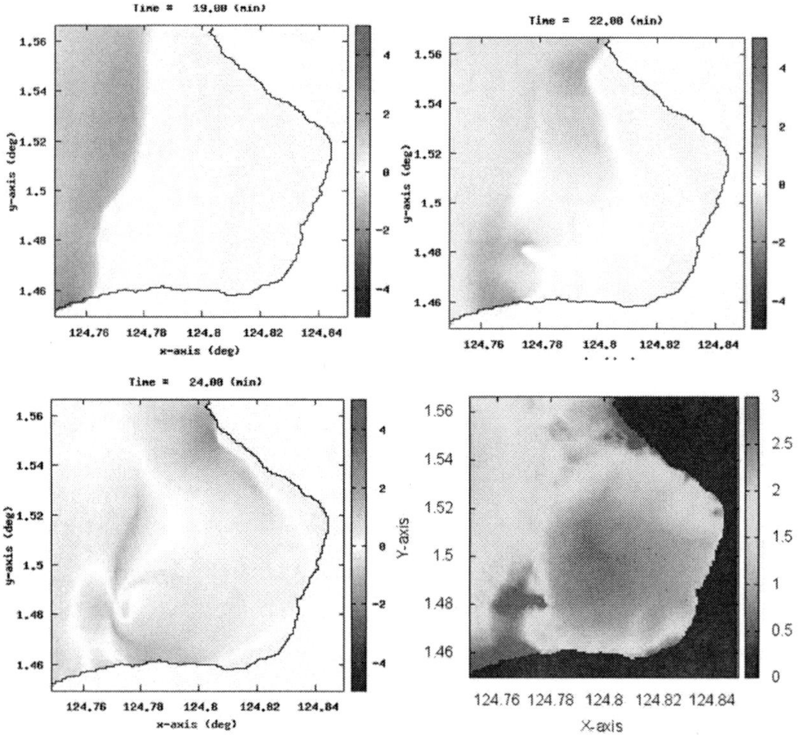

Figure 3. Tsunami wave propagation at layer 4 for 19 minutes, 22 minutes, 24 minutes and maximum run up on Manado Beach

The simulation results in Figure 3 are the same as the previous picture, but in this condition the area is smaller so that the affected area specifically can be seen more clearly. From the picture it can be seen that the high level (run up maximum) of tsunami waves that can occur in Manado bays is around 1.5-3 meters

To see clearly what the height of the tsunami in some places on the Manado coast, there are several Tide Gauges (TG) which are shown in the following Figure 4.

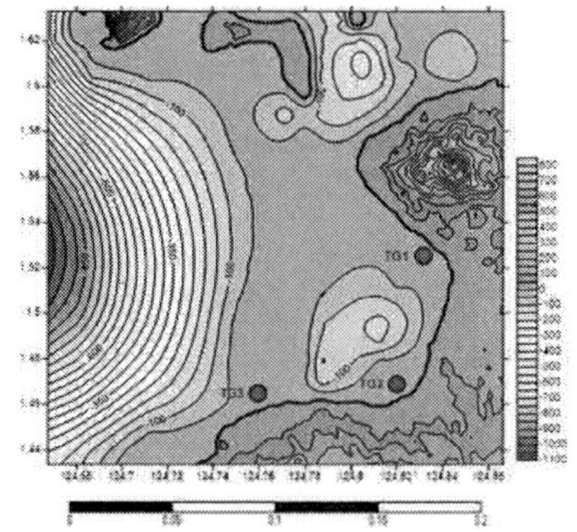

Figure 4. Positioning of Tide Gauges (TG)

The recording results for each Tide Gauge are the tsunami propagation time and wave height as shown in the following Figure 5.

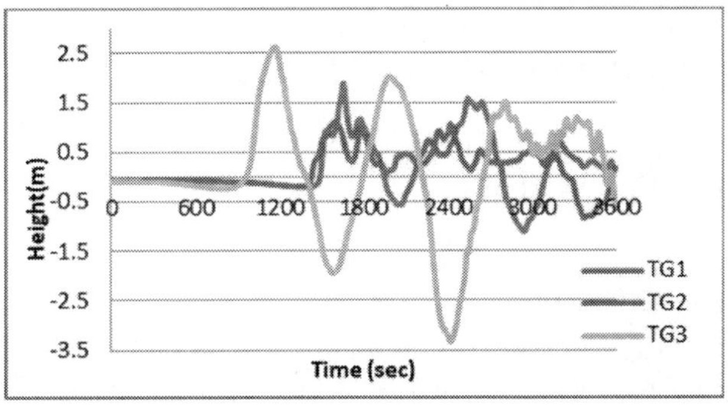

Figure 5. Time Series of Manado Bays Tsunami Wave Propagation, shows the location of each Tide Gauge

5. Conclusion

This model successfully simulates the propagation of tsunami waves due to earthquakes in North Sulawesi using the TUNAMI-N2 model. The northern coast region of the north arm Sulawesi is very vulnerable to the tsunami disaster. The height of the tsunami that can occur in this region is around 1-10 meters. Especially for beaches Manado bays the tsunami wave period ranges from 20 to 24 minutes after an earthquake with a height of around 1.5-3 meters. This tsunami height can be greater than 3 meters considering that the Manado beach is in the shape of a bay where tsunami wave amplification can occur.

References

[1] Horspool N, Pranantyo I, Griffinl J, Latief H, Natawidjaja, Kongko W, Cipta A, Bustaman, Anugrah, and Thio H K 2014 *Natural Hazards Earth System. Sciences* **14** 3105–3122.

[2] Aydan O 2008 Seismic and Tsunami Hazard Potential in Indonesia with a special emphasis on Sumatra Island *Journal of The School of Marine Science and Technology* **6** 19-38.

[3] Latief H, Puspito N T, Imamura F 2000 *Journal of Natural Disaster Science* **22(1)**, 25-43.

[4] Lovholt F, Glimsdal S, Habitz C, Horspool N, Smebye H, de Bono D, and Nadim F 2012 *Journal of Geophysical Research.*B **117** 03910

[5] Parwanto N B, Oyama T 2013 *International Journal of Disaster Risk Reduction.* **7** 122-141.

[6] Lovholt F D, Kuhn H, Bungum C B, Harbitz C, and Glimsdal S 2012 *Journal of Geophysical Research.* **117** 1-19.

[7] Imamura F, Yalciner A C and Ozyurt G 2006 *Tsunami Modelling Manual* (*TUNAMI model*). (Japan: Tohoku University)

[8] Satake K and Imamura, F 1995 *Journal Physics Earth* **43** 259-277.

[9] Lalli F, Bruschi A, Cassese M L, Lotti A, Pesarino V 2016 *Journal of Operational Oceanography* **9** 215–222.

IOP Conf. Series: Materials Science and Engineering **567** (2019) 012006 doi:10.1088/1757-899X/567/1/012006

Mathematical model of enzymatic reactions of coconut coir substrate treated by alkaline and ionic liquid into the reducing sugars using power series

H F Sangian[1], A Widjaja[2]

[1]Jurusan Fisika, Universitas Sam Ratulangi Manado, Sulawesi Utara, Indonesia 95115
[2]Jurusan Teknik Kimia, Institut Teknologi Sepuluh Nopember Surabaya, Jawa Timur, Indonesia 60111

Corresponding author: arief_w@chem-eng.its.ac.id

Abstract. The research purposed at formulating the kinetics model of reducing sugars liberated from coconut coir dust treated with alkaline, ionic liquid, alkaline followed by ionic liquid employing a power series. The method was successful to predict the sugar concentration released from the substrate in a biocatalytic reaction employing cellulose, and a mixture of cellulase+xylanase at any time. The yield of sugar released enzymatically was modeled as an equation in power series in which the results were comparable to those of the experimental data. The biggest constant rate obtained was of 0.0200000 h^{-1} for an application of a mixture of enzymes on the conversion of NaOH-treated substrate into the reducing sugars. Generally, the trend of the modelled graph was close to the experiment but the error was quietly wide in the initial times since the conversion rate was very high. The lines were an agreement when the time was above 8h for all pre-treatments (native-, 1%NaOH-, IL-, and 1%NaOH+IL substrates). It was indicative the power series can be used to solve a differential equation generated from a rate equation.

1. Introduction

Currently, the scientists have been interested in the lignocellulosic materials because they can be converted into sugars and biofuels and are available abundantly on earth especially, in the tropical countries [1][2]. The study showed that prior to conversion into more valuable materials, lignocellulose should be treated that purposed to decline the lignin content and to modify the cellulosic structure [3]. The conversion evaluation of lignocellulose treated by chemical or physical pre-treatment has been reported by many investigators [4][5][6].

The kinetic study is an important aspect of the chemical reaction in which it can be predicted an amount of the product in the extended range of time. The poplar wood has been liquefied into the reducing sugar by using the wet oxidation pre-treatment and the yield of product was formulated in the form of the mathematical model. The equation obtained was successful in predicting the reducing sugar yielded [7]. The transformation of sugarcane bagasse into levulinic acid enhanced by acid was reported by authors [8]. The kinetic formulas of enzymatic hydrolysis were written in some equations depending on the temperatures and sulfuric acid concentrations employed.

The kitchen waste was converted to the bioethanol by using acid and hot water treatments whereby the kinetics of the biochemical reaction was modeled in the simple form [9]. Authors [10] reviewed the kinetic model of enzymatic hydrolysis of many substrates. The results found the simultaneous hydrolysis and yeasting in continuous mode gave the biggest yield. The mathematical model of anaerobic digestion in the fermenter in producing biogas was studied previously [11]. The studies

IOP Conf. Series: Materials Science and Engineering **567** (2019) 012006 doi:10.1088/1757-899X/567/1/012006

discovered as described that through a kinetic mathematical model, the biochemical process inside the reactor would be understood deeply and the prediction of product yield was well known.

This study reported the simple kinetic model of bioconversion of coconut coir dust treated by ionic liquid 1-ethyl-3-methylimidazolium dimethyl phosphate ([mmim][dmp]) and alkaline using the power series. The IL employed in this study was synthesized in the author's Lab and the result obtained through the kinetic model was compared to that of the experimental data. The kinetic model was formulated after the reducing sugars were obtained through enzymatic hydrolysis with a catalyst loading was constant.

2. Materials and Methods

2.1. Chemicals

The coconut coir was obtained from a copra farmer in Minahasa Regency, North Sulawesi Indonesia. The biomass was milled until the size of 100-120 mesh employed an apparatus the Retsch GmbH Rheinische Strade 362/2781, Haan Germany. The ionic liquid [MMIM][DMP], cellulase, and xylanase were purchased from Sigma-Aldrich, St. Louis, MO, USA. The procedures for all steps started with pretreatments and solids characterization were published previously [3][4][6].

2.2. Enzymatic reaction

The enzymatic reaction to convert substrate into sugars was conducted by adapting a previous report (Zhang and Bao, 2017). The sodium acetate set at pH 3, 30 mL, and 0.1M was employed as a suspension and was mixed with a 1 gram of (120 mesh) inside Ellen-Meyer flask. The temperature of the reactor was kept at 60 °C by being stirred for 48h. The sugars started liberating from the substrate when 0.2ml cellulase or cellulase+xylanase was added to the reactor and if needed the citric acid was also added to maintain pH at 3. The sample as much as 0.4ml was taken periodically from the mixture and dissolved with 1.8 ml of pure water as well as 3 ml DNS. The concentration of sugars obtained was measured by employing a DNS method [12].

2.3. The enzymatic reaction model

The mathematical model, which was proposed, followed and modified from an established equation as proposed previously by authors [7][13]. The first, lignocellulose converted into the reducing sugars could be modeled as

$$coconut\ coir \rightarrow sugars \tag{1}$$

The rate formulation was defined as follows:

$$-r = km_S{}^n \tag{2}$$

Where, r is reaction rate (*g/s*); k is the reaction constant (unit depends on the reaction order), m_S is the substrate mass (g) and n is the reaction order. In this study, since the catalyst loading is constant, the influence of *cellulase* was absorbed in k parameter and the temperature reaction was kept in 60°C and the variable r was the decrease rate of substrate mass with respect to time or $\frac{dm_S}{dt}$.

To solve Eq. (2) was assumed that the order of enzymatic hydrolysis was an order one, $n=1$ as proposed by investigators [13][14]. The mass of the initial substrate and reducing sugar were assigned by m_O and m_{RS}. The substrate mass could be written as $m_S = m_O - m_{RS}$ so the differential equation of Eq. (2) was modified as:

$$\frac{dm_{RS}}{dt} + km_{RS} = km_O \tag{3}$$

The solution of Eq. (3) was solved in two ways by the analytical method and mathematical series. The solutions of Eq. (3) were two parts, a homogeneous solution in which the right side was zero and a particular solution that was easy to obtain and will be elaborated widely in results and discussion below. By employing a variable separation, the mass of reducing sugar (m_{RS}) can be written as

$$m_{RS} = m_O\left(1 - e^{-kt}\right) \tag{4}$$

IOP Conf. Series: Materials Science and Engineering **567** (2019) 012006 doi:10.1088/1757-899X/567/1/012006

The solution of Eq. (3) could be expanded in form of mathematical series by assuming the k is much less than unity as follows:

$$m_{RS} = \sum_{n=0}^{N} a_n t^n = a_o + a_1 t + a_2 t^2 + a_3 t^3 + a_4 t^4 + \cdots \tag{5}$$

The advantage of using the power series method is that we do not need to solve the differential equation analytically which is more difficult by employing the conventional way. The using of power series technique only needs the ability to equate the parameters which attach to the similar order and then quantities are made as function k and a_o.

The first work was to find the solution of the homogeneous part by equalizing zero on the right side in Eq. (3). The Eq. (5) was derived with respect to time obtained Eq. (6) as follows:

$$\frac{dm_{RS}}{dt} = n \sum_{n=1}^{N} a_n t^{n-1} = a_1 + 2a_2 t + 3a_3 t^2 + 4a_4 t^3 + \cdots \tag{6}$$

After Eq. (6) is substituted into Eq. (3), the new expression will be obtained the following:

$$a_1 + 2a_2 t + 3a_3 t^2 + 4a_4 t^3 + \cdots = -k(a_o + a_1 t + a_2 t^2 + a_3 t^3 + a_4 t^4 + \cdots) \tag{7}$$

Now, parameters $a_1, a_2, a_3, a_4 \ldots\ldots a_n$ are attributed with quantities k and a_o by equalizing the terms both sides which have the same power and the results are shown below.

$$\begin{aligned} a_1 &= -ka_0 \\ a_2 &= \frac{k^2}{1x2} a_0 \\ a_3 &= -\frac{k^3}{1x2x3} a_0 \\ a_4 &= \frac{k^4}{1x2x3x4} a_0 \\ a_5 &= -\frac{k^5}{1x2x3x4x5} a_0 \\ a_6 &= \frac{k^6}{1x2x3x4x5x6} a_0 \\ &\vdots \\ &a_n;\ a_m \end{aligned} \tag{8}$$

The eq. (8) shows that the parameter is split into two parts, even and odd series as presented in Eq. (9) and (10).

$$a_n = \sum_{n=0}^{N} \frac{k^{2n} a_o}{(2n)!} \tag{9}$$

$$a_m = -\sum_{m=0}^{N} \frac{k^{2m+1} a_o}{(2m+1)!} \tag{10}$$

The particular part is found through supposing that m_{RS} = *constant* whose value is a m_o after being substituted for Eq. (3). The general solution obtained is an addition to the homogeneous- and particular parts are written as:

$$m_{RS} = m_o + \sum_{n=0}^{N} \frac{(kt)^{2n} a_o}{(2n)!} + \sum_{m=0}^{N} \frac{(kt)^{2m+1} a_o}{(2m+1)!} \tag{11}$$

The Eq. (11) can be expanded in the form of

$$m_{RS} = m_O + a_o \left[1 - (kt) + \frac{1}{2!}(kt)^2 - \frac{1}{3!}(kt)^3 + \frac{1}{4!}(kt)^4 - \frac{1}{5!}(kt)^5 + \cdots\right] \tag{12}$$

The experiment was observed that at $t=0$, $m_{RS}=0$, the a_o found is a negative of the initial mass of the substrate, so the Eq. (12) could be written as

$$m_{RS} = m_O\left\{1 - \left[1 - (kt) + \frac{1}{2!}(kt)^2 - \frac{1}{3!}(kt)^3 + \frac{1}{4!}(kt)^4 - \frac{1}{5!}(kt)^5 + \cdots\right]\right\}$$

$$Yield = \left\{\left[(kt) - \frac{1}{2!}(kt)^2 + \frac{1}{3!}(kt)^3 - \frac{1}{4!}(kt)^4 + \frac{1}{5!}(kt)^5 + \cdots\right]\right\} \tag{13}$$

The Eq. (13) derived from the power series is actually similar to the Eq. (4) that is obtained by solving the differential equation of Eq. (3) by an analytical method [15][16]. The fittings the data from the experiment and model were carried out to obtain the rate constant k using the solver *software* that is available in an excel program and would be elaborated the following.

Table 1. The comparative yields of sugar released from treated substrates (in g sugar/g substrate) with experiment and kinetic model data vs time course for employing the cellulase.

t(h)	No Pretreatment		1%NaOH-		4%NaOH-		IL-		1%NaOH+IL-	
	Exp. Yield	Model Yield	Exp. Yield	Model Yield	Exp. Yield	Model Yield	Exp. Yield	Model Yield	Exp. Yield	Model Yield
0	0.0000	0.0000	0.0000	0.0000	0.0200	0.0000	0.0000	0.0000	0.0000	0.0000
1	0.0500	0.0030	0.1100	0.0064	0.0900	0.0056	0.0700	0.0047	0.1099	0.0074
2	0.0600	0.0060	0.1100	0.0128	0.1000	0.0112	0.0700	0.0093	0.1394	0.0148
4	0.0500	0.0119	0.1200	0.0254	0.1000	0.0224	0.0900	0.0185	0.1361	0.0293
6	0.0600	0.0178	0.1400	0.0379	0.1200	0.0334	0.1000	0.0276	0.1485	0.0436
8	0.0600	0.0237	0.1500	0.0501	0.1300	0.0442	0.1100	0.0367	0.1498	0.0577
10	0.0700	0.0295	0.1500	0.0623	0.1300	0.0550	0.1100	0.0456	0.1675	0.0717
12	0.0700	0.0354	0.1500	0.0743	0.1400	0.0656	0.1200	0.0545	0.1661	0.0854
18	0.0800	0.0526	0.1600	0.1093	0.1400	0.0968	0.1200	0.0806	0.1866	0.1253
24	0.0800	0.0695	0.1600	0.1430	0.1500	0.1269	0.1200	0.1060	0.2065	0.1634
30	0.0900	0.0860	0.1800	0.1755	0.1600	0.1560	0.1400	0.1307	0.2041	0.1999
36	0.0900	0.1023	0.1900	0.2067	0.1700	0.1842	0.1500	0.1547	0.2162	0.2348
42	0.1000	0.1184	0.1900	0.2367	0.1700	0.2114	0.1400	0.1781	0.2068	0.2682
48	0.1000	0.1341	0.1800	0.2656	0.1600	0.2377	0.1400	0.2008	0.2001	0.3002

3. Results and Discussion

In the present work elaborated the yield of the reaction kinetic model of the substrate treated by an ionic liquid, alkaline and, alkaline followed by ionic liquid converting into the reducing sugars after conducting a fitting process and compared to the experimental data which are presented in Table 1 and

2. Fig. 1 shows the yield model of reducing sugar liberated from treated substrates (coconut coir) using a single enzyme, cellulase. If compared to the experimental yield that the error of the initial reaction was quite big since the sugar rate released from the substrate for pre-treatment was relatively high [17][18]. The assumption that the reaction for enzymatic hydrolysis is the first order is necessary to be evaluated for this case. As time increases, however, the experimental- and kinetic model yields tend to high accuracy that they are coming to an agreement.

The experimental data of sugar yields (in g sugar/g substrate) released from solids at 1h for original-, 1%NaoH-, 4%NaOH, IL, and 1%NaOH+IL substrates are 0.0500, 0.1100, 0.1000, 0.0700, and 0.10991, while the model yields are 0.003, 0.0064, 0.0056, 0.0047, and 0.0074, respectively. When time course is approaching 12h, the experimental and model yields decrease their differences. The experimental and model yields liberated from native coconut coir and substrates treated by 1%NaoH-, 4%NaOH, IL and 1%NaOH+IL at 12h are 0.0700; 0.0354, 0.1500; 0.0743, 0.1400; 0.0656, 0.1200; 0.0545, and 0.16613; 0.0854, respectively. The final time course the two yields resulted to close figures as follows: 0.1000; 0.1341, 0.1800; 0.2656, 0.1600; 0.2377, 0.1400; 0.2008, and 0.2001; 0.3002, respectively.

The data shows that the combination method, NaOH+IL pre-treatment gives the biggest yields of the sugar obtained from both an experimental and the model hydrolysis that was comparable to previous reports by the other authors.

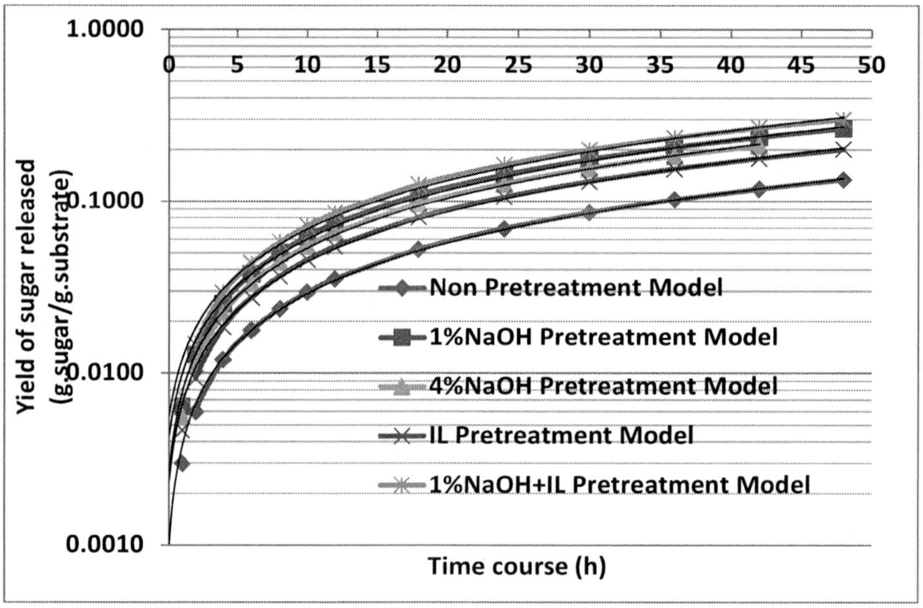

Figure 1. The kinetic model of reducing sugars yields released from substrates with respect to the time course employing a single cellulose

The second place of sugar yield found is the NaOH pre-treatment and then followed by IL technique and the least is the original substrate. It was found that NaOH pre-treatment could increase significantly the sugar released from substrate since lignin and hemicellulose dissolved into alkaline and the porosity of the substrate increased [19]. When alkaline concentration inclined to 4%, the sugars liberated decreased slightly. The increase of alkaline concentration started liquefying cellulose and hemicellulose as the source of sugar. It meant that the more alkaline added can be decreasing the yield of sugar obtained since cellulose is decomposed into the water-soluble substances.

Table 2 shows the yields of sugar (experiment and model) liberated from coconut coir treated by 1%NaoH-, 4%NaOH, IL, and 1%NaOH+IL versus time course using a mixture of cellulase+xylanase.

Meanwhile, the sugar yield of the kinetic model which is fit from experimental data is presented in Fig. 2. Generally, the behaviour of the kinetic model of enzymatic hydrolysis using mixture enzymes as shown in the figure is similar to that of employing single cellulase as described previously. The real difference is that the yield of sugar freed enzymatically from substrate treated all methods and employing mixture enzyme is higher than that of applying a single enzyme (cellulase). The significant increase occurred since the synergy of cellulase trims the 1,4-β-connections into glucose while xylanase broke down the hemicellulose rings becoming the xylose monomers [20]. The sugar yield of a model for an original solid using mixture of enzymes at 48h is of 0.2066 that is higher than that of cellulase recorded at 0.1800 g sugar/g substrate. The sugar from the model and hydrolyzed of 1%NaOH-treated solid gave of 0.3061 g sugar/g substrate that increased from 0.2656 using cellulase. While NaOH+IL pretreatment resulted in the biggest yield modelled observed at 0.3760 employing a mixture of enzymes if compared to that of single cellulase obtained at 0.3002 g sugar.g substrate.

Table 2. The comparative yields of sugar released from treated substrates (in g sugar/g substrate) with experiment and kinetic model data vs time course for employing the mixture cellulase and xylanase.

t(h)	No Pretreatment		1%NaOH-		4%NaOH-		IL-		1%NaOH+IL-	
	Exp. Yield	Model Yield	Exp. Yield	Model Yield	Exp. Yield	Model Yield	Exp. Yield	Model Yield	Exp. Yield	Model Yield
0	0.0000	0.0000	0.0000	0.0000	0.0000	0.0000	0.0000	0.0000	0.0000	0.0000
1	0.0593	0.0048	0.1212	0.0076	0.0900	0.0064	0.2340	0.0146	0.1338	0.0098
2	0.0766	0.0096	0.1557	0.0151	0.1136	0.0128	0.2669	0.0289	0.1742	0.0195
4	0.0843	0.0191	0.1633	0.0300	0.1214	0.0254	0.2991	0.0570	0.1954	0.0385
6	0.0851	0.0285	0.1730	0.0446	0.1379	0.0378	0.2943	0.0843	0.2216	0.0572
8	0.0897	0.0378	0.1783	0.0591	0.1446	0.0501	0.3153	0.1107	0.2259	0.0756
10	0.1054	0.0471	0.1707	0.0733	0.1494	0.0622	0.2912	0.1365	0.2434	0.0936
12	0.1257	0.0562	0.1770	0.0873	0.1509	0.0742	0.2943	0.1614	0.2507	0.1112
18	0.1473	0.0831	0.1942	0.1280	0.1635	0.1092	0.3100	0.2321	0.2527	0.1621
24	0.1459	0.1093	0.1983	0.1670	0.1722	0.1429	0.3328	0.2968	0.2541	0.2100
30	0.1419	0.1347	0.2180	0.2042	0.1780	0.1753	0.3394	0.3561	0.2527	0.2552
36	0.1423	0.1594	0.2173	0.2397	0.1964	0.2065	0.3530	0.4104	0.2525	0.2979
42	0.1444	0.1833	0.2080	0.2736	0.1851	0.2365	0.3490	0.4601	0.2441	0.3381
48	0.1360	0.2066	0.1917	0.3061	0.1737	0.2653	0.3168	0.5057	0.2322	0.3760

Table 3 presents the constant rate of reaction kinetic model of sugars released from substrates that were pretreated by 1%NaoH-, 4%NaOH, IL, and 1%NaOH+IL and compared to non-pretreatment written in the first row. The constant rates of sugar from native coconut coir employing cellulase and mixture of enzymes are of 0.0029999 and 0,0048220 h^{-1}. When solid was treated with 1%NaOH, the constant rate increased to 0.0064311 (cellulase) and 0.0076116 h^{-1} (cellulase+xylanase). If alkaline concentration inclined to 4%, the constant rate went down to 0.006000 h^{-1} for two enzymatic. The decrease was similar to the experimental data obtained which was caused by the decomposing of cellulose into water-soluble substances. The combination method (NaOH+IL), the constant rate of an enzymatic reaction increased to 0.0074400 (cellulase) and 0.02 h^{-1} (cellulase+xylanase) that were the highest rate obtained. The significant increase was primarily caused by the dissolution of lignin and hemicellulose because of alkaline treatment and then the change in substrate crystallinity after the IL application [21]. The substrate used in this study was coconut coir dust that had a lignin composition of around 41% [22]. The combination method, alkaline followed by ionic liquid was the best option for high lignin biomass before converting it into sugars, or bioethanol.

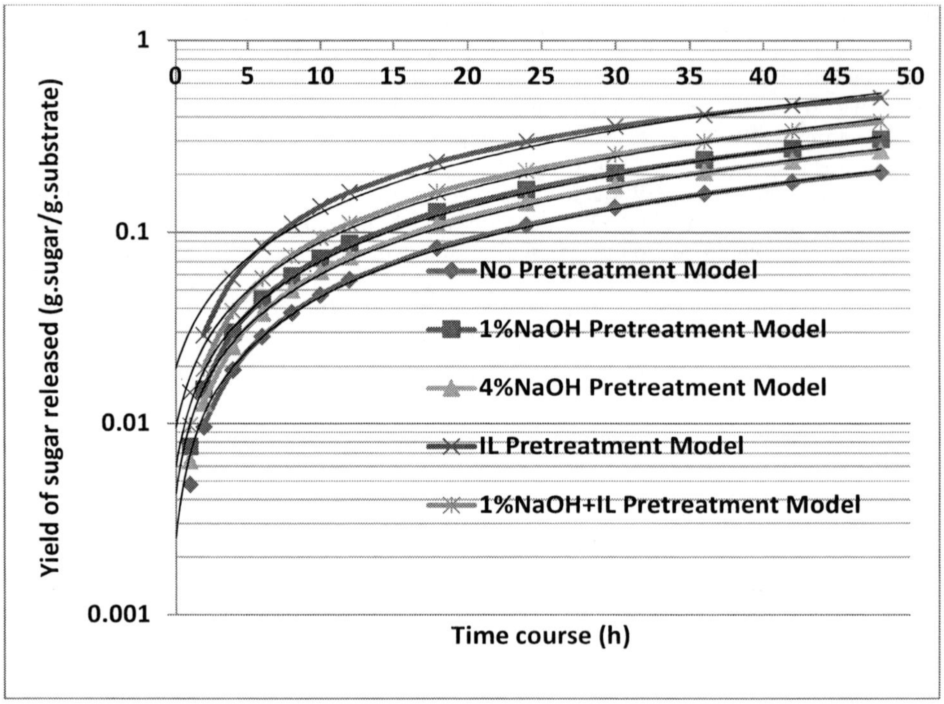

Figure 2. The kinetic model of reducing sugars yields released from substrates with respect to the time course employing a mixture of cellulase and xylanase.

Table 3. The constant rate the chemical reaction of sugar released from substrates using cellulase and mixture cellulase+xylanase

No	Pretreatment	The constant rate (1/h)	
		cellulase	Cellulase+xylanase
1	No-	0.0029999	0.0048220
2	1%NaOH-	0.0064311	0.0076116
3	4%NaOH-	0.0060000	0.0060000
4	IL-	0.0046700	0.0146717
5	1%NaOH+IL-	0.0074400	0.0200000

The regressions show that the parameters of determination for employing cellulase were calculated and found from 0.9874 (no-pretreatment), 0.8364 (1%NaOH-), 0.7966 (4%NaOH-), 0.8066 (IL-), and 0.8127 (1%NaOH+IL-). While those employing mixture of enzymes declined to 0.6111, 0.4161, 0.5572, 0.3499, and 0.34419, respectively. The low R^2 values are indicative that the employing of a mixture of enzymes is harder to predict than that of single cellulase.

4. Conclusions

The power series were successful to predict the amounts of sugar liberated from lignocellulose treated by an ionic liquid, alkaline, alkaline followed by ionic liquid and compared to the experimental data. It was discovered that the power series equation derived from the general reaction rate was similar to that of the conventional method. The constant rates of an enzymatic reaction of treated substrates employing the mixture of enzymes were generally higher than those of applying the single cellulase. The determination parameters employing the mixture of enzymes were lower than that of cellulase. It meant that the sugar yielded enzymatically using cellulase+xylanase was more difficult to be predicted.

Acknowledgments
The authors would like to give thanks very much for the Higher Education Department of Indonesia Government for financial assistance and the Biochemical Engineering Lab of Institute Technology Sepuluh Nopember (ITS) Surabaya that gave authors wide access in conducting many experiments.

References
[1] Widjaja A, Agnesty S Y Sangian H F and Gunawan S 2015 *Bulletin of Chemical Reaction Engineering & Catalysis* **10** 70-77
[2] Sangian H F and Widjaja A 2017 *BioResources* **12** (4) 8030-8046
[3] Sangian H F, Ranggina D, Ginting G M, Purba A A, Gunawan S and Widjaja A 2015 *Scientific Study & Research Chemistry & Chemical Engineering, Biotechnology, Food Industry* **16** (1) 013 – 027
[4] Sangian H, Kristian J, Rahma S, Agnesty S Y., Gunawan S and Widjaja A 2015 *Journal of Engineering and Technological Sciences* **47** (2) 137-148
[5] Toor S S, Rosendahl L and Rudolf A A 2011 *Energy* **36** 2328-2342
[6] Sangian H F, Junaidi K, Sukmawati R, Hellen K D, Debra A P, Silvya YA, Setiyo G and Arief W 2015 *Bulletin of Chemical Reaction Engineering & Catalysis* **10** (1) 8-22
[7] Liu S, Fang G, Wang Q, Deng Y and Han S 2011 *Bioresources* **6** (4) 4229-4237
[8] Girisuta B, Dussan K, Haverty D, Leahy J J and Hayes M H B 2013 *Chemical Engineering Journal* **217** 61–70
[9] Cekmecelioglu D and Uncu O N 2013 *Waste Management* **33** (2013) 735–739
[10] Morales-Rodriguez R, Meyer A S, Gernaey K V and Sin G 2011 *Bioresource Technology* **102** 1174–1184
[11] Lübken M, Gehring T and Wichern M 2010 *Appl Microbiol Biotechnol* 85 1643–1652
[12] Miller G L 1959 *Analytical Chemistry* **31**(3) 426-428
[13] Levine S E, Fox J M, Blanch H W and Clark D S 2010 *Biotechnology and Bioengineering* **107** (1) 37-51
[14] Garrote G, Domınguez H and Parajo J C 2001 *Process Biochemistry* **36** 571–578
[15] Rice G R and Do DD 1995 *Applied mathematics and Modeling for Chemical Engineering.* Jhon Wiley & Sons Inc
[16] Boas M 1983 *Mathematical Methods in the Physical Sciences* Jhon Wiley & Sons Inc
[17] Aguilar R, Ramırez J A, Garrote G, Vazquez M 2002 *Journal of Food Engineering* 55 309–318
[18] Howell J A and Stuck J D 1975 *Biotechnology and Bioengineering* **XVII** 873-893
[19] Wu H, Dai X, Zhou S L, Gan,Y Y, Xiong Z Y, Qin Y H, Ma J, Yang L, Wu Z K, Wang T L, Wang W G, Wang C W 2017 *Bioresource Technology* **241** 70–74.
[20] Song H T, Gao Y, Yang Y M, Xiao W J, Liu S H, Xia W C, Liu Z L, Yi L and Jiang, Z B 2016 *Bioresource Technology* **219** 710-715
[21] McIntosh S and Vancov T 2011 *Biomass and Bioenergy* **35** 3094-3103
[22] Van Dam J E G, Van den Oever M J A, Keijsers E R P, Van der Putten J C, Anayron C, Josol F and Peralta A 2006 *Ind. Crops Prod.* 24, 96-104

IOP Conf. Series: Materials Science and Engineering **567** (2019) 012007 doi:10.1088/1757-899X/567/1/012007

Measuring the tourist destination images based on service quality using fuzzy inference system

M Wullur[1], I N Sutapa[2]

[1]Management Department, Sam Ratulangi University, INDONESIA
[2]Industrial Engineering Department, Petra Christian University, INDONESIA

Corresponding author: mantapa@petra.ac.id

ABSTRACT. The article discusses the measurement of tourist destination images perceived by travelers based on five dimensions of the service quality, i.e. tangibles, reliability, responsiveness, empathy, and assurance. Measurement problems occur when a traveller assesses the quality of a tourism destination service subjectively with vague boundaries and perceives the image of a tourism destination to vary. To address these issues, we designed an inference model using Mamdani's fuzzy inference system. The results of this study are quantitative assessments of the image of tourism destinations by various travellers based on qualitative perceptions of the quality of service experienced by the traveller.

1. Introduction

Increasing the competitiveness of the tourism industry can create new sources for regional economic growth. North Sulawesi, as a case study, has a great potential in the tourism sector which can be utilized to increase economic growth. However, from the field survey, it is found that the infrastructure, and accessibility to tourist objects are inadequate. Environmental cleanliness, tourist comfort, and safety in the tourist destinations are still far from expectations. Tourist destinations are generally not managed properly in order to look attractive and sustainably preserved. Human resources as the main players in the tourism industry, are not competent enough [1]. If these weaknesses are not properly addressed for improvements, the potential for increasing economic growth will be difficult to achieve.

The tourism industry is a very complex service industry [2], involving many players from hotel or accommodation businesses, travel agents, restaurants, food-beverage retailers, transportation businesses, the government, and local communities. All of these players serve travelers who have various expectations, experiences, lifestyles and abilities. All travelers are expected to have memorable experiences and satisfying services so they will stay longer and spend more money. To satisfy these travelers, the most important component of the tourism industry is the image of the tourist destination [3]. The image of the destination is mainly related to the quality of services carried out in an integrated and systematic manner by all tourism players. Managing the tourism service industry, which involves many players in serving the various desires and needs of the travelers, requires an integrated management system approach. One management system that can be applied is Total Quality Management. In this study, several components of TQM are considered, as the initial step of the study is limited to determining the measurement or indicator of service quality and its relation to the measurement or indicator of the image of the tourism destination. This is important to study, as it gets an overview of the travelers' expectations and perceptions based on their experience in some tourism objects or tourist destinations. This service quality measurement can be used by decision-makers as a monitoring and evaluation tool in improving service quality in the tourism service industry [4].

Furthermore, data which are measures or indicators of service quality experienced by travelers are generally not abundant and incompletely delivered unsystematically to tourism decision makers. Only a few travelers want to express their experiences in some tourist destinations, and, if there is, usually a few data can be conveyed. Lots of data are scattered only as a moving story. This is because there are not many officers and medias to collect the data from the traveler. If there are some data collection medias, they are only for some limited needs of the specific company, so not much data are shared for the needs of the other players [5, 6]. The data submitted by the traveler are generally subjective and qualitative, namely in the form of words or statements that are vague or inexact. Besides, with the increasing numbers of the tourists, the types and forms of the data can be variously diverse, in line with the number of the tourists [7].

2. Result and Discussion

2.1. Fuzzy Inference System

Fuzzy inference system (FIS) is a process of formulating the mapping from a given input to an output using fuzzy logic [8, 9]. The mapping provides a foundation from which decision can be made. FIS is sometimes called fuzzy reasoning, used in a fuzzy rule, to determine the rule outcome from the given rule input information. Fuzzy rules represent modelling knowledge or experience. Fuzzy rules are needed to compute the outcome for output variables in consequence of the rule when specific information is assigned to input variables in the rule antecedent. The main components of FIS include fuzzification interface, inference engine and defuzzification [9]. A basic structure of FIS comprising three components and rules can be seen in Figure 1.

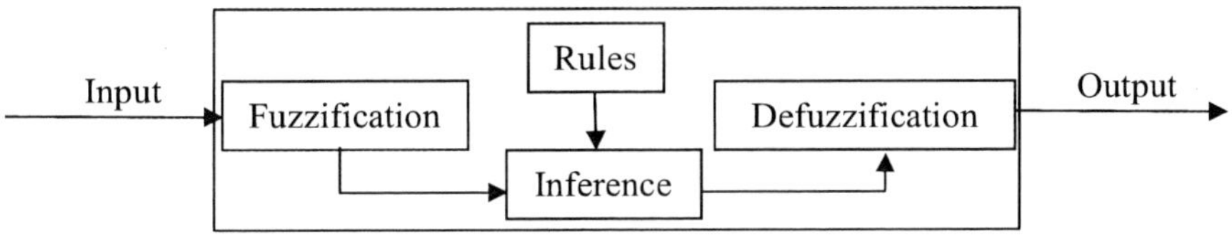

Figure 1. Basic Structure of a Fuzzy Inference System

FIS can be envisioned as involving a knowledge base and a processing stage. The knowledge base provides some membership functions and fuzzy rules needed for the process. In the processing stages, crisp numerical variables are the inputs of the system. These variables are passed through a fuzzification stage which are transformed into linguistic variables, then become the fuzzy input for the inference engine. This fuzzy input is transformed by the rules of the inference engine to a fuzzy output. A defuzzification stage, then, changes these linguistic results into numerical values that become the output of the system [9]. Creating a decision using FIS involves several steps.

The steps in FIS are created to develop an inference for the tourism destination image (output) based on service quality dimensions (input). Five dimensions of service quality, namely tangible, reliability, responsiveness, assurance, and empathy may influence the tourism destination image. In Table 1, the indicators for each dimension of the service quality and the tourism destination image are identified.

Mamdani inference system [10] allows a system to take in a set of input values, and applies a set of fuzzy rules to those values to derive a quantitative output value (Figure 2). The data from the service quality dimension become the inputs of the Mamdani inference, and the output of the system is the tourism destination image. Figure 2 illustrates the processes to transform inputs to an output.

IOP Conf. Series: Materials Science and Engineering **567** (2019) 012007 doi:10.1088/1757-899X/567/1/012007

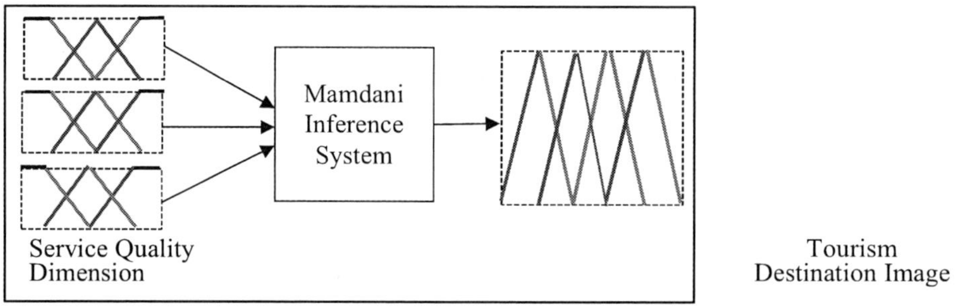

Figure 2. Transformation from the Input to the Output of the System

2.2. Output and Input of the System

The output of the inference system is the tourist destination image. Based on [11, 12], the data for the system outputs are taken from the answers to the questions according to the indicators of image, i.e. the destination environment is attractive, the tourist destinations are easy to access, the travelers enjoy the style of buildings in the destination, there is a good transportation system in the destination, the local residents of the destination are friendly, and the local residents of the destination are excellent in welcoming the travelers. The data for the system inputs are taken from the indicator of service quality dimension (Table 1), the dimension and its indicators are derived from a previous research [2, 3, 13]. The data can be collected from the back pockets of an airplane, airport waiting rooms, hotel lobbies, restaurants, tourist information offices, and some other resources, such as via questionaries' to the travelers, checklist buttons at hotels/restaurant toilets.

Based on the defined functional and operational system characteristics, the input and output data, from now on referred to as system variables are needed to fuzzify. The following steps are performed to obtain the tourist destination image based on service quality dimensions.

Table 1. Service Quality's Indicators for Input Variable

Dimension	Indicators
Tangible	The infrastructure is designed with high-quality standards.
	Modern and technologically relevant vehicles are available.
	There is enough security guards and provided security and comfort.
	The meals are of high quality.
	The accommodation and facilities are appealing and in sound design.
	The physical appearance of the hotel is tidy and clean.
	The physical appearance of the tours escort is tidy and clean.
	There is free Wi-Fi internet with large bandwidth.
	There is an adequate supply of electricity.
	There is an adequate HP charger facility.
	There are enough volunteer photographers for a selfie or groupie.
	Tourist information officers are easy to find.
Responsiveness	Tourism servicer shows sincere interest in problem-solving.
	Tourism servicer provides adequate and clear information about the service they deliver
	Tourism servicer is able to fulfil requests promptly on time.
	Tourism servicer provides full information regarding the entertainment offered.
	Tourism servicer shows sincere willingness and interest in helping/assisting.
	Tourism servicer provides advice on how to best utilize free time.
Empathy	Services offered are supplied by pleasant and friendly personnel.
	My exceptions and special needs are met as expected.
	Personal safety is considered in a significant aspect of every service provided.
	Local people care about and pay attention to traveler needs.
	Tourism destinations pay attention to the needs of disable travelers
Assurance	Well-Trained, customer-oriented personnel serves the travelers
	The level of service quality reinforces traveler's confidence in the service provided.
	Detailed, experienced, and competent tour/hotel escorts are provided to facilitate travelers.
	Staff communicates with fluently and in an understandable manner.
	All food and beverages sold are certified according to international standards.
	There is an OK sign from the results of the food and beverage audit conducted by the authorities.
	Every traveling infrastructure has been audited regularly and has passed the authorized institution.
Reliability	Directions and signs are available properly
	Directions and signs easily guide me in finding the needed locations.
	Services delivered are correct from the first time.
	Services are delivered as promised to travelers.
	Scheduled tours are on time.
	No troubles occur with the service providers during my stay.
	Information from taxi drivers/tourist information officers/hoteliers/the tour escorts is detailed, clear, and very helpful.
	Food and beverages sold are guaranteed safe and healthy.

2.3. *Fuzzy Membership Function of System Variables*

System variables are fuzzified to obtain a fuzzy membership function. The system recognizes the input and output variables and defines its membership. In this research, all membership functions for the input variables (service quality dimension) are defined in three linguistic terms, '*High*', '*Medium*', and '*Low*'. The output variable (tourism destination image) is defined in five linguistic terms, '*Very High*', '*High*', '*Medium*', and '*Low*', and '*Very Low*'. The experienced travelers formulate membership functions for each variable. Following are the membership functions for each variable after normalized on [0, 1] intervals. The membership functions (MF) for '*Low*', '*Medium*', and '*High*' fuzzy set of input variables (equation 1) are approximated by a trapezoidal function $MF_i = \mu_i(x) = tz(a, b, c, d)$, where *i={L='Low', M='Medium', H='High'},* and for output variable (equation 2) is a triangular function $MF_j = \mu_j(y) = ta(a, b, c)$, where *j={VL='Very Low', L='Low', M='Medium', H='High', VH='Very High'}.*

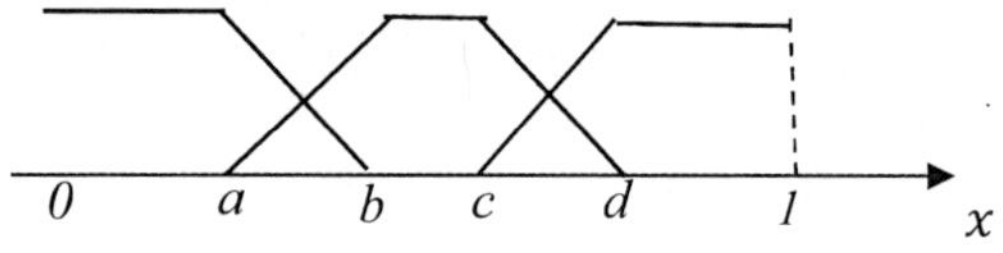

Figure 3. Membership Function for Input Variables

Where

$$\mu_L(x) = tz(0,0,a,b) = \begin{cases} 1 & 0 \le x < a \\ \frac{b-x}{b-a} & a \le x < b, \end{cases} \qquad \mu_H(x) = tz(c,d,1,1) = \begin{cases} \frac{x-c}{d-c} & c \le x < d \\ 1 & d \le x \le 1 \end{cases}$$

$$\mu_M(x) = tz(a,b,c,d) = \begin{cases} \frac{x-a}{b-a} & a \le x < b \\ 1, & b \le x < c \\ \frac{d-x}{d-c} & c \le x < d \end{cases} \tag{1}$$

For the output variable:

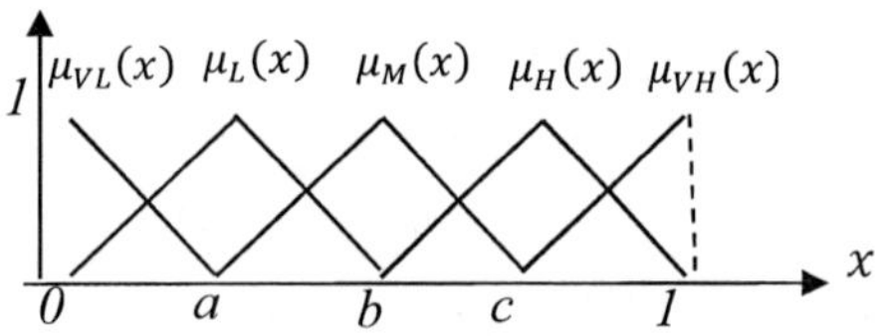

Figure 4. Membership Function for Output Variable

$$\mu_{VL}(x) = ta(0,0,a) = \frac{a-x}{a} \quad 0 \le x < a, \qquad \mu_{VH}(x) = ta(c,1,1) = \frac{x-c}{1-c} \quad c \le x < 1$$

$$\mu_L(x) = ta(0,a,b) = \begin{cases} \frac{x-a}{a} & 0 \le x < a \\ \frac{b-x}{b-a} & a \le x < b, \end{cases}$$

$$\mu_M(x) = ta(a,b,c) = \begin{cases} \frac{x-a}{b-a} & a \le x < b \\ \frac{c-x}{c-b} & b \le x \le c \end{cases}$$

$$\mu_H(x) = ta(b,c,1) = \begin{cases} \frac{x-b}{c-b} & b \le x < c \\ \frac{1-x}{1-c} & c \le x \le 1 \end{cases} \tag{2}$$

2.4. Fuzzy Rules Development

Fuzzy rules are developed from data conveyed by travelers after they visit a tourism object or a tourist destination. The forms and types of information are varied according to their different experiences, knowledge, and expectations. Fuzzy rules are defined as the *IF-THEN* rules to describe the system behavior. The rules are designed to describe the causality relationship among the indicators of service

IOP Conf. Series: Materials Science and Engineering **567** (2019) 012007 doi:10.1088/1757-899X/567/1/012007

quality dimension and the tourist destination images (equation 3).

For example:

> *IF* the direction sign for the tourism object is very clear (high) *AND* the highway infrastructure is very good (high), *AND* the tourism destination is very comfortable (high) *THEN* the traveler will give a positive image of the tourism destination (very high).

In general, the travelers' ratings can be stated based on several dimensions of service quality, as follows:

> *IF* an *tangible* indicator is 'fuzzy set *i*' *AND* an *reliability* indicator is 'fuzzy set *j*' *AND* ··· *THEN the image of a tourism destination* is 'fuzzy set *k*'. (3)

Where fuzzy set *i, j* can be categories as '*Low*'. '*Medium*', or '*High*', while the fuzzy set *k* of output variable can be categories as '*Very Low*', '*Low*'. '*Medium*', '*High*', or '*Very High*'.

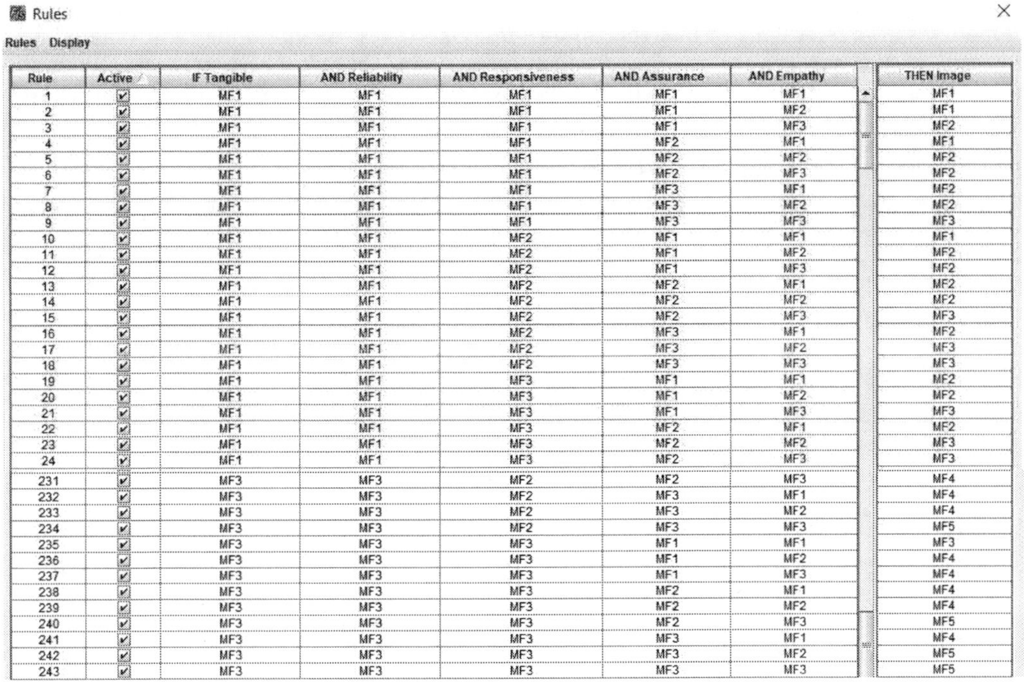

Rules

Rules Display

Rule	Active	IF Tangible	AND Reliability	AND Responsiveness	AND Assurance	AND Empathy	THEN Image
1	✓	MF1	MF1	MF1	MF1	MF1	MF1
2	✓	MF1	MF1	MF1	MF1	MF2	MF1
3	✓	MF1	MF1	MF1	MF1	MF3	MF2
4	✓	MF1	MF1	MF1	MF2	MF1	MF1
5	✓	MF1	MF1	MF1	MF2	MF2	MF2
6	✓	MF1	MF1	MF1	MF2	MF3	MF2
7	✓	MF1	MF1	MF1	MF3	MF1	MF2
8	✓	MF1	MF1	MF1	MF3	MF2	MF2
9	✓	MF1	MF1	MF1	MF3	MF3	MF3
10	✓	MF1	MF1	MF2	MF1	MF1	MF1
11	✓	MF1	MF1	MF2	MF1	MF2	MF2
12	✓	MF1	MF1	MF2	MF1	MF3	MF2
13	✓	MF1	MF1	MF2	MF2	MF1	MF2
14	✓	MF1	MF1	MF2	MF2	MF2	MF2
15	✓	MF1	MF1	MF2	MF2	MF3	MF3
16	✓	MF1	MF1	MF2	MF3	MF1	MF2
17	✓	MF1	MF1	MF2	MF3	MF2	MF3
18	✓	MF1	MF1	MF2	MF3	MF3	MF3
19	✓	MF1	MF1	MF3	MF1	MF1	MF2
20	✓	MF1	MF1	MF3	MF1	MF2	MF2
21	✓	MF1	MF1	MF3	MF1	MF3	MF3
22	✓	MF1	MF1	MF3	MF2	MF1	MF2
23	✓	MF1	MF1	MF3	MF2	MF2	MF3
24	✓	MF1	MF1	MF3	MF2	MF3	MF3
231	✓	MF3	MF3	MF2	MF2	MF3	MF4
232	✓	MF3	MF3	MF2	MF3	MF1	MF4
233	✓	MF3	MF3	MF2	MF3	MF2	MF4
234	✓	MF3	MF3	MF2	MF3	MF3	MF5
235	✓	MF3	MF3	MF3	MF1	MF1	MF3
236	✓	MF3	MF3	MF3	MF1	MF2	MF4
237	✓	MF3	MF3	MF3	MF1	MF3	MF4
238	✓	MF3	MF3	MF3	MF2	MF1	MF4
239	✓	MF3	MF3	MF3	MF2	MF2	MF4
240	✓	MF3	MF3	MF3	MF2	MF3	MF5
241	✓	MF3	MF3	MF3	MF3	MF1	MF4
242	✓	MF3	MF3	MF3	MF3	MF2	MF5
243	✓	MF3	MF3	MF3	MF3	MF3	MF5

Figure 5. Fuzzy Rules of the System

As a simulation, each dimension of service quality (input) uses an indicator and the tourist destination image (output) is expressed as an indicator. Table 5 shows 243 fuzzy rules that are obtained from all possible indicators for service quality dimensions and an indicator for a tourist destination image. To shorten each fuzzy set, it is used MF1=*Low*, MF2=*Medium*, and MF3=*High* for input variables (service quality dimension), and for output variable (Image) MF1=*Very Low*, MF2=*Low*, MF3=*Medium*, MF4=*High*, and MF5=*Very High*.
Example of fuzzy rule:

Rule 1: *IF* an indicator in the tangible dimension is *Low AND* an indicator in the reliability dimension is *Low AND* an indicator in the responsiveness dimension is *Low AND* an indicator in the assurance dimension is *Low AND* an indicator in the empathy dimension is *Low THEN* the tourism destination image will be *Very Low*.

⋮

Rule 24: *IF* an indicator in the tangible dimension is *Low AND* an indicator in the reliability dimension is *Low AND* an indicator in the responsiveness dimension is *High AND* an indicator in the assurance dimension is *Medium AND* an indicator in the empathy dimension is *High THEN* the tourism destination image will be *Medium*.

The inference rules set the premise to create an output, then the output needs to defuzzify to obtain a crisp value.

2.5. Defuzzification

Defuzzification step is needed to convert all input data into three linguistic terms (*Low, Medium*, or *High*) that can be used to observe the tourist destination images (*Very Low, Low, Medium, High*, or *Very High*). The defuzzification process transforms the fuzzy set into the crisp value that is meaningful to the decision maker.

Figure 6 shows the results of inference of 243 *IF-THEN* rules. Based on the crisp value of each dimension of the service quality, fuzzification is done into a red fuzzy set according to the corresponding fuzzy set, then connected with the *AND* logic which is operationalized with the *MIN* operator.

$$\mu_{A_1}(x) \cap \mu_{A_2}(x) \cap \cdots \cap \mu_{A_n}(x) = MIN[\mu_{A_1}(x), \mu_{A_2}(x), \cdots, \mu_{A_n}(x)] \quad (4)$$

Where $A_1, A_2, \cdots, A_n$ are fuzzy set for indicators of each service quality dimension.

Furthermore, the combination of all rules is operationalized with logic *OR* or *MAX*, the result is a total fuzzy set that describes the resulting image.

$$\mu_{B_1}(y) \cup \mu_{B_2}(y) \cup \cdots \cup \mu_{B_m}(y) = MAX[\mu_{B_1}(y), \mu_{B_2}(y), \cdots, \mu_{B_m}(y)] \quad (5)$$

Where $B_1, B_2, \cdots, B_m$ are fuzzy set for tourism destination image as a result of each rule.

Finally, defuzzification is performed by calculating the center point or centroid, resulting in a tourist destination image in the form of crisp number, i.e. a quantitative score between [0, 1].

$$Centroid = \frac{\int_0^1 \mu_B(y) y dy}{\int_0^1 \mu_B(y) dy} \quad (6)$$

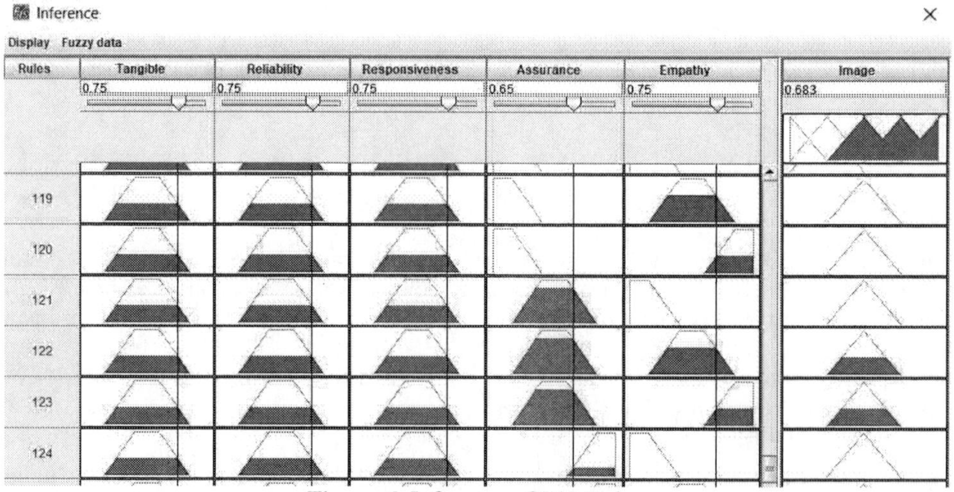

Figure 6. Inference of 243 Rules

Figure 6 can be used as a detection and monitoring tool by decision-makers, who are engaged in the tourism industry, to find out the image of the tourist services provided based on the quality of tourist destination services, which are expressed by travelers from their experiences and expectations.

Based on the 243 fuzzy rules above, the simulation can be done to determine the impact of changes in each indicator of the service quality on the image of a tourist destination. From these simulations, it can be inferred which indicator has a significant impact on improving the image of the tourist destinations. Furthermore, changes can also be seen in 3 dimensions or 2 dimensions (Figure 7). As an example, the changes in the image of tourist destinations can be observed due to changes in an indicators of tangible and reliability dimensions, when other indicators of the service quality are constant. It appears that, when services on the tangible and reliability dimensions increase, the image also increases.

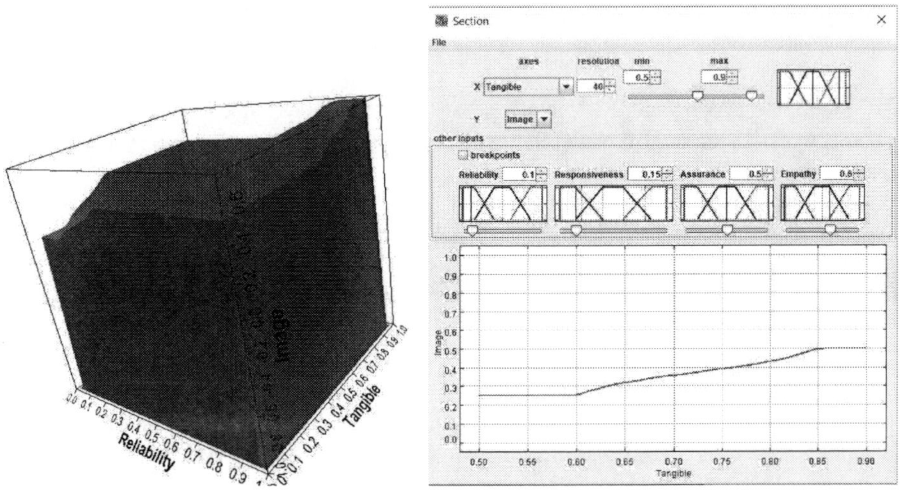

Figure 7. Visualize The Tourism Destination Image in 3D and 2D

3. **Conclusion**

The image of the tourist destinations is important in increasing the number of travelers to visit. Increasing visitors have an impact on the regional economic growth of the tourism area. The increased tourist destination image is a strategic effort that must be carried out by all tourism players. One of the efforts to improve the image is by improving the service quality. To find out the service quality that needs to be prioritized, it is compulsory to understand the indicators of the service quality. This initial research has identified several measures/indicators that are in line with the tourist destination services. This study also produces reasoning rules to improve the image of the tourist destinations based on several indicators of the service quality dimensions using a fuzzy inference system. This research is still in the early stage, and there are several weaknesses, as it has not yet produced reasoning rules using all possible indicators of the service quality dimensions. In the future, it needs to be developed continuously the reasoning rules resulting from interviews or direct observation of travelers or from data collection of various social medias. Then, it needs to be integrated automatically with government-owned information systems, and distributed to tourism players in real time through online media.

References

[1] ------, 2016, *Tourism Industry Report, North Sulawesi Province*. Survey Team of Potency and Competitiveness of North Sulawesi Tourism Industry in Facing MEA (in Bahasa Indonesia edition), PSE-KP, Gadjah Mada University.

[2] M. A. Bhat, (2012), "Tourism Service Quality: A Dimension-specific Assessment of SERVQUAL", *Global Business Review*, 13 (2), pp. 327–337.

[3] M. N. Akroush, L. E. Jraisat, D. J. Kurdieh, R. N. Al-Faouri, and L. T. Qatu, (2016), "Tourism service quality and destination loyalty –the mediating role of destination image from international tourists' perspectives", *Tourism Review*, 71 (1), pp. 18-44.

[4] J. de Ona, R. de Ona, L. Eboli, and G. Mazzulla, (2016), "Index numbers for monitoring transit service quality", *Transportation Research Part A: Policy and Practice*, 84, pp. 18-30.

[5] S. Farias, F. Aguiar, M. Kovacs, and F. Sales, (2013), "Destination image on the web: evaluation of Pernambuco's official tourism destination websites", *Business Management Dynamics*, 2 (10), pp. 35-48.

[6] A. Said, A. Shuib, N. Ayob, and F. Yaakub, (2013), "An evaluation of service quality from visitors' perspectives: the case of Niah National Park in Sarawak", *International Journal of Business and Society*, 14 (1), pp. 61-78.

[7] Sreekumar, S. Mahapatra, and S. S. Mahapatra, (2015), Service Quality of Indian Banks: A Fuzzy Inference System Approach, *Asian Academy of Management Journal*, 20 (2), pp. 59–80.

[8] L. Zadeh, (1996), "Fuzzy logic computing with words", *IEEE Transactions on Fuzzy Systems*, 4, pp. 103–111.

[9] N. Sabri, S. A. Aljunid, M. S. Salim, R. B. Badlishah, R. Kamaruddin, M. F. A. Malek, (2013), "Fuzzy Inference System: Short Review and Design", *International Review of Automatic Control*, 6 (4), pp. 441-449.

[10] E.H. Mamdani and S. Assilian, (1975), "An Experiment in Linguistic Synthesis with a Fuzzy Logic Controller", *International Journal of Man-Machine Studies*, 7, pp. 1–13.

[11] G. Hankinson, (2005), "Destination brand images: a business tourism perspective", *Journal of Services Marketing*, 19 (1), pp. 24-32.

[12] S. Kim, S. Holland, and H. Han, (2013), "A structural model for examining how destination image, perceived value, and service quality affect destination loyalty: a case study of Orlando", *International Journal of Tourism Research*, 15 (1), pp. 313-328.

[13] A. Parasuraman, V. A. Zeithaml, and L. L. Berry, (1988). "SERVQUAL: a multiple-item scale for measuring consumer perceptions of service quality", *Journal of Retailing*, 64 (1), pp. 12-40.

IOP Conf. Series: Materials Science and Engineering **567** (2019) 012008 doi:10.1088/1757-899X/567/1/012008

Rainfall prediction using backpropagation algorithm optimized by Broyden-Fletcher-Goldfarb-Shanno algorithm

S Anam[1]

[1] Department of Mathematics, Faculty of Sciences, Brawijaya University, Malang 65145, Indonesia

Corresponding author: syaiful@ub.ac.id

Abstract. An extreme climate change results in a long dry season and an extreme rainfall results the losses in various areas of life. Rainfall prediction becomes an important thing for planning in many life sectors. Many prediction methods have been proposed, such as Autoregressive Integrated Moving Average (ARIMA) and Artificial Neural Networks (ANN). ANN has some advantages compared with the ARIMA model. Backpropagation algorithm is one of the ANN which has been successfully used in various fields. However, the performance of the backpropagation algorithm depends on the architecture and the optimization method used. The standard backpropagation algorithm optimized by gradient descent method works slowly to get a small error. The Broyden–Fletcher–Goldfarb–Shanno (BFGS) algorithm works faster than gradient descent method. For this reason, this paper proposes the rainfall prediction using the backpropagation algorithm optimized by the BFGS algorithm. From the experiment results, it can be shown that the backpropagation algorithm optimized by the BFGS algorithm gives better result compared with the standard backpropagation algorithm for rainfall prediction. The big number of neuron hidden causes overfitting and the small number of neuron hidden make the worst accuracy. Choosing the right learning rate will produce better accuracy.

1. Introduction

Climate is one of the important factors of agricultural productivity [1]. Climate change will give many impacts in agricultural and livestock production, hydrologic balances and others. Agriculture is heavily dependent on climate. Climate depends on rainfall, sunshine hours, temperature, relative humidity and length of the drought period, which are called climatic factors. The climatic factors result in cycle-to-cycle variability of crops production. The uncontrollable natures of climate factors are changing over time which affects the agricultural, social, economic and environmental sustainability of a country. Climate change also influences potentially the earth's biological systems and human health [2]. Temperature, humidity, and rainfall patterns are the meteorological factors that influence transmission intensity of infectious diseases Developing country are expected to face a host of health effects due to climate change, including water-borne and vector-borne diseases such as malaria and dengue.

Global warming causes climate change in almost all parts of the world. An extreme climate change results in a long dry season and an extreme rainfall results the losses in various areas of life. High rainfall can cause floods, disease outbreaks and health problems, transportation problems, landslide, and crop failure, while low rainfall has a negative impact on the agricultural sector, i.e. land drought and crop failure [1,2,3].

The weather or rainfall is influenced by several factors, namely temperature, relative humidity, air pressure, wind speed, total cloud cover, and sun exposure. The weather or rainfall prediction becomes an important thing for planning in many life sectors. Many prediction methods for rainfall prediction have been proposed, such as Autoregressive Integrated Moving Average (ARIMA) [4]. Artificial Neural Network (ANN) is an alternative method for prediction. The ARIMA model works on the assumption of stationary, while ANN can work without assumptions. It means that the ANN method is more flexible compared with the ARIMA model. Adebiyi et al. in 2014 have proven that the ANN model is superior over the ARIMA model [5].

ANN model is also able to acquire knowledge even though there is no certainty, generalize and extract from a particular data pattern, create a pattern of knowledge through self-organizing. One of the popular ANN models that have been successfully used is backpropagation neural networks. It has been successfully used in various fields. However, the performance of the backpropagation neural networks depends on the architecture and the optimization method used.

The standard backpropagation neural networks use the gradient descent method for optimizing the error. The gradient descent method is very slow to get a small error. Quasi-Newton method is one of the good methods to solve the minimization problem. The size of the gradient change is constructed well enough to result in superlinear convergence. The quasi-Newton method has better convergence compared with the gradient descent method. Broyden–Fletcher–Goldfarb–Shanno (BFGS) algorithm is one of type quasi-Newton methods which is a good iterative method for obtaining a local optimum. This method has also been proven to be better than gradient descent when applied to the neural network [6]. For this reason, this paper proposes the rainfall prediction by using the backpropagation neural networks optimized by Broyden–Fletcher–Goldfarb–Shanno algorithm. This research uses the average temperature, average relative humidity, and average wind speed for predicting rainfall. An automatic rainfall prediction based on weather data will realize a better life.

2. Related Works

This section will describe some theories and the previous research related to this research, such as rainfall, BFGS algorithm, and neural networks.

2.1. Rainfall

Rainfall is defined as the amount of water that falls on a flat ground surface during a certain period measured by unit height (mm) above the horizontal surface if it does not occur evaporation, runoff, and infiltration. One rainfall millimeters (1 mm), meaning in one-meter area square on a flat place where water is high one millimeter.

Rainfall is very important in determining the yield of crop cultivation. Increasing rainfall in an area raises the potential for flooding. Conversely, if there is a decrease in rainfall in one area, will cause drought [7]. According to Wilson (1993), influencing factors a lot of rainfall is humidity, air pressure, temperature and wind speed [8].

2.2. BFGS Algorithm

Quasi-Newton method is an alternative method that can be used for solving optimization problems. This method was successfully used for minimizing errors on artificial neural networks [9]. Quasi-Newton method is a method that is used when the calculation of the Hessian matrix is difficult or time-consuming. This method has a rapid convergence when compared with the method of gradient descent. One of quasi-Newton method types is the BFGS method. To solve the problem of minimization of the Eq. (1), the BFGS method can be used for solving it.

$$\min f(\mathbf{x}_k) \tag{1}$$

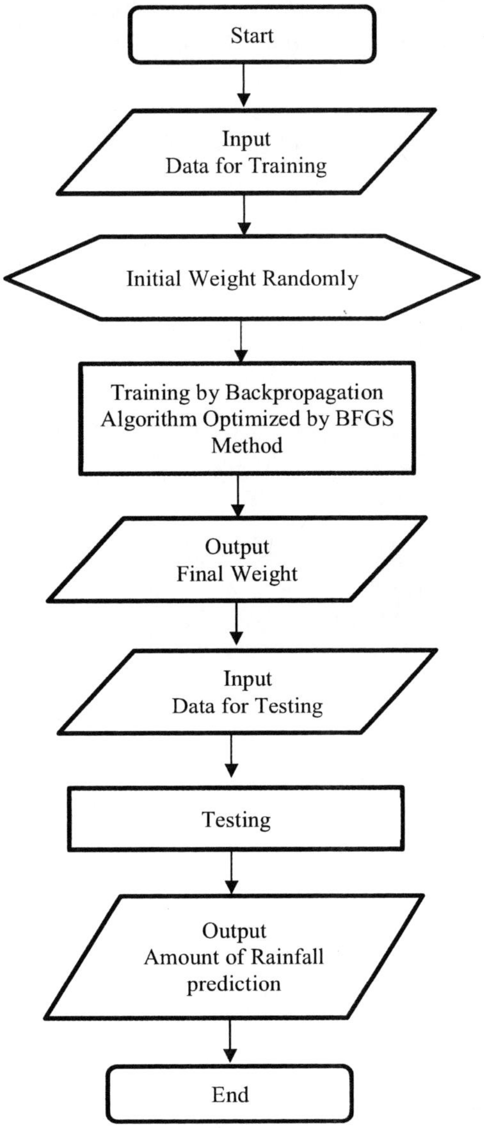

Figure 1. Rainfall prediction system by backpropagation neural network optimized by the BFGS method.

The BFGS algorithm can be seen in Algorithm 1 [6].

Algorithm 1: Algorithm of BFGS method

a. Input $\mathbf{x}_0$, ε (stopping criteria) and k_{max}.

b. Set an initial iteration k = 0 and **B= I** where I is an identity matrix.

c. Calculate $f(\mathbf{x}_k)$

d. While ($\|\nabla(\mathbf{x}_k)\|>\varepsilon$) or ($k<=k_{max}$) do
   1. Calculate $\mathbf{d}_k = -\mathbf{B}_k f(\mathbf{x}_k)$, where $\mathbf{d}_k$ is a generating direction.
   2. Select α that is able to minimize $f(\mathbf{x}_k + \alpha_k \mathbf{d}_k)$
   3. $\mathbf{x}_{k+1} = \mathbf{x}_k + \alpha_k \mathbf{d}_k$
   4. Calculate $\mathbf{B}_{k+1}$ by using Eq. (2).

$$\mathbf{B}_{k+1} = \mathbf{B}_k - \frac{\mathbf{B}_k\mathbf{s}_k(\mathbf{B}_k\mathbf{s}_k)^T}{\mathbf{s}_k^T\mathbf{B}_k\mathbf{s}_k} + \frac{\mathbf{y}_k\mathbf{y}_k^T}{\mathbf{y}_k^T\mathbf{s}_k} + \phi_k\left[\mathbf{s}_k^T\mathbf{B}_k\mathbf{s}_k\right]\mathbf{v}_k\mathbf{v}_k^T \tag{2}$$

where $\phi_k \in [0,1]$, $\mathbf{s}_k = \mathbf{x}_{k+1} - \mathbf{x}_k$,

$\mathbf{y}_k = \nabla f(x_{k+1}) - \nabla f(x_k)$ and $\mathbf{v}_k = \left[\frac{\mathbf{y}_k}{\mathbf{y}_k^T\mathbf{s}_k} - \frac{\mathbf{B}_k\mathbf{s}_k}{\mathbf{s}_k^T\mathbf{B}_k\mathbf{s}_k}\right]$.

   5. k = k + 1

e. End while

2.3. Backpropagation Neural Networks

Artificial neural network (ANN) is an information processing system that has characters such as neural networks in the human brain. There are two types of learning in ANN to modify their weight, namely supervised learning and unsupervised learning. Backpropagation learning model is a learning model used for multilayer networks. The optimization method commonly used in backpropagation learning is the gradient descent method. In this method, the weight of the network moves along the negative direction of the gradient [6].

3. Proposed Method

The system of the proposed method is used to predict the amount of rainfall as shown in figure 1. By using this system, the amount of rainfall in an area can be predicted based on the factors of the rainfall. The first step, the training data of rainfall which are the previous month's rainfall, average temperature, average relative humidity, and average wind speed are inputted in the system. The initial weight of the backpropagation neural networks is generated randomly. The next step is that the backpropagation

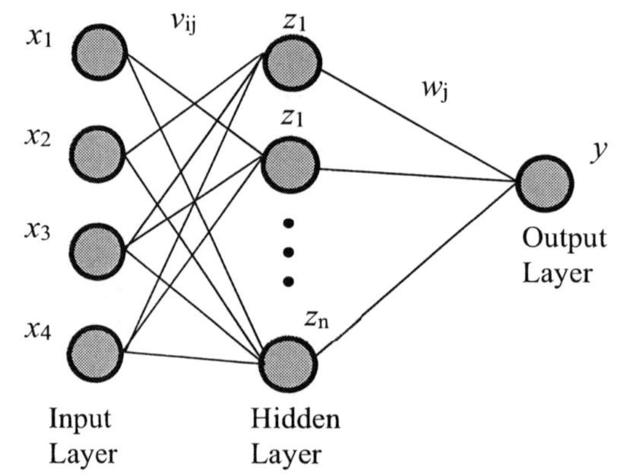

Figure 2. Architecture of backpropagation neural network.

neural networks will do training based on the training data. In this step, the weight of the neural networks will be improved by minimizing an error between the output of networks and target. The error is minimized by the BFGS method. The final weight will be used for testing step. In the testing step, the data test is inputted and the output is a prediction of the amount of rainfall.

The hypothesis in this study is that the previous month's rainfall, average temperature, average relative humidity and average wind speed in the previous month will affect rainfall this month. The data input of this program/system is the average rainfall in the previous month, average temperature, average relative humidity, and average wind speed.

The network architecture is developed by using the number of variables that affect many factors that influence rainfall. These variables include the average rainfall in the previous month average temperature, average relative humidity, average wind speed, and rainfall. Consequently, the variables used as network inputs are four variables which are the previous month average temperature (x_1), average temperature (x_2), average relative humidity (x_3), average wind speed (x_4), while the variables used as network output are the amount of rainfall (y). The network architecture is shown in figure 2. It has a three-layer, they are the input layer, the hidden layer and the output layer as shown in figure 2. Conversely for the number of hidden in the network is determined by trial (trial and error). The aim is to find out which architecture is most suitable for accurately recognizing patterns. Therefore, in this thesis an architectural model is taken, including: 3-5-1, 3-20-1, 3-50-1, 3-100-1, and 3-150-1. The meaning of architectural model 3-5-1 is that the network is built by 3 neurons in the input layer and 1 neuron in the output layer while the number of neurons in the hidden layer is 5 neurons.

The training of network by the backpropagation algorithm has three phases which are the feedforward of the input data, the backpropagation error, and the weights adjustment. The backpropagation algorithm is as in Algorithm 2.

Algorithm 2: Backpropagation neural networks

Step 0. Initialize the weights of the neural networks by random values and t=1 (iteration t^{th}).

Step 1. While error> min error or iteration < max iteration, do step 2-8

Step 2. For each data, do step 3-8

Feedforward :

Step 3. Every input unit x_i, i=1,2,...,n gets an input x_i and distributes it to all units in the hidden layer.

Step 4. Every hidden unit $(z_j, j = 1,2, \dots, p\ )$ calculates its weighted $(v_{ij}(t))$ by an input signal as in Eq.(3)

$$\boldsymbol{z\_in_j = v_{0j}(t) + \textstyle\sum_i x_i\, v_{ij}(t)}, \tag{3}$$

where v_{0j} is a bias on the hidden unit j.

Calculate its output signal by an activation function as in Eq.(4) or Eq.(5)

$$\boldsymbol{z_j = f(z\_in_j)} \tag{4}$$

or

$$z_j = \frac{1}{1+ex^{(-z\_in_j)}} \tag{5}$$

then sends z_j to all units in the output layer.

Step 5. Every output unit $(y_{\_k}, k = 1,2, \dots, m\ )$ is calculated by considering its weighted input signals, it can be shown as in Eq.(6),

$$y\_in_k = w_{0k}(\boldsymbol{t}) + \sum_{j=1}^{P} z_j\, w_{jk}(\boldsymbol{t}) \tag{6}$$

where $w_{0k}\ (\boldsymbol{t})$ is a bias on the output unit k.

Therefore, the output signal is calculated by an activation function as in Eq. (7) or Eq.(8)

$$y_k = f(y\_in_k) \tag{7}$$

$$\text{or } y_k = \frac{1}{1+\exp^{(-y\_in_k)}} \tag{8}$$

Backpropagation of error :

Step 6. Every output unit ($y_k, k = 1,2,\dots,m$) accepts a target pattern matching to the input data, and calculates its error as in Eq.(9),

$$\delta_k = (t_k - y_k)f'(y_{in_k}) \tag{9}$$

where , δ_k is error correction weight for w_{jk}

Step 7. Every hidden unit ($z_j, j = 1,2,\dots,p$) counts its $\delta_k$as in Eq.(10),

$$\delta\_in_j = \sum_{k=1}^{m} \delta_k w_{jk} \tag{10}$$

$$\delta_j = \delta\_in_j f'(z\_in_j) \tag{11}$$

Update weights and biases:

Step 8. Every output unit ($y_k, k = 1,2,\dots,m$) improves its bias and weights $j = 1,2,\dots,p$ as in Eq.(12)

$$w_{jk}(t+1) = w_{jk}(t) + \alpha B_k(t)(\delta_k z_j(t)) \tag{12}$$

where $B_k(t)$ is an aproximation of Hessian matrix. In this case, the approach to the Hessian matrix is updated by formulation in Eq (13).

$$B_k(t+1) = B_k(t) + \frac{\Delta(-\delta_k z_j(t))\Delta(-\delta_k z_j(t))^T}{\Delta(-\delta_k z_j(t))^T \Delta w_{jk}(t)} + \frac{(-\delta_k z_j(t))(-\delta_k z_j(t))^T}{(-\delta_k z_j(t))^T \left(\frac{\Delta w_{jk}(t)}{\alpha}\right)} \tag{13}$$

Each hidden unit ($Z_j, j = 1,2,\dots,p$) improves its bias and weights $i = 1,2,\dots,n$ as in Eq.(14).

$$v_{ij}(t+1) = v_{ij}(t) + \alpha A_j(t)(\gamma_j x_i(t)) \tag{14}$$

where $A_j(t)$ is defined as in Eq (15).

$$A_j(t+1) = A_j(t) + \frac{\Delta(-\gamma_j x_i(t))\Delta(-\gamma_j x_i(t))^T}{\Delta(-\gamma_j x_i(t))^T \Delta v_{ij}(t)} + \frac{(-\gamma_j x_i(t))((-\gamma_j x_i(t))^T}{((-\gamma_j x_i(t))^T \left(\frac{\Delta v_{ij}(t)}{\alpha}\right)} \tag{15}$$

Step 9. Check the stopping creteria.

The activation functions used in this paper are the sigmoid function for the hidden layer and the identity function for the output layer. The learning rate used is static. The learning rate is intended to see how the influence given to ANN's ability to recognize patterns. In this case, the several learning rate (α) valaus are used which are 0.001, 0.005, 0.01, 0.1 and 0.2 to analyze the effect of the learning rate.

The stopping criteria is the maximum iteration amount of 1000. It means that when the neural networks learning has reached the number of iterations, the learning process will stop, if the number of iteration is less then 1000, the learning will continue.

4. Result and Discussion

For the evaluation of the system of rainfall prediction, the data used were taken from Karangploso Meteorology Climatology and Geophysics Council Station, Malang Regency. The number of data is 47 data from 2014 until 2017. The data are divided into data training and testing data. The number of data training is 40 data (85% from data total) and the number of data training is 7 data (15% from data total). For implementing the system, the hardware used is Core i5 processor, 1.70 GHz, 2. 4.0 GHz RAM, 500 GB hard drive and the software used is MATLAB R2013a. The input data of the neural networks are the previous month's rainfall, the average temperature, the average relative humidity, and the average wind speed and the output of the neural networks is rainfall this month.

Table 1 and 2 show the comparison of the performance of the BFGS and gradient descent methods to optimize the backpropagation neural networks for the different architecture. They show that the BFGS method has better accuracy than the gradient descent. For a small number of neurons hidden, the BFGS method and the gradient descent almost have the same computational time. The number of the hidden layer should not be too small and too many. The big number of neuron hidden causes overfitting, it means that the accuracy for data training is too good but the accuracy for data

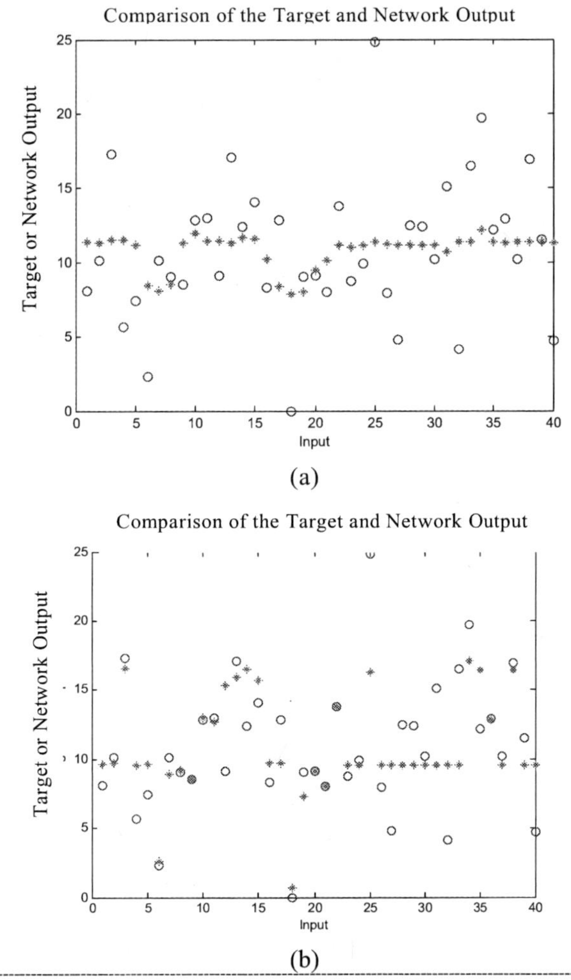

Figure 3. Comparison of the target and network output for training data. (a). the standard backpropagation (b) the backpropagation algorithm optimized by Broyden–Fletcher–Goldfarb–Shanno algorithm

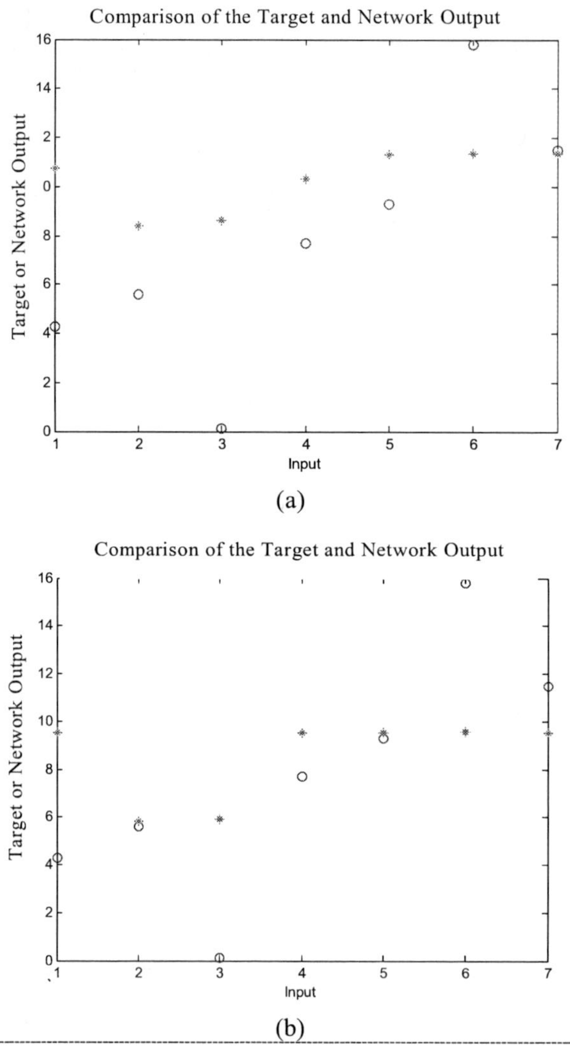

Figure 4. Comparison of the target and network output for testing data. (a). the standard backpropagation (b) the backpropagation algorithm optimized by Broyden–Fletcher–Goldfarb–Shanno algorithm.

test is worst. Table 3 and 4 show the accuracy comparison of the BFGS method and the gradient descent method for predicting the amount of rainfall. It can be shown that the best learning is 0.01.

The backpropagation algorithm optimized by BFGS algorithm results in the best MSE (Mean Square Error) of 10.09 for training data and 20.13 for testing data, while the standard backpropagation algorithm results in MSE of 16.38 for training data and 18.98 for testing data. Figure 3 (a) and (b) show the comparison of the rainfall truth data, and the prediction results by the standard backpropagation algorithm and the backpropagation algorithm optimized by BFGS algorithm for training data, respectively. Figure 4 (a) and (b) show the comparison of the rainfall truth data, and the prediction results by the standard backpropagation algorithm and the backpropa-

Table. 1. Evaluation results by using data training for different architectures with learning rate 0.005.

Architecture	MSE (Means Square Error)		Computational Time	
	Gradient Descent	BFGS	Gradient Descent	BFGS
3-5-1	22.29	**22.29**	10.02	**8.03**
3-20-1	20.38	**13.96**	**11.46**	14.55
3-50-1	18.86	**7.85**	**10.14**	30.40
3-100-1	17.47	**0.32**	**10.02**	24.65
3-150-1	16.38	**0.32**	**11.73**	69.88

Table. 2. Evaluation results by using data test for different architecture with learning rate 0.005.

Architecture	MSE (Means Square Error)	
	Gradient Descent	BFGS
3-5-1	**31.74**	**31.74**
3-20-1	27.10	**23.76**
3-50-1	23.71	**20.13**
3-100-1	**18.98**	109.41
3-150-1	**25.17**	126.38

Table. 3. Evaluation results by using data training for different learning rate with 50 neurons hidden.

Learning rate	MSE (Means Square Error)		Computational Time	
	Gradient Descent	BFGS	Gradient Descent	BFGS
0,001	19.37	**12.094**	**9.94**	18.56
0,005	20.38	**13.96**	**11.46**	14.55
0,01	18.84	**10.09**	**9.98**	14.95
0,1	19.23	**10.25**	**7.71**	31.92
0,2	18.69	**10.87**	**10.99**	31.09

Table. 4. Evaluation results by using data test for different learning rate with 50 neurons hidden.

Learning rate	MSE (Means Square Error)	
	Gradient Descent	BFGS
0,001	27.73	**19.27**
0,005	27.10	**23.76**
0,01	27.64	**19.41**
0,1	24.90	**23.48**
0,2	26.08	**23.24**

gation algorithm optimized by the BFGS algorithm for testing data, respectively. From the figures 3 and 4, it can be seen that the backpropagation algorithm optimized by BFGS algorithm gives better result compared with the standard backpropagation algorithm for predicting the amount of rainfall.

5. Conclusion

The learning rate, the neurons number of the hidden layer and the maximum number of iterations should be determined. It can be concluded that the performance of the backpropagation neural network depends on the number of neurons in the hidden layer and the learning used. The number of the hidden layer should not be too small and too many. The big number of neuron hidden causes over fitting and the small number of neuron hidden make the worst accuracy. When the number of neurons in the hidden layer increases, the computational time also increases. The increasing of computational time does not increase significantly for the gradient descent, but the computational time for the BFGS method increase significantly. Learning rates that are too small or too large will result in poor accuracy. Therefore, choosing the right learning rate will produce better accuracy. The backpropagation algorithm optimized by the BFGS algorithm gives better result compared with the standard backpropagation algorithm for predicting the amount of rainfall.

Acknowledgments
We would like to say our greatest thanks to Brawijaya University who has supported to join this conference.

References
[1] Aydinalp C and Cresser M S 2008 The effects of global climate change on agriculture American-Eurasian *J. Agric. & Environ. Sci.* **3** (5) pp 672-676.
[2] Dhara V R, Schramm P J and Luber G 2013 Climate change & infectious diseases in India: Implications for health care providers *Indian J Med Res* **138** pp 847-852
[3] Xu F, He Z, Sha Z, Zhuang L and Sun W 2013 Assessing the impact of rainfall on traffic operation of urban road network *Procedia - Social and Behavioral Sciences* **96** pp 82-89
[4] Eni D and Adeyeye F J 2015 Seasonal ARIMA modeling and forecasting of rainfall in Warri Town, Nigeria *Journal of Geoscience and Environment Protection* **3** pp 91-98
[5] Adebiyi A A, Adewumi A O and Ayo C K 2014 Comparison of ARIMA and artificial neural networks models for stock price prediction *Journal of Applied Mathematics* **2014** pp1-7
[6] Anam S, Yanti I and Widodo A 2010 Penggunaan metode quasi Newton pada jaringan syaraf tiruan backpropagation *Jurnal Teknika* **2** (1) pp 43-56
[7] Novelina P and Brodjol S S U 2016 Pemodelan data curah hujan di Kabupaten Banyuwangi dengan metode ARIMA dan radial basis function neural network *Jurnal Sains dan Seni ITS* **5** (2) pp 2337-3520
[8] Wilson E M 1993 *Hidrologi Teknik*, Erlangga Jakarta

IOP Conf. Series: Materials Science and Engineering **567** (2019) 012009 doi:10.1088/1757-899X/567/1/012009

A novel 3-D chaotic system with line equilibrium: dynamical analysis, coexisting attractors, offset boosting control and circuit design

A Sambas[1], S Vaidyanathan[2], S Zhang[3], Mujiarto[1], Sukono[4], M Mamat[5], Subiyanto[6]

[1]Department of Mechanical Engineering, Universitas Muhammadiyah Tasikmalaya, Indonesia
[2]Research and Development Centre, Vel Tech University, Avadi, Chennai, India
[3]School of Physics and Opotoelectric Engineering, Xiangtan University, Hunan, China
[4]Department of Mathematics, Faculty of Mathematics and Natural Sciences, Universitas Padjadjaran, Indonesia.
[5]Faculty of Informatics and Computing, Universiti Sultan Zainal Abidin, Kuala Terengganu, Malaysia
[6]Department of Marine Science, Faculty of Fishery and Marine Science, Universitas Padjadjaran, Indonesia.

Corespoding author: _acengs@umtas.ac.id

Abstract. A 3-D new chaotic system with five nonlinearities is proposed in this paper. A novel feature of our chaotic system is that there is no linear term in it. We also show that the chaotic system consists of equilibrium points on the z-axis (line equilibrium) as well as two equilibrium points on the (x, y)-plane. The dynamical properties of the new chaotic system are described in terms of phase portraits, bifurcation diagram, Lyapunov exponents, coexisting attractors, coexisting bifurcation and offset boosting control. Finally, an electronic circuit realization of the new chaotic system is presented in detail to confirm the feasibility of the theoretical chaotic model.

1. Introduction

Chaotic systems are characterized by their high sensitivity to small changes in initial conditions [1-2] and they have many applications in science and engineering such as weather systems [3-4], ecology [5], neurons [6-7], biology [8-10], cellular neural networks [11-12], chemical reactors [13-14], oscillators [15-20], robotics [21-24], encryption [25-30], finance systems [31-32], circuits [33-45], etc.

In the chaos literature, there is good interest in finding chaotic systems with infinite number of equilibrium points such as line equilibrium [46-50], square equilibrium [51], ellipse equilibrium [52], conch equilibrium [53], circle equilibrium [54], heart-shaped equilibrium [55], etc.

In this research paper, we report the finding of a new chaotic system with equilibrium points on the $z-$axis (line equilibrium) as well as two equilibrium points on the $(x,y)-$plane. We describe the phase plots of the chaotic system and do a rigorous dynamic analysis by finding bifurcation diagrams, Lyapunov exponents, equilibrium points, etc. Bifurcation analysis gives valuable information about the chaotic systems [56-61].

Section 2 describes the new chaotic system, its phase plots and equilibrium points. Section 3 describes the dynamic analysis of the new chaotic system. Section 4 depicts an electronic circuit realization of the new chaotic system. Section 5 draws the main conclusions.

Methodology of this work can be detailed as follows. Section 2 describes a dynamical analysis of the properties of the new 3-D chaotic system reported in this work. We calculate the equilibrium of the system and show that the system has line equilibrium. We display the phase portraits of the system. Section 3 describes a detailed bifurcation analysis of the new 3-D chaotic system including multi-stability and coexisting attractors. Section 4 gives a circuit simulation for the new 3-D chaotic system.

2. A new chaotic system with line equilibrium

In this work, we report a new 3-D system given by the dynamics

$$\begin{cases} \dot{x} = z\ \mathrm{sign}(y) \\ \dot{y} = x|x| - y|y| \\ \dot{z} = a|x| - bxy \end{cases} \tag{1}$$

Where x, y, z are state variables and a, b are positive constants.

In this paper, we show that the dynamical system (1) is *chaotic* for the parameter values

$$a = 0.6,\ b = 1.6 \tag{2}$$

For numerical simulations, we take the initial values of the system (4) as $X(0) = (0.2, 0.2, 0.2)$.

Figure 1 shows the phase portrait of the new system (4) for $(a,b) = (0.6, 1.6)$ and initial conditions $X(0) = (0.2, 0.2, 0.2)$. Figures 1 (a)-(c) show the 2-D phase plots of the new chaotic system (1) in (x, y), (y, z), (x, z) coordinate planes, while Figure 1 (d) shows the 3-D phase plot of the new chaotic system (1).

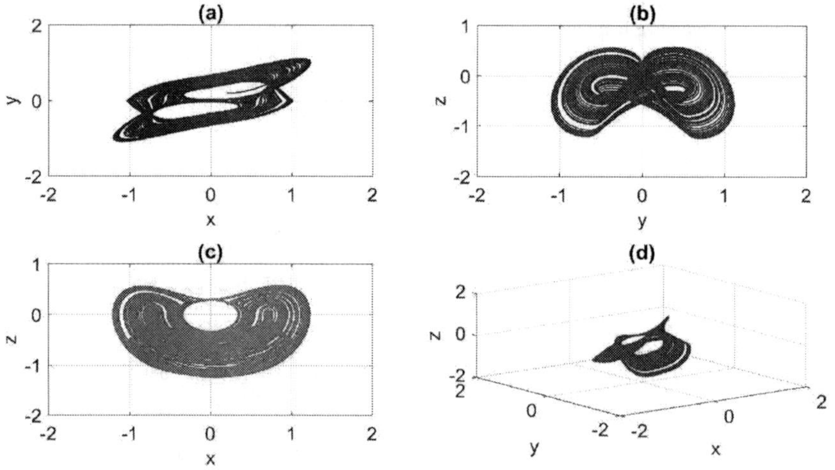

Figure 1. Plots of the new chaotic system (1) for $(a,b) = (0.6, 1.6)$ and $X(0) = (0.2, 0.2, 0.2)$

We take the parameters a and b as in the chaotic case (2), *i.e.* $(a,b) = (0.6, 1.6)$.

The equilibrium points of the new chaotic system (1) are obtained by solving the system:

$$z\ \mathrm{sign}(y) = 0 \tag{3a}$$

$$x|x| - y|y| = 0 \tag{3b}$$

$$a|x| - bxy = 0 \tag{3c}$$

Solving the equations (3), we obtain the set of equilibrium points

$$S=\{(x,y,z)\in R^3 : x=0, y=0\}\cup\{(0.3750,0.3750,0),(-0.3750,-0.3750,0)\}.$$

Thus, the new chaotic system (1) has the whole of z-axis as its equilibrium points as well as the two points (0.3750, 0.3750, 0) and (-0.3750,-0.3750, 0) on the $(x, y)-$plane.

3. Numerical Study

3.1 Bifurcation and Chaos

We fix $b = 0.6$ and vary a in the range of [0, 5]. The bifurcation diagram for varying a and the related graphs of Lyapunov exponents are provided in Figure 2 (a), (b). Obviously, from the bifurcation diagram, one can get that the system is in periodic state in the beginning, then goes into chaos. In addition, there is a big periodic window. Similarly, from Figure 3 (a), one can see that the system changes from period to chaos in the whole region except for a large period-3 window in the region of [1.1, 1.3]. Note that the bifurcation diagram and the Lyapunov exponents match well with each other.

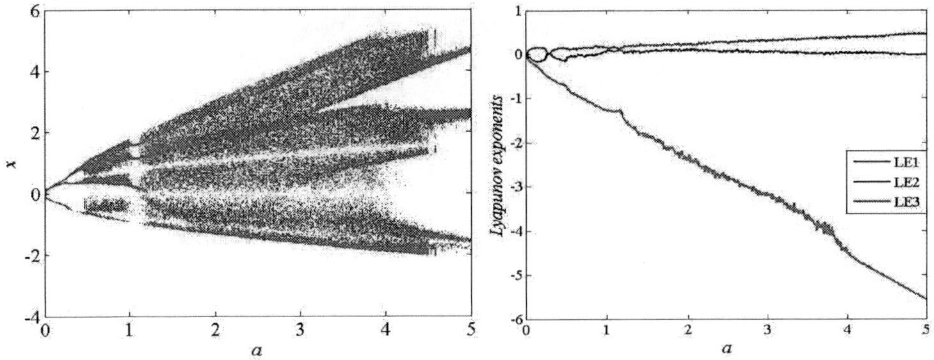

(a) (b)

Figure 2 (a) Bifurcation diagram of system (1) versus the parameter a for $b = 0.6$ and initial conditions $(x(0), y(0), z(0)) = (0.2, 0.2, 0.2)$; **(b)** Lyapunov spectrum of system (1) when varying the parameter a for $b = 0.6$, and initial conditions $(x(0), y(0), z(0)) = (0.2, 0.2, 0.2)$.

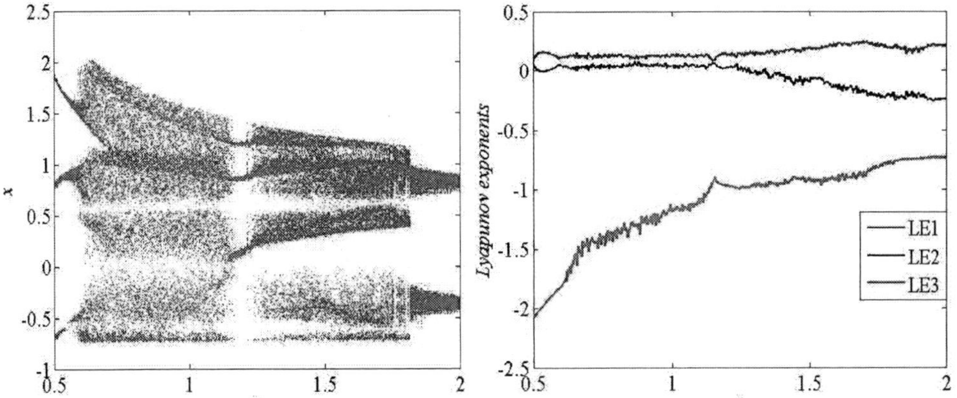

(a) (b)

Figure 3 (a) Bifurcation diagram of the system (1) versus the parameter b for $a = 1.6$ and initial conditions $(x(0), y(0), z(0)) = (0.2, 0.2, 0.2)$; **(b)** Lyapunov spectrum of the system (1) when varying the parameter b for $b = 1.6$, and initial conditions $(x(0), y(0), z(0)) = (0.2, 0.2, 0.2)$.

3.2 Coexisting attractor

In this work, the bifurcation diagrams of the system (1) versus a [0, 5] are shown in Figure 4, where the blue color starts from the initial conditions (0.2 0.2 0.2) and the red color starts from the initial conditions (-0.2 -0.2 0.2), respectively. As can be seen from the bifurcation diagram, there exists coexisting attractors in the very narrow regions of [0.25, 0.5] and [4.5, 5]. The coexisting chaotic attractors can be seen in Fig. 5a, when $a = 4.5$ and the coexisting periodic attractors can be seen in Fig. 5b, when $a = 0.25$.

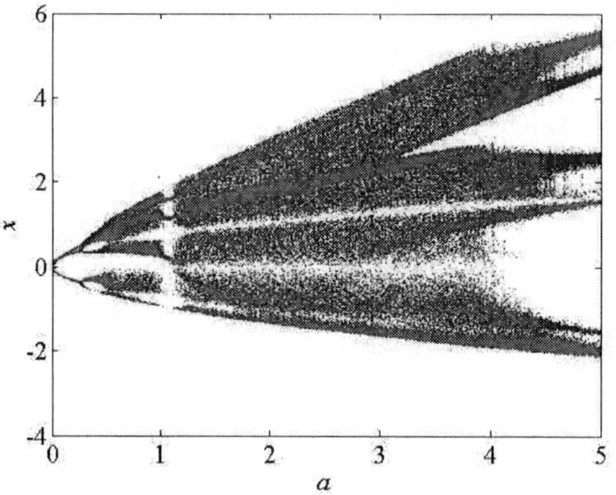

Figure 4 The bifurcation diagrams of the system (1) with a from 0 to 5 for $b = 1.6$. condition: (x (0) , y (0) , z (0)) = (0.2 0.2 0.2) (*blue*), (x (0) , y (0) , z (0)) = (-0.2 -0.2 0.2), (*red*)

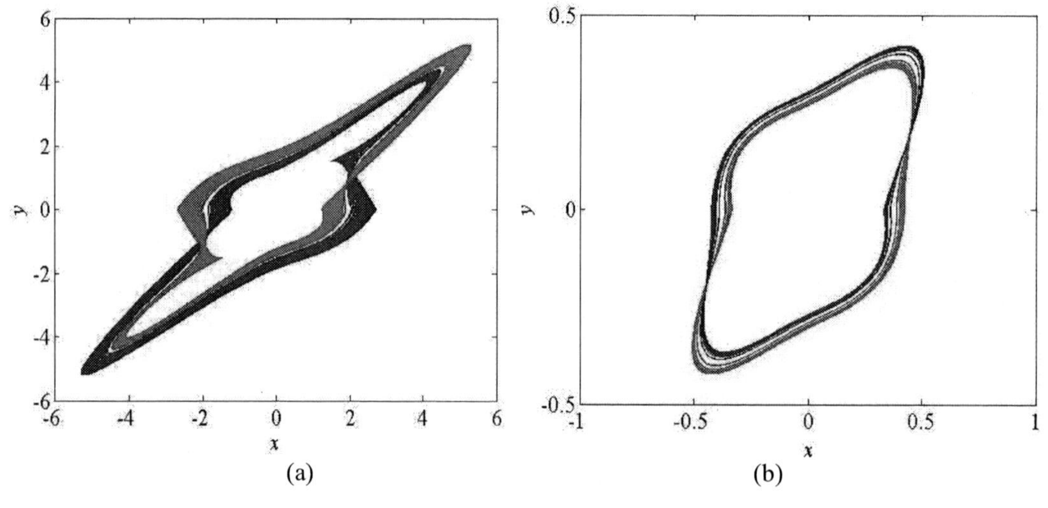

Figure 5 Phase portraits of the system (1) displayed in the x–y plane when changing the value of parameter a: (a) the coexisting chaotic attractors for a = 4.5
(b) the coexisting periodic attractors for $a = 0.25$.

3.3 Offset boosting control

Clearly, the state variable z appears only once in the first equation of the system. Therefore, we can control the state variable z conveniently. The state variable z is offset-boosted by replacing z with $z + k$, in which k is a constant. The system can be rewritten as

$$\begin{cases} \dot{x} = sign(y)(z+k) \\ \dot{y} = x|x| - y|y| \\ \dot{z} = a|x| - bxy \end{cases} \tag{4}$$

Consequently, the chaotic signal z can be transformed from a bipolar signal to a unipolar signal when varying the control parameter k. When increasing the boosting controller k, the chaotic signal z is boosted from a bipolar signal to a unipolar one as illustrated in Figure 6. Phase portraits of system (1) are adjusted according to the boosting controller as illustrated in Figure 7.

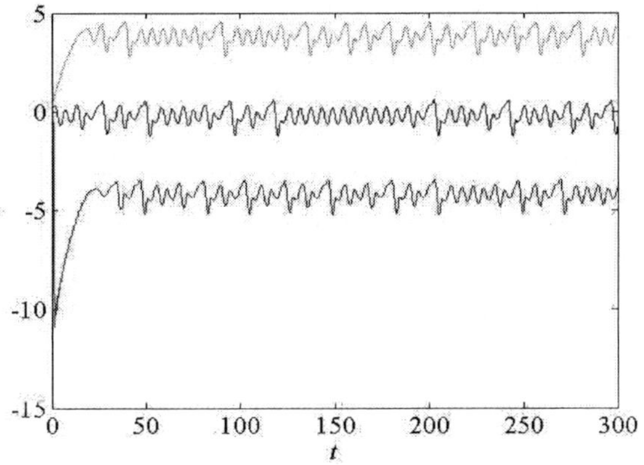

Figure 6 The signal z with different values of the offset boosting controller k: $k = 0$ (blue colour); $k = 4$ (red colour); $n = -4$ (green colour).

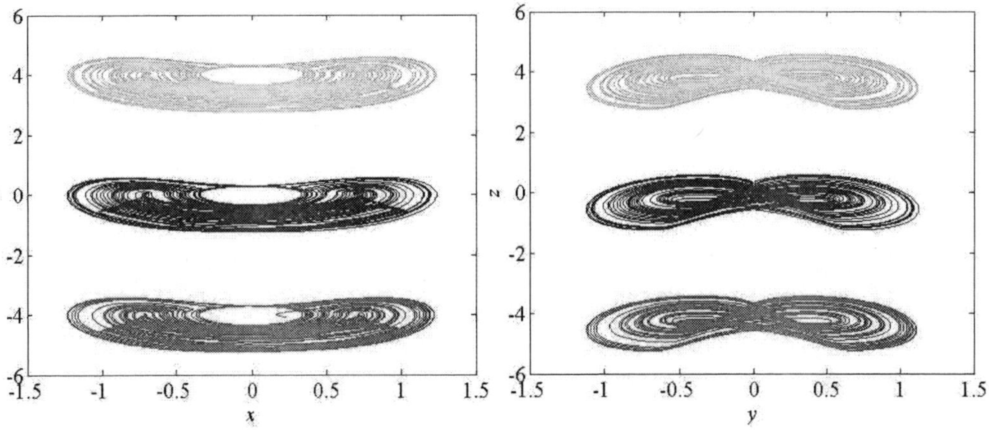

Figure 7 Phase portraits in different planes and different values of the offset boosting controller k: (a) x - z plane, (b) y - z plane for $k = 0$ (blue colour), $k = 4$ (red colour), $k = -4$ (green colour).

4. Circuit implementation of the new chaotic system

In this work, the electronic circuit design of signum nonlinearity (1) is presented. The electronic circuit is designed in MultiSIM platform. The signum nonlinearity which detailed electronic circuit is depicted in Figure 8 consists of two resistors and two operational amplifiers. A detailed analysis of the signum circuit can be found in [62].

The circuital equations of the designed new chaotic system are given by

$$\begin{cases} \dot{x} = \dfrac{1}{10C_1R_1} z\,sign(y) \\ \dot{y} = \dfrac{1}{10C_2R_2} x|x| - \dfrac{1}{10C_2R_3} y|y| \\ \dot{z} = -\dfrac{1}{C_3R_4}|x| - \dfrac{1}{10C_3R_5} xy \end{cases} \quad (5)$$

Where x , y , and z are the voltages across the capacitors C_1, C_2 and C_3, respectively. The circuit has been implemented by using MultiSIM $R_1 = R_2 = R_3 = 40$ kΩ, $R_4 = 66.67$ kΩ, $R_9 = 1.425$ MΩ, $R_5 = 25$ kΩ, $R_6 = R_7 = R_8 = R_{10} = R_{11} = R_{12} = R_{13} = R_{14} = R_5 = R_{16} = R_{17} = R_{18} = R_{19} = R_{20} = R_{21} = 100$ kΩ, $C_1 = C_2 = C_3 = 1$ nF. Obtained MultiSIM results in Figure 9 indicate that the circuit exhibits chaotic attractors. The MultiSIM results (see Figure 9) based simulations are carried out to confirm the results of theoretical analysis (see. Figure 1).

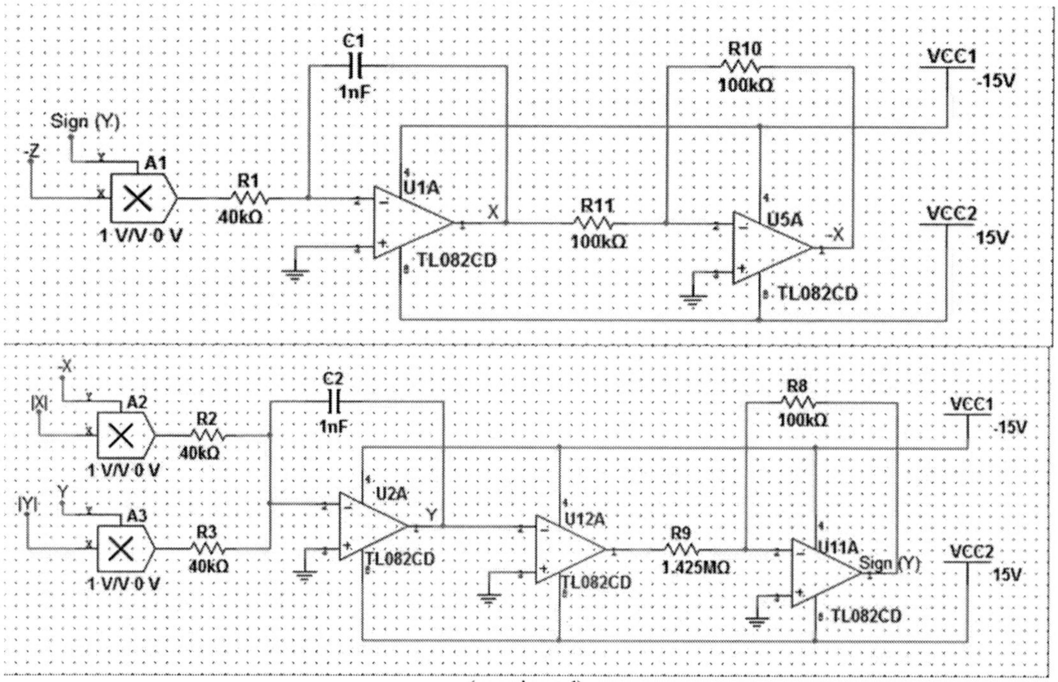

(continued)

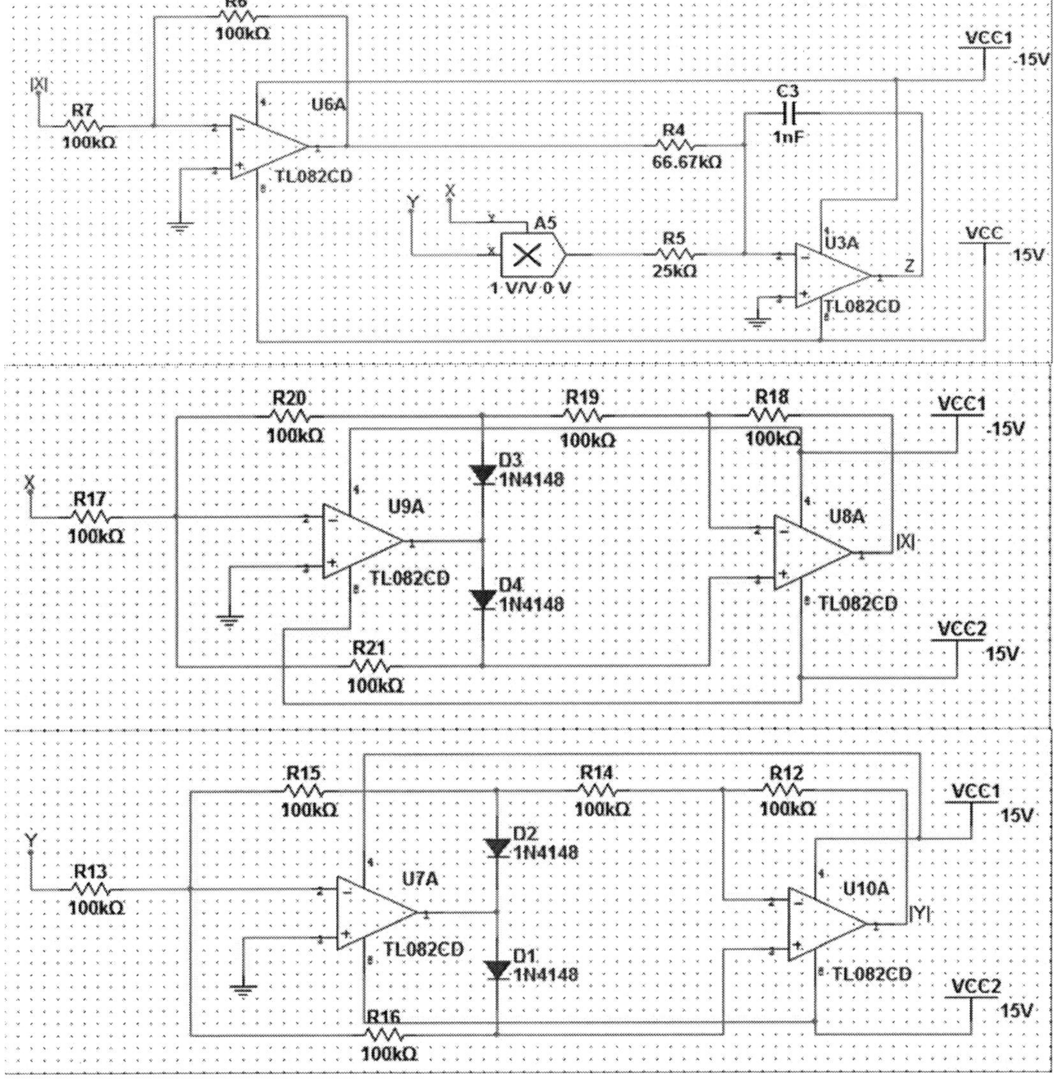

Figure 8 The electronic circuit schematic of new chaotic system (1)

IOP Conf. Series: Materials Science and Engineering **567** (2019) 012009 doi:10.1088/1757-899X/567/1/012009

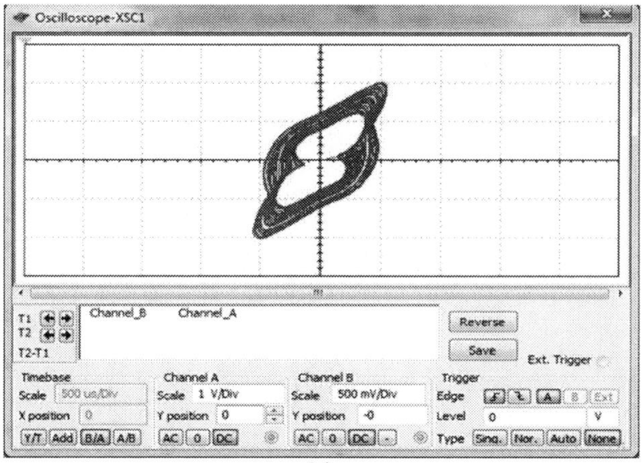

(a)

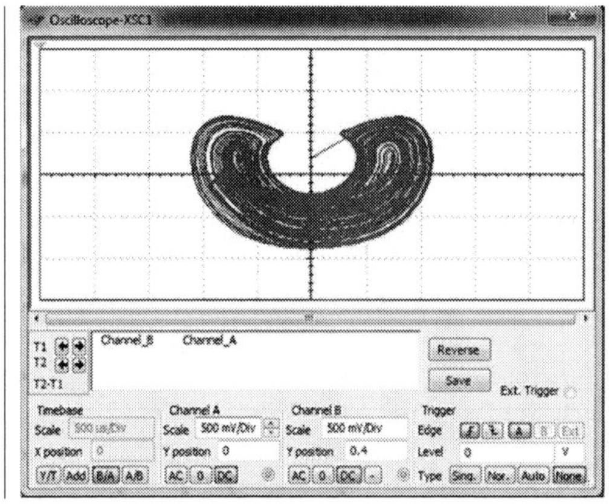

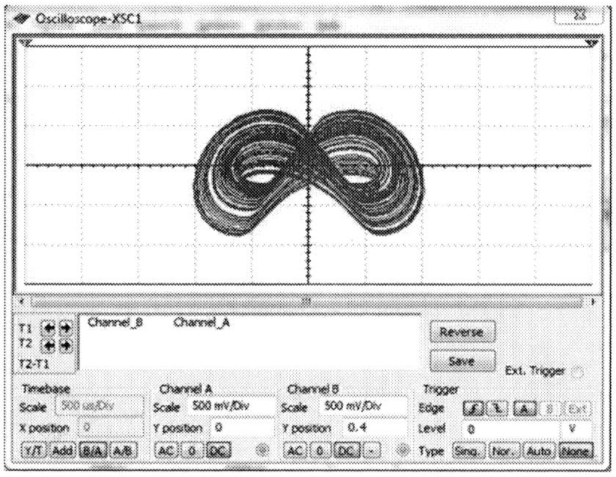

(b) (c)

Figure 9 MultiSIM chaotic attractors of the new chaotic system (1)
(a) x_1- x_2 plane (b) x_2- x_3 plane and (c) x_1-x_3 plane.

5. Conclusions

A new chaotic system with five nonlinearities was announced in this paper. Our proposed chaotic system has a novel feature that there is no linear term in it. We also showed that the chaotic system has equilibrium points on the z-axis (line equilibrium) as well as two equilibrium points on the (x, y)-plane. The dynamical properties of the new chaotic system were analyzed in terms of phase portraits, bifurcation diagram, etc. Finally, an electronic circuit realization of the new chaotic system was displayed in detail to confirm the feasibility of the theoretical chaotic model.

References

[1] Vaidyanathan S and Volos C 2017 *Advances and Applications in Chaotic Systems* (Berlin: Springer)
[2] Alligood K T, Sauer T D and Yorke J A 2000 *Chaos: An Introduction to Dynamical Systems* (Berlin: Springer)

[3] Vaidyanathan S, Azar A T, Rajagopal K, Sambas A, Kacar S and Cavusoglu U 2018 International Journal of Simulation and Process Modelling **13** 281-296
[4] Vaidyanathan S, Volos C K, Rajagopal K, Kyprianidis I M and Stouboulos I N 2015 *Journal of Engineering Science and Technology Review* **8** 74-82
[5] Vaidyanathan S 2015 *International Journal of PharmTech Research* **8** 974-981
[6] Vaidyanathan S 2015 *International Journal of PharmTech Research* **8** 1-11
[7] Vaidyanathan S 2015 *International Journal of PharmTech Research* **8** 117-127
[8] Tomita K 1982 *Journal of Theoretical Biology* **99** 111-118
[9] Vaidyanathan S 2015 *International Journal of PharmTech Research* **8** 106-116
[10] Vaidyanathan S 2015 *International Journal of PharmTech Research* **8** 156-166
[11] Njitacke Z T and Kengne J 2018 *AEU-International Journal of Electronics and Communications* **93** 242-252
[12] Vaidyanathan S 2015 *International Journal of PharmTech Research* **8** 946-955
[13] Saad M, Safieddine A and Sultan R 2018 *Journal of Physical Chemistry A* **122** 6043-6047
[14] Vaidyanathan S 2015 *International Journal of ChemTech Research* **8** 159-171
[15] Hellen E H and Volkov E 2018 *Communications in Nonlinear Science and Numerical Simulation* **62** 462-479
[16] Vaidyanathan S and Rasappan S 2011 *Communications in Computer and Information Science* **131** 585-593
[17] Pisarchik A N, Huerta-Cuellar G and Kulp C W 2018 *Communications in Nonlinear Science and Numerical Simulation* **62** 134-145
[18] Vaidyanathan S 2015 *International Journal of Modelling, Identification and Control* **23** 380-392
[19] Vaidyanathan S 2012 *Lecture Notes of the Institute for for Computer Sciences, Social-Informatics and Telecommunications Engineering* **85** 124-133,
[20] Xu Q, Zhang Q, Jiang T, Bao B and Chen M 2018 *Circuit World* **44** 108-114
[21] Vaidyanathan S, Sambas A, Mamat M and Sanjaya W S M 2017 *Archives of Control Sciences* **27** 541-554
[22] Singh J P, Lochan K, Kuznetsov N V and Roy B K 2017 *Nonlinear Dynamics* **90** 1277-1299
[23] Wang Y, Mou Y and Zhang J 2018 *Journal of Harbin Engineering University* **39** 584-593
[24] Mansour S M B, Sundarapandian V and Naceur S M 2016 *International Journal of Control Theory and Applications* **9** 37-54
[25] Akgul A, Moroz I, Pehlivan S and Vaidyanathan S 2016 *Optik* **127** 5491-5499
[26] Vaidyanathan S, Akgul A, Kacar S and Cavusoglu U 2018 *European Physical Journal Plus* **133** 46
[27] Vaidyanathan S 2015 *Kyungpook Mathematical Journal* **55** 563-586
[28] Vaidyanathan S and Rajagopal K 2017 *International Journal of Simulation and Process Modelling* **12** 165-178
[29] Dou Y, Liu X, Fan H and Li M 2017 *Optik* **145** 456-464
[30] Vaidyanathan S, Sambas A, Mamat M and Sanjaya W S M 2017 *International Journal of Modelling, Identification and Control* **28** 153-166
[31] Idowu B A, Vaidyanathan S, Sambas A, Olusola O I and Onma O S 2018 *Studies in Systems, Decision and Control* **133** 271-295

[32] Tacha O I, Volos C K, Kyprianidis I M, Stouboulos I N, Vaidyanathan S and Pham V T 2016 *Applied Mathematics and Computation* **276** 200-217
[33] Volos C K, Pham V T, Vaidyanathan S, Kyprianidis I M and Stouboulos I N 2015 *Journal of Engineering Science and Technology Review* **8** 142-151
[34] Daltzis P, Vaidyanathan S, Pham V T, Volos C, Nistazakis E. andTombras G. 2018 *Circuits, Systems, and Signal Processing* **37** 613-615
[35] Sambas A, Vaidyanathan S, Mamat M and Mada Sanjaya W S 2018 *Studies in Systems, Decision and Control* **133** 365-373
[36] Vaidyanathan S, Jafari S, Pham V T, Azar A T and Alsaadi F E 2018 *Archives of Control Sciences* **28** 239-254
[37] Wang X, Vaidyanathan S, Volos C, Pham V T and Kapitaniak T 2017 *Nonlinear Dynamics* **89** 1673-1687
[38] Sambas A, Mamat M, Vaidyanathan S, Mohamed M A, Mada Sanjaya W S and Mujiarto 2018 *WSEAS Transactions on Systems and Control* **13** 345-352
[39] Bao B, Xu L, Wang N, Bao H, Xu Q and Chen M 2018 *AEU-International Journal of Electronics and Communications* **94** 26-35
[40] Sambas A, Mamat M, Vaidyanathan S, Mohamed M A and Mada Sanjaya W S 2018 *International Journal of Engineering and Technology* **7** 1245-1250
[41] Volos C, Maaita J O, Vaidyanathan S, Pham V T, Stouboulos I and Kyprianidis I 2017 *IEEE Transactions on Circuits and Systems II: Express Briefs* **64** 339-343
[42] Pham V T, Jafari S, Volos C, Giakoumis A, Vaidyanathan S and Kapitaniak T 2016 *IEEE Transactions on Circuits and Systems II: Express Briefs* **63** 878-882
[43] Mamat M, Vaidyanathan S, Sambas A, Mujiarto, Sanjaya W S M and Subiyanto 2018 *IOP Conference Series: Materials Science and Engineering* **332** 012033
[44] Lien C H, Vaidyanathan S, Sambas A, Sukono, Mamat M, Sanjaya W S M and Subiyanto 2018 *IOP Conference Series: Materials Science and Engineering* **332** 012010
[45] Vaidyanathan S, Sambas A, Sukono, Mamat M, Gundara G, Sanjaya W S M and Subiyanto 2018 *IOP Conference Series: Materials Science and Engineering* **332** 012048
[46] Zhang S, Zeng Y, Li Z, Wang M and Xiong L 2018 *Chaos: An Interdisciplinary Journal of Nonlinear Science* **28** 013113
[47] Kingni S T, Pham V T, Jafari S and Woafo P 2017 *Chaos, Solitons and Fractals* **99** 209-218
[48] Jafari S and Sprott J C 2013 *Chaos, Solitons and Fractals* **57** 79-84
[49] Vaidyanathan S 2016 *Studies in Computational Intelligence* **636** 471-494
[50] Li C and Sprott J C 2014 *Physics Letters A* **378** 178-183
[51] Gotthans T, Sprott J C and Petrzela J 2016 *International Journal of Bifurcation and Chaos* **26** 1560137
[52] Pham V T, Jafari S, Wang X and Ma J 2016 *International Journal of Bifurcation and Chaos* **26** 1650069
[53] Mamat M, Vaidyanathan S, Sambas A, Mohamed M A, Sampath S and Sanjaya W S M 2018 *International Journal of Engineering and Technology* **7** 1410-1414
[54] Gotthans T and Petrzela J 2015 *Nonlinear Dynamics* **81** 1143-1149
[55] Pham V T, Jafari S and Volos C 2017 *Optik* **131** 343-439
[56] Zhang S, Zeng Y, Li Z, Wang M, Zhang X and Chang D 2018 *International Journal of Dynamics and Control* **23** 1-12
[57] Zhang S, Zeng Y and Li Z 2018 *Chinese Journal of Physics* **56** 793-806
[58] Zhang S, Zeng Y C and Li Z J 2018 *J. Computational and Nonlinear Dynamics* **13** 1-10
[59] Zhang S, Zeng Y, Li Z, Wang M and Xiong L 2018 *Pramana* **90** 63.
[60] Wang L, Zhang S, Zeng Y C and Li Z J 2018 *Electronics Letters* **52** 1008-1010
[61] Zhang S, Zeng Y C, Li Z J, Wang M J and Xiong L 2018 *Chaos* **28** 013113
[62] Pham V T, Volos C, Jafari S and Kapitaniak T 2018 *Journal of Crcuits, Systems and Computers*, **27** 1850066.

IOP Conf. Series: Materials Science and Engineering **567** (2019) 012010 doi:10.1088/1757-899X/567/1/012010

Determination of the status of utilization and effort of little tuna (*Euthynnus affinis*) caught in the North Bolaang-Mongondow waters North Sulawesi

J S Kekenusa[1], S B Rondonuwu[2], M S Paendong[1]

[1]Jurusan Matematika, Universitas Sam Ratulangi Manado, Sulawesi Utara, Indonesia 95115
[2]Jurusan Biologi, Universitas Sam Ratulangi Manado, Sulawesi Utara, Indonesia 95115

Corresponding author's e-mail: johnskekenusa@unsrat.ac.id

Abstract. Little tuna (*Euthynnus affinis*) needs tobe managed well because even as a renewable natural resources, but can undergo depletion or extinction. One of the approach inthe management of the resources is by modelling. The analysis was performed aimingto get the best estimated for the surplus production model to determine the maximum sustainable yields (MSY), utilization level, and effort level of little tuna. The data of catch and fishing effort little tuna collected from the Marine and Fisheries Service of the North Bolaang-Mongondow Regency and the North Sulawesi Province. Best Surplus Production Model, which is used to assess the potential of little tuna is ***Schaefer Model***. Optimal effort (E_{MSY}) of **482 trips** per year, with catchesof optimal C_{MSY} **465.13 tons** per year. The effort level for 2017 is **114.10 %,** which shows the inefficiency of effort, the utilization level of **125.99 %,** showing occur overfishing.

1. Introduction

Little tuna (*Euthynnus affinis*) classified as pelagic fishery resource is important and one of the non-oil export commodity in North Sulawesi. Little tuna in North Sulawesi (including North Bolaang-Mongondow waters) in 2016 reached 30,000 tons per year, with a value of about 100 billion rupiahs [1]. Research on little tuna gererally discuss the exploitation in increase production, not much research on the status of utilization (including aspects of sustainability and efficiency) resources. Data on the level of utilization of the fish resources are very important, as it will determine whether the resource use is less tan optimal, optimal, or excessive. Excessive utilization of fish resources would threaten its sustainability. By knowing the level of resource utilization on the little tuna, is expected tobe done in a planned and suatainable management.

The simplest model of the dynamics of fish populations is Surplus Production Model (SPM), by treating the fish as a single biomass that can not be divided, which is subject to the rules of simple increases and decreases in biomass. This model, commonly used in the assessment of fish stocks using only the data of catch and fishing effort generally available.

This study aims to get the best SPM, as well as knowing how much the result of maximum sustainable yields (MSY), utilization level, and the level of effort of little tuna in the North

Bolaang-Mongondow waters.

2. Materials and Methods

The primary and secondary data of little tuna catching is collected from the North Bolaang-Mongondow waters. Production and fishing effort data collected from the Marine and Fisheries Service of North Bolaang-Mongondow Regency and North Sulawesi Province during years 2008-2017. Data (variables) used for the analysis of the surplus production model is the data of the Catch (**C_t**) per year, fishing effort (**E_t**) per year, and **CPUE** (Catch per Unit of Effort)

The models estimator who analyzed and evaluated are: **Schaefer, Fox, Schnute, Walter-Hilborn, Clarke-Yoshimoto-Pooley (CYP)**. Base on the results of **statistical evaluation (sign suitability of regression coefficient, the value of R^2, the validation value, and significancy of the regression coefficient of models**), we get the **"best"** as estimator. From the best of model can be calculated C_{MSY} value, utilization level, and the level of effort of little tuna.

The simplest model of the dynamics of fish populations is a surplus production model that treats the fish population as a single biomass that can not be divided, which is subject to the simple rules of the rise and decline. The production model is dependent on the amount of four kinds, namely: biomass population at a given time t (B_t), catches for a certain time t (C_t), fishing effort at a certain time t (E_t), and the natural growth rate constant (r) [2]. This model was first developed by Schaefer, who was initially the same as the form of logistic growth model. According to Coppola and Pascoe [3], equation surplus consists of several constants that are affected by natural growth, the ability of fishing gear, and carrying capacity. Constants allegedly using models of biological parameter estimators of surplus production equation, namely the model: Equilibrium Schaefer, Schaefer Disequilibrium, Schnute, and Walter - Hilborn. Based on the four models were selected the most appropriate or best fit of the estimation of others. According to Sparre and Venema [4], formulas surplus production model is valid only if the slope parameter (b) is negative, which means the addition of fishing effort will lead to a decrease in the catch per fishing effort. If the parameter b positive value, then it can not be done estimating the optimum amount of stock and effort, but it can only be concluded that the addition of fishing effort is still possible to increase the catch.

Prediction of optimum fishing effort (E_{opt}) and the maximum sustainable catch (C_{MSY}) approached the surplus production model. Between the catch per unit of effort (CPUE) and fishing effort can be either linear or exponential relationship [5]. Surplus Production Model consists of two models, namely basic model of Schaefer (linear relationship) and the Gompertz model developed by Fox with forms exponential relationship [5].

The procedures for estimates of surplus production models follows these equations:

Surplus production models first developed by **Schaefer**, who was initially the same as the form of logistic growth model. The model is as follows:

$$\frac{dB_t}{dt} = G(B_t) = r B_t \left(1 - \frac{B_t}{K}\right) \tag{1}$$

This equation does not include the effect of the catching, so Schaefer wrote back to :

$$\frac{dB_t}{dt} = r B_t \left(1 - \frac{B_t}{K}\right) - C_t \tag{2}$$

K is the carrying capacity of the marine environment, and C_t is the catch that can be written as:

$$C_t = q E_t B_t \tag{3}$$

q is catchability, and Et indicates fishing effort. This equation can be written as:

$$\frac{C_t}{E_t} = q\ B_t = \text{CPUE} \tag{4}$$

From the differential equation (2), the optimum catchment can be calculated at the time $\frac{dB_t}{dt}$ = 0, also called settlement at the point of balance (equilibrium), in the form of:

$$r\ B_t\,(1 - \frac{B_t}{K}) - C_t = 0, \quad \text{or}$$

$$C_t = r\ B_t\,(1 - \frac{B_t}{K}) = q\ E_t\ B_t \tag{5}$$

From equation (3) and (5), find value of B_t obtained as follows:

$$B_t = K\,(1 - \frac{qE_t}{r}) \tag{6}$$

So that equation (5) becomes:

$$C_t = q\ K\ E_t\ (1 - \frac{qE_t}{r})$$

$$= q\ K\ E_t - \frac{q^2K}{r}\ E_t^2 \tag{7}$$

Equation (7) is simplified further by Schaefer becomes:

$$\frac{C_t}{E_t} = a - b\ E_t, \quad \text{or}$$

$$C_t = a\ E_t - b\ E_t^2 \tag{8}$$

while the $a = q\ K$ and $b = \frac{q^2K}{r}$

This linear relationship is used widely for calculating C_{MSY} through the determination of the first derivative of C_t with E_t to find optimal solutions, both to catch and fishing effort. The first derivative of C_t to Et is : $\frac{dC_t}{dE_t} = a - 2b\ E_t$, in order to obtain the alleged E_{opt} (optimum fishing effort) and C_{MSY} (maximum sustainable yields) respectively :

$$E_{opt} = \frac{a}{2b} = \frac{r}{2q} \tag{9}$$

by entering the value of E_{opt} in equation (8), will be obtained C_{MSY} as follows:

$$C_{MSY} = a\ E_t - b\ E_t^2$$

$$= a\,(\frac{a}{2b}) - b\,(\frac{a}{2b})^2$$

IOP Conf. Series: Materials Science and Engineering **567** (2019) 012010 doi:10.1088/1757-899X/567/1/012010

$$= \frac{a^2}{4b}$$

by substituting a = qK and b = $\frac{q^2K}{r}$ will be obtained,

$$C_{MSY} = \frac{a^2}{4b} = \frac{q^2K^2}{4q^2K/r} = \frac{rK}{4} \quad (10)$$

The values of a and b are estimated by the least squares method approach that is commonly used to estimate the coefficient of a simple regression equation. Furthermore, by including the value of E_{opt} in equation (6) is obtained optimum biomass (B_{MSY}) as follows:

$$\begin{aligned} B_{MSY} &= K - \frac{Kq}{r} E_{opt} \\ &= K - \frac{Kq}{r} \left(\frac{r}{2q}\right) \\ &= K - \frac{K}{2} \\ &= \frac{K}{2} \end{aligned} \quad (11)$$

The values of the parameter q, K, and r can be calculated using the Fox algorithm, as referenced in Sularso [6], as follows:

$$q_t = \ln\left[\left[\left(zU_t^{-1} + \frac{1}{b}\right) / \left(zU_{t+1}^{-1} + \frac{1}{b}\right)\right]\right] / (z) \quad (12)$$

where $z = -(a / b) / E^*$, $E^* = (E_t + E_{t+1}) / 2$, $U_t = \frac{C_t}{E_t}$ and the value of q is the geometric mean of the value of q_t. From the values of a, b, and q, can then be calculated values of K and r.

Model of Fox has several characteristics that are different from the model Schaefer, that it biomass growth following the Gompertz growth model [7]. The relation of CPUE with effort (E) follows a negative exponential pattern :

$$C_t = E_t \, . \, \exp(a - b\, E_t) \quad (13)$$

Efforts optimum is obtained by equating the first derivative of Ct to E_t equal to zero and find :

$$E_{opt} = \frac{1}{b} \quad (14)$$

The maximum sustainable yields of catch (C_{MSY}) is obtained by inserting the value of the optimum effort into equation (13), and obtained:

$$C_{MSY} = \frac{1}{b} e^{a-1} \quad (15)$$

Model of Schnute [8], suggests another version of the surplus production model is dynamic and deterministic. Schnute method is considered as a modification of the model in the form of discrete Schaefer (Roff, 1983, referred by Tinungki) [9],

$$\ln\left(\frac{U_{t+1}}{U_t}\right) = r - \frac{r}{qK}\left(\frac{U_t + U_{t+1}}{2}\right) - q\left(\frac{E_t + E_{t+1}}{2}\right)$$

$$= a - b\left(\frac{U_t + U_{t+1}}{2}\right) - c\left(\frac{E_t + E_{t+1}}{2}\right) \quad (16)$$

where a = r, $b = \frac{r}{qK}$, and c = q, is the regression coefficient estimators.

Walter and Hilborn (1976) referred by Tinungki [9], to develop other types of surplus production model, known as the regression model. Walter - Hilborn Model, using a simple differential equation, by the following equation :

$$\frac{U_{t+1}}{U_t} - 1 = r - \frac{r}{Kq} U_t - q E_t$$

$$= a - b U_t - c E_t \quad (17)$$

where a = r, $b = \frac{r}{Kq}$, and c = q, is the regression coefficient estimators.

Estimation of biological parameters for the surplus production model can also be done through estimation techniques proposed by Clarke, Yoshimoto, and Pooley (CYP) [9, 10]. The parameters which allegedly is r, K, and q, the model is expressed as follows:

$$\ln(U_{t+1}) = \left(\frac{2r}{2+r}\right)\ln(qK) + \frac{2-r}{2+r}\ln(U_t) - \frac{q}{2+r}(E_t + E_{t+1}) \quad (18)$$

where : $a' = \frac{2r}{2+r}$, $a = a'\ln(qK)$, $b = \frac{2-r}{2+r}$, $c = \frac{q}{2+r}$

thus equation (18) can be written in the form :

$$\ln(U_{t+1}) = a' \ln(qK) + b\ln(U_t) - c(E_t + E_{t+1})$$

$$= a + b\ln(U_t) - c(E_t + E_{t+1}) \quad (19)$$

3. Results and Discussion

Catches of little tuna fisheries in the North Bolaang-Mongondow waters fluctuate from year to year. Data catching in 2008-2017, are presented in Table 1.

From the analysis of regression, equation for **Schaefer Model** : $\frac{C_t}{E_t} = 1.929 - 0.002\ E_t$, with a coefficient of determination (R^2) = 0.734 and a significance level of $p < 0.05$. Thus, a production model estimator catches Schaefer model according to the equation (8) is: $\mathbf{C_t = 1.929\ E_t - 0.002\ E_t^2}$.

From the results of the regression analysis for **Fox Model** :

$\text{Ln}\ \frac{C_t}{E_t} = 0.733 - 0.001\ E_t$, with $R^2 = 0.731$ ($p < 0.05$). Estimates of catches corresponding to the model Fox equation (13) : $C_t = E_t . e^{(0.733 - 0.001 E_t)}$

Years	Catch (ton), C_t	Effort (trip), E_t	CPUE = $\frac{C_t}{E_t}$ (ton/trip)
2008	401.0	240	1.6708
2009	391.0	300	1.3033
2010	457.0	480	0.9521
2011	400.9	400	1.0022
2012	541.6	484	1.1190
2013	531.0	528	1.0057
2014	524.7	527	0.9956
2015	471.0	480	0.9813
2016	509.0	525	0.9695
2017	586.0	550	1.0655
Mean	**481.320**	**451.400**	**1.1065**

Table 1. Total catch, fishing efforts, and CPUE (Catch per Unit of Efforts) of Little tuna in North Bolaang-Mongondow waters 2008-2017

Source : *Calculated from the Marine and Fisheries Service and Landing Station of Fish in North Bolaang-Mongondow*

For **Schnute model** according to equation (16), obtained regression equation:

$$\ln\left(\frac{U_{t+1}}{U_t}\right) = 0.055 - 0.337\left(\frac{U_t + U_{t+1}}{2}\right) + 0{,}01\left(\frac{E_t + E_{t+1}}{2}\right)$$

with $R^2 = 0.457$, and all the regression coefficient was significant ($p > 0.05$).

In **Walter-Hilborn Model** using equation (17) derived regression equation:

$$\frac{U_{t+1}}{U_t} - 1 = 1.009 - 0.651\, U_t - 0.001\, E_t$$

With $R^2 = 0.831$ and not all regression coefficients were significant ($p < 0.05$). In the regression equation **CYP Model**, according to equation (19):

$$\ln(U_{t+1}) = 0.325 + 0.074 \ln(U_t) - 0.000320(E_t + E_{t+1})$$

with $R^2 = 0.464$, and the regression coefficient are not significant ($p > 0.05$). The results of calculations for validation surplus production model of 5 models is presented is summarized in Table 2.

Table 2. Results of the surplus production model validation

	Model Schaefer	**Model Fox**	**Model Schnute**	**Model Walter-Hilborn**	**Model CYP**
Sign Suitability	**Appropriate**	**Appropriate**	**Not Appropriate**	**Appropriate**	**Appropriate**
R^2 Value	**0.734**	0.731	0.457	0.831	0.464
Validation Value	**0.1019**	0.2258	13.6514	0.2214	0.9871
Significance Coefficient	**Significant**	Significant	Not Significant	Not all significant	Not Significant

From the results of the calculations in Table 2, it appears that the most appropriate is **Schaefer model** with the R^2 value is quite large (R^2 = 0.734) and validation (residual value) is smallest. Schaefer model obtained values of a = 1.929 and b = 0.002, with equation (9) and (10) can be calculated optimum value of Effort (E_{opt}) and the maximum sustainable catch (C_{MSY}) as follows:

$$E_{opt} = \frac{a}{2b} = \frac{1.929}{2(0.002)} = 482.25 \approx 482 \text{ trips per year.}$$

$$C_{MSY} = \frac{a^2}{4b} = \frac{1.929^2}{4(0.002)} = 465.13 \text{ tons per year.}$$

This means that in order to preserve the bonito fisheries resources technically and biologically, in a year the number of units should not exceed 482 trips. To preserve the Little tuna resources in the waters North Bolaang-Mongondow, the maximum of fish that can be caught at 465.13 tons per year. Furthermore, from the value of E_{opt} and C_{MSY} can be calculated fishing effort levels and utilization level of bonito for a particular year for example in 2017, as follows:

$$\text{The level of effort in 2017} = \frac{E_{2017}}{E_{opt}} \text{ x } 100\%$$

$$= \frac{550}{482} \text{ x } 100\% = 114.10\,\%$$

$$\text{The utilization level in 2017} = \frac{C_{2017}}{C_{MSY}} \text{ x } 100\%$$

$$= \frac{586.0}{465.13} \text{ x } 100\% = 125.99\,\%.$$

From the calculation, it turns out Little tuna fishing effort at the North Bolaang-Mongondow waters in 2017, greater than the maximum sustainable level of effort. This shows that fishing effort is not efficient. The utilization level for the year 2017, is more than optimum level, its mean a sign of *overfishing* (catch-over). The same result of Little tuna fishing effort and utilization level at the Talaud waters shows not efficient and overfishing [11], at Manado waters [12], also at Bitung waters [13].

This study describes the use of some **statistical criteria** in selecting the best surplus production model. By applying some statistical criteria in selecting a surplus production model, will obtain better results. Researchers in the field of fisheries get guidelines for setting selection criteria for surplus production models, as well as avoiding the direct application of one model in analyzing the surplus production model in a waters.

4. Conclusions and Recommendations

4.1 Conslusions

1. The surplus production model that can be used to examine the catch of little tuna in the North Bolaang-Mongondow waters is Schaefer model, by the equation : $C_t = 1.929\ E_t - 0.002\ E_t^2$.
2. The maximum sustainable yield of little tuna C_{MSY} is 465.13 tons per year, obtained at the of fishing effort E_{MSY} 482 trips per year. For the year 2017 the amount of 125.99 % utilization level is a sign of overfishing, with the level effort 114.10 % indicating inefficiencies in fishing effort.

4.2 Recommendations

1. In applying surplus production model in a waters location, not only directly using one particular model, but should use some of the models are chosen base on statistical criteria.

These criteria involve, among others : suitability sign of the coefficient of models, coefficient of determination (R^2), the value of validation, and the significance of the regression coefficients.

2. There are indications occur overfishing and the presence of the inefficiency of fishing effort of little tuna in the North Bolaang-Mongondow waters, recommended immediate supervisión by competent intitutions to handle this issue.

Appendix

Appendix 1. Regression analysis of Surplus Production Model of little tuna data in North Bolaang-Mongondow waters

1. Model Schaefer

Model Summary

Model	R	R Square	Adjusted R Square	Std. Error of the Estimate
1	.857[a]	.734	.701	.1224453

a. Predictors: (Constant), Et

ANOVA[a]

Model		Sum of Squares	df	Mean Square	F	Sig.
1	Regression	.331	1	.331	22.058	.002[b]
	Residual	.120	8	.015		
	Total	.451	9			

a. Dependent Variable: Ct_per_Et

b. Predictors: (Constant), Et

Coefficients[a]

Model		Unstandardized Coefficients		Standardized Coefficients	T	Sig.
		B	Std. Error	Beta		
1	(Constant)	1.929	.179		10.757	.000
	Et	-.002	.000	-.857	-4.697	.002

a. Dependent Variable: Ct_per_Et

2. Model Fox

Model Summary

Model	R	R Square	Adjusted R Square	Std. Error of the Estimate
1	.855[a]	.731	.697	.0970864

a. Predictors: (Constant), Et

ANOVA[a]

Model		Sum of Squares	df	Mean Square	F	Sig.
1	Regression	.205	1	.205	21.748	.002[b]
	Residual	.075	8	.009		
	Total	.280	9			

a. Dependent Variable: Ln_CtperEt

b. Predictors: (Constant), Et

Coefficients[a]

Model		Unstandardized Coefficients		Standardized Coefficients	t	Sig.
		B	Std. Error	Beta		
1	(Constant)	.733	.142		5.157	.001
	Et	-.001	.000	-.855	-4.663	.002

a. Dependent Variable: Ln_CtperEt

3. Model Schnute

Model	R	R Square	Adjusted R Square	Std. Error of the Estimate
1	.676[a]	.457	.276	.1253970

a. Predictors: (Constant), EttambahsatutambatEtperdua, UttambahsatutambahUtperdua

ANOVA[a]

Model		Sum of Squares	df	Mean Square	F	Sig.
1	Regression	.079	2	.040	2.522	.160[b]
	Residual	.094	6	.016		
	Total	.174	8			

a. Dependent Variable: LnUttambahsatuperUt
b. Predictors: (Constant), EttambahsatutambatEtperdua, UttambahsatutambahUtperdua

Coefficients[a]

Model		Unstandardized Coefficients		Standardized Coefficients	t	Sig.
		B	Std. Error	Beta		
1	(Constant)	.055	1.130		.049	.963
	UttambahsatutambahUtperdua	-.337	.597	-.369	-.564	.593
	EttambahsatutambatEtperdua	.001	.001	.326	.498	.636

a. Dependent Variable: LnUttambahsatuperUt

4. Model Walter-Hilborn

Model Summary

Model	R	R Square	Adjusted R Square	Std. Error of the Estimate
1	.912[a]	.831	.775	.0498222

a. Predictors: (Constant), Et, Ct_per_Et

ANOVA[a]

Model		Sum of Squares	df	Mean Square	F	Sig.
1	Regression	.073	2	.037	14.766	.005[b]
	Residual	.015	6	.002		
	Total	.088	8			

a. Dependent Variable: UttambahsatuperUtkurangsatu
b. Predictors: (Constant), Et, Ct_per_Et

Coefficients[a]

Model		Unstandardized Coefficients		Standardized Coefficients	t	Sig.
		B	Std. Error	Beta		
1	(Constant)	1.009	.329		3.070	.022
	Ct_per_Et	-.651	.161	-1.468	-4.047	.007
	Et	-.001	.000	-.693	-1.911	.105

a. Dependent Variable: UttambahsatuperUtkurangsatu

5. **Model CYP**

Model Summary

Model	R	R Square	Adjusted R Square	Std. Error of the Estimate
1	.681[a]	.464	.285	.0831340

a. Predictors: (Constant), EttambahEttambahsatu, Ln_CtperEt

ANOVA[a]

Model		Sum of Squares	df	Mean Square	F	Sig.
1	Regression	.036	2	.018	2.597	.154[b]
	Residual	.041	6	.007		
	Total	.077	8			

a. Dependent Variable: LnUttambahsatu
b. Predictors: (Constant), EttambahEttambahsatu, Ln_CtperEt

Coefficients[a]

Model		Unstandardized Coefficients		Standardized Coefficients	t	Sig.
		B	Std. Error	Beta		
1	(Constant)	.325	.348		.932	.387
	Ln_CtperEt	.074	.321	.141	.232	.824
	EttambahEttambahsatu	.000	.000	-.554	-.909	.399

a. Dependent Variable: LnUttambahsatu

Appendix 2. Validation of Surplus Production Model of little tuna

Years	C_t (tons)	E_t (trips)	Validation: Abs($\frac{C_t - \hat{C}_t}{C_t}$)				
			Schaefer	**Fox**	**Schnute**	**Walter-Hilborn**	**CYP**
2008	401.0	240	.0197	.0201	5.1685	.2930	.9916
2009	391.0	300	.0178	.1830	7.7111	.1644	.9893
2010	457.0	480	.1265	.3527	15.8021	.1465	.9855
2011	400.9	400	.1412	.3920	12.6905	.0666	.9861
2012	541.6	484	.1319	.1463	13.7002	.2793	.9876
2013	531.0	528	.1212	.2206	16.4307	.2653	.9863
2014	524.7	527	.0125	.2341	16.5566	.2563	.9861
2015	471.0	480	.0934	.3125	15.3621	.1719	.9859
2016	509.0	525	.2219	.2699	16.9143	.2331	.9857
2017	586.0	550	.0197	.1270	16.1783	.3382	.9870
Mean	**481.320**	**451.400**	**0.1019**	**0.2258**	**13.6514**	**0.2214**	**0.0871**

1. Schaefer Model : $\hat{C}_t = 1.929E_t - 0{,}002\,E_t^{\,2}$
2. Fox Model : $\hat{C}_t = E_t.e^{(0.733-0{,}001\,E_t)}$

3. Schnute Model : $\hat{Y} = a - b X_1 - c X_2 = 0.055 - 0.337 X_1 + 0{,}001 X_2$

$r = a = 0.055 \quad q = c = 0{,}01 \quad b = \dfrac{r}{Kq} = 0.337$

$$K = \frac{r}{bq} = \frac{0.055}{(0.337)(0.01)} = 16.3204$$

$$\hat{C}_t = Kq\,E_t - \frac{Kq^2}{r} E_t^{\,2} = 0.1632\,E_t - 0.0297\,E_t^2$$

4. Walter – Hilborn Model :

$$\hat{Y} = a - b X_1 - c X_2 = 1.009 - 0.651 X_1 - 0{,}001 X_2$$

$r = a = 1.009 \quad q = c = 0.001 \quad b = \dfrac{r}{Kq} = 0.651$

$$K = \frac{r}{bq} = \frac{1.009}{(0.651)(0{,}001)} = 1549.9232$$

$$\hat{C}_t = Kq\,E_t - \frac{Kq^2}{r} E_t^{\,2} = 1.5499\,E_t - 0.001636\,E_t^2$$

5. CYP Model : $\hat{Y} = a + b X_1 - c X_2 = 0.325 + 0.074 X_1 - 0.000320 X_2$

$$r = \frac{2(1-b)}{1+b} = \frac{2(1-0.074)}{1+0.074} = 1.7244 \quad q = -c(2-r) = 0.000320(2-1.7244) = 0.00008819$$

$$Q = \frac{a(2+r)}{2r} = \frac{0.325(2+1.7244)}{2(1.7244)} = 0.3509$$

$$K = \frac{e^Q}{q} = \frac{e^{0.3509}}{0.00008819} = 161.0500$$

$$\hat{C}_t = Kq\,E_t - \frac{Kq^2}{r} E_t^{\,2} = 0.0142\,E_t - 0.000000726\,E_t^2$$

References

[1] [DKP] Dinas Kelautan dan Perikanan Provinsi Sulawesi Utara. 2017. Statistik Perikanan Tangkap Provinsi Sulawesi Utara Tahun 2016.

[2] Boer, M., dan K.A. Azis. 1995. Prinsip-prinsip Dasar Pengelolaan Sumberdaya Perikanan Melalui Pendekatan Bio-Ekonomi. Jurnal Ilmu-ilmu Perairan dan Perikanan III(2):109-119.

[3] Coppola G., and S. Pascoe. 1996. A Surplus Production Model with a non-linear Catch-Effort Relationship. (Research Paper 105) Center for the Economics and Managemant of Aquatic Resources University of Portsmouth.

[4] Sparre , P. and S.C. Venema. 1999. Introduksi Pengkajian Stok Ikan Tropis. Buku 1 Manual. (Terjemahan J. Widodo. I.G.S. Merta, S. Nurhakim, dan M. Badrudin). Pusat Penelitian dan Pengembangan Perikanan, Badan Penelitian dan Pengembangan Pertanian (Kerjasama dengan Organisasi Pangan dan Pertanian Perserikatan Bangsa-bangsa). Jakarta. 438 hal.

[5] Gulland, J.A. 1983. Fishing and Stock of Fish at Iceland. Mui.`Agric. Fish Food, Invest. (Ser.2) 23(4): 52 – 70.

[6] Sularso, A. 2005. Alternatif Pengelolaan Perikanan Udang di Laut Arafura. Disertasi (Tidak Dipublikasikan). Sekolah Pascasarjana Institut Pertanian Bogor. Bogor. 130 hal.

[7] Fox, W.W. 1970. An Exponential Surplus Yield Model for Optimazing Exploited Fish

Population. Trans. Am. Fish Soc. 99(1):80-88.
[8] Schnute, J. 1977. Improved Estimates from the Schaefer Production Models : Theoretical Considerations : J. Fish. Res. Board Can., 34:583-663.
[9] Tinungki, G. M. 2005. Evaluasi Model Produksi Surplus dalam Menduga Hasil Tangkapan Maksimum Lestari untuk Menunjang Pengelolaan Perikanan Lemuru Di Selat Bali. Disertasi (Tidak Dipublikasikan). Sekolah Pascasarjana Institut Pertanian Bogor. Bogor. 207 hal.
[10] Fauzi, A., dan S. Anna. 2005. Pemodelan Sumberdaya Perikanan dan Kelautan untuk Analisis Kebijakan. PT. Gramedia Pustaka Utama, Jakarta
[11] Kekenusa, J. S., Marline S. Paendong, Winsy Ch. D. Weku, and Sendy B. Rondonuwu. 2015. Determination of the Status of Utilization and Management Scenarios Bonito (*Auxis rochei*) Caught in the Talaud Waters North Sulawesi. Sci. J. of Applied Mathematics and Statistics 2015; 3(2):39-46.
[12] Kekenusa, J. S., S. B. Rondonuwu, dan M. S. Paendong. 2017. Penentuan Status Pemanfaatan dan Skenario Pengelolaan Ikan Tongkol (*Auxis rochei*) di Perairan Manado – Sulawesi Utara. Makalah Hasil Penelitian yang disampaikan pada Seminar Nasional MIPA-*net* tanggal 25 Agustus 2017.
[13] Kekenusa, J. S., Sendy B. Rondonuwu, and Marline S. Paendong. 2018. Determination of the Status of Utilization and Effort of Little Tuna (*Euthynnus affinis*) Caught in the Bitung Waters North Sulawesi. International Journal of ChemTech Research 2018; 11(02):340-354.

IOP Conf. Series: Materials Science and Engineering **567** (2019) 012011 doi:10.1088/1757-899X/567/1/012011

Discrete optimal capital investment and financing policies in fishery resource harvesting with reserve area

M N Alatas[1], T Bakhtiar[1], F Hanum[1]

[1]Division of Operation Research, Department of Mathematics, IPB University (Bogor Agricultural University), Kampus IPB Dramaga, Bogor 16680, INDONESIA

Corresponding author: s.n.alatas@gmail.com

Abstract. This research discusses a problem in renewable resource (fish) harvesting on single-owner fishery where the dynamics of growth are expressed by system of difference equations. The model involves three state variables (stock of equity, resource stock in reserve zone and unreserved zone) and two control variables (effort rate and dividend payout rate). Discrete Pontryagin maximum principle was applied to obtain optimality conditions which must be satisfied by the variables. Further, the forward-backward sweep method was used to determine the numerical solution of optimal control model. Three scenarios of interest rate applied to find out the response of company in managing harvesting effort and dividend payout rate were investigated.

1. Introduction

A fishery is all about catching, maintenance, and cultivation of fish. The characteristic of rightful ownership of fishery resource is a common property that leads to an open-access fishery [1][2]. In open-access fishery, profit tends to decrease and stocks may be extinct. As the result of that, the tendency of common property and open-access fishery shift to be a single firm has noticed in Canada, New Zealand, Finland, and Iceland [3]. Furthermore, to prevent overfishing that can lead to the extinction of stocks, there are many efforts have been applied including the development of mathematical approaches to make best decisions.

Clark *et al.* discussed optimal harvesting at the renewable resource (fish) for a fish company with irreversible investment assumption [4]. Jørgensen and Kort discussed optimal harvesting in renewable resource, which allowed the firm to have a debt or lend with the interest rate depend on the firm's cumulative amount of debt or lending level and replaced irreversible investment assumption with perfectly reversible investment assumption [5]. Dubey *et al.* introduced and analyzed model of resource (fish) with allocation zone, that is reverse zone and non-reserve zone [6]. The harvesting allowed at the non-reserve zone. The three models above are continuous types, but in the real world, the continuous model is less realistic because the decision will be made at a discrete and probably periodic time.

In this work we present an optimal harvesting model of a renewable natural resource. We offer the discrete-time version of Jørgensen and Kort of optimal investment and finance in renewable resource harvesting and that of Dubey *et al.* of fishery resource dynamic with reserve area [5][6]. In this model, we employed three state variables (stock of equity, resource stock in reserve zone and non-reserve zone) and two control variables (effort rate and dividend payout rate) were introduced. It is shown that company has a debt when the capital or stock of equity is not enough for harvesting, and when the capital or stock of equity is enough for harvesting, the company will stop to debt and start to give a lend. In the meantime the company is always make dividend payment.

2. Optimal control problem

Let $S(k)$ denotes the stock of fishery resource at non-reserve area by time k, $V(k)$ represents the stock of fishery resource at reserve area by time k and $X(k)$ refers to the stock of equity at time k, where $k = 1,2,\dots,N$. The discrete dynamical model of fishery resource management at reserve area that combine the harvesting effort, capital investment and financing policies can be given as in model (1) below:

$$\begin{aligned} S(k+1) - S(k) &= \alpha S(k)\left(1 - \frac{S(k)}{M}\right) - r_1 S(k) + r_2 V(k) - qE(k)S(k), \\ V(k+1) - V(k) &= \beta V(k)\left(1 - \frac{V(k)}{L}\right) + r_1 S(k) - r_2 V(k), \\ X(k+1) - X(k) &= (pqS(k) - c - ma)E(k) - C\big(B(k)\big) - D(k), \end{aligned} \tag{1}$$

In the above model, we define by $E(k)$ and $D(k)$ the harvesting effort rate and dividend payout rate by time k, respectively. The cost of fund C is assumed to be a function of the level of debt/lending by time k, $B(k)$. Thus implicitly we have $B(k) = mE(k) - X(k)$, where m is the unit cost of harvesting effort. When $X > mE$ $(B < 0)$ the firm has excess equity that leads to the firm to give a lending, and when $X < mE$ $(B > 0)$ the cost of harvesting effort more than equity that leads to the firm to debt. In model (1), $\alpha, \beta, r_1, r_2, p, q, c, M$ and L are all parameters.

The control problem is to find control variables E and D such that maximized the present value of dividend payout over an infinite horizon $J = \int_0^\infty D(t)e^{-it}dt$ that can further be expressed by (2) as follows.

$$\max J = \sum_{k=0}^{\infty} (1+i)^{-k}\big[(pqS(k) - c - ma)E(k) - C\big(B(k)\big) - iX(k)\big], \tag{2}$$

where i is the market interest rate. In this problem it is assumed that all control variables are bounded, i.e., $E_{min} \le E(k) \le E_{max}$ and $D_{min} \le D(k) \le D_{max}$ for all time k. It is also assumed that $S(0) = S_0, V(0) = V_0, X(0) = X_0$ and $S(N), V(N), X(N)$ are all free.

We take into account two intervention strategies, namely fish harvesting and dividend payment. Three cases were considered relating to the choice of cost of fund functions (3)-(5):

Linear

$$C\big(B(k)\big) = zB(k), \tag{3}$$

piecewise linear

$$C\big(B(k)\big) = \begin{cases} z_1 B(k) & ;\ B(k) \le 0 \\ z_2 B(k) & ;\ B(k) > 0, \end{cases} \tag{4}$$

and quadratic

$$C\big(B(k)\big) = \begin{cases} z_1 B(k) & ;\ B(k) \le 0 \\ z_1 B(k) + z_3 B^2(k) & ;\ B(k) > 0. \end{cases} \tag{5}$$

Here, z_1, z_2 and z_3 are positive constants related to interest rate for either debt or lend. In the linear case it is obvious that the interest rate is constant and the same for debt or lending. For the piecewise linear case the interest rate is constant but different for debt or lending. For the quadratic case, the interest rate is constant for lending and quadratic for debt that means the interest rate will increase related to the debt level.

3. Optimality conditions

We generate the necessary condition for the optimal control by using the discrete version of Pontryagin Maximum Principle [7]. We need to define the Hamiltonian H based on (2) subject to system state (1) to obtain the necessary conditions:

$$\begin{aligned} H = & \frac{1}{(1+i)^k}\left[(pqS(k) - c - ma)E(k) - C\big(B(k)\big) - iX(i)\right] \\ & + P_1(k)\left[\alpha S(k)\left(1 - \frac{S(k)}{M}\right) - r_1 S(k) + r_2 V(k) - qE(k)S(k)\right] \\ & + P_2(k)\left[\beta V(k)\left(1 - \frac{V(k)}{L}\right) + r_1 S(k) - r_2 V(k)\right] \\ & + P_3(k)\left[(pqS(k) - c - ma)E(k) - C\big(B(k)\big) - D(k)\right], \end{aligned} \tag{6}$$

with $P_1(k)$, $P_2(k)$ and $P_3(k)$ are adjoin functions to be optimally selected. The optimality conditions insisted by Pontryagin Maximum Principle are: (i) $H_E = 0, H_D = 0$, (ii) $P_i(k) - P_i(k-1) = -H_{x_i}$, where $x_i \in \{S, V, X\}$, and (iii) $x_i(k+1) - x(k) = f_i$, where f_i is the right hand side of system state (1), respectively.

The optimal system of equation (1) is reclaimed by applying the third condition of Pontryagin Maximum Principle, i.e., by taking the appropriate partial derivative of Hamiltonian (6) with respect to the associated state variables. Furthermore, we are able to determine the optimal control variables E^*, D^* by imposing the first principle, i.e., $H_E = 0, H_D = 0$. The optimal controls for every cases of cost of fund functions are given as follow (7)-(9).

Linear:

$$E^* = \begin{cases} E_{max} & ; \ \sigma > 0 \\ E_{min} & ; \ \sigma < 0' \end{cases} \tag{7}$$

with

$$\sigma = \left(\frac{1}{(1+i)^k} + P_3(k)\right)\left[(pqS(k) - c - ma) - z\right] - P_1(k)\big(qS(k)\big).$$

Piecewise linear:

$$E^* = \begin{cases} E_{1max} & ; \ B(k) \le 0, \sigma_1 > 0 \\ E_{1min} & ; \ B(k) \le 0, \sigma_1 < 0 \\ E_{2max} & ; \ B(k) > 0, \sigma_2 > 0 \\ E_{2min} & ; \ B(k) > 0, \sigma_2 < 0. \end{cases} \tag{8}$$

with

$$\sigma_i = \left(\frac{1}{(1+i)^k} + P_3(k)\right)\left[(pqS(k) - c - ma) - z_i\right] - P_1(k)\big(qS(k)\big).$$

Quadratic:

$$E^* = \begin{cases} E_{1\,max} & ; \ B(k) \le 0, \sigma > 0 \\ E_{1\,min} & ; \ B(k) \le 0, \sigma < 0 \\ E_{2\,min} & ; \ B(k) > 0, E_2{}^* < E_{2\,min} \\ E_2{}^* & ; \ B(k) > 0, E_{2\,min} \le E_2{}^* \le E_{2\,max} \\ E_{2\,max} & ; \ B(k) > 0, E_2{}^* > E_{2\,max} \end{cases} . \tag{9}$$

with

$$\sigma = \left(\frac{1}{(1+i)^k} + P_3(k)\right)(pqS(k) - c - ma - z_1) - qP_1(k)S(k),$$

and

$$E_2^* = \frac{1}{m}\left(\frac{A - z_1}{2z_3} + X(k)\right).$$

where

$$A = pqS(k) - c - ma - \frac{qP_1(k)S(k)}{(1+i)^{-k} + P_3(k)}.$$

Since Hamiltonian is linear with respect to control function D, then the optimal control D^* for every cases is given by the following bang-bang control (10):

$$D^* = \begin{cases} D_{max} & ; \ \tau > 0 \\ D_{min} & ; \ \tau < 0' \end{cases} \tag{10}$$

with

$$\tau = -P_3(k).$$

The second principle of Pontryagin's Maximum is stated as follows. Given optimal controls E^*, D^* and corresponding solutions to the system states S^*, V^*, X^* that maximized objective function (2), there exists functions P_1, P_2 and P_3 that satisfying the system of adjoin (11) below

$$\begin{aligned} P_1(k-1) &= P_1(k)\left[\alpha\left(1 - \frac{2S(k)}{M}\right) - r_1 - qE(k) + 1\right] \\ &\quad + \left(\frac{1}{(1+i)^k} + P_3(k)\right) pqE(k) + P_2(k)r_1, \\ P_2(k-1) &= P_1(k)r_2 + P_2(k)\left[\beta\left(1 - \frac{2V(k)}{L}\right) - r_2 + 1\right], \\ P_3(k-1) &= \frac{1}{(1+i)^k}\left[C'\big(B(k)\big) - i\right] + P_3(k)\left[C'\big(B(k)\big) + 1\right]. \end{aligned} \tag{11}$$

Since $S(N), V(N), X(N)$ are all free, the following transversality conditions (12) must be satisfied by adjoin functions:

$$P_1(N) = 0, P_2(N) = 0, P_3(N) = 0. \tag{12}$$

4. Results and discussion

In this section, we discuss the numerical solution and the optimal control of the optimal system. An iterative forward method is used for solving the state system and the iterative backward method is used for solving the adjoin system. We implement this procedure based on the parameter values.

Table 1 present a value of parameter used in the numerical simulation of the model. Due to lack of data, we make a hypothesis for the parameter value within realistic range. The initial value for the state are given by $S_0 = 7500$, $V_0 = 9000$, and $X_0 = 10$.

We start comparing the case of the optimal harvesting with three kind of cost of fund function: linear, piecewise linear and quadratic. In Figure 1, the fish population shown in blue line and the red line is environmental carrying capacity. It is shown that under control the carrying capacity of environment in both zones are maintained. The opposite situation happened in unreserved zone when no control was applied. In Figure 2, we shown the optimal harvesting with respect to the type of cost of fund function: linear, piecewise linear and quadratic, respectively. At the end, the value of optimal harvesting is similar, that is in the maximum level of harvesting $E = 19$. In Figure 3, we shown that the company always make dividend payment with the results are similar for every cases. In Figure 4, we shown the value of equity and debt/lending level, respectively. The value of equity and debt/lending level is similar for every cases, that in the beginning the company have negative equity which mean expenditure more than income, and the company have a debt, at the some point the equity will be positive and the company have a lend. Figures 5-7 depict the adjoin functions and their satisfactions on the final time conditions (8). While, Figures 8-10 illustrate the switching functions σ and τ.

Table 1. Value of parameter

Parameter	Value	Unit
α	0.5	% per year
β	0.6	% per year
r_1	0.2	% per year
r_2	0.25	% per year
M	416304	Ton
L	450000	Ton
q	0.00456	% per ship
p	0.1	IDR 10^7/ton
c	3	IDR 10^7/ship
a	0.05	% per year
m	10	IDR 10^7/ship
i	0.08	% per year
z_1	0.03	% per year
z_2	0.11	% per year
z_3	0.0002	% per year
E_{min} (Linear)	2	Ship per year
E_{max} (Linear)	19	Ship per year
$E_{1\,min}$ (Piecewise linear)	11	Ship per year
$E_{1\,max}$ (Piecewise linear)	19	Ship per year
$E_{2\,min}$ (Piecewise linear)	2	Ship per year
$E_{2\,max}$ (Piecewise linear)	10	Ship per year
$E_{1\,min}$ (Quadratic)	2	Ship per year
$E_{1\,max}$ (Quadratic)	19	Ship per year
$E_{2\,min}$ (Quadratic)	2	Ship per year
$E_{2\,max}$ (Quadratic)	30	Ship per year
D_{min}	0	IDR 10^7
D_{max}	100	IDR 10^7

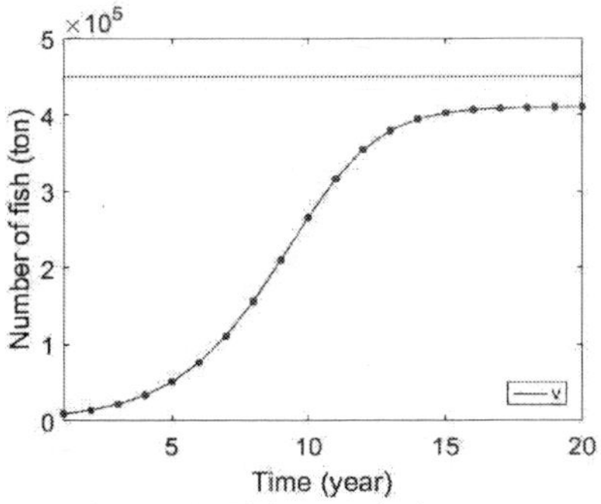

Figure 1. Fish population at reserve area (left) and non-reserve area (right)

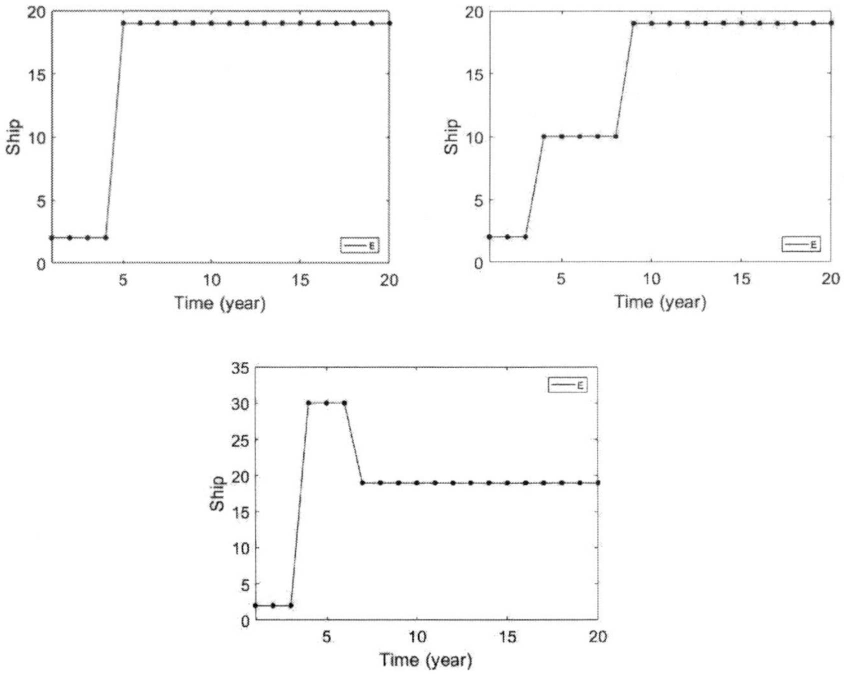

Figure 2. Optimal harvesting for linear, piecewise linear and quadratic cost of fund functions respectively

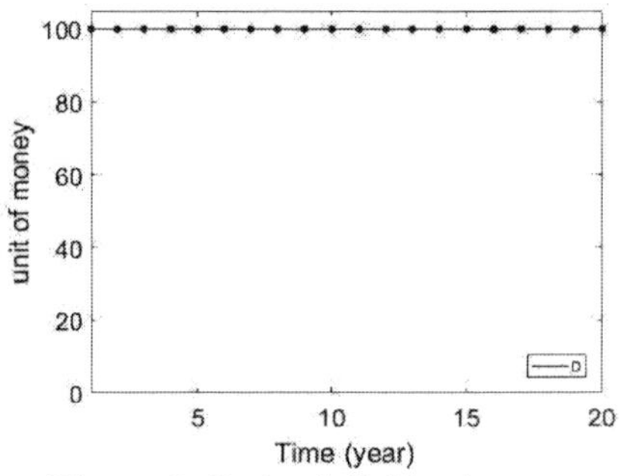

Figure 3. Optimal dividend payment

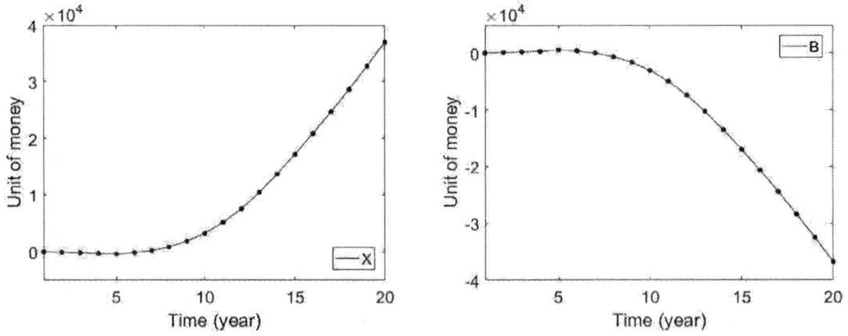

Figure 4. Equity (left) and Debt/lending level (right)

IOP Conf. Series: Materials Science and Engineering **567** (2019) 012011 doi:10.1088/1757-899X/567/1/012011

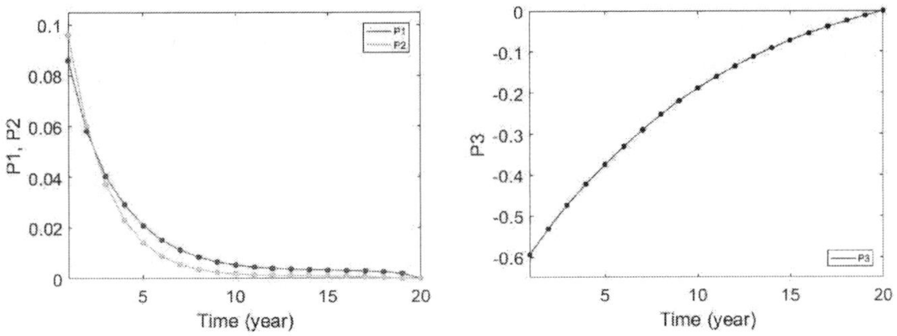

Figure 5. Adjoin function (linear case)

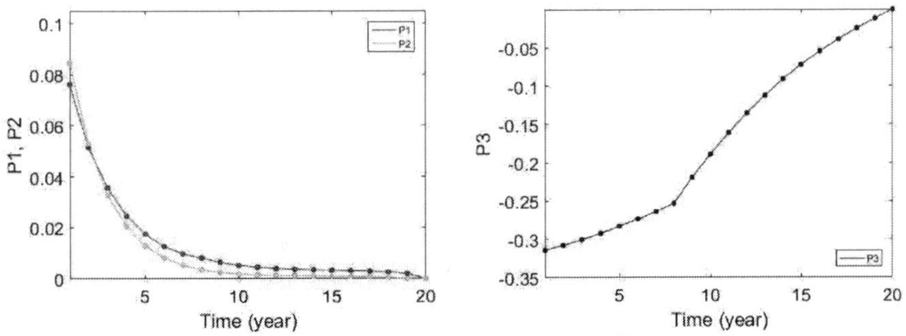

Figure 6. Adjoin function (piecewise linear case)

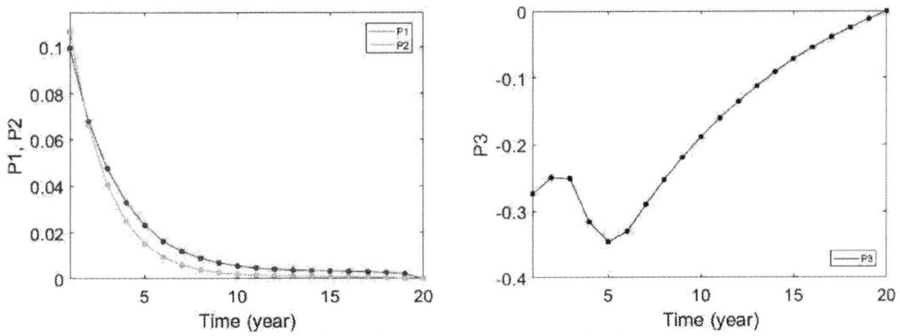

Figure 7. Adjoin function (quadratic case)

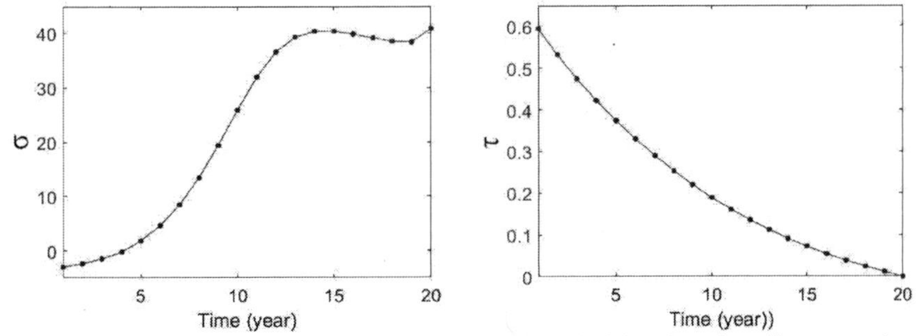

Figure 8. Harvesting and dividend payment switching function (linear case)

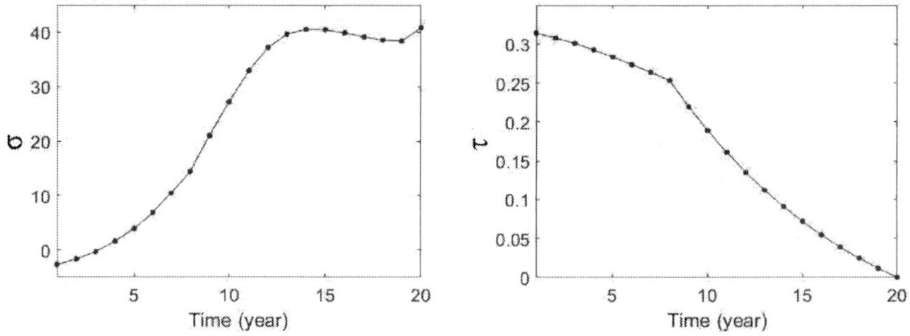

Figure 9. Harvesting and dividend payment switching function (piecewise linear case)

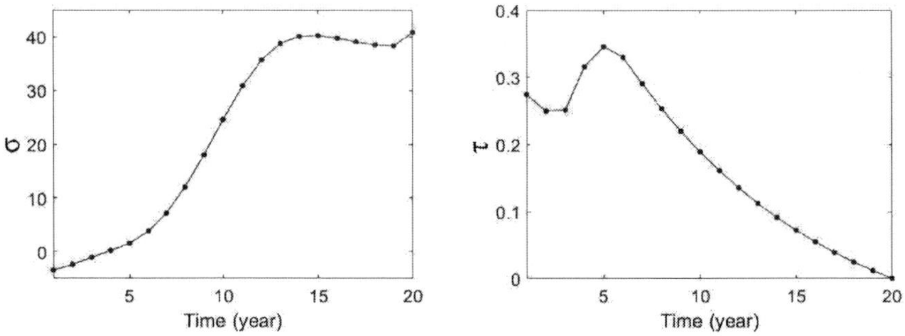

Figure 10. Harvesting and dividend payment switching function (quadratic case)

5. Conclusion

In this work, we have discussed a simple mathematical model of optimal harvesting in renewable resource (fish) involved capital investment and financing policies, with reserve and non-reserve area of harvesting. Based on numerical simulation, the resource without harvesting that leads to over population exceed the environmental carrying capacity so the harvesting effort is needed.

Optimal harvesting effort and dividend payout was determine by discrete version of Pontryagin maximum principle. At the time $k = 1$ for every cases, the firm need to have a debt for doing harvesting and the firm has to pay the interest, at the time $k = 7$, $k = 9$, $k = 7$ for linear, piecewise linear and quadratic case, respectively, the firm has paid off the debt and start giving a lend and receive an interest payment that leads to increase the value of equity dramatically, in the meantime the company is always make dividend payment

Acknowledgments

Authors gratefully acknowledge financial support provided by Ministry of Research, Technology and Higher Education of the Republic of Indonesia under grant No. 011/SP2H/LT/DRPM/IV/2017 through the PUPT Scheme of Bogor Agricultural University.

References

[1] Charles AT 2001 *Sustainable Fishery System* (Iowa: Blackwell Science)
[2] Harris JM and Roach B 2015 *Environmental and Natural Resource Economics A contemporary Approach*. (New York: Routledge)
[3] Novak A, Kaitala V, Feichtiger G 1996 Resource leasing and optimal periodic capital investment. *Mathematical Methods of Operation Research,* 43: 47-67

IOP Conf. Series: Materials Science and Engineering **567** (2019) 012011 doi:10.1088/1757-899X/567/1/012011

[4] Clark CW, Clarke FH and Munro GR 1979 The optimal exploitation of renewable resource stocks: problems of irreversible investment. *Econometrica*, **47(1)**: 25–47
[5] Jørgensen S and Kort PM 1997 Optimal investment and finance in renewable resource harvesting. *Journal of Economic Dynamics and Control*, **21**: 603–630
[6] Dubey B, Chandra P and Sinha P 2003 A model for fishery resource with reserve area. *Nonlinear Analysis: Real World Application*, **4**: 625–637
[7] Tu PNV 1984 *Introductory Optimization Dynamics Optimal Control with Economics and Management Science Applications* (New York: Springer)

IOP Conf. Series: Materials Science and Engineering **567** (2019) 012012 doi:10.1088/1757-899X/567/1/012012

A hyperchaotic hyperjerk system with four nonlinearities, its dynamical analysis and circuit realization

Mujiarto[1], S Vaidyanathan[2], S Zhang[3], A Sambas[1], Sukono[4], A S Praiwi[5], Subiyanto[6]

[1]Department of Mechanical Engineering, Universitas Muhammadiyah Tasikmalaya, Indonesia
[2]Research and Development Centre, Vel Tech University, Avadi, Chennai, India
[3]School of Physics and Opotoelectric Engineering, Xiangtan University, Hunan, China
[4]Department of Mathematics, Faculty of Mathematics and Natural Sciences, Universitas Padjadjaran, Indonesia.
[5]Faculty of Teacher Training and Education, Universitas Muhammadiyah Tasikmalaya, Indonesia
[6]Department of Marine Science, Faculty of Fishery and Marine Science, Universitas Padjadjaran, Indonesia.

Corresponding author: acengs@umtas.ac.id

Abstract. A new four-dimensional hyperchaotic hyperjerk system with four nonlinearities is proposed in this paper. The dynamical properties of the new hyperjerk system are described in terms of phase portraits, Lyapunov exponents, Kaplan-Yorke dimension, dissipativity, etc. Also, a detailed dynamical analysis of the new hyperjerk system has been carried out with bifurcation diagram and Lyapunov exponents. As an engineering application, an electronic circuit realization of the new hyperchaotic hyperjerk system is designed via MultiSIM to confirm the feasibility of the theoretical hyperchaotic hyperjerk model.

1. Introduction

Hyperjerk systems are special types of mechanical chaotic systems arising in chaos literature [1-2]. Chaotic systems are very useful in many applications in science and engineering such as weather systems [3-4], ecology [5], neurons [6-7], biology [8-10], cellular neural networks [11-12], chemical reactors [13-14], oscillators [15-20], robotics [21-24], encryption [25-30], finance systems [31-32], circuits [33-45], secure communication [46-50], etc.

In physics, a hyperjerk ODE can be written as the high-order dynamics

$$\frac{d^n x}{dt^n} = \varphi\left(x, \frac{dx}{dt}, \ldots, \frac{d^{n-1}x}{dt^{n-1}}\right), \quad (n \geq 4) \tag{1}$$

In (1), $x(t)$ stands for the *displacement*, $\frac{dx}{dt}$ the *velocity*, $\frac{d^2x}{dt^2}$ the *acceleration*, $\frac{d^3x}{dt^3}$ the *jerk* and higher-order derivatives are called as *hyperjerk* terms.

Thus, we call the ODE (1) as the hyperjerk differential equation.

For qualitative analysis, it is convenient to express the hyperjerk ODE (1) in a system form.

Using phase variables, we can express the hyperjerk differential equation (1) as follows:

$$\begin{cases} \dot{x}_1 & = & x_2 \\ \dot{x}_2 & = & x_3 \\ \vdots & \vdots & \vdots \\ \dot{x}_n & = & \varphi(x_1, x_2, \ldots, x_n) \end{cases} \tag{2}$$

Jerk systems are special cases of hyperjerk systems when $n = 3$. Thus, jerk systems can be described by the following general system of differential equations.

$$\begin{cases} \dot{x}_1 = x_2 \\ \dot{x}_2 = x_3 \\ \dot{x}_3 = \varphi(x_1, x_2, x_3) \end{cases} \tag{3}$$

Many jerk systems have been reported in the chaos literature [51-58]. Jerk systems have important applications in mechanical engineering [1-2]. Some famous jerk systems can be cited as Sprott systems [51], Li system [52], Elsonbaty system [53], Coullet system [54], Kengne system [55], Vaidyanathan systems [56-60], etc.

In the literature, many hyperjerk systems have been reported by many scientists [61-68]. Some popular hyperjerk systems are Chlouverakis system [61], Munmuangsaen system [62], Daltzis system [63], Wang system [64], Pham system [65], Vaidyanathan systems [66-70], etc.

In this research paper, we report the finding of a new hyperchaotic hyperjerk system with four nonlinearities. We describe the phase plots of the hyperjerk system and do a rigorous dynamic analysis by finding bifurcation diagrams, Lyapunov exponents, etc. Bifurcation analysis is very useful to understand the special properties of chaotic and hyperchaotic systems [71-76].

Section 2 describes the new hyperchaotic hyperjerk system, its phase plots and Lyapunov exponents. Section 3 describes the dynamic analysis of the new hyperchaotic hyperjerk system. Furthermore, an electronic circuit realization of the new chaotic system is presented in detail in Section 4. The circuit experimental results of the new hyperjerk system in Section 4 agreement with its numerical simulations via MATLAB obtained in Section 2. Section 5 draws the main conclusions.

2. A new hyperchaotic hyperjerk system

In this work, we report a new 4-D hyperjerk system given by the dynamics

$$\begin{cases} \dot{x}_1 = x_2 \\ \dot{x}_2 = x_3 \\ \dot{x}_3 = x_4 \\ \dot{x}_4 = -x_1 - x_2 - ax_3 + b(|x_2| + |x_3|) - cx_1^4 x_4 - dx_2^2 \end{cases} \tag{4}$$

where x_1, x_2, x_3, x_4 are state variables and a, b, c, d are positive constants.

In this paper, we show that the hyperjerk system (1) is *hyperchaotic* for the parameter values

$$a = 3.6, \quad b = 0.02, \quad c = 3, \quad d = 0.05 \tag{5}$$

For numerical simulations, we take the initial values of the hyperjerk system (4) as $X(0) = (0.1, 0.1, 0.1, 0.1)$.

Figures 1-4 show the 2-D projections of the new hyperjerk system (4) in (x_1, x_2), (x_2, x_3), (x_3, x_4) and (x_1, x_4) coordinate planes, respectively.

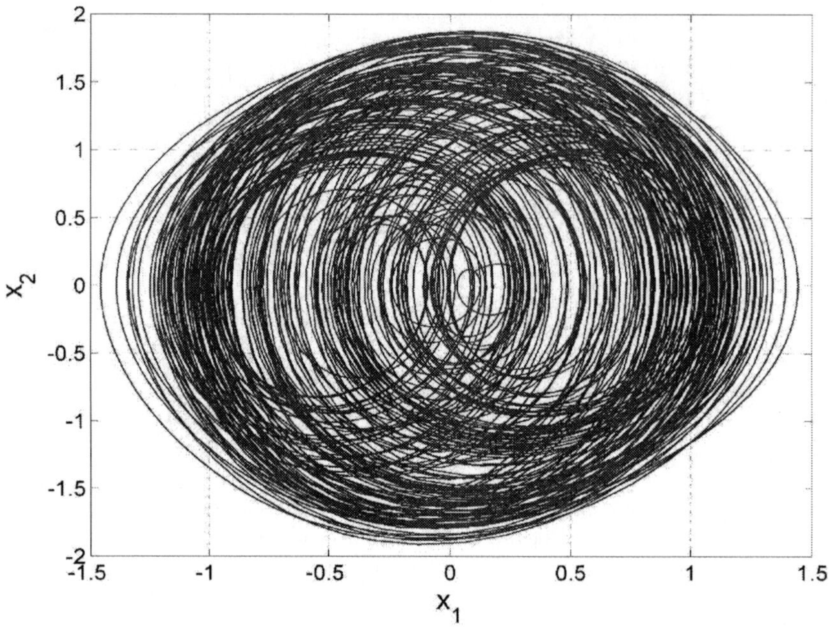

Figure 1. 2-D plot of the hyperchaotic hyperjerk system (4) in the (x_1, x_2) plane for $X_0 = (0.1, 0.1, 0.1, 0.1)$ and $(a, b, c, d) = (3.6, 0.02, 3, 0.05)$

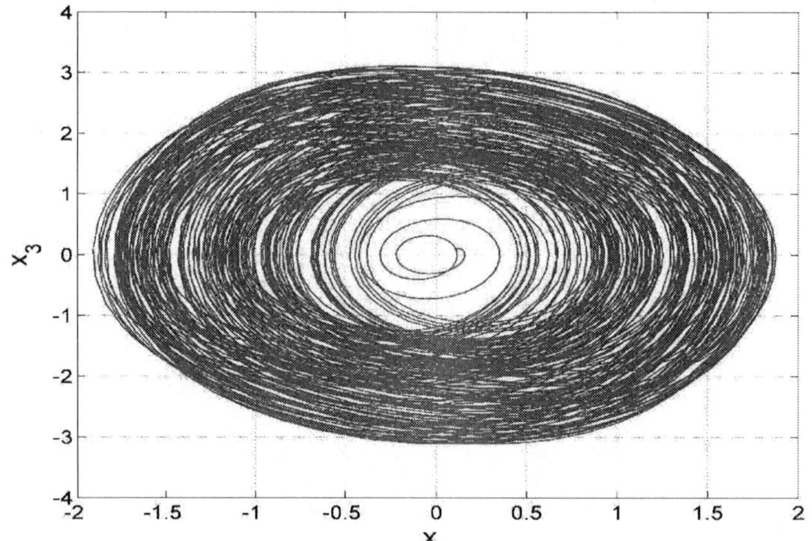

Figure 2. 2-D plot of the hyperchaotic hyperjerk system (4) in the (x_2, x_3) plane for $X_0 = (0.1, 0.1, 0.1, 0.1)$ and $(a, b, c, d) = (3.6, 0.02, 3, 0.05)$

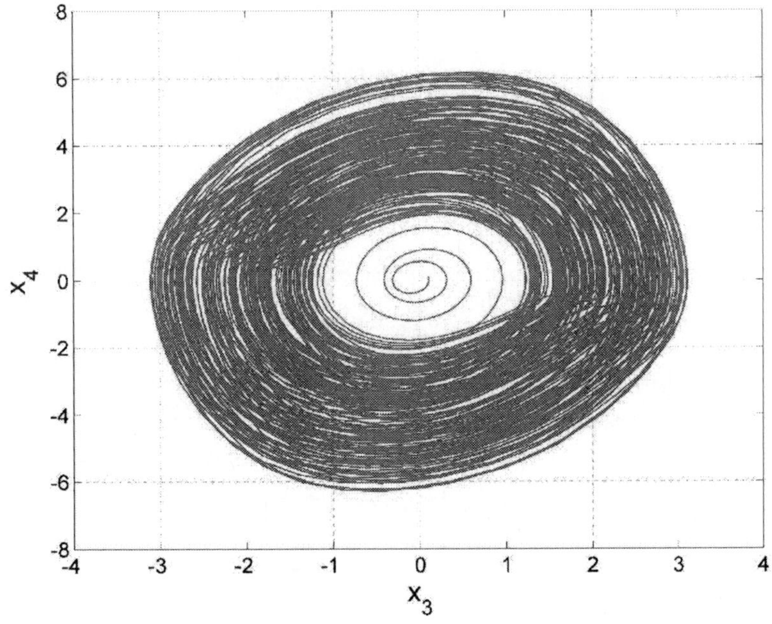

Figure 3. 2-D plot of the hyperchaotic hyperjerk system (4) in the (x_3, x_4) plane for $X_0 = (0.1, 0.1, 0.1, 0.1)$ and $(a, b, c, d) = (3.6, 0.02, 3, 0.05)$

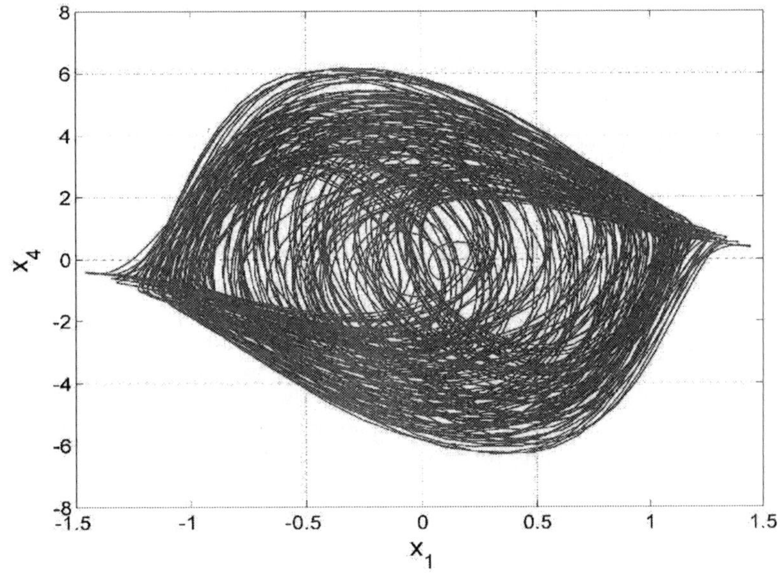

Figure 4. 2-D plot of the hyperchaotic hyperjerk system (4) in the (x_1, x_4) plane for $X_0 = (0.1, 0.1, 0.1, 0.1)$ and $(a, b, c, d) = (3.6, 0.02, 3, 0.05)$

For the rest of this section, we take the values of the parameters as in the hyperchaotic case (5), *i.e.* $(a,b,c,d) = (3.6, 0.02, 3, 0.05)$.

The equilibrium points of the new hyperchaotic hyperjerk system (4) are obtained by solving the system of equations

$$x_2 = 0 \tag{6a}$$

$$x_3 = 0 \tag{6b}$$

$$x_4 = 0 \tag{6c}$$

$$-x_1 - x_2 - ax_3 + b(|x_2| + |x_3|) - cx_1^4 x_4 - dx_2^2 = 0 \tag{6d}$$

From (6a), (6b) and (6c), we deduce that $x_2 = x_3 = x_4 = 0$.

Substituting these in (7d), we obtain $-x_1 = 0$. This gives $x_1 = 0$.

Hence, $E_0 = (0,0,0,0)$ is the unique equilibrium of the chaotic jerk system (4).

The Jacobian matrix of the new hyperjerk system (4) at $E_0 = (0,0,0,0)$ is obtained as

$$J = \begin{bmatrix} 0 & 1 & 0 & 0 \\ 0 & 0 & 1 & 0 \\ 0 & 0 & 0 & 1 \\ -1 & -1 & -3.6 & 0 \end{bmatrix} \tag{7}$$

The Jacobian J has the spectral values $\lambda_{1,2} = 0.1604 \pm 1.8395j$, $\lambda_{3,4} = -0.1604 \pm 0.5172j$

This shows that the equilibrium point E_0 is a saddle-focus and unstable.

For the parameter values as in the hyperchaotic case (5) and the initial state $X_0 = (0.1, 0.1, 0.1, 0.1)$, the Lyapunov exponents of the new jerk system (4) are determined using Wolf's algorithm as

$$LE_1 = 0.1344, \; LE_2 = 0.0411, \; LE_3 = 0, \; LE_4 = -1.2929 \tag{8}$$

The hyperjerk system (4) is hyperchaotic since it has 2 positive Lyapunov exponents. Thus, the system (4) exhibits a self-excited strange hyperchaotic attractor. Also, we note that the sum of the Lyapunov exponents in (8) is negative. This shows that the hyperjerk system (4) is dissipative.

The Kaplan-Yorke dimension of the hyperjerk system (4) is determined as

$$D_{KY} = 3 + \frac{LE_1 + LE_2 + LE_3}{|LE_4|} = 3.1357, \tag{9}$$

which indicates the high complexity of the hyperchaotic hyperjerk system (4).

3. Bifurcation Analysis for the New Hyperchaotic Hyperjerk System

In this section, we describe a bifurcation analysis for the new hyperjerk system (4) introduced in Section 2. Bifurcation analysis is an important topic for studying chaotic systems [71-76].

Here, we select a and c as the control parameters and fix others.

We fix $b = 0.02$, $c = 3$, $d = 0.05$, the initial condition $X_0 = (0.1, 0.1, 0.1, 0.1)$ and vary a in the region of $[3.6, 4.6]$.

Obviously, from the bifurcation diagram and the Lyapunov exponents shown in Figure 5, one can get that the hyperjerk system (4) depicts hyperchaos in the region of [3.6, 3.95]; then the system gets into chaos and finally converts into periodic orbits. In addition, there is quasi-periodic behavior in the parameter range. Some sample results are shown in Figures 6-9.

IOP Conf. Series: Materials Science and Engineering **567** (2019) 012012 doi:10.1088/1757-899X/567/1/012012

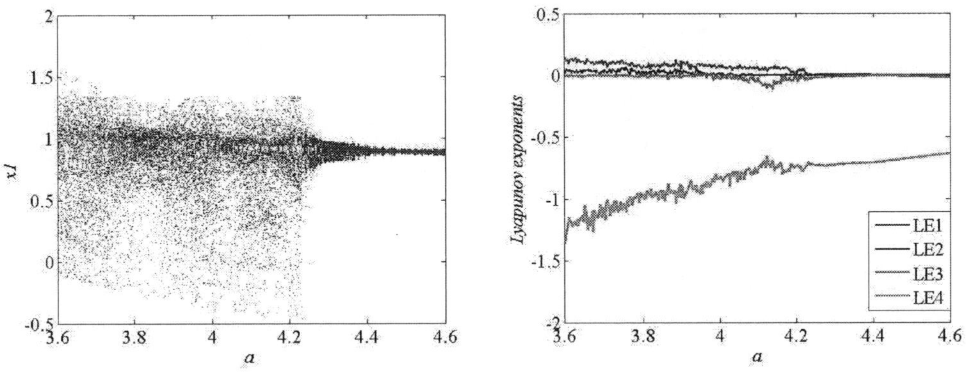

Figure 5. Bifurcation diagram and Lyapunov exponents of the new hyperjerk system (4), where we fix $b = 0.02$, $c = 3$, $d = 0.05$ and the initial conditions (0.1, 0.1, 0.1, 0.1)

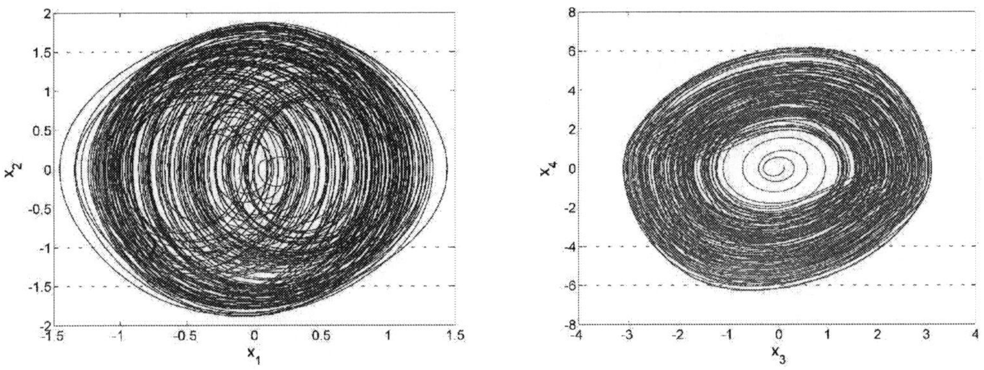

Figure 6. Phase plots of the new hyperjerk system (4), where we fix $b = 0.02$, $c = 3$, $d = 0.05$ and the initial conditions (0.1, 0.1, 0.1, 0.1). When $a = 3.6$, the system (5) depicts hyperchaos.

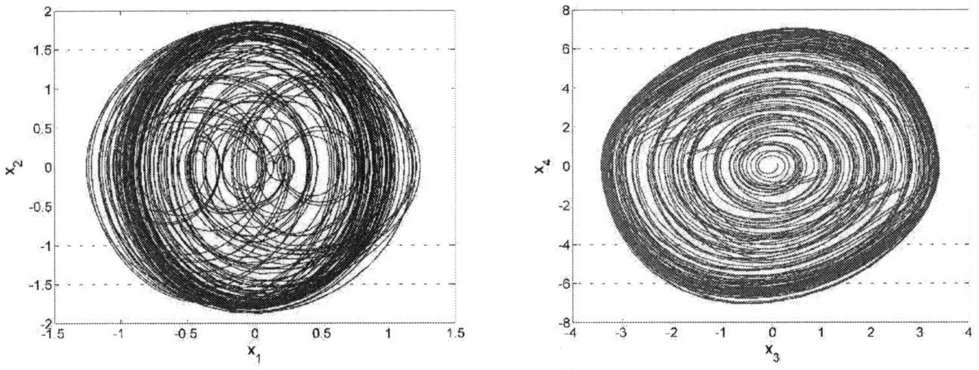

Figure 7. Phase plots of the new hyperjerk system (4), where we fix $b = 0.02$, $c = 3$, $d = 0.05$ and the initial conditions (0.1, 0.1, 0.1, 0.1). When $a = 4$, the system (4) depicts chaos.

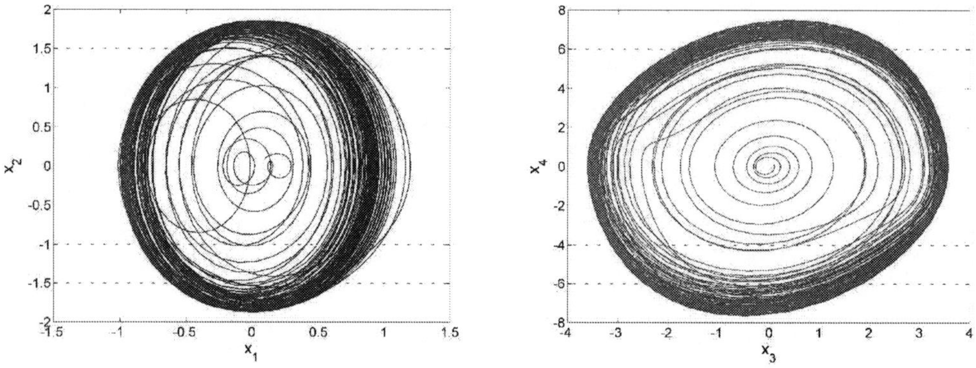

Figure 8. Phase plots of the new hyperjerk system (4), where we fix $b = 0.02$, $c = 3$, $d = 0.05$ and the initial conditions (0.1, 0.1, 0.1, 0.1). When $a = 4.25$, the system (4) depicts quasi-period motion.

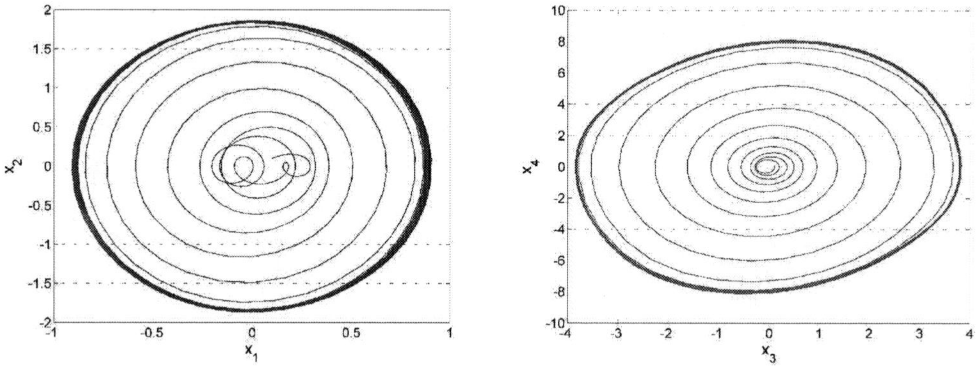

Figure 9. Phase plots of the new hyperjerk system (4), where we fix $b = 0.02$, $c = 3$, $d = 0.05$ and the initial conditions (0.1, 0.1, 0.1, 0.1). When $a = 4.6$, the system (4) depicts periodic orbit.

We fix $a = 3.6$, $b = 0.02$, $d = 0.05$, the initial conditions (0.1, 0.1, 0.1, 0.1) and vary c in the region of [3,6]. The constant Lyapunov exponent behavior, meaning the values of the Lyapunov exponents keep invariable when the parameters vary in a certain range, has been reported in some chaotic systems [76]. From the Lyapunov exponent spectrum shown in Figure 10, one can see that the hyperjerk system (4) displays that all the values of the Lyapunov exponents are unchanged and moreover, the value of the maximum Lyapunov exponent keeps invariable and positive when the control parameter c increases in the region of [3, 6]. That means the system exhibits robust chaos behavior, which is very important for real-world applications.

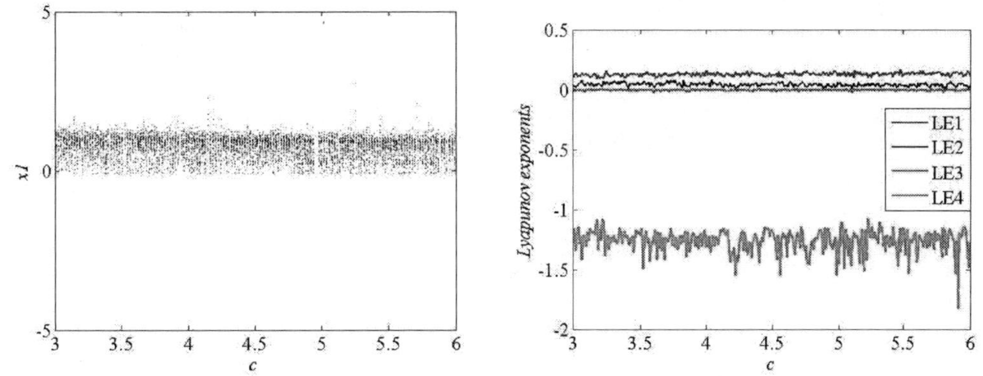

Figure 10. Bifurcation diagram and Lyapunov exponents of the new hyperjerk system (4), where we fix $a = 3.6$, $b = 0.02$, $d = 0.05$ and the initial conditions (0.1, 0.1, 0.1, 0.1)

4. Circuit Implementation of the New Hyperchaotic Hyperjerk System

In this section, we design an electronic circuit based on the hyperjerk system (4) in MultiSIM software. A circuit design containing three channels with respect to the variables x_1, x_2, x_3, x_4 of system (4) is given in Figure 11. The circuit includes simple electronic elements such as resistors, multipliers, capacitors, op-amps and diodes.

In this study, a linear scaling is considered as follows:

$$\begin{cases} \dot{x}_1 = x_2 \\ \dot{x}_2 = x_3 \\ \dot{x}_3 = \dfrac{x_4}{2} \\ \dot{x}_4 = -2x_1 - 2x_2 - 2ax_3 + 2b(|x_2| + |x_3|) - 16cx_1^4 x_4 - dx_2^2 \end{cases} \tag{10}$$

By applying Kirchhoff's laws to this circuit, its dynamics are presented by the following circuital equations:

$$\begin{cases} \dot{x}_1 = \dfrac{1}{C_1R_1} x_2 \\ \dot{x}_2 = \dfrac{1}{C_2R_2} x_3 \\ \dot{x}_3 = \dfrac{1}{C_3R_3} x_4 \\ \dot{x}_4 = -\dfrac{1}{C_4R_4} x_1 - \dfrac{1}{C_4R_5} x_2 - \dfrac{1}{C_4R_6} x_3 + \dfrac{1}{C_4R_7} |x_2| + \dfrac{1}{C_4R_8} |x_3| - \dfrac{1}{C_4R_9} x_1^4 x_4 - \dfrac{1}{C_4R_{10}} x_2^2 \end{cases} \tag{11}$$

The values of components in Figure 11 are chosen to match the parameters of new hyperjerk system (4) as follows: $R_1 = R_2 = 400$ kΩ, $R_3 = 800$ kΩ, $R_4 = R_5 = 200$ kΩ, $R_6 = 55.55$ kΩ, $R_7 = R_8 = 10$ MΩ, $R_9 = 8.33$ kΩ, $R_{10} = 8$ MΩ, $R_{11} = R_{12} = R_{13} = R_{14} = R_{15} = R_{16} = R_{17} = R_{18} = R_{19} = R_{20} = R_{21} = R_{22} = R_{23} = R_{24} = R_{25} = R_{26} = R_{27} = R_{28} = R_{29} = R_{30} = 100$ kΩ, $C_1 = C_2 = C_3 = C_4 = 1$nF. The circuit simulations of the phase plots are displayed in Figs 12 (a)-(d), which show the chaotic attractors in x_1- x_2 plane, x_2- x_3

IOP Conf. Series: Materials Science and Engineering **567** (2019) 012012 doi:10.1088/1757-899X/567/1/012012

plane, x_3- x_4 plane and x_1- x_4 plane, respectively. respectively. As can be seen from the MultiSIM outputs in Figure 12 and numerical simulation in Figures 1-4, the results are similar.

5. Conclusions

A new four-dimensional hyperchaotic hyperjerk system with four nonlinearities was announced in this paper. The dynamical properties of the new hyperjerk system are described in terms of phase portraits, Lyapunov exponents, Kaplan-Yorke dimension, dissipativity, etc. Also, a detailed dynamical analysis of the new hyperjerk system was done with bifurcation diagram and Lyapunov exponents. Furthermore, an electronic circuit realization of the new hyperchaotic hyperjerk system was carried out via MultiSIM to confirm the feasibility of the theoretical hyperchaotic hyperjerk model.

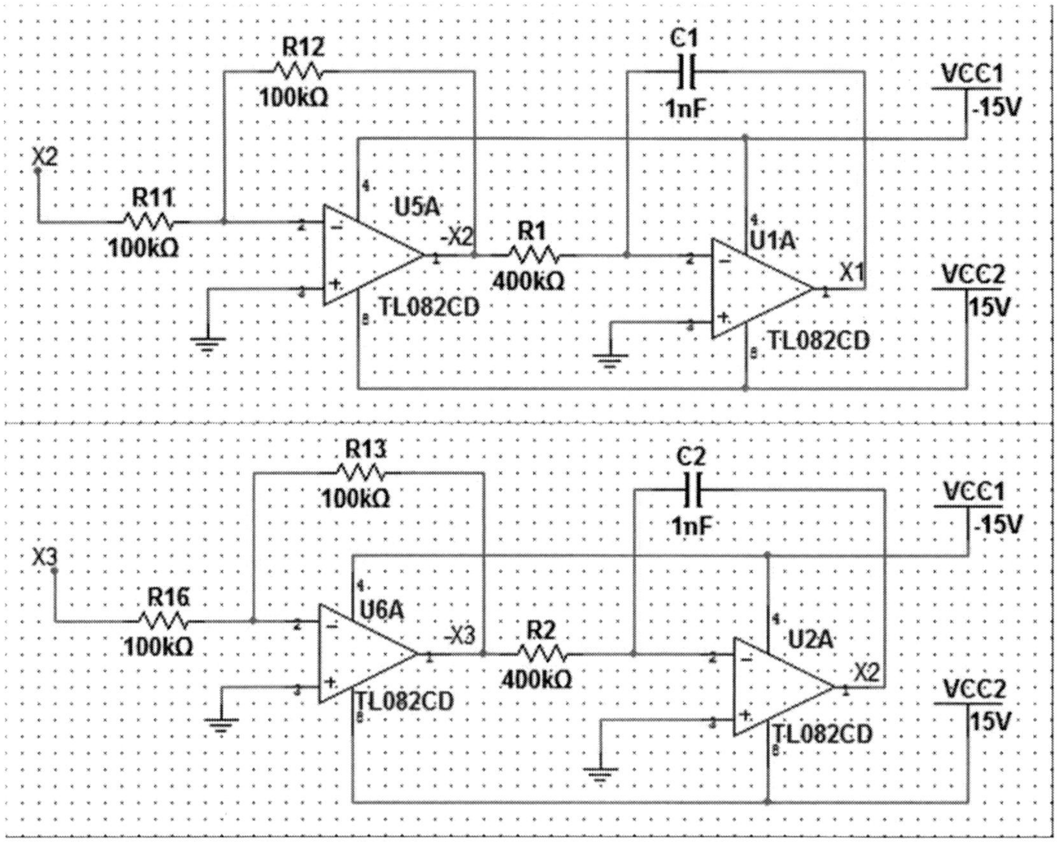

IOP Conf. Series: Materials Science and Engineering **567** (2019) 012012 doi:10.1088/1757-899X/567/1/012012

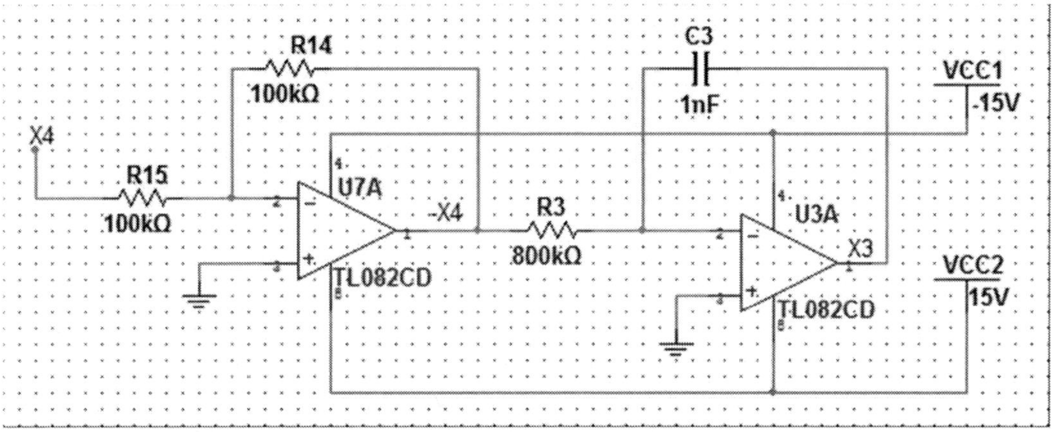

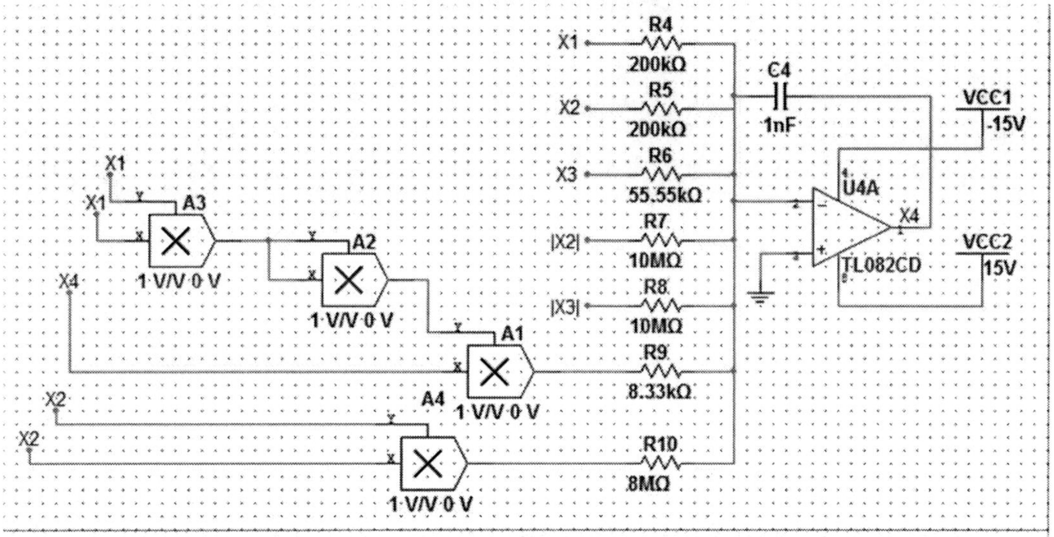

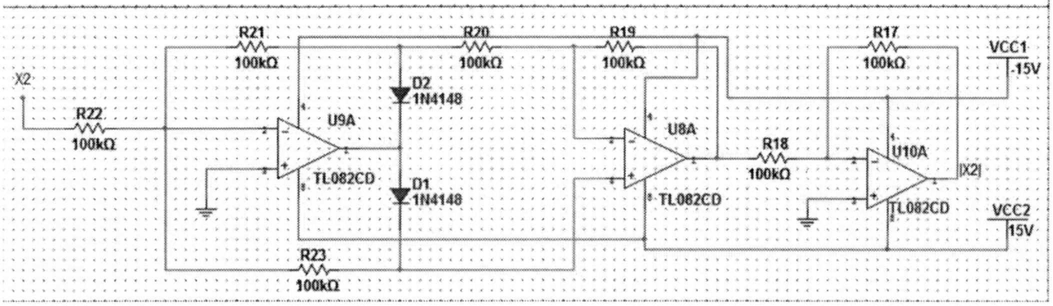

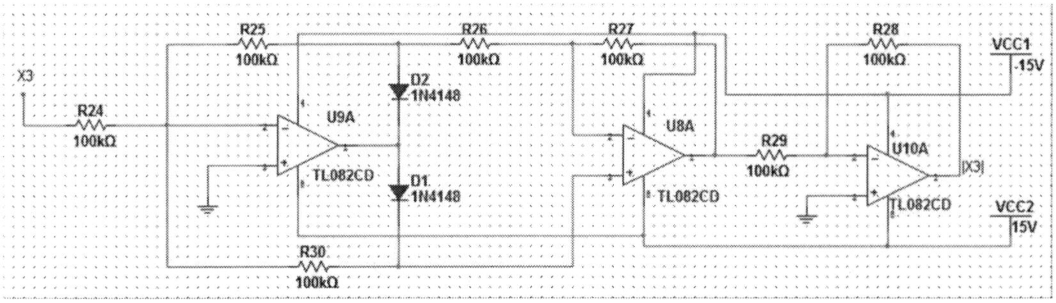

Fig. 11 Circuit design for the proposed new four-dimensional hyperchaotic hyperjerk system (4)

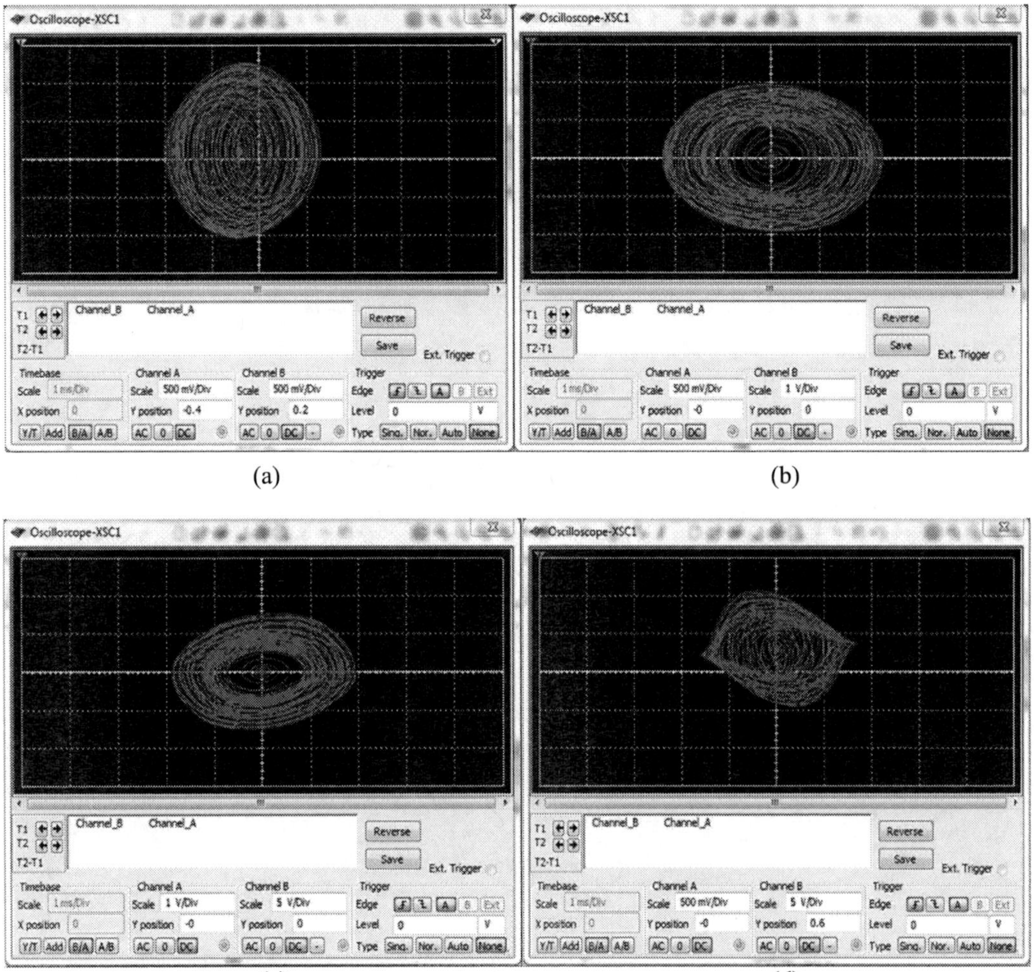

(a) (b)

(c) (d)

Fig. 12 MultiSIM chaotic attractors of the new four-dimensional hyperchaotic hyperjerk system (4) (a) x_1- x_2 plane, (b) x_2- x_3 plane, (c) x_3- x_4 plane and (d) x_1- x_4 plane.

IOP Conf. Series: Materials Science and Engineering **567** (2019) 012012 doi:10.1088/1757-899X/567/1/012012

References

[1] Vaidyanathan S and Volos C 2017 *Advances and Applications in Chaotic Systems* (Berlin: Springer)

[2] Pham V T, Vaidyanathan S, Volos C and Kapitaniak T 2018 *Nonlinear Dynamical Systems with Self-Excited and Hidden Attractors* (Berlin: Springer)

[3] Vaidyanathan S, Azar A T, Rajagopal K, Sambas A, Kacar S and Cavusoglu U 2018 International Journal of Simulation and Process Modelling **13** 281-296

[4] Vaidyanathan S, Volos C K, Rajagopal K, Kyprianidis I M and Stouboulos I N 2015 *Journal of Engineering Science and Technology Review* **8** 74-82

[5] Vaidyanathan S 2015 *International Journal of PharmTech Research* **8** 974-981

[6] Vaidyanathan S 2015 *International Journal of PharmTech Research* **8** 1-11

[7] Vaidyanathan S 2015 *International Journal of PharmTech Research* **8** 117-127

[8] Tomita K 1982 *Journal of Theoretical Biology* **99** 111-118

[9] Vaidyanathan S 2015 *International Journal of PharmTech Research* **8** 106-116

[10] Vaidyanathan S 2015 *International Journal of PharmTech Research* **8** 156-166

[11] Njitacke Z T and Kengne J 2018 *AEU-International Journal of Electronics and Communications* **93** 242-252

[12] Vaidyanathan S 2015 *International Journal of PharmTech Research* **8** 946-955

[13] Saad M, Safieddine A and Sultan R 2018 *Journal of Physical Chemistry A* **122** 6043-6047

[14] Vaidyanathan S 2015 *International Journal of ChemTech Research* **8** 159-171

[15] Hellen E H and Volkov E 2018 *Communications in Nonlinear Science and Numerical Simulation* **62** 462-479

[16] Vaidyanathan S and Rasappan S 2011 *Communications in Computer and Information Science* **131** 585-593

[17] Pisarchik A N, Huerta-Cuellar G and Kulp C W 2018 *Communications in Nonlinear Science and Numerical Simulation* **62** 134-145

[18] Vaidyanathan S 2015 *International Journal of Modelling, Identification and Control* **23** 380-392

[19] Vaidyanathan S 2012 *Lecture Notes of the Institute for for Computer Sciences, Social-Informatics and Telecommunications Engineering* **85** 124-133,

[20] Xu Q, Zhang Q, Jiang T, Bao B and Chen M 2018 *Circuit World* **44** 108-114

[21] Vaidyanathan S, Sambas A, Mamat M and Sanjaya W S M 2017 *Archives of Control Sciences* **27** 541-554

[22] Singh J P, Lochan K, Kuznetsov N V and Roy B K 2017 *Nonlinear Dynamics* **90** 1277-1299

[23] Wang Y, Mou Y and Zhang J 2018 *Journal of Harbin Engineering University* **39** 584-593

[24] Mansour S M B, Sundarapandian V and Naceur S M 2016 *International Journal of Control Theory and Applications* **9** 37-54

[25] Akgul A, Moroz I, Pehlivan S and Vaidyanathan S 2016 *Optik* **127** 5491-5499

[26] Vaidyanathan S, Akgul A, Kacar S and Cavusoglu U 2018 *European Physical Journal Plus* **133** 46

[27] Vaidyanathan S 2015 *Kyungpook Mathematical Journal* **55** 563-586

[28] Vaidyanathan S and Rajagopal K 2017 *International Journal of Simulation and Process Modelling* **12** 165-178

[29] Dou Y, Liu X, Fan H and Li M 2017 *Optik* **145** 456-464

[30] Vaidyanathan S, Sambas A, Mamat M and Sanjaya W S M 2017 *International Journal of Modelling, Identification and Control* **28** 153-166

[31] Idowu B A, Vaidyanathan S, Sambas A, Olusola O I and Onma O S 2018 *Studies in Systems, Decision and Control* **133** 271-295

[32] Tacha O I, Volos C K, Kyprianidis I M, Stouboulos I N, Vaidyanathan S and Pham V T 2016 *Applied Mathematics and Computation* **276** 200-217
[33] Volos C K, Pham V T, Vaidyanathan S, Kyprianidis I M and Stouboulos I N 2015 *Journal of Engineering Science and Technology Review* **8** 142-151
[34] Daltzis P, Vaidyanathan S, Pham V T, Volos C, Nistazakis E. andTombras G. 2018 *Circuits, Systems, and Signal Processing* **37** 613-615
[35] Sambas A, Vaidyanathan S, Mamat M and Mada Sanjaya W S 2018 *Studies in Systems, Decision and Control* **133** 365-373
[36] Vaidyanathan S, Jafari S, Pham V T, Azar A T and Alsaadi F E 2018 *Archives of Control Sciences* **28** 239-254
[37] Wang X, Vaidyanathan S, Volos C, Pham V T and Kapitaniak T 2017 *Nonlinear Dynamics* **89** 1673-1687
[38] Sambas A, Mamat M, Vaidyanathan S, Mohamed M A, Mada Sanjaya W S and Mujiarto 2018 *WSEAS Transactions on Systems and Control* **13** 345-352
[39] Mamat M, Vaidyanathan S, Sambas A, Mohamed M A, Sampath S and Sanjaya W S M 2018 *International Journal of Engineering and Technology* 7 1410-1414
[40] Sambas A, Mamat M, Vaidyanathan S, Mohamed M A and Mada Sanjaya W S 2018 *International Journal of Engineering and Technology* **7** 1245-1250
[41] Volos C, Maaita J O, Vaidyanathan S, Pham V T, Stouboulos I and Kyprianidis I 2017 *IEEE Transactions on Circuits and Systems II: Express Briefs* **64** 339-343
[42] Pham V T, Jafari S, Volos C, Giakoumis A, Vaidyanathan S and Kapitaniak T 2016 *IEEE Transactions on Circuits and Systems II: Express Briefs* **63** 878-882
[43] Mamat M, Vaidyanathan S, Sambas A, Mujiarto, Sanjaya W S M and Subiyanto 2018 *IOP Conference Series: Materials Science and Engineering* **332** 012033
[44] Lien C H, Vaidyanathan S, Sambas A, Sukono, Mamat M, Sanjaya W S M and Subiyanto 2018 *IOP Conference Series: Materials Science and Engineering* **332** 012010
[45] Vaidyanathan S, Sambas A, Sukono, Mamat M, Gundara G, Sanjaya W S M and Subiyanto 2018 *IOP Conference Series: Materials Science and Engineering* **332** 012048
[46] Khorashadizadeh S and Majidi M H 2017 *AEU-International Journal of Electronics and Communications* **82** 37-44
[47] Xu G, Xu J, Xiu C, Liu F and Zang Y 2017 *Neurocomputing* **227** 108-112
[48] Zaher A A 2018 *Computers and Electrical Engineering* **71** 77-92
[49] Zaher A A and Abu-Rezq 2011 *Communications in Nonlinear Science and Numerical Simulation* **16** 3721-3737
[50] Naderi B and Kheiri H 2016 *Optik* **127** 2407-2412
[51] Sprott J C 1997 *American Journal of Physics* **65** 537-543
[52] Li C and Sprott J C and Xing H 2016 *Physics Letters A* **380** 1172-1177
[53] Elsonbaty A and El-Sayed A M A 2017 *Nonlinear Dynamics* **90** 2637-2655
[54] Coullet P, Tresser C and Arneodo A 1979 *Physics Letters A* **72** 268-270
[55] Kengne J, Njikam S M and Folifack Signing V R 2018 *Chaos, Solitons and Fractals* **106** 201-213
[56] Vaidyanathan S 2015 *International Journal of Modelling, Identification and Control* **23** 380-392
[57] Vaidyanathan S and Azar A T 2016 *Studies in Fuzziness and Soft Computing* **337** 581-607
[58] Vaidyanathan S 2017 *Archives of Control Sciences* **27** 409-439
[59] Vaidyanathan S 2017 *Studies in Computational Intelligence* **709** 393-417
[60] Vaidyanathan S and Azar A T 2016 *Studies in Fuzziness and Soft Computing* **337** 349-376
[61] Chlouverakis K E and Sprott J C 2006 *Chaos, Solitons and Fractals* **28** 739-746
[62] Munmuangsaen B and Srisuchinwong B 2011 *Chaos, Solitons and Fractals* **44** 995-1003
[63] Daltzis P, Vaidyanathan S, Pham V T, Volos C, Nistazakis E and Tombras G 2018 *Circuits Systems and Signal Processing* **37** 613-635

[64] Wang X, Vaidyanathan S, Volos C, Pham V T and Kapitaniak T 2017 *Nonlinear Dynamics* **89** 1673-1687
[65] Pham V T, Vaidyanathan S, Volos C, Wang X and Hoang D V 2017 *Studies in Computational Intelligence* **701** 59-80
[66] Vaidyanathan S, Jafari S, Pham V T, Azar A T and Alsaadi F E 2018 *Archives of Control Sciences* **28** 239-254
[67] Vaidyanathan S *2017 Studies in Computational Intelligence* **701** 393-423
[68] Vaidyanathan S 2016 *Archives of Control Sciences* **26** 311-338
[69] Vaidyanathan S 2016 *Studies in Computational Intelligence* **635** 185-209
[70] Vaidyanathan S 2016 *International Journal of Control Theory and Applications* **9** 257-278
[71] Zhang S, Zeng Y, Li Z, Wang M, Zhang X and Chang D 2018 *International Journal of Dynamics and Control* **23** 1-12
[72] Zhang S, Zeng Y and Li Z 2018 *Chinese Journal of Physics* **56** 793-806
[73] Zhang S, Zeng Y C and Li Z J 2018 *J. Computational and Nonlinear Dynamics* **13** 1-10
[74] Zhang S, Zeng Y, Li Z, Wang M and Xiong L 2018 *Pramana* **90** 63
[75] Wang L, Zhang S, Zeng Y C and Li Z J 2018 *Electronics Letters* **52** 1008-1010
[76] Zhang S, Zeng Y C, Li Z J, Wang M J and Xiong L 2018 *Chaos* **28** 013113

IOP Conf. Series: Materials Science and Engineering **567** (2019) 012013 doi:10.1088/1757-899X/567/1/012013

Convergence of binomial tree methods to Black Scholes model on determining stock option prices

Riaman[1], K Parmikanti[1], I Irianingsih[1], K Joebaedi[1], S Supian[1]

[1]Department of Mathematics, Universitas Padjadjaran, 45363, INDONESIA.

Corresponding author: riaman@unpad.ac.id

Abstract. There are many alternative investment tools that can be used to be the choice of investors. One of them is a derivative product. Derivative products that are more widely known and traded on financial markets are options. Option is a contract or agreement between two parties to buy or sell an instrument. To minimize risk, it is necessary to determine the option price by determining the fair price of the option. This option pricing can be done by using the Binomial Tree method and the Black Scholes method. Some factors that influence options are stock prices, strike prices, maturity, volatility, and interest rates. This paper discusses the European option pricing call on the shares of Bank Central Asia (BCA) with the Binomial Tree method and the Black Scholes Method. From the results of the research, it is found that the Binomial Tree method will converge to the Black Scholes method if the time partition increases.

1. Introduction

The development of the investment in the world is now increasingly rapid and growing. This matter is indicated by the increasing number of alternative investment tools used to be the choice of investors. One of them is derivative products. Derivative products are financial instruments whose value depends on the value of underlying assets can also be used to manage and minimize losses caused by fluctuations in the prices of underlying assets [1]. Derivative products are widely known and traded in financial markets, namely options. Options as one of the financial products that are widely assessed mathematically, both deterministically and stochastically.

Option is a contract or agreement between two parties, whereby the contract holder has the right not the obligation to buy or sell an underlying asset from the seller to a certain asset at a certain price and time [2]. Based on the type of option rights are divided into two, namely the purchase option (call option) and the selling option (put option). Buy options give the buyer the right to buy a particular stock at a predetermined price at any time during the contract, while the selling option gives the buyer the right to sell a certain price at a predetermined price at any time during the contract [3]. Options are divided into two types based on the type of time period of implementation, namely the type option

America and European type options. The American type option is an option that is done at or before maturity, while the European type option is an option that is done at maturity [4].

In buying shares that are more expensive than the market or selling stocks that are cheaper than the market, it is necessary to determine the option price to minimize risk, namely by determining the fair price of the option. This option pricing can be done in several ways, including the Binomial Tree Method and the Black and Scholes Method [5] (Boddie et al. 2014)

The Binomial Tree Method was first introduced by Cox, Ross, and Rubinstein in 1979. This method reveals that stock price movements in the market have two possibilities, namely up and down [6]. Meanwhile, the Black Scholes method has been published since 1973 by Fisher Black and Myron Scholes in "*The Pricing of Option and Corporate Liabilities*". This method can only be done on the European type option pricing which can only be done at maturity, and has the assumption that no dividends will be paid during the option period, there is not transaction fees, risk free and constant interest rates over time, and the stock price follows geometric Brown motion [7].

2. Methodology

Option contracts are an agreement for option holders to buy an underlying asset on a certain date (maturity date) and at a certain price level (strike price). An option holder can make a decision to buy or sell an option in a particular market situation. In the European type call option, the option holder can ignore the option if the stock price at maturity is lower than the purchase price because it does not provide benefits for option holders, this condition can be called a strike price. Conversely, the option holder will sell shares at a strike price that is higher than the market price, the option holder will benefit by buying a low price share then selling it at a high price to the writer (seller). Writer must buy shares from the option holder who has bought the selling option as a transaction risk.
Option contracts based on the rights granted are as follows:

On the buy option (*call option*):
1. In-the-money: strike price is less than the stock price at the time of transaction ($K<S_t$).
2. At-the-money: strike price equals the stock price at the time of the transaction ($K = S_t$).
3. Out-of-the-money: strike price is greater than the stock price at the time of transaction ($K>S_t$)

On the sell option (*put option*):
1. In-the-money: strike price is greater than the stock price at the time of transaction ($K>St$).
2. At-the-money: strike price equals the stock price at the time of the transaction ($K = S_t$).
3. Out-of-the-money: strike price is less than the stock price at the time of transaction ($K<S_t$).

2.1 Lemma Itô

Lemma Itô is a stochastic analogy of the chain rules in ordinary differential equations to solve a stochastic integral. Lemma Itô was discovered by a mathematician, namely K. Itô in 1951.
Suppose the variable X follows the Itô process

$$dX = a\,(X,\,t)\;dt + b\,(X,\,t)\;dWt \tag{1}$$

Where dWt is the Wiener process and a, b is a function of X, t. The variable X has a drift level a and a variance level of b^2. Lemma Itô shows the functions, G from X and t follows the process:

$$dG = \left(\frac{\partial G}{\partial x}a + \frac{\partial G}{\partial t} + \frac{1}{2}\frac{\partial^2 G}{\partial x^2}b^2\right)dt + \frac{\partial G}{\partial x}b\;dW_t \tag{2}$$

dWt is a Wiener in the equation (2). G follows the Itô process, using the drift rate

$$\frac{\partial G}{\partial x}a + \frac{\partial G}{\partial t} + \frac{1}{2}\frac{\partial^2 G}{\partial x^2}b^2$$

Standard deviation of changes in a short period of time Δt in stock prices and Parameter μ is the level of expectation of stock returns, σ is the volatility of stock prices.

$$dS = \mu S dt + \sigma S dW \tag{3}$$

μ and σ are constant, from lemma Itó follows the process of Itó then the function G and S is

$$dG = \left(\frac{\partial G}{\partial S} \mu S + \frac{\partial G}{\partial t} + \frac{1}{2} \frac{\partial^2 G}{\partial S^2} \sigma^2 \sigma S^2 \right) dt + \frac{\partial G}{\partial S} \sigma S dW_t \tag{4}$$

The solution of equation (4) is

$$S_t = S_0 \exp\left\{ \left(\mu + \frac{\sigma^2}{2} \right) T + \sigma W_t \right\} \tag{5}$$

2.2 Volatility

Stock price volatility is the movement of the rise and fall of stock prices on the stock exchange in the annual period used to measure the level of risk of a stock. Volatility values are at positive intervals between 0 and infinity ($0 \leq \sigma \leq \infty$). The higher the level of volatility, the higher the level of rise and fall of stock returns obtained.Volatility can be calculated using the formula:

$$\sigma = \sqrt{\frac{1}{n-1} \sum_{i=1}^{n} (R_i - \bar{R}_i)^2 k} \tag{6}$$

or with the following formula:

$$\sigma = \sqrt{\frac{1}{n-1} \sum_{i=1}^{n} R_i^2 - \frac{1}{n(n-1)} \left(\sum_{i=1}^{n} R_i \right)^2 k} \tag{7}$$

2.3 Binomial Tree Method

In the Binomial method, the option price is calculated by looking for the present value of the expected profit from using the option (payoff) when the option is used at the loan deadline. Stock price and execution price are needed in calculating payoff. For a simple stock price binomial method, the time interval is $[0, T]$ divided into N sub intervals, which are the same length as the points for $0 = t_0 < t_1 < t_2 < \cdots < t_n = T$, where $t_1 = i\Delta t$ for ($i = 0, 1, 2, ..., N$), $\Delta t = T/N$ and $S_t = S_{ti}$ is stock price at time t_i.

Assumptions:

a. In time interval Δt the stock price can go up or down to $S_t \rightarrow S_u$ or $S_t \rightarrow S_d$ with $0 < d < 1 < u$
b. The opportunity for stock prices to rise P (up) $= p$ and P (down) $= (1 - p)$
c. There is no dividend payment for a certain period of time.
d. Expectation of stock price return equal to risk free interest r. So that the stock price S_t moves randomly S_i and t_i becomes S_{i+1} and at the time ti this means:

$$E(S_{i+1}) = S_i e^{\mu \Delta t} \tag{8}$$

In the Binomial method, when $t_1 = i\Delta t$ there will be $i + 1$ the stock price that might occur:

$$S_{i,j} = S_0 u^j d^{i-j}, i = 0,1,2,3, \dots, N \text{ dan } j = 0,1,2,3, \dots, i \tag{9}$$

The equation above is not recursive, so it requires an equation, so the calculation does not take long.

$$E(S_N) = S_0 e^{N\mu\Delta t} \tag{10}$$

To calculate the initial stock price of the matured stock price using the discount $e - \mu\Delta t$ from the above equation with the formula:

$$S_0 = e^{-\mu\Delta t} E(S_1) \tag{11}$$

$$S_{ij} = e^{-\mu\Delta t}(pS_{i+1,j+1} + qS_{i,j+1}) \tag{12}$$

So that the value of the European call option is at the time ti that is the average of the current option value $ti + 1$ that is $\{SiN\}$, i = 0,1,2, ... N. Thus, the payoff value of a call option is

$$C_{iN} = maks(S_{ij} - K, 0) \tag{13}$$

The backward binomial tree method for calculating call options at t = 0, is:

$$C_{ij} = e^{-\mu\Delta t}\, E(C_{iN} e^{\mu\Delta t}) \tag{14}$$

$$C_{ij} = e^{-\mu\Delta t}(pC_{i+1,j+1} + (1-p)C_{i+1,j}) \tag{15}$$

2.4 Black Scholes method

The Black-Scholes method is a method used to determine option values. In the financial world, this model was developed by Fischer Black and Myron Scholes in 1973. This method is limited because only can be used to determine the European type option value that is executed at maturity, while this model does not apply to American type options, because the option can be executed at any time until the due date.There are several assumptions for formulating option values developed by Fisher Black and Myron Scholes:

1. The distribution of stock prices is lognormal and the range of returns on stocks is constant.
2. There are no transaction fees and taxes in buying or selling shares or options.
3. The type of option used is the type of European option, where the option can only be executed when the maturity date.
4. There is no possibility of the act of buying low-priced securities in a market and at the same time selling it at a higher price in a different market, so as to gain profits without risk.
5. Short-term risk-free interest rates are known and their value is constant.
6. Trading of underlying assets is continuous.
7. There is no prohibition on Short selling.

The European type Call option prices determined by the Black Scholes formula are as follows:

$$C = S_0 N(d_1) - Ke^{-\mu T} N(d_2) \tag{16}$$

Where

$$d_1 = \frac{\ln\left(\frac{S_0}{K}\right) + \left(\mu + \frac{\sigma^2}{2}\right)T}{\sigma\sqrt{t}} \tag{17}$$

$$d_2 = d_1 - \sigma\sqrt{T} \tag{18}$$

where $N(d_1)$ is the cumulative density function of the normal distribution of d_1, and $N(d_2)$ is the normal distributive cumulative density function of d_2.

3. Results and Discussion

The data used in this study is secondary data, namely the historical data of stock prices from PT. Bank Central Asia Tbk. (BCA) with a period of January 5, 2017 - January 5 2018. The stock price used is the daily closing price (close price) and the option price calculated is only the European type call option. With a risk-free interest rate of 0.065 or 6.5% with an initial stock price of15,675.00 IDR, a strike price of 500.00 IDR, a maturity of 3 months and the stock price is assumed to follow the lognormal distribution and no dividends are taken into account. By using the data return of the stock price of PT. Bank Central Asia Tbk. (BBCA) is calculated to find the volatility value as follows:

Table 1. Stock Closing Price Data

Date	*Close Price* (IDR)
January 5, 2017	15675
January 6, 2017	15600
January 9, 2017	15350
⋮	⋮
January 5, 2018	22250

By using the equation (7), the stock closing return price is calculated. The expected return of the stock closing price is 0.0014. The calculation results of $R_t - R_{\bar{t}}$ can be seen in Table 2.

Table 2. The calculation results $Rt{-}R\bar{\bar{t}}$

t	Stock Closing Price (IDR)	R_t	$R_t{-}R_{\bar{t}}$	$(R_t{-}R_{\bar{t}})2$
1	15600	-0.0048	-0,0062	0.00004
2	15350	-0.0162	-0.0176	0.00031
3	15400	0.0033	0.0019	0.0000
4	15300	-0.0065	-0.0079	0.00006
⋮	⋮	⋮	⋮	⋮
255	22250	0.0011	-0.0003	0.00000

And the volatility is $\sigma = 0.1846$. So, obtained the value of σ used in this study is equal to 0.1846 or 18.46% means that the stock price movement of PT. Bank Central Asia Tbk. (BCA) of 18.46% during the period of January 5, 2017 - January 5, 2018.

3.1 Simulation of Stock Option Price Determination Using the Binomial Tree Method

In determining the stock option price using the Binomial Tree method, the first step is to find the value of $E(S_N)$, as follows:

$$E(S_N) = S_0 e^{N\mu\Delta t}$$

Table 3. The calculation results$E(S_N)$,C_{ij} dan C

t	Stock Closing Price (IDR)	K	C_{iN}	C_{ij}	$E(S_N)$
1	15600	500	15179	15175,13	15679
2	15350	550	15132,99	15125,28	15682,99
3	15400	600	15086,99	15075,46	15686,99
4	15300	650	15040,99	15025,66	15690,99
⋮	⋮	⋮	⋮	⋮	⋮
255	22.250,00	13200	3527,718	3305,709	16727,72

So, obtained the European type call option value using the Binomial Tree Method with $K = 500$ and the 3 month maturity is 15,175.13 IDR.

3.2 Simulation of Stock Option Pricing Using the Black Scholes Method

In determining the stock option price using the Black Scholes method, the first step to look for is to calculate the value of d_1 and d_2. Using equations (18) and (19), it is obtained as follows:

$$d_1 = \frac{\ln\left(\frac{S_0}{K}\right) + \left(\mu + \frac{\sigma^2}{2}\right)T}{\sigma\sqrt{T}}$$

Table 4. The calculation results$N(d_1)$dan $N(d_2)$

T	S_0	d_1	d_2	$N(d1)$	$N(d2)$
1	15675	37.54847	37.45617	1.00000000	1.00000000
2	15675	36.51585	36.42356	1.00000000	1.00000000
3	15675	35.57315	35.48086	1.00000000	1.00000000
4	15675	34.70595	34.61366	1.00000000	1.00000000
⋮	⋮	⋮	⋮	⋮	⋮
255	15675	2.08407	1.991773	0.98142322	0.976901994

Furthermore, the option to buy the Black Scholes method is determined as follows:

$C = S_0N(d_1) - Ke - \mu TN\ (d_2)$

$C = 15675\ (1) - 500.e - 0.065\ (0.25)\ (1)$

$C = 15675 - (491.9406595)\ (1)$

$C = 15675 - 491.9406595$

$C = 15183.0593$

So, we get the European type call option value, using the Black Scholes Method, with $K = 500$ and 3 months maturity amounting to 15,183.06 IDR.

4. Conclusion

From the results of the study, it was found that the stock price at the end of the contract is greater than the strike price, this means $St > K$, so that some conclusions are obtained, namely: In the Binomial Tree method, it is possible for investors to exercise their rights and will get a profit equal to the difference between the share price and the strike price when selling the call option of 15,175.13 IDR. In the Black Scholes model, it is possible for investors to use their rights and will benefit as much as the difference between the share price and the strike price, if they sell the call option of 15,183.06 IDR. From the results of the analysis in the discussion of this study it can be concluded that the calculation of the call option price with the Binomial tree method converges close to the calculation of the Black-Scholes model. Based on the value of the Call option price on the Binomial tree method of 15175.13 IDR and the Black Scholes method call option price of 15,183.06 IDR. The greater time partition to n in the binomial tree method, the option value will converge to the option value of the Black-Scholes method.

Acknowledgments

The Authors would like to thank to DRPMI, directorate of research and community service, Padjadjaran University, who gave funding for this research and dissemination of this paper via RFU.

References

[1] Roy, S dan Ferdiansyah, T .2002. Sekuritas derivatif: madu atau racun?. Jakarta: Salemba Empat.
[2] Bugnevicius, J dan Radakhrishnan, P (2004). Option Valuation With The Black-Scholes Model. London: London School Of Economics.
[3] Bezberodov, 2016, An Introduction to Financial Option Valuation. United Kingdom: Cambridge University Press.
[4] Hull, C Jhon.2009. Option,Futures and Other Derivatives. Seventh Edition. New Jersey : Pearson Pretince Hall, International Edition.

[5] Bodie, dkk. 2014. Invesment Tenth Edition. New York : Mc Graw-Hill Education.
[6] Cox, dkk. 1979. Option pricing.A Simplified Approach Journal of Financial Economics.
[7] Black, F. Scholes, M. 1973. The Pricing of Options and Corporate Liabilities, The Journal of Political Economy, 81 (3).
[8] Higham,Desmond J,. 2004. An Introduction to Financial Option Valuation. United Kingdom: Cambridge University Press.

IOP Conf. Series: Materials Science and Engineering **567** (2019) 012014 doi:10.1088/1757-899X/567/1/012014

Optimization of coral reef conservation based on coral larval dispersion using hamiltonian path (case study: Bunaken National Park)

S E Chungdinata[1], C E J C Montolalu[1*], M L Mananohas[1], R Tumilaar[1], N Nainggolan[1], A J Rindengan[1], Y A R Langi[1]

[1]Department of Mathematics, Faculty of Mathematics and Natural Sciences, Sam Ratulangi University, INDONESIA

*Corresponding author: chriestelly@yahoo.com

Abstract. Bunaken sea is one of the fisheries source for Manado and around it. But unfortunately, some studies have shown that the coral reefs in some parts of Bunaken Sea have been deteriorating. Some of the causes are coral retrieval, illegal fishing using explosive and poison, marine pollution, and sea temperature increase. Deteriorated reef can be recognized through the color becomes white or known as coral bleaching. In order to preserve coral reefs, there are some factors that can be learned. One of the factors is the location of the healthy coral reefs and can produce coral larval to help recovery process of the deteriorated reefs. Based on some research, coral larval dispersion will be modelled using Graph Theory that is objects connectivity, especially the Hamiltonian Path. Reef spots are represented by vertices and dispersion paths are represented by edges. Based on research, the first location that must be conserved is M23 because it gives good impact and has high possibilities to help recovery of the other 16 locations. Research about determine the coral larval dispersion path can be used to help coral reef conservation to minimize the time and budget, also can support the development of marine, fisheries, and tourism.

1. Introduction

Bunaken National Park is measured to have an area about 75,265 ha. There are 5 islands in this national park, those are Naen, Bunaken, Manado Tua, Siladen, and Mantehage Island, also small islands around it [1].

There are three islands adjacent each other, namely Bunaken Island, Manado Tua Island, and Siladen Island. Bunaken National Park geographically include in “Coral Triangle”. It’s because the “coral triangle” connects the Papua, Philippines, and Pacific Ocean. And in this “Coral Triangle” there are thousands of fish species who’s live inside the coral reef.

The diversity of fish and coral reef species inside the Bunaken National Park are the background from Manado City Government to make this area as Marine Tourism area and as National Park by the Central Government. Bunaken Sea is one of the fisheries source for Manado and around it. But unfortunately, some studies have shown that the coral reefs in some parts of Bunaken Sea have been deteriorating [2].

IOP Conf. Series: Materials Science and Engineering **567** (2019) 012014 doi:10.1088/1757-899X/567/1/012014

Some of the causes are coral retrieval, illegal fishing using explosive and poison, marine pollution, and sea temperature increase. Deteriorated coral reef can be recognized through the color. The color of the deteriorated coral reef becomes white or known as coral bleaching.

In order to preserve coral reefs, there are some factors that can be learned. One of the factors is the location of the healthy coral reefs and can produce coral larval to help recovery process of the deteriorated reefs [3]. The return of a coral reef ecosystem to a functional state after mass bleaching mortality will depend on successful reproduction and recolonization by remaining corals and by corals from outside the ecosystem [4]. This research is centered on coral larval dispersion from each healthy reefs spot and see if its path reach the deteriorated reef. If coral larval can reach the deteriorated reef, then there are possibilities that they can help the recovery of the declined reefs.

Based on some research [5] and [6], formed a theoretical foundation that benthic marine organism affected by the ocean currents and larvals.

Based on some research [7] and [8], coral larval dispersion will be modeled using Graph Theory that is Hamiltonian Path. In Graph Theory, especially the objects connectivity. Reef spots are represented by vertices and dispersion paths are represented by edges.

There are 17 sustainable development goals, and one of it is life below water. So, this research can help the development on marine sciences, fisheries, and tourism.

Based on the background explained above, the problem to be solved is how to optimize the model of coral reef conservation based on coral larval dispersion?

This study aims to optimize the conservation of the coral reef based on coral larval dispersion using the Hamiltonian Path. The result of study is to help coral reef conservation policies to minimize the time and budget for the government in Indonesia, especially Sulawesi Utara.

2. Graph Theory

Definition 2.1 : A graph $G = (V, E)$ consists of V, a nonempty set of vertices (or nodes) and E, a set of edges. Each edge has either one or two vertices associated with it, called its endpoints. And edge is said to connect its endpoints [9].

2.1. Connectivity

A walk in a graph is a sequence (not necessarily distinct) vertices $v_1, v_2, \dots, v_k$ such that $v_i v_{i+1} \in E$ for $i = 1, 2, \dots, k-1$. Such a walk is somethimes called a $v_1 - v_k$ walk, and v_1 and v_k are the end vertices of the walk. If the vertices in a walk are distinct, then the walk is called a path [10].

2.2. Hamiltonian Path

Definition 2.2 : A simple path in a graph G that passes through every vertex exactly once is called a Hamilton path. That is, the simple path $x_0, x_1, \dots, x_{n-1}, x_n$ in the graph $G = (V, E)$ is a Hamilton path if $V = \{x_0, x_1, \dots, x_{n-1}, x_n\}$ and $x_i \neq x_j$ for $0 \leq i \leq j \leq n$ [9].

3. Research Methodology

Data used in this paper are secondary data obtained from [11] and [12]. Method used in this paper is literature review and the sources used are obtained from books, journals, and articles from the internet. And the steps are:

1. Collect the data,
2. Determine the location of coral reefs and coral dispersal paths,
3. Determine the vertices as the location of coral reefs and the edges as the coral dispersal paths,
4. Make the Hamiltonian Path from the vertices and edges which has been determined,
5. Determine the most optimal location (vertex) from the Hamiltonian Path.

IOP Conf. Series: Materials Science and Engineering **567** (2019) 012014 doi:10.1088/1757-899X/567/1/012014

4. Result and Discussions

4.1. Dispersion Model using Hamiltonian Path

Based on ocean current data from Badan Informasi Geospasial (BIG) [11] and coral reefs data from Reef Base [12], a graph which represents reefs locations as vertices and coral larval dispersal paths as edges can be constructed. Each square on the map (Figure 1) is represented as a vertex so there are 49 vertices formed in the graph.

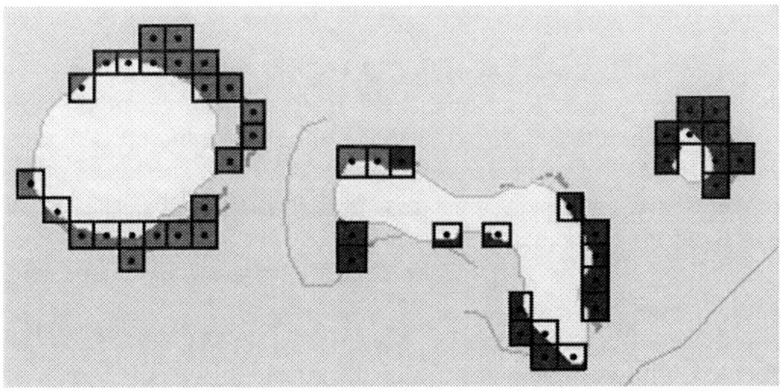

Figure 1. Vertices based on map

Coral reefs locations which were represented as vertices are reefs in three islands: Manado Tua, Bunaken, and Siladen. As can be seen in Figure 1, all the coral reefs in Manado Tua have been in high threat. In Bunaken, almost all its reefs have been in very high threat. While in Siladen, all its reefs have been in very high threat.

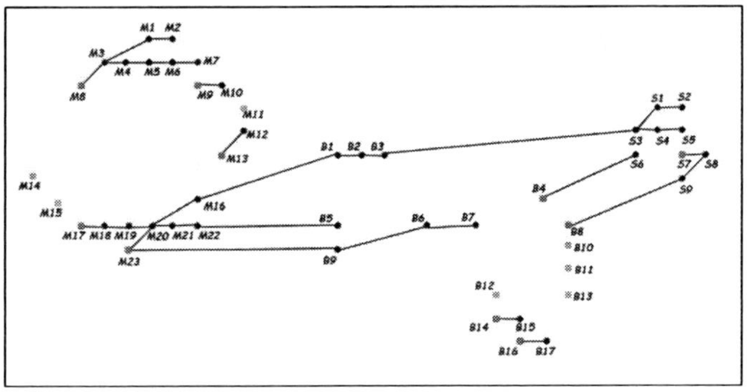

Figure 2. Graph from location of coral reefs

● : Source ● : Sink ●: Isolated Vertex

15 hamiltonian paths plotted are:

1. M8 → M3 → M4 → M1 → M2
2. M8 → M3 → M4 → M5 → M6 → M7
3. M9 → M10
4. M13 → M12
5. M17 → M18 → M19 → M20 → M21 → M22 → B5
6. M17 → M18 → M19 → M20 → M16 → B1 → B2 → B3 → S3 → S1 → S2
7. M23 → M20 → M16 → B1 → B2 → B3 → S3 → S1 → S2

8. M23 → M20 → M15 → B1 → B2 → B3 → S3 → S4 → S5
9. M23 → M20 → M21 → M22 → B5
10. M23 → B9 → B6 → B7
11. B4 → B6
12. B8 → S9 → S8
13. B14 → B15
14. B16 → B17
15. S7 → S8

From the graph formed on Figure 2, out of the 49 vertices, 10 vertices were chosen as sources, i.e. spots for the first act for conservation locations. Reefs restoration might be done in these 10 sources as initial acts in order to provide healthy reefs in these locations. Once the reefs in these locations restored, they can produce healthy larval and release them to ocean currents. The larval released will then use the ocean currents as transportation and reach other reefs in their paths. When they reach the damaged reefs, they will help restore the reefs in sexual reproduction. The reefs in the larval paths that absorb larval were then set as Sinks. From Figure 2, it can be seen that there are 33 sinks indicated in the graph modeled.

From the graph modeled with 10 sources, i.e. M8, M9, M13, M17, M23, B4, B8, B14, B16 and S7, 15 hamiltonian path can be plotted. The paths are 15 dispersal paths of the larval that were assumed can help the restoration process of the 33 sinks, i.e. M1, M2, M3, M4, M5, M6, M7, M10, M12, M16, M18, M19, M20, M21, M22, B1, B2, B3, B5, B6, B7, B9, B15, B17, S1, S2, S3, S4, S5, S6, S8, and S9.

4.2. Deteriorated Coral Reef Recovery Analysis

It can be assumed that by putting the 10 sources in beginning acts of conservation, the 33 sinks can be restored too. Thus, the number of reefs that can be cured will be 42 locations.

There are some spots in which although the reefs have been restored would not significantly affect other locations because there is a small possibilities that the ocean currents will connect them directly. Reefs spots which are not affected or affect other spots were then identified as Isolated Vertices. These isolated vertices are M14, M15, M11, B10, B11, B13, and B12. Thus, these 7 locations have to be preserved one by one distinctly.

In order to optimize conservation time and budget, spot that has to be conserved first is M23. This will affect 16 other locations in the same path. The second spot that has to be preserved is M17 because this will affect other 15 locations in the same path. The next one is M8 since it will affect 7 other locations. Consequently, B8 has to be preserved for it affects 2 other locations. Whereas, locations that affect to 1 location only are M9, M13, B4, B14, B16, and S7. Eventually, the last spots to be conserved are the isolated vertices for they do not significantly influence restoration process in other locations.

5. Conclusion

Based on analysis result and literature review we can conclude that the first location that must be conserved is M23 because it gives good impact and has high possibilities to help recovery of the other 16 locations.

6. Recommendation

Recommendation in this paper is we need more research and further development about locations of the coral reef, species of coral reef, sea current based on season, and coral health level. When doing conservation, we need cooperation between university, government especially Ministry of Tourism, Ministry of Maritime Affairs and Fisheries, and investors. I hope this paper can support the development of marine, fisheries, and tourism. Research about determine the coral larval dispersion path can be used to help coral reef conservation to minimize the time and budget.

IOP Conf. Series: Materials Science and Engineering **567** (2019) 012014 doi:10.1088/1757-899X/567/1/012014

References

[1] Universitas Terbuka 2018 Sekilas Sulawesi Utara (http://www.manado.ut.ac.id, acessed 3 March 2018).
[2] Coral Watch 2018 Pemutihan Terumbu Karang (https://id.coralwatch.org/bleaching-event-indonesia, acessed 3 March 2018).
[3] Rani C 2003 Perikanan dan terumbu karang yang rusak: bagaimana mengelolanya? *Jurnal Bionatura* **5** 2 97 – 111.
[4] Westmacott S, Teleki K, Wells S and West J 2000 *Management of Bleached and Severely Damaged Coral Reefs* Gland and Cambridge: IUCN.
[5] Roughgarden J, Gaines S and Possingham H 1988 Recruitment dynamics in complex life cycles *Science* **241** 4872 1460-1466.
[6] Possingham H and Roughgarden J 1990 Spatial population dynamics of a marine organism with a complex life cycle *Ecology* **71** 3 973-985.
[7] Thomas C, Lambrechts J, Wolanski E, Traag V, Blondel V, Deleersnijder E and Hanert E 2014 Numerical modelling and graph theory tools to study ecological connectivity in the great barrier reef *Ecological Modelling* **272** 160–174.
[8] Treml E, Halpin P, Urban D and Pratson F 2008 Modeling population connectivity by ocean currents, a graph-theoretic approach for marine conservation *Landscape Ecology* **23** 19-36 (https://link.springer.com, acessed 3 March 2018).
[9] Rosen K 2012 *Discrete Mathematics and Its Applications 7th Edition* New York: McGraw-Hill.
[10] Harris J, Hirst J and Mossinghoff M 2008 *Combinatorics and Graph Theory 2nd Edition* New York: Springer.
[11] Badan Informasi Geospasial (http://tides.big.go.id, acessed 3 March 2018).
[12] Reef Base (http://reefgis.reefbase.org, acessed 3 March 2018).

IOP Conf. Series: Materials Science and Engineering **567** (2019) 012015 doi:10.1088/1757-899X/567/1/012015

Expansion of the investment portfolio performance assessment model based on value-at-risk using a time series approach

Sukono[1], D Susanti[1], E S Hasbullah[1], Y Hidayat[2], Subiyanto[3]

[1]Department of Mathematics, Faculty of Mathematics and Natural Sciences, Universitas Padjadjaran, INDONESIA
[2]Department of Statistics, Faculty of Mathematics and Natural Sciences, Universitas Padjadjaran, INDONESIA
[3]Department of Marine Sciences, Faculty of Fishery and Marine Sciences, Universitas Padjadjaran, INDONESIA

Corresponding author: sukono@uanpad.ac.id

Abstract. Portfolio performance assessment needs to be carried out before or after the investment decision is taken, in order to minimize the possibility of risk loss. This paper discusses the expansion of the investment portfolio performance appraisal model based on Value-at-Risk, where the analyzed stock returns on mean and volatility is non-constant.The aim is to increase the likelihood of achieving investment objectives by investors. In this paper the mean is estimated using autoregressive moving average models, while the non-constant volatility is estimated using generally autoregressive conditional heteroscedastic models. The estimator's ofmean and non-constant volatility are then used for the analysis of investment portfolio optimization. Portfolio optimization issues are followed based on the basic framework of the Mean-Value-at-Risk model. The solution to the investment portfolio optimization problem is done by using the Lagrange multiplier technique and the Kuhn-Tucker method.Assessment of investment portfolio performance is based on Reward to Value-at-Risk, which is then used to compare the two investment portfolios A and B are analyzed. The results of the analysis show that portfolio A has better performance than portfolio B. So it is recommended to investors to choose an investment portfolio A, to achieve a better level of success.

1. Introduction

In dealing with risky investments, investors must make a decision to choose an efficient portfolio that has better performance. To make a decision, an assessment of an efficient portfolio needs to be done. The performance assessment of an efficient portfolio can be carried out before or after the investment decision is taken [1][2][3]. Efficient portfolio performance evaluation is to increase the likelihood of achieving investor goals. In conditions of uncertainty, investors cannot choose investment opportunities only by considering the level of profit offered. Investors need to consider the element of risk [4][5]. Therefore, the assessment of investment performance will be based on the level of profit and risk [6][7].

Portfolio risk is the possibility of a level of profit deviating from what is expected. Therefore, certain dispersal measures are often used as a measure of risk [8][9]. Standard deviation or variance is often used as a measure of investment portfolio risk [10]. However, many loss risk events exceed the standard deviation or variance.Therefore, the idea arises to measure risk using a quantile, or more popularly called Value-at-Risk (*VaR*) [10][11]. The amount of *VaR* depends on the average parameter value and volatility of a stock return, as well as the probability of possible risk of loss. Stock returns often have a non-constant mean and volatility, and even have long memory effects [12][13].

This paper aims to analyze portfolio performance measurement based on *VaR* risk measure. Average and volatility as constituents in *VaR* will be analyzed using a time series approach. The average is estimated using autoregressive fractionally integrated moving average (ARFIMA) models. Non-constant volatility is estimated using the generally autoregressive conditional heteroscedastic (GARCH) models.For efficient portfolio selection is based on the Mean-VaR portfolio optimization model. While to measure portfolio performance is done using the Reward to Value-at-Risk (*RVaR*) approach. The application of this method is used to analyze ten stocks traded in the capital market in Indonesia. The aim is to compare and choose a portfolio that has better performance than other portfolios. This study is useful for investors to find out the performance of two investment portfolios, where risk is measured by Value-at-Risk, where data follows a time series pattern.

2. Methodology

Suppose P_{it} and r_{it} successively stated prices and stock returns i ($i = 1,...,N$ and N the number of stocks analyzed), at the time t ($t = 1,...,T$, T data observation period). Stock returns r_{it} calculated using formula $r_{it} = \ln(P_{it} / P_{it-1})$ [5; 14]. The return data model estimation is then performed on mean and volatility as follows.

2.1 *Modeling of mean and volatility*

Modeling of mean. Identification of long memory effects on stock returns data r_{it}. Identification is done using the rescale (R / S) method or the Geweke and Porter-Hudak (GPH) methods. Estimation of fractional differentiation parameters d_i ($i = 1,...,N$ and N the number of stocks analyzed), carried out using the maximum likelihood method [11; 13]. Confidence interval $(1-\alpha)100\%$ for d_i is $\hat{d}_i - z_{\alpha/2}.\sigma_{d_i} < d_i < \hat{d}_i + z_{1-\alpha/2}.\sigma_{d_i}$ with $\hat{d}_i$ is estimator of d_i, and $z_{\alpha/2}$ standard normal distribution percentile if given a level of significance α. Suppose d_i fractional differentiation to be tested by the hypothesis. Suppose also σ_{d_i} standard deviation of d_i. Hypothesis testing is carried out against $H_0 : \hat{d}_i = 0$ and $H_1 : \hat{d}_i \neq 0$ use $z_{d_i} = d_i / \sigma_{d_i}$. Test criteria are reject H_0 if value $z_{d_i} < z_{\alpha/2}$ atau $z_{d_i} > z_{1-\alpha/2}$ [12; 14].

Fractional differentiation processes are defined as:

$$(1-B)^{d_i} r_{it} = a_{it}, \; -0.5 < d_i < 0.5 ; \tag{1}$$

with $\{a_{it}\}$ is residual white noise series, and B stated the backshift operator. If fractional differentiation series $(1-B)^{d_i} r_{it}$ follow the model of ARMA (p,q), then r_{it} called the autoregressive fractionally integrated moving average degree process p, d and q, or ARFIMA(p,d,q) [14]. Model equation of ARMA(p,q) is:

$$r_{it} = \psi_{i0} + \sum_{g=1}^{p} \psi_{ig} r_{it-g} + a_{it} + \sum_{h=1}^{q} \theta_{ih} a_{it-h} , \tag{2}$$

with ψ_{i0} constants, and ψ_{ig} $(g=1,...,p)$,and θ_{ih} $(h=1,...,q)$ parameter coefficient of the mean stock return model i $(i=1,...,N$ and N the number of stocks analyzed). Assumed $\{a_{jt}\}$ residual white noise sequence with zero mean and variance $\sigma^2_{a_j}$ [14][15].

The stages of the average modeling process include: (i) Model identification, (ii) Estimation of parameters, (iii) Test diagnosis, and (iv) Prediction [14].

Modeling of volatility. Stock return volatility modeling is performed using generalized autoregressive conditional heteroscedastic (GARCH) models. Suppose μ_{it} and σ^2_{it} successive mean and volatility of stock returns i $(i=1,...,N$ and N the number of stocks analysed), at the time t ($t=1,...,T$ and T data observation period). Residual a_{it} the above has an equation $a_{it}=r_{it}-\mu_{it}$ [11; 14]. Volatility σ^2_{it} will follow the GARCH model with degrees m and n or written as GARCH(m,n), when:

$$a_{it}=\sigma_{it}\varepsilon_{it}\,,\quad \sigma^2_{it}=\alpha_{i0}+\sum\nolimits_{k=1}^{m}\alpha_{ik}a^2_{it-k}+\sum\nolimits_{l=1}^{n}\beta_{il}\sigma^2_{it-l}+\varepsilon_{it}\,, \tag{3}$$

with α_{i0} constants, and α_{ik} $(k=1,...,m)$,and β_{il} $(l=1,...,n)$ parameter coefficient of stock return volatility model i $(i=1,...,N$ and N the number of stocks analyzed). Assumed $\{\varepsilon_{it}\}$ squence of random variables are mutually independent and have identical distributions (iid) with an average of 0 and variance 1 $\alpha_{i0}>0$, $\alpha_{ik}\geq 0$, $\beta_{il}\geq 0$, and $\sum\nolimits_{k=1}^{\max(m,n)}(\alpha_{ik}+\beta_{ik})<1$ [14][16].

Stages of the volatility modeling process include: (i) Estimation of the average model, (ii) ARCH effect test, (iii) Model identification, (iv) Estimation of the volatility model, (v) Test diagnosis, and (vi) Prediction [14].

Using the mean model (2) and volatility (3), the prediction is carried out aimed at calculating the mean $\hat{\mu}_{it}=\hat{r}_{iT}(1)$ and variance $\hat{\sigma}^2_{it}=\hat{\sigma}^2_{iT}(1)$, namely the prediction of the l-step forward after the time period to T [14].

2.2 *Portfolio modal and Value-at-Risk*

In the formation of an investment portfolio w, will relate to the proportion of funds allocated to each of the shares analyzed. Suppose w_i is the proportion of funds allocated to stocks i, where $\sum\nolimits_{i=1}^{N}w_i=1$, then portfolio return can be expressed as:

$$R_{wt}=\sum_{i=1}^{N}w_iR_{it}\,, \tag{4}$$

where R_{wt} portfolio return w at time t, and N the number of stocks in the formation of a portfolio[10; 14].

Based on (5), the mean (expectation) portfolio is obtained with weights w_i can be stated as:

$$\hat{\mu}_{wt}=\sum_{i=1}^{N}w_i\hat{\mu}_{it} \tag{5}$$

While portfolio variance can be expressed as:

$$\sigma^2_{wt}=\sum_{i=1}^{N}w_i^2\sigma^2_{it}+2\sum_{i=1}^{N}\sum_{j=1}^{N}w_i\,w_j\sigma_{ij}\;;\; i\neq j \tag{6}$$

where $\sigma_{ij}=Cov(r_{it},r_{jt})$.

Suppose the amount of the initial investment is one unit, and the level of significance of the risk of loss is α , then VaR for portfolios with weights w_i is:

$$VaR_{wt} = z_\alpha \hat{\sigma}_{wt} - \hat{\mu}_{wt} \tag{7}$$

where z_α is the percentile value of a standard normal distribution with a level of significance α [10; 11; 14].

2.3 Mean-Vary Portfolio Optimization

Suppose that the vector values of expectations and covariance matrices are given successively by: $\boldsymbol{\mu}^T = (\hat{\mu}_{1t},...,\hat{\mu}_{it})$, with $\hat{\mu}_{it} = E[R_{it}]$, $i = 1,...,N$, and $\boldsymbol{\Sigma} = (\hat{\sigma}_{ij})_{i,j=1,...,N}$, with $\hat{\sigma}_{ij} = Cov(r_{it}, r_{jt})$, $i, j = 1,...,N$. Refer to the previous discussion, the weight of stock returns in a portfolio $\mathbf{w}^T = (w_1,...,w_N)$, where $\sum_{i=1}^N w_i = 1$ or $\mathbf{e}^T\mathbf{w} = 1$ with $\mathbf{e} = (1,...,1)^T$ vector with one-on-one element. Referring to equation (5) can be rewritten as

$$\mu_{wt} = E[R_{wt}] = \boldsymbol{\mu}^T\mathbf{w}, \tag{8}$$

and equation (6) is rewritten as:

$$\sigma_{wt}^2 = Var(R_{wt}) = \mathbf{w}^T\boldsymbol{\Sigma}\mathbf{w}. \tag{9}$$

Use the level of significance α, the percentile z_α obtained from the standard normal distribution table. So the Value-at-Risk investment portfolio equation (13) can be rewritten as:

$$VaR_{wt} = z_\alpha \sigma_{wt} - \mu_{wt} = z_\alpha(\mathbf{w}^T\boldsymbol{\Sigma}\mathbf{w})^{1/2} - \boldsymbol{\mu}^T\mathbf{w}. \tag{10}$$

A portfolio $\mathbf{w}^*$ *called (Mean-VaR) efficiently if there is no portfolio* $\mathbf{w}$ *with* $\mu_{wt} \geq \mu_{wt^*}$ *and* $VaR_{wt} < VaR_{wt^*}$ [3; 7; 9]. To get an efficient portfolio, the objective function is determined maximally $\{2\tau\mu_{wt} - VaR_{wt}\}$, $\tau \geq 0$ where τ is investor risk tolerance. So, for investors with risk tolerance $\tau \geq 0$ must solve optimization problems [13; 5]:

$$\begin{aligned} &\text{maks}\{2\tau\boldsymbol{\mu}^T\mathbf{w} - z_\alpha(\mathbf{w}^T\boldsymbol{\Sigma}\mathbf{w})^{1/2} + \boldsymbol{\mu}^T\mathbf{w}\} \\ &\quad \text{kendala } \mathbf{e}^T\mathbf{w} = 1 \end{aligned} \tag{11}$$

Because of the covariance matrix $\boldsymbol{\Sigma}$ semidefinite positive, the objective function is quadratic concave. Therefore, (12) is a concave quadratic optimization problem. The Lagrange function is given by:

$$L(\mathbf{w}, \lambda) = (2\tau + 1)\boldsymbol{\mu}^T\mathbf{w} - z_\alpha(\mathbf{w}^T\boldsymbol{\Sigma}\mathbf{w})^{1/2} + \lambda(\mathbf{e}^T\mathbf{w} - 1).$$

Using the Kuhn-Tucker theorem, the optimality requirement is:

$$\partial L / \partial \mathbf{w} = (2\tau + 1)\boldsymbol{\mu} - z_\alpha \boldsymbol{\Sigma}\mathbf{w} / (\mathbf{w}^T\boldsymbol{\Sigma}\mathbf{w})^{1/2} + \lambda\mathbf{e} = 0 \text{ dan } \partial L / \partial \lambda = \mathbf{e}^T\mathbf{w} - 1 = 0. \tag{12}$$

Based on algebraic calculations, if for example $A = \mathbf{e}^T\boldsymbol{\Sigma}^{-1}\mathbf{e}$, $B = (2\tau + 1)\,(\boldsymbol{\mu}^T\boldsymbol{\Sigma}^{-1}\mathbf{e} + \mathbf{e}^T\boldsymbol{\Sigma}^{-1}\boldsymbol{\mu})$ and $C = (2\tau + 1)^2(\boldsymbol{\mu}^T\boldsymbol{\Sigma}^{-1}\boldsymbol{\mu}) - z_\alpha^2$, then the ABC formula is obtained:

$$\lambda = \{-B + (B^2 - 4AC)^{1/2}\} / 2A \tag{13}$$

For $\tau \geq 0$, solving equation (12) obtained a portfolio with weight vector $\mathbf{w}^*$

$$\mathbf{w}^* = \frac{(2\tau + 1)\boldsymbol{\Sigma}^{-1}\boldsymbol{\mu} + \lambda\boldsymbol{\Sigma}^{-1}\mathbf{e}}{(2\tau + 1)\mathbf{e}^T\boldsymbol{\Sigma}^{-1}\boldsymbol{\mu} + \lambda\mathbf{e}^T\boldsymbol{\Sigma}^{-1}\mathbf{e}} \tag{14}$$

If vector $\mathbf{w}^*$ substituted into equations (8) and (10), then the average portfolio return and optimum Value-at-Risk will be obtained [2; 9].

2.4 Reward-to Value-at-Risk (RVaR)

Because in this case portfolio risk is measured based on Value-at-Risk (*VaR*), the size of the Reward to Volatility (*RVAR*) portfolio performance is expanded to *RVaR*. Meaning *RVaR* measures the comparison between portfolio risk premiums and Value-at-Risk (VaR). The mathematical equation of $RVaR$ is:

$$RVaR = \frac{\mu_{wt} - \mu_f}{VaR_{wt}}. \tag{15}$$

Where μ_{wt} mean portfolio return w at time t, μ_f mean of risk-free asset return, and VaR_{wt} Value-at-Risk portfolio w at time t. Measures of good portfolio performance are determined based on the greatest value of $RVaR$ [11].

3. Result and Analysis

The analyzed stock data is accessed through the website http://www.finance.go.id//. The data consists of 10 (ten) selected shares, for the period January 2, 2015 up to June 4, 2018, which includes shares: INDF, DEWA, AALI, LSIP, ASII, TURB, HDMT, BMRI, UNTR, BBRI . Next, it is called sequentially S_1 up to S_{10}. Share prices include the opening price, the highest price, the lowest price, and the closing price, but only the closing price is closed.

3.1 The results of mean and volatility modeling

In this section ten shares are analyzed S_1 up to S_{10}. The analysis starts with calculating the return of each stock, then identifying the long memory effect, estimating the average model and the volatility model.

Identification of long memory effects. To identify long memory effects, it is done by estimating fractional differentiation parameters d_i $(i = 1,...,10)$ in equation (1). Estimates were made using the Gewek and Porter-Hudak methods, with the help of software R. The estimation results obtained fractional differentiation values $\hat{d}_1$ =0.333742; $\hat{d}_7$ = 0.016421 and $\hat{d}_{10}$ = -0.062398. Based on the results of the hypothesis test shows that stocks S_1, S_7 and S_{10} there is a significant long memory effect, with fractional differentiation $\hat{d}_1$, $\hat{d}_7$ and $\hat{d}_{10}$. Whereas other stocks do not have long memory effects. Results of fractional differentiation estimates for stocks S_1 up to S_{10} given in Table-1 column $\hat{d}_i$.

Mean model estimates. In this section Eviews 9 software is used for estimating the mean model. The fractional stock return data will be estimated for the mean model. The first stage is the identification and estimation of the mean model. Identification was carried out through the fractional function (ACF) and partial autocorrelation function (PACF) samples. Based on the ACF and PACF patterns, tentative models are possible for the return data of each stock S_1 up to S_{10} determined. Then the model estimation is done by referring to equation (2). From the model estimation and diagnostic test can be obtained a significant mean model of stocks S_1 up to S_{10}, the results are given in Table-1 below.

Estimation of volatility models. In this section Eviews 9 software is also used to estimate the volatility model. First, the detection of the existence of an element of autoregressive conditional heteroscedasticity (ARCH) against residuals a_{it} $(i = 1,...,10)$ from the mean model. Detection is done using the ARCH-LM test method. Detection results indicate that the calculation value of χ^2 (obs * *R-Square*) on each stock S_1 up to S_{10} generate a probability of 0.0000 or 5% smaller, which means there are elements of ARCH.

Table 1. The results of estimation of mean and volatility models of stocks S_1 up to S_{10}

Stocks (S_i)	Time Series Model	**Frac.Dif.** ($\hat{d}_i$)	**Mean** ($\hat{\mu}_{it}$)	**Variance** ($\hat{\sigma}_{it}^2$)
S_1	ARFIMA(1, $\hat{d}$,0)-GARCH(1,1)	0.333742	0.015399	0.002643
S_2	ARFIMA(2, $\hat{d}$,2)-ARCH(1)-M	0	0.039007	0002797
S_3	ARFIMA(0, $\hat{d}$,1)-GARCH(3,3)	0	0.003315	0.001331
S_4	ARFIMA(1, $\hat{d}$,1)-GARCH(1,1)	0	0.008672	0.001921
S_5	ARFIMA(0, $\hat{d}$,1)-GARCH(1,1)	0	-0.000262	0.001873
S_6	ARFIMA(1, $\hat{d}$,1)-FIGARCH(1,1)	0	0.022085	0.001181
S_7	ARFIMA(1, $\hat{d}$,1)-GARCH(1,1)	0.016421	0.003564	0.001073
S_8	ARFIMA(0, $\hat{d}$,1)-EGARCH(1,1)	0	0.001594	0.001411
S_9	ARFIMA(2, $\hat{d}$,2)-TGARCH(1,1)	0	0.020709	0.013362
S_{10}	ARFIMA(0, $\hat{d}$,1)-GARCH(1,1)	-0.062398	-0.000865	0.001237

Second, identification and estimation of volatility models are carried out. The volatility model used is the generalized autoregressive conditional heteroscedasticity (GARCH) model referring to equation (3). Based on correlogram residual squares a_{it}^2 $(i=1,...,10)$, set a tentative volatility model for each stock S_1 up to S_{10}. Estimation of the volatility model is carried out simultaneously with the average model.

In the volatility modeling process it is also shown that based on the ARCH-LM test, residuals ε_{it} $(i=1,...,10)$ from the volatility model of each stock S_1 up to S_{10} is white noise. The estimation results of the average model and volatility for stocks S_1 up to S_{10} given in Table-1 column of "Time Series Model". Furthermore, the mean and volatility equations are used to estimate values $\hat{\mu}_{it}=\hat{r}_{it}(1)$ and $\hat{\sigma}_{it}^2=\hat{\sigma}_{it}^2(1)$ $(i=1,...,10)$; is a recursive 1-step forward prediction. The results for each stock S_1 up to S_{10} given in Table-1 column of $\hat{\mu}_{it}$ and $\hat{\sigma}_{it}^2$.

3.2 The results of portfolio optimization A and B

In this section analyzes two portfolios, namely portfolios A and B. portfolio of stocks that were analyzed include stocks S_1 up to S_{10}, which is divided into two portfolios. Portfolio A consists of stocks S_1 up to S_5, while portfolio B consists of stocks S_6 up to S_{10}.

Portfolio Optimization A

A portfolio of five stocks, namely stock S_1 up to S_5. Estimator based on the mean stock return S_1 up to S_5, in Table-1 the average vector is arranged as follows:

$$\boldsymbol{\mu}_A^T = (0.015399\ \ 0.039007\ \ 0.003315\ \ 0.008672\ \ -0.000262)$$

Portfolio A consists of five stocks, so the unit vector $\mathbf{e}^T=(1\ \ 1\ \ 1\ \ 1\ \ 1)$. Next, use a volatility estimator in Table-1 and a return covariance estimator between stocks S_1 up to S_5 covariance matrix is formed as:

$$\boldsymbol{\Sigma}_A = \begin{pmatrix} 0.002643 & -2.49868\times10^{-7} & -9.11399\times10^{-9} & -3.79169\times10^{-8} & -3.92689\times10^{-8} \\ -2.49868\times10^{-7} & 0.002797 & 8.90245\times10^{-8} & -1.27711\times10^{-7} & -2.23112\times10^{-8} \\ -9.11399\times10^{-9} & 8.90245\times10^{-8} & 0.001331 & 9.74434\times10^{-8} & 6.30434\times10^{-7} \\ -3.79169\times10^{-8} & -1.27711\times10^{-7} & 9.74434\times10^{-8} & 0.001921 & 7.96921\times10^{-8} \\ -3.92689\times10^{-8} & -2.23112\times10^{-8} & 6.30434\times10^{-7} & 7.96921\times10^{-8} & 0.0001873 \end{pmatrix}$$

Based on the matrix $\boldsymbol{\Sigma}_A$, then the inverse matrix can be calculated $\boldsymbol{\Sigma}_A^{-1}$.

Optimization is carried out based on portfolio problems in equation (11). Next, vector $\boldsymbol{\mu}^T$ and $\mathbf{e}^T$ and matrix $\boldsymbol{\Sigma}_A^{-1}$, used to calculate the optimum weight vector using equation (14). Where risk tolerance τ on condition $\tau \geq 0$ in portfolio optimization here is simulated by taking several values that meet the requirements $\mathbf{e}^T\mathbf{w} = 1$. Assuming short sale is not permitted, taking the risk tolerance value is only for value $0 \leq \tau \leq 0.486$. This is due to the risk tolerance value $\tau > 0.486$ produce a negative weight.

For each risk tolerance value $0 \leq \tau \leq 0.486$ generate portfolio mean return $\hat{\mu}_A$ and the level of risk VaR_A different. Curved lines between pairs $\hat{\mu}_A$ and VaR_A it forms an efficient surface. Where the mean minimum portfolio return is 0.015736 with a minimum VaR of 0.019629, and the mean highest portfolio return is 0.025191 with a maximum VaR of 0.025022.

Ratio between $\hat{\mu}_A$ and VaR_A the biggest is 0.90099912 or obtained when risk tolerance τ =0.486. Ratio between $\hat{\mu}_A$ and VaR_A continue to experience an increase in risk tolerance intervals $0 \leq \tau \leq 0,486$. Based on Mean-VaR portfolio optimization analysis, the optimal portfolio composition of stocks S_1 up to S_5 produce a weight vector $\mathbf{w}^T$ = (0.21701 0.51389 0.09856 0.17038 0.00016). Where the composition of the optimal portfolio produces $\hat{\mu}_A$= 0.025191 with VaR_A= 0.025022 which is also a maximum portfolio.

Portfolio Optimization B

Portfolio B consists of five stocks, namely stocks S_6 up to S_{10}. From the average estimator of stock returns S_6 up to S_{10}, in Table-1 the average vector is arranged as follows:

$$\boldsymbol{\mu}_B^T = (0.022085\ \ 0.003564\ \ 0.001594\ \ 0.020709\ \ -0.000865)$$

Portfolio B consists of five stocks, so the unit vector $\mathbf{e}^T$ = (1 1 1 1 1). Next, from the volatility estimator in Table-1 and the return covariance estimator between stocks S_6 up to S_{10} covariance matrix are formed as follows:

$$\boldsymbol{\Sigma}_B = \begin{pmatrix} 0.001181 & 2.05888\times10^{-7} & 2.63104\times10^{-7} & 1.60945\times10^{-7} & 1.60497\times10^{-7} \\ 2.05888\times10^{-7} & 0.001073 & -2.65181\times10^{-8} & -4.58792\times10^{-8} & -2.38116\times10^{-8} \\ 2.63104\times10^{-7} & -2.65181\times10^{-8} & 0.001411 & 7.18467\times10^{-7} & 7.62223\times10^{-7} \\ 1.60945\times10^{-7} & -4.58792\times10^{-8} & 7.18467\times10^{-7} & 0.0013362 & 6.47969\times10^{-7} \\ 1.60497\times10^{-7} & -2.38116\times10^{-8} & 7.62223\times10^{-7} & 6.47969\times10^{-7} & 0.001233 \end{pmatrix}$$

Based on the matrix $\boldsymbol{\Sigma}_B$, then the inverse matrix can be calculated $\boldsymbol{\Sigma}_B^{-1}$.

The portfolio optimization problem is based on equation (11). Using vectors $\boldsymbol{\mu}^T$ and $\mathbf{e}^T$ and matrix $\boldsymbol{\Sigma}_B^{-1}$, Optimal weight vector is calculated using equation (14). Risk tolerance τ on condition $\tau \geq 0$ in portfolio optimization here is simulated by taking several values that meet the requirements $\mathbf{e}^T\mathbf{w} = 1$. Assuming short sale is not permitted, taking the risk tolerance value is only for value $0 \leq \tau \leq 0.409$. Because for the risk tolerance value $\tau > 0.409$ produce a negative weight.

Each risk tolerance value $0 \leq \tau \leq 0.409$produce $\hat{\mu}_B$ and VaR_B different. Curved lines between pairs $\hat{\mu}_B$ and VaR_B it forms an efficient surface. Where the mean minimum portfolio return is generated $\hat{\mu}_B$ = 0.013436 with a minimum risk level VaR_B= 0.014542. The highest mean return portfolio is 0.018832 with a maximum VaR of 0.017021.

Ratio between $\hat{\mu}_B$ and VaR_B the biggest is 0.9509312 or obtained when risk tolerance τ =0.409. Ratio between $\hat{\mu}_B$ and VaR_B continue to experience an increase in risk tolerance intervals $0 \leq \tau \leq 0{,}409$. Based on Mean-VaR portfolio optimization analysis, the optimal portfolio composition of stocks S_6 up to S_{10} is a stock weight vector $\mathbf{w}^T$ = (0.46850 0.09989 0.04210 0.38933 0.00018). Where the composition of this optimal portfolio produces $\hat{\mu}_B$ = 0.018832 dan VaR_B= 0.017021 which is also $\hat{\mu}_B$ and VaR_B maximum.

3.3 Performance of portfolio A and B

In this section an assessment of the performance of portfolios A and B is carried out, the aim is to determine the performance of each portfolio at various risk tolerances that meet both portfolios. Performance appraisal is carried out referring to equation (15). Calculation $RVaR_A$ for portfolio A and $RVaR_B$ for portfolio B, the results is summarized in the following Table-2. Based on the RVaR values presented in Table 2, it appears that for the risk tolerance value $0 \leq \tau \leq 0.50$ the values $RVaR_A$ greater than values $RVaR_B$.

Table 2. The values of $RVaR_A$ and $RVaR_B$

No	τ	$RVaR_A$	$RVaR_B$
1	0.00	0.666721	0.387433
2	0.05	0.697884	0.405700
3	0.10	0.728374	0.423855
4	0.15	0.757918	0.441876
5	0.20	0.786218	0.459663
6	0.25	0.812914	0.477149
7	0.30	0.835589	0.494258
8	0.35	0.859494	0.510899
9	0.40	0.878353	0.526912
10	0.45	0.893241	0.542234
11	0.50	**0.900935**	0.556655
12	0.55	-	0.570065
13	0.60	-	0.582166
14	**0.658**	-	**0.594192**

For more details, RVaR values in Table-2 can be presented in graphical form as given by Figure 1.

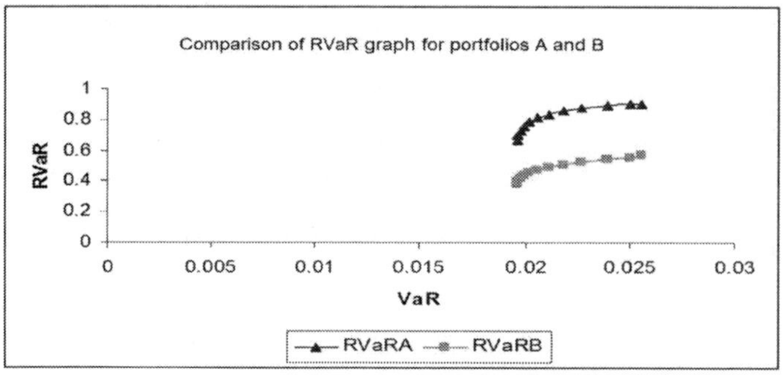

Figure 1. Graph of $RVaR_A$ and $RVaR_B$

In Figure-1 it is also seen that for the risk tolerance value $0 \leq \tau \leq 0.50$ graph of $RVaR_A$ always above than the graphof $RVaR_B$. This situation shows that portfolio A performs better than portfolio B. Therefore, it is advisable for investors to choose portfolio A which consists of stocks S_1, S_2, S_3, S_4 and S_5.

4. Conclusion

In this paper a discussion has been carried out on the expansion of the investment portfolio performance assessment model based on Value-at-Risk using a time series approach. Based on the identification of the long memory effect shows that stock returns S_1, S_7 and S_{10} there is an element of long memory. Average modeling and non-constant volatility indicate that stocks return S_1, S_7 and S_{10} following the ARFIMA-GARCH model; S_2 ARMA-ARCH-M model; S_4, S_4 and S_5 ARMA-GARCH model; S_6 ARMA-FIGARC model; S_8 ARMA-EGARCH model; and S_9 ARMA-TGARCH model. The average model and volatility are used to estimate the average values and variances of each stock. The portfolio A is composed of S_1 up to S_5, whereas portfolio B consists of S_6 up to S_{10}. Portfolio optimization is formed based on the Mean - Value-at-Risk model. Portfolio performance appraisal has been carried out based on the Reward to Value-at-Risk approach, and the results show that portfolio A has better performance than portfolio B. Based on the analysis of investment portfolios which include stocks S_1 up to S_{10}, investors are recommended to choose portfolio A. The limitation of this study is that the comparison of the performance of the investment portfolio is only valid if the risk is measured by Value-at-Risk, and the data follows a time series pattern.

Acknowledgements

Acknowledgments are conveyed to the Rector, Director of Directorate of Research, Community Involvement and Innovation, and the Dean of Faculty of Mathematics and Natural Sciences, Universitas Padjadjaran, with whomthe Internal Grant Program of Universitas Padjadjaran was made possible to fund this research.The grant is a means of enhancing research and publication activities for researchers at Universitas Padjadjaran.

References

[1] Hadi, A.S., Naggar, A.A.E., and Bary, M.N.A. 2016. New Model and Method for Portfolios Selection. Applied Mathematical Sciences, Vol. 10, 2016, no. 6, pp. 263–288. doi: 10.12988/ams.2016.58541.

[2] Soeryana, E., Halim, N.B.A., Sukono, Rusyaman, E., and Supian, S. 2016. Mean-Variance Portfolio Optimization on Some Stocks by Using Non Constant Mean and Volatility Models Approaches. *Proceedings of the International Conference on Industrial Engineering and Operations Management*, Vol. 8-10 March 2016, pp. 3124-3131.

[3] Sukono, Hidayat, Y., Lesmana, E., Putra, A.S., Napitupulu, H., and Supian, S. 2018. Portfolio Optimization by Using Linear Programing Models Based on Genetic Algorithm. *IOP Conf. Series: Materials Science and Engineering*, 300 (2018) 012001 doi:10.1088/1757-899X/300/1/012001.

[4] Napitupulu, H., Sukono, Mohd, I.B., Y Hidayat, Y., and Supian, S. 2018. Steepest Descent Method Implementation on Unconstrained Optimization Problem Using C++ Program. *IOP Conf. Series: Materials Science and Engineering*, 332 (2018) 012024 doi:10.1088/1757-899X/332/1/012024.

[5] Sukono, Susanti, D., Najmia, M., Lesmana, E., Napitupulu, H., Putra, A.S., and Supian, S. 2018. Analysis of Stock Investment Selection Based on CAPM Using Covariance and Genetic Algorithm Approach. *IOP Conf. Series: Materials Science and Engineering*, 332 (2018) 012046 doi:10.1088/1757-899X/332/1/012046.

[6] Hasbullah, E.S., Suyudi, M., Halim, N.B.A., Sukono, Gustaf, F., and Putra, A.S. 2018. A Comparative Study of Three Pillars System and Banking Methods in Accounting Long-Term Purposes of Retiree in Indonesian Saving Account. *IOP Conf. Series: Materials Science and Engineering,* 332 (2018) 012017 doi:10.1088/1757-899X/332/1/012017.

[7] Soeryana, E., Halim, N.B.A., Sukono, Rusyaman, E., and Supian, S. 2017. Mean-Variance Portfolio Optimization by Using Non Constant Mean and Volatility Based on the Negative Exponential Utility Function. *AIP Conference Proceedings* **1827**, 020042 (2017); doi: 10.1063/1.4979458.

[8] Soeryana, E., Halim, N.B.A., Sukono, Rusyaman, E., and Supian, S. 2017. Mean-Variance Portfolio Optimization by Using Time Series Approaches Based on Logarithmic Utility Function. *IOP Conf. Series: Materials Science and Engineering* 166 (2017) 012003 doi:10.1088/1757-899X/166/1/012003.

[9] Sukono, Sidi, P., Susanti, D., and Supian, S. 2017. Quadratic Investment Portfolio Without a Risk Free Asset Based on Value-at-Risk. *Journal of Engineering and Applied Sciences,* 12 (19): 4846-4850.

[10] Balibey, M. and Turkyilmaz, S. 2014. Value-at-Risk Analysis in the Presence of Asymmetry and Long Memory: The Case of Turkish Stock Market. *International Journal of Economics and Financial Issues,* Vol. 4, No. 4, 2014, pp.836-848.

[11] Sukono, Lesmana, E., Susanti, D., and Napitupulu, H. 2017. Estimating the Value-at-Risk for Some Stocks at the Capital Market in Indonesia Based on ARMA-FIGARCH Models. *IOP Conf. Series: Journal of Physics: Conf. Series,* 909 (2017) 012040 doi :10.1088/1742-6596/909/1/012040.

[12] Korkmaz, T., Cevic, E.I. & Ozatac, N. (2009). Testing for Long Memory in ISE Using ARFIMA-FIGARCH Model and Structural Break Test. *International Research Journal of Finance and Economics.* Issue 26, 1450-2887.

[13] Turkyilmaz, S. and Balibey, M. 2014. Long Memory Behavior in the Returns of Pakistan Stock Market: ARFIMA-FIGARCH Models. *International Journal of Economics and Financial Issues,* Vol. 4, No. 2, 2014, 400-410.

[14] Tsay, R.S. 2005. *Analysis of Financial Time Series.* Second Edition. Hoboken, New Jersey: John Wiley & Sons, Inc.

[15] Sukono, Hidayat, Y., Suhartono, Sutijo, B., Bon, A.T., and Supian, S. 2016. Indonesian Financial Data Modeling and Forecasting By Using Econometrics Time Series and Neural Network. *Global Journal of Pure and Applied Mathematics.*ISSN 0973-1768 Volume 12, Number 4 (2016), pp. 3745-3757.

[16] Stancu, S. Empirical Results of Modeling EUR/RON Exchange Rate Using ARCH, GARCH, EGARCH, TARCH and PARCH Models. *Romanian Statistical Review,* 1/2017, pp. 57-72.

IOP Conf. Series: Materials Science and Engineering **567** (2019) 012016 doi:10.1088/1757-899X/567/1/012016

Book embedding of 3-crossing-critical graphs with rational average degree between 3.5 and 4

B Pinontoan[1], J Titaley[1], C E J C Montolalu[1]

Mathematics Department, Sam Ratulangi University, Manado, INDONESIA

Corresponding author: bpinonto@unsrat.ac.id

Abstract. We consider the graphs $P(m, n)$ with average degree between 3.5 and 4, made up of m identical pieces together with n identical pieces glued together in a circular fashion, such that in any drawing of $P(m, n)$ on the plane, there are exactly three pairwise edges crossings, and when deleting any edge of the graph, the number of crossings of the remaining graph decreases. We then embed $P(m, n)$ into a book such that the vertices are put on a line called the spine and the edges are put on half-planes called the pages, which have the spine as their common boundary, without crossings. In this paper, we show that the minimal number of pages needed to embed $P(m, n)$ into a book is three.

1. Introduction

A graph G is a pair (N, E) where N is a non-empty set of vertices and E is a set of edges. We consider only simple graphs, i.e. without parallel edges and loops. The *degree* of a vertex v is the number of edges incident with v. A path from v_1 to v_n in graph G is an alternating vertex and edge sequence $v_1\ e_1\ v_2\ e_2\ v_3\ e_3 \ldots e_n\ v_n$, where edge $e_k = v_k v_{k+1}$ $(1 \leq k \leq n)$. Since we only consider simple graphs, we denote path from v_1 to v_n only by sequence of vertices $v_1 v_2 v_3 \ldots v_n$,. Paths P_1 and P_2 are disjoint if there is no intersection between P_1 and P_2. The *average degree* of a graph G is the sum of degrees of all vertices of G divided by the number of vertices of G. A drawing of a graph G in a plane is a mapping of G to the plane such that vertices are mapped to distinct points of the plane and edges are mapped into arcs connecting the corresponding points. A crossing in a drawing of graph G is a point in the plane not corresponding with a vertex of G, where two arcs intersect. The *crossing number* $cr(G)$ of graph G is the minimal number of crossings among all possible drawings of graph G in the plane. The crossing number problem was introduced by Tutte [1]. A graph G with $cr(G) = k$ is *k-crossing-critical* if deleting any edge e of G makes $cr(G\text{-}e) < k$. Recent studies on crossing-critical graphs are conducted in [2], [3], and [4].

The *book with k* pages is the topological space B_k that consists of a line called the *spine* and k half-planes called the *pages*, all having the spine as their common boundary. A *k-page* book embedding of a graph G is an embedding of G into book in which the vertices are on the spine, and each edge is contained in one page without crossing. Given a graph G, the minimum k such that G can be embedded in a *k-page* book is the *pagenumber* $\rho(G)$ of graph G [5]. Figure 1(a) shows the complete bipartite graph $K_{3,3}$ and Figure 1(b) shows its embedding into 3-page book.

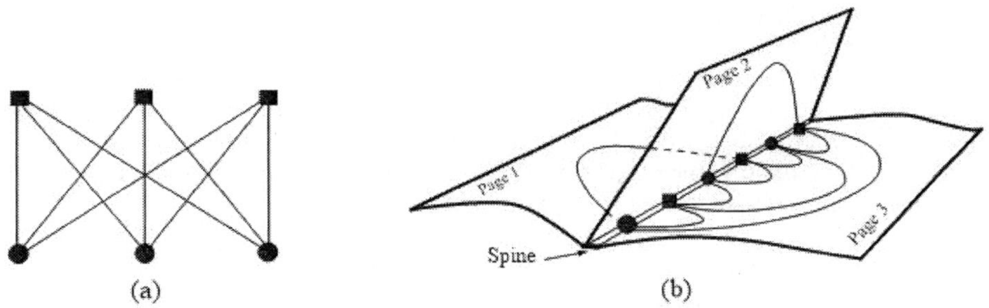

Figure 1. Complete bipartite graph $K_{3,3}$ and its embbeding into 3-page book.

The pagenumber problem was introduced by Kainen [6] and attracts attentions since Chung, Leighton and Rosenberg [7] have pointed out its applications to Very Large Scale Integration (VLSI) design. In general, given a graph *G*, it is not easy to determine the pagenumber of *G*; the pagenumber problem is NP-Complete [7].

Pinontoan and Richter [8] introduced the notion of *tile* which is a small piece of graph with certain conditions and can be glued together in a circular fashion to build an infinite family of graphs. They gave as example the infinite family of the $\binom{2h+3}{2}$-crossing-critical graphs, with positive integer *h*, having rational average degree $r \in (3.5, 4)$. In this paper, we set $h = 0$ and get 3-crossing-critical graphs $P(m, n)$, for $0 \leq m, n$ and $3 \leq m + n$, which are made up by gluing of *m* copies of a tile and *n* copies of other tile, where one of the tile is twisted, in a circular fashion. We embed $P(m, n)$ into book and show that $\rho(P(m, n)) = 3$.

2. Infinite Family of 3-crossing-critical graphs with rational average degree between 3.5 and 4.

In this section we recall the notion of *tile* which was introduced by Pinontoan and Richter [8] and define the infinite family of 3-crossing-critical graphs having rational average degree between 3.5 and 4.

2.1. Tiles

A *tile* is a 3-tuple (G, L, R) where *G* is a connected graph, the *left-wall L* and the *right-wall R* are finite sequence of vertices of *G*. A *tile drawing* of tile $T = (G, L, R)$ is the drawing of *G* on the unit square $[0, 1] \times [0, 1]$ such that the vertices of *L* occur in the line $\{0\} \times [0, 1]$ in the decreasing order, the vertices of *R* occur in the line $\{1\} \times [0, 1]$ also in the decreasing order, and only the vertices $L \cup R$ occur on the boundary $\{0, 1\} \times [0, 1]$. The *tile crossing number tcr*(*T*) of tile *T* is the smallest number of crossings in any tile drawing of *T*. A tile *T* is *planar* if $tcr(T) = 0$, otherwise non-planar. The *twist T'* of $T = (G, L, R)$ is the tile obtained by reversing the order of *R*.

Figure 2 shows the tile $S = ((\{a, b, c, d, e, f\}, \{ab, ae, bc, bd, cf, de, df\}), (a, c), (e, f))$, the tile $T = ((\{x, y, z, u, v\}, \{xz, xu, yz, uz, vz, yv\}), (x, y), (u, v))$, the twist $S' = ((\{a, b, c, d, e, f\}, \{ab, ae, bc, bd, cf, de, df\}), (a, c), (f, e))$ of *S*, and the twist $T' = ((\{x, y, z, u, v\}, \{xz, xu, yz, uz, vz, yv\}), (x, y), (v, u))$ of *T*. Note that the tiles *S* and *T* are planar, whereas their twists *S'* and *T'* respectively are not planar with tile crossing number $tcr(S') = 3$ and $tcr(S') = 3$.

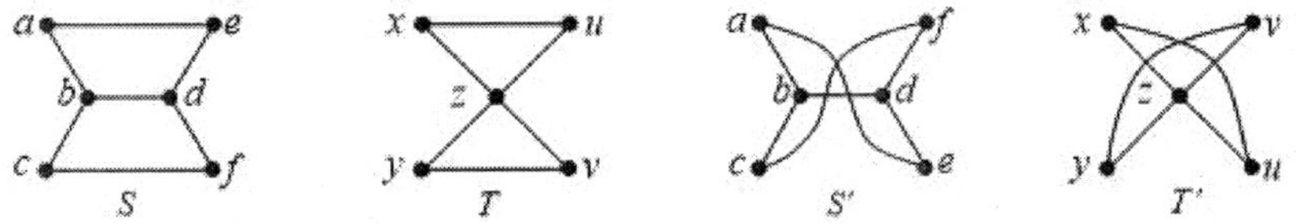

Figure 2. Tiles *S* and *T* together with their twists *S'* and *T'*.

A tile $S = (G_1, L_1, R_1)$ is compatible with tile $T = (G_2, L_2, R_2)$ if $|R_1| = |L_2|$. If $S = (G_1, L_1, R_1)$ is compatible with $T = (G_2, L_2, R_2)$, then *catenation* of tiles S and T is $ST = (G_1 \cup G_2, L_1, R_2)$ where R_1 is identified with L_2. A tile is self-*compatible* if it is compatible to itself. If T is a *self-compatible* tile, then T^n is the catenation of n copies of T in a linear fashion and $o(T^n)$ is the catenation of n copies of T where the last and the first copy of T also put in catenation to get a circular fashion of arrangement. Let T be a self-compatible tile, then $\otimes(T^n) = o(T^{n-1}T')$, the *twist circular* catenation of n copies of T. Note that T^n is a tile, whereas $o(T^n)$ and $\otimes(T^n)$ are graphs.

Figure 3 shows the catenation ST of tiles S and T where the vertices e and f of S are identified with respectively the vertices x and y of T, and catenation $ST'T$ of tiles S, twist T' of T, and tile T. In Figure 3, we can see that there are three pair-disjoint paths in $ST'T$, namely {*aeyv*, *cfxu*}, {*aeyv*, *cbdfz'xzu*}, and {*abdez'xzu*, *cfxu*}.

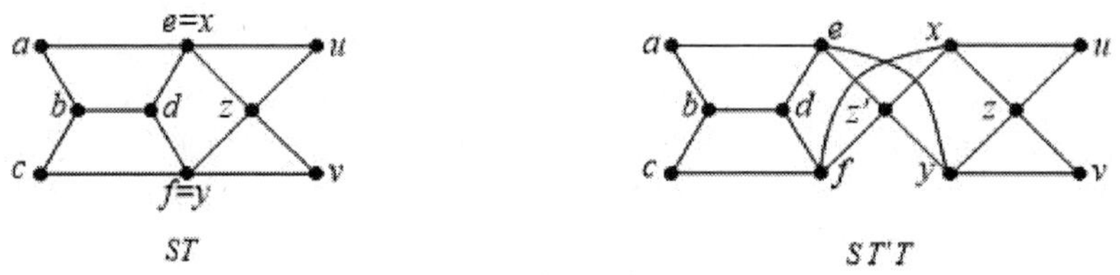

Figure 3. Catenations ST and $ST'T$.

2.2. The infinite family P(m, n)

Pinontoan and Richter [8] defined an infinite family of the $\binom{2h+3}{2}$-crossing-critical graphs, with h is positive integer. We set h to zero and define, for $0 \leq m, n$ and $3 \leq m + n$, the graph $P(m, n) = \otimes(S^mT^n)$ where S and T are the tiles in Figure 2, i.e. the twist circular catenation of m copies of tile S and n copies of tiles n. More precisely, for $0 \leq m, n$ and $3 \leq m + n$, the graph $P(m, n)$ has vertices $a_1, b_1, c_1, d_1, a_2, b_2, c_2, d_2, \ldots, a_m, b_m, c_m, {}_m, x_1, y_1, z_1, x_2, y_2, z_2, \ldots, x_n, y_n$, and z_n, and the edges a_kb_k, b_kc_k, b_kd_k $(1 \leq k \leq m)$, $a_ka_{k+1}, c_kc_{k+1}, d_ka_{k+1}, d_kc_{k+1}$ $(1 \leq k \leq m-1)$, x_kz_k, y_kz_k $(1 \leq k \leq n)$, $x_kx_{k+1}, y_ky_{k+1}, z_kx_{k+1}, z_ky_{k+1}$ $(1 \leq k \leq n-1)$, d_mx_1, d_my_1, x_nc_1, and y_ma_1. The twist can be put as S' the twist of S or T' the twist of T as pictured in Figure 2.

Figure 4 shows $P(2, 4)$ with the twist S' of S and with the twist T' of T.

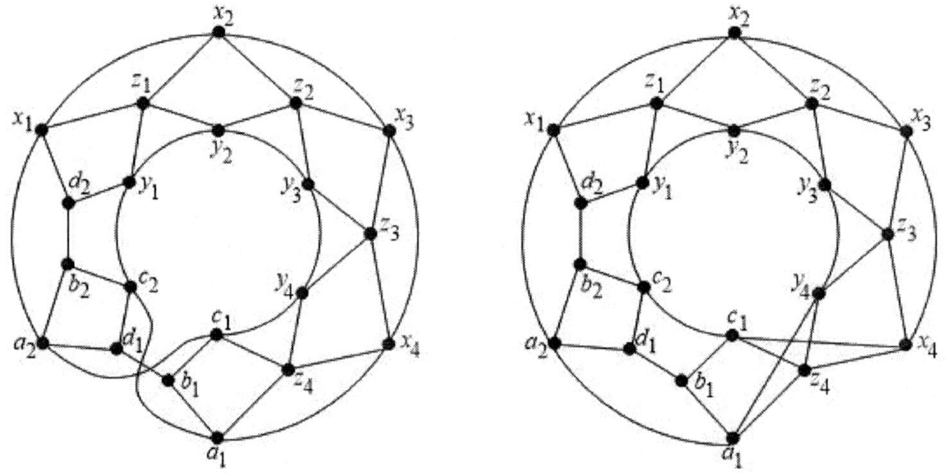

Figure 4. Graph $P(2, 4)$ with twist S' and with twist T'.

IOP Conf. Series: Materials Science and Engineering **567** (2019) 012016 doi:10.1088/1757-899X/567/1/012016

2.3. P(m, n) is 3-crossing-critical with rational average degree between 3.5 and 4

As shown in Subsection 2.1, there are three pair-disjoint paths in *ST'T*. Similarly, there are three pair-disjoint paths in *SS'S*, *TT'T*, and *TS'S*. Hence, using the tile mechanism for calculation crossing numbers established in [8], we conclude that $cr(P(m, n) = 3$.

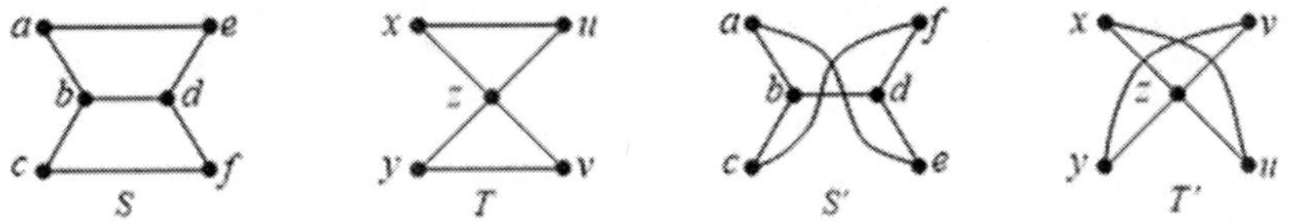

Consider the *S'* of tile *S* in Figure 2. Obviously, deleting the edges *ae*, *cf*, end *bd* decreases the number of crossings. If the edge *ab* is deleted, then the edge *cf* can be rerouted only to cross *ab* and does not cross *bd*, and hence there are only two crossings. Similarly, by symmetry, deleting other edges of *S'* will decrease the number of crossings. Now consider *T'*. Obviously deleting any of the edges involved in the crossings will decrease the number of crossings. The edges *yz* and *uz* are in symmetry with the egdes *xz* and *vz* and hence deleting any of these edges will also decrease the number of crossings. Thus, deleting any edge in $P(m, n)$ will decrease $cr(P(m, n))$. Therefore, $P(m, n)$ is 3-crossing-critical graph.

The average degree of $P(m, n)$ is

$$\frac{4\big(2m+3(n+1)\big)+3(2m)}{4m+3n+3}=\frac{14m+12n+12}{4m+3n+3}.$$

Given any rational number $r \in (3.5, 4)$, we have $4rm + 3rn + 3r = 14m + 12n + 12$ which is equivalent to

$$m=\frac{(3n+3)(4-r)}{4r-14}.$$

Hence, for any rational number $r \in (3.5, 4)$, there are many choices for n to get integer m. Therefore, $P(m, n)$ is 3-crossing-critical graph with rational average degree in the interval (3.5, 4).

3. Book Embeddings of *P*(*m*, *n*)

In this section, we prove the main result.

Theorem. Let $0 \le m, n$ and $3 \le m + n$. Then $\rho(P(m, n)) = 3$.

Proof. Let $0 \le m, n$ and $3 \le m + n$. We showed that $cr(P(m, n) = 3$ and so $P(m, n)$ is not planar and hence it needs at least three pages to embed into a book. Thus $\rho(P(m, n)) \ge 3$.

Now we show that $\rho(P(m, n)) \le 3$ by drawing the embedding of $P(m, n)$ into a 3-page book, as follows. Put on the page 1 the following edges: a_kb_k, b_kc_k, b_kd_k $(1 \le k \le m)$, a_ka_{k+1}, d_ka_{k+1} $(1 \le k \le m-1)$, x_kz_k, y_kz_k $(1 \le k \le n)$, x_kx_{k+1}, z_kx_{k+1} $(1 \le k \le n-1)$, a_mx_1, and d_mx_1. On page 2, put the following edges c_kc_{k+1}, d_kc_{k+1} $(1 \le k \le m-1)$, y_ky_{k+1}, z_ky_{k+1} $(1 \le k \le n-1)$, c_my_1, d_my_1, c_1z_n, and a_1z_n. Finally, put the edges c_1x_{n+1} and a_1y_{n+1} on page 3. Therefore $\rho(P(m, n)) = 3$.

Figure 5 shows a 3-page book embedding of graph $P(2, 4)$.

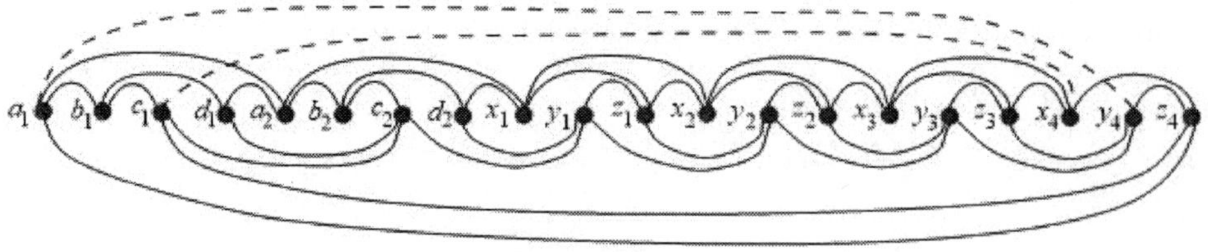

Figure 5. Book embedding of $P(2, 4)$.

Acknowledgments
This research was funded by Sam Ratulangi University research grant SK No. 897/UN12/LL2018 and Contact No. 1863/UN12.13/LT/2018.

References

[1] Tutte W T 1970 Toward a Theory of Crossing Numbers. *J. Comb. Th.* **8** 45-53.
[2] Beaudou L, Hernández–Véelez C and Salazar G 2013 Making a graph crossing-critical by multiplying its edges, *Electronic J. of Combinatorics* **20**, Research Paper 61, 14
[3] Bokal B, Oporowski B, Richter R B, and Salazar G 2016 Characterizing 2-crossing-critical graphs. *Advances in Applied Mathematics* **74** 23-208.
[4] Dvořák Z, Hliněný P and Mohar B 2018 Structure and generation of crossing-critical graphs. *Submitted.*
[5] Balogh J and Salazar G 2015 Book embeddings of regular graphs. *SIAM J. on Discrete Mathematics* **29** 811-22.
[6] Kainen P C 1974 An introduction to topological graph theory, in *Graphs and Combinatorics*, Ed R Bari and F Harary (New York: Springer) 76-108.
[7] Chung F R T, Leighton F T and Rosenberg A L 1987 Embedding graphs in books: a layout problem with applications to VLSI design, *SIAM J. on Algebraic and Discrete Methods* **8** 33-58.
[8] Pinontoan B and Richter R B 2003 Crossing numbers of sequences of graphs II: planar tiles. *J. Graph Theory* **42** 332-41.

On model of supply chain using max-plus algebra

D P W Putra[1]

[1]Department of Mathematics Education, Universitas Sanata Dharma, INDONESIA.

Corresponding author: dewa@usd.ac.id

Abstract. Supply chain scheduling is used to manage the stability of the distribution of a commodity. In this paper, we will discuss the modelling of rice commodity supply chains using max-plus algebra. The rice distribution system data was obtained from the Yogyakarta Regional Logistics Bureau (Bulog) and its subsidiary company PT JPL. We get the supply chain model in the form of a matrix over max-plus algebra.

1. Introduction

Indonesia has been known as agricultural country. Rice is most important agricultural commodity. The demand of rice must be fulfilled by the supply of rice.Chain between supply-distribution-demand of rice has to stable. Supply chain is defined as a group of inter-connected participating companies that add value to a stream of transformed inputs from their source of origin to the end products or services that are demanded by designated end-customers [1]. There are four intrinsic flow of supply chain, i.e. material flow, infomation flow, finace flow, and comercial flow.

Indonesian Bureau of Logistics (Bulog) is a company which distibute rice in Indonesia. Bulog has 26 regional division who all of them has the same task to keep stability of rice commodity in Indonesia. Bulog was helped by it's subsidiary company, PT Jasa Prima Logistik (JPL) to distribute the commodity to people (customers). Bulog as a supplier and JPL as distributor cooperate to make the price of these commodity relatively stable. Bulog and JPL must make sure the distribution of the commodities are effective and efficient. Bulog made a schedule to distribute the commodity based on the demand of the customers. JPL help Bulog to prepare trucks to loading the commodity.

In [2][3][4] supply chain model of rice use causality diagram of supply chain, linear programming and LINDO, and goal programming model respectively. All the model use social variable approach and linear quantification.

Max-plus algebra is the system of algebra with two operation maximization and addition. The structure of this system is semiring [5]. Some work of tranportation such as scheduling at bus station, railway station, and airport using max-plus algebra [5][6]. Max-plus algebra also use to model supply chain of oil feedstock between two fuel terminal using tanker [6][7].

In this paper we will discuss of model supply chain of rice that distribute from Bulog and JPL using max-plus algebra.

2. Pleminaries

Definition 2.1 Max-Plus Algebra [2]

Define a set $\mathbb{R}_{max} = \mathbb{R} \cup \{\varepsilon\}$, *where* $\mathbb{R}$ *is the set of real number and* $\varepsilon := -\infty$. *For every* $a, b \in \mathbb{R}_{max}$ *define operations* $\oplus$ and $\otimes$ *by*

$$a \oplus b := \max\{a, b\} \ \ and \ \ a \otimes b := a + b$$

The set $\mathbb{R}_{max}$ *with operations* $\oplus$ and $\otimes$ *is called Max-Plus Algebra and denoted by* $\Re = (\mathbb{R}_{max}, \oplus, \otimes)$. Clearly, $a \oplus \varepsilon = a$ and $a \oplus 0 = a$, for all $a \in \mathbb{R}_{max}$. The structure of $\Re$ is semiring with unity. Unit element in $\Re$ is ε and neutral element is $e = 0$. Furthermore, $\Re$ is commutative semiring.

Definition 2.2 Supremum on Max-Plus Algebra
If set $\{a_i : i \in \mathbb{R}\}$ is countable set, with $a_i \in \mathbb{R}_{max}$ then

$$\bigoplus_{i\in\mathbb{R}} a_i = \bigoplus_{i=1}^{\infty} a_i = \sup_{i\in\mathbb{R}} a_i.$$

Definition 2.3 Power on Max-Plus Algebra
Given $a \in \mathbb{R}_{max}$ *and* $n \in \mathbb{R}$, *where* $\mathbb{R}$ *is the set of natural number, and define*

$$a^{\otimes n} := \underbrace{a + a + a + \cdots + a}_{n \text{ times}}$$

If $n = 0$, *we define* $a^{\otimes n} = e$.
We will generalize definition 2.2 for the power of $a \in \mathbb{R}_{max}$ to arbitrary $x \in \mathbb{R}$. We have,

$$a^{\otimes x} = xa$$

for all $x \in \mathbb{R}$.

Next, we will introduce matrix over max-plus algebra and basic operation on these matrix. We denote the set of all matrix $m \times n$ over max-plus algebra by $\mathbb{R}_{max}^{m\times n}$. The element of matrix $M \in \mathbb{R}_{max}^{m\times n}$ can be written as,

$$M = \begin{bmatrix} a_{11} & a_{12} & \cdots & a_{1n} \\ a_{21} & a_{22} & \cdots & a_{2n} \\ \vdots & \cdots & \ddots & \vdots \\ a_{m1} & a_{m2} & \cdots & a_{mn} \end{bmatrix}$$

where $a_{ij} \in \mathbb{R}_{max}$ for $0 < i \le m$ and $0 < j \le n$.
The sum of two matrices $A, B \in \mathbb{R}_{max}^{m\times n}$ denoted by $A \oplus B$, is defined by

$$[A \oplus B]_{ij} = a_{ij} \oplus b_{ij} = \max(a_{ij}, b_{ij})$$

and for the $A \in \mathbb{R}_{max}^{m\times p}$ and $B \in \mathbb{R}_{max}^{p\times n}$ product of two matrices $A \otimes B$ is defined by

$$[A \otimes B]_{ij} = \bigoplus_{j=1}^{p} a_{ij} \otimes b_{jk} = \max_{j\in p}\{a_{ij} + b_{jk}\}$$

The zero matrix E is the matrix whose has entry $[\mathrm{E}]_{ij} = \varepsilon$ for all i and j. The identity matrix $I \in \mathbb{R}_{max}^{n\times n}$ is the matrix whose has entry

$$[I]_{ij} = \begin{cases} 0, & i = j \\ \varepsilon, & i \ne j \end{cases}.$$

Clearly, the structure $(\mathbb{R}_{max}^{n\times n}, \oplus, \otimes)$ is semiring, with zero element E and unity I. The nth power of matrix $A \in \mathbb{R}_{max}^{m\times n}$ is defined by

$$A^{\otimes n} = \underbrace{A \otimes A \otimes \cdots \otimes A}_{n \text{ times}}$$

If $n = 0$ then $A^{\otimes n} = I^{n\times n}$.

Base on definition 2.2 and definition of the power of matrix, we will define supremum of power of matrix by

$$A^{+} = \bigoplus_{i=1}^{\infty} A^{\otimes i}$$

The entry of A^{+} is $\left[A^{+}\right]_{ij} = \sup\left\{\left[A^{k}\right]_{ij} : k \geq 0\right\}$.

Structure of max-plus algebra is semiring, which not always has an invers respect to addition. It means that we will find some problem to solve equation of linear equation $A \otimes x = b$. The solution of this equation does not always exists. But, inequality $A \otimes x \leq b$ always has solution. We will find out the greatest solution for vector x such that satisfy the inequality $A \otimes x \leq b$. This solution is called principal solution and denoted by $x^*(A,b)$. The principal solution of this inequality is given the by the theorem bellow.

Theorem 2.4
Suppose $A \in \mathbb{R}_{max}^{m \times n}$ is the matrix over max-plus algebra which every column contains at least one element does not equal to ε and $b \in \mathbb{R}_{max}^{m}$, then

$$\left[x^*(A,b)\right]_j = \min\left\{b_i - a_{ij} : i \in m \text{ and } a_{ij} > \varepsilon\right\}.$$

Theorem 2.4 has not given the solution of equation $A \otimes x = b$ but only the principal solution for inequality $A \otimes x \leq b$. The next lemma will give relationship between the solution existence of linear equation and the principal solution.

Lemma 2.5
If equation $A \otimes x = b$ has solution then the principal solution of $A \otimes x \leq b$ is the solution of $A \otimes x = b$.

For example, we have system of linear equation with two variables over max-plus algebra

$$\begin{cases} 3 \otimes x_1 \oplus 1 \otimes x_2 = 2 \\ 0 \otimes x_1 \oplus 4 \otimes x_2 = 5 \end{cases}$$

Matrix form of this system is $A \otimes x = b$, where $A = \begin{bmatrix} 3 & 1 \\ 0 & 4 \end{bmatrix}$, $x = \begin{bmatrix} x_1 \\ x_2 \end{bmatrix}$, and $b = \begin{bmatrix} 2 \\ 5 \end{bmatrix}$.

We have the principal solution for $A \otimes x \leq b$ is $x^*(A,b) = \begin{bmatrix} -1 \\ 1 \end{bmatrix}$. The principal solution also satisfy system $A \otimes x = b$. Hence, the solution of this system of linear equation is $x_1 = -1$ and $x_2 = 1$.

3. Method

The supply chain that will be modeled is supply chain of rice commodity in Yogyakarta, Indonesia. Data supply chain distribution was obtained by the Head of Supplier, Operations and Public Services Division of Yogyakarta Regional Division through interviews. Data related to consumer demand for rice is obtained from JPL's rice distribution realization data in June 2018. We conducted a literature study to develop mathematical models using max-plus algebra from the rice distribution supply chain in Yogyakarta.

IOP Conf. Series: Materials Science and Engineering **567** (2019) 012017 doi:10.1088/1757-899X/567/1/012017

4. Result and Discussion

Supply chain of rice in Bulog Divre Yogyakarta as follow the figure.

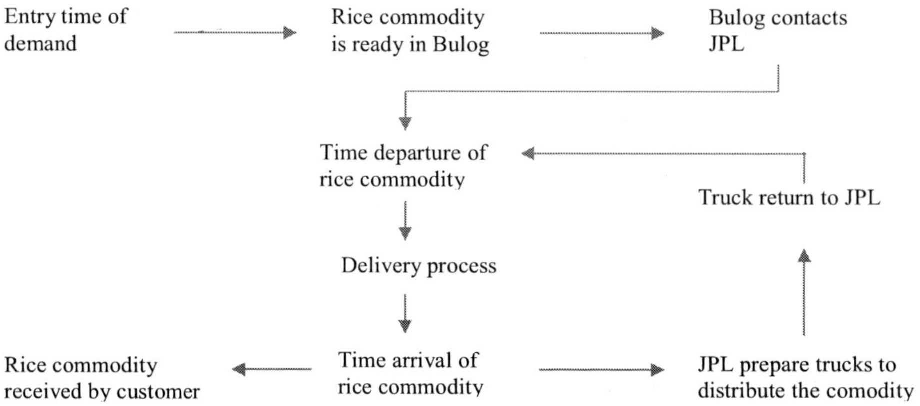

Figure 1. Supply Chain of Rice in Bulog Divre Yogyakarta

If n is the number of truck then system of equation in max-plus algebra as follows.

$$\begin{cases} t_1(k) = p_r \otimes t_2(k-n) \otimes p_w \oplus u(k) \\ t_2(k) = p_d \otimes t_1(k) \\ y(k) = t_2(k) \end{cases} \tag{1}$$

with $t_1(k)$ and $t_2(k)$ are the time of rice departure from Bulog and the time of rice arrive to customer. Time spent by the truck to delivery is $p_d(k)$ and return to Bulog is $p_r(k)$. JPL need time to prepare the trucks $p_w(k)$. Finally, $u(k)$ and $y(k)$ are entry time of rice demand and time of rice supply arrive to customers. The model of equation system (1), in the matrix form as follows

$$\vec{t}(k) = A_0 \otimes \vec{t}(k) \oplus A_1 \otimes \vec{t}(k-n) \oplus B_0 \otimes u(k) \tag{2}$$

$$y(k) = C \otimes \vec{t}(k) \tag{3}$$

where,

$$\vec{t}(k) = \begin{bmatrix} t_1(k) \\ t_2(k) \end{bmatrix},\ A_0 = \begin{bmatrix} \varepsilon & \varepsilon \\ p_d & \varepsilon \end{bmatrix}, A_1 = \begin{bmatrix} \varepsilon & p_r \otimes p_w \\ \varepsilon & \varepsilon \end{bmatrix},\ B_0 = \begin{bmatrix} e \\ \varepsilon \end{bmatrix}, \text{ and } C = [\varepsilon \ \ e].$$

We substitute $\vec{t}(k)$ in equation (2) with $\vec{t}(k)$ in left hand side, and we have

$$\vec{t}(k) = A_0 \otimes \left[A_0 \otimes \vec{t}(k) \oplus A_1 \otimes \vec{t}(k-n) \otimes B_0 \otimes u(k) \right] \otimes A_1 \otimes \vec{t}(k-n) \otimes B_0 \otimes u(k)$$

$$= A_0^{\otimes 2} \otimes \vec{t}(k) \otimes \left(A_0 \otimes A_1 \otimes \vec{t}(k-n) \oplus A_1 \otimes \vec{t}(k-n) \right) \oplus \left(A_0 \otimes B_0 \otimes u(k) \oplus B_0 \otimes u(k) \right) \tag{4}$$

Next, we substitute again $\vec{t}(k)$ in equation (4) with $\vec{t}(k)$ in left hand side of equation (2) and we have

$$\vec{t}(k) = A_0^{\otimes 3} \otimes \vec{t}(k) \oplus \bigoplus_{i=0}^{2} A_0 \otimes A_1 \otimes \vec{t}(k-n) \oplus \bigoplus_{i=0}^{2} A_0 \otimes B_0 \otimes u(k) \tag{5}$$

Analog, we substitute $\vec{t}(k)$ for n times for $n \to \infty$, and we have

$$\vec{t}(k) = A_0^{\otimes n} \otimes \vec{t}(k) \oplus \bigoplus_{i=0}^{n-1} A_0 \otimes A_1 \otimes \vec{t}(k-n) \oplus \bigoplus_{i=0}^{n-1} A_0 \otimes B_0 \otimes u(k) \tag{6}$$

We have $A_0^{\otimes i} = E$ for $i \geq 2$ and

$$\bigoplus_{i=0}^{n-1} A_0 \otimes A_1 \otimes \vec{t}(k-n) = I \oplus \left(\bigoplus_{i=1}^{\infty} A_0^{\otimes i} \otimes A_1 \right) \vec{t}(k-n) \text{ and}$$

$$\bigoplus_{i=0}^{n-1} A_0 \otimes B_0 \otimes u(k) = I \oplus \left(\bigoplus_{i=1}^{\infty} A_0 \otimes B_0 \right) u(k)$$

Hence, we can simplify equation (2) became

$$\vec{t}(k) = A \otimes \vec{t}(k-n) \oplus B \otimes u(k), \tag{7}$$

where $A = \left[\bigoplus_{i=0}^{\infty} A_0^{\otimes i} \right] \otimes A_1$, $B = \left[\bigoplus_{i=0}^{\infty} A_0^{\otimes i} \right] \otimes B_0$

We will determine matrix form of equation (3) by substitute $\vec{t}(k)$ in equation (7). Now, we have

$$y(k) = C \otimes \left(A \otimes \vec{t}(k-n) \oplus B \otimes u(k) \right) \tag{8}$$

Based on equation (7), we can conclude $\vec{t}(k-\alpha n) = A \otimes \vec{t}(k-(\alpha+1)n) \oplus B \otimes u(k-\alpha n)$ with $\alpha \in \cup\{0\}$. We repeat the substitute until n times for $n \to \infty$ and we have

$$y(k) = C \otimes A^{\otimes \alpha} \otimes \vec{t}(k-\alpha n) \oplus \left(\bigoplus_{i=0}^{\infty} C \otimes A^{\otimes i} \otimes B \otimes u(k-in) \right) \tag{9}$$

Then $y(k) = \bigoplus_{i=0}^{\infty} C \otimes A^{\otimes i} \otimes B \otimes u(k-in)$, since $k - \alpha n < n$.

$$Y(k) = H \otimes U(k) \tag{10}$$

where H is lower triangular matrix with $[H]_{ij} = CA^{i-j}B$. Equation (7) and (10) are the models of supply chain of rice commodity using max-plus algebra.

Simulation Model

We used data of rice distribution in district of Kulon Progo, Yogyakarta on June 2018 to simulate the model of supply chain using Max-Plus Algebra.

The truck can carry maximum loads about 6,000 kg. For the simulation, we used the schedule of distribution on 21 June 2018, which is rice distributed to Pengasih and Kokap. Total demand was 95,970 kg. Truck spent 2.5 hour from Bulog to Pengasih and took 2 hour from Pengasih to Bulog. JPL need an hour to prepare the truck before used everyday. We estimated JPL need 8 truck to distribute the rice.

Table 1. Demand and Schedule Distribution of rice in Kulon Progo

No	Sub-District of Kulon Progo	Total of Demand (Kg)	Schedule of Distribution
1	Pengasih	52,480	21 June 2018
2	Kokap	43,490	
3	Panjatan	37,860	25 June 2018
4	Sentolo	57,140	
5	Girimulyo	28,100	26 June 2018
6	Kalibawang	39,820	
7	Wates	36,100	
8	Samigaluh	37,570	27 June 2018
9	Nanggulan	36,350	
10	Temon	23,920	
11	Galur	32,620	28 June 2018
12	Lendah	47,760	
	Total	473,210	

(Source Bulog Divre Yogyakarta June 2018)

Table 2. Simulation Model on Schedule Distribution of Rice on 21 June 2018

Truck	Sub-District	Schedule Distribution	Vol (K Kg)	Time Delivery (hours)	Time Return (hours)	Prepare in JPL (hours)	Time Departure	Time Arrive	Decision
1	Pengasih	21-Jun-18	6	2.5	2	1	7	10.5	On-time
2	Pengasih	21-Jun-18	6	2.5	2	1	7	10.5	On-time
3	Pengasih	21-Jun-18	6	2.5	2	1	7	10.5	On-time
4	Pengasih	21-Jun-18	6	2.5	2	1	7	10.5	On-time
5	Pengasih	21-Jun-18	6	2.5	2	1	7	10.5	On-time
6	Pengasih	21-Jun-18	6	2.5	2	1	7	10.5	On-time
7	Pengasih	21-Jun-18	6	2.5	2	1	7	10.5	On-time
8	Pengasih	21-Jun-18	6	2.5	2	1	7	10.5	On-time
1	Pengasih & Kokap	21-Jun-18	5.99	3	2.2	0	10.5	15.5	On-time
2	Kokap	21-Jun-18	6	2.5	2	0	10.5	15	On-time
3	Kokap	21-Jun-18	6	2.5	2	0	10.5	15	On-time
4	Kokap	21-Jun-18	6	2.5	2	0	10.5	15	On-time
5	Kokap	21-Jun-18	6	2.5	2	0	10.5	15	On-time
6	Kokap	21-Jun-18	6	2.5	2	0	10.5	15	On-time
7	Kokap	21-Jun-18	5.99	2.5	2	0	10.5	15	On-time
8	Kokap	21-Jun-18	5.99	2.5	2	0	10.5	15	On-time

References

[1] Karo, Natalia Br, Analisis Optimasi Distribusi Beras Bulog di Provinsi Jawa Barat, *Jurnal OE*, Vol. VII, No. 3, 2015, pp. 252-270.

[2] Garside, Annisa Kesy and Hasyim Yusuf Asjari, Simulasi Ketersediaan Beras di Jawa Timur, *Jurnal Ilmiah Teknik Industri*, Vol. 14 No. 1, 2015, pp. 47-58.

[3] Karo, Natalia Br, Analisis Optimasi Distribusi Beras Bulog di Provinsi Jawa Barat, *Jurnal OE*, Vol. VII, No. 3, 2015, pp. 252-270.

IOP Conf. Series: Materials Science and Engineering **567** (2019) 012017 doi:10.1088/1757-899X/567/1/012017

[4] Trisilawaty, Cory, Marimin, Noer Azam Achsani, Analisis Optimasi Rantai Pasok Beras dan Penggunaan Gudang di Perum Bulog Divre DKI Jakarta, *Pangan,* Vol. 20, No. 2, 2011, pp. 177-195.

[5] Heidergott, Bernd, G.J. Olsder, and J. Van der Woude, *Max Plus at Work (Modeling and Analysis of Synchronized Systems: A Course on Max Plus Algebra and Its Application)*, Princeton University Press, New Jersey, 2006.

[6] Subiono and Kistosil Fahim, On Computing Supply Chain Scheduling Using Max-Plus Algebra, *Applied Mathematical Science*, Vol. 10, 2016, pp. 477-486.

[7] Mufidah, Shofiyatul and Subiono, Model Rantai Pasok Mengunakan Petri Net dan Aljabar Max Plus dengan Mempertimbangkan Prioritas Transisi, *Proceeding: Seminar Nasional Pendidikan Matematika Ahmad Dahlan*, 2014, pp. 985-1002.

IOP Conf. Series: Materials Science and Engineering **567** (2019) 012018 doi:10.1088/1757-899X/567/1/012018

2-cut splicing and 4-cut splicing on DNA molecule

I Aisah[1], P R E Jayanti[1], A K Supriatna[1]

Department of Mathematics, Faculty of Mathematics and Natural Sciences, Universitas Padjadjaran Bandung, Indonesia

Corresponding author: isah.aisah@unpad.ac.id

Abstract. Graph has many applications both in mathematics and various other disciplines especially to describe model problem. Graph can be used to describe the process of splicing system in DNA molecule. Splicing system is a mathematical model of the system in the action of restriction enzyme in DNA molecule. Splicing language is a language constructed by splicing system. In this paper we discussed about 2 cut splicing and 4 cut splicing on DNA molecule. It had been recognized that at the time of splicing, the graph takes the semi graph structure and it generates the words which leads to the Parikh concepts. Since a DNA which consist of (A, C, G, T) can be seen as a word, thus the modelling can be done within the framework of formal language theory. After the splicing process, the main element of the DNA molecule is mapped into the binary alphabet (a, b). The language formed from the binary alphabet is mapped into a matrix form.

1. Introduction

Splicing system is a mathematical model of the system in restriction actions on DNA molecules. The concept of splicing systems is introduced as a tool to represent string restructuring that occurs during linear polymer deviation in the presence of specific enzyme activity [1].

DNA or *deoxyribonucleic acid* is a molecule that carries the genetic information in living cell. DNA sequences are three-dimensional objects are more easily described using graphs [2]. So that the main structure of the DNA molecule (A, G, T, C) can be mapped into the binary alphabet (a, b). The Parikh matrix is a type of upper triangular matrix. The Parikh matrix results from string mapping formed from the binary alphabet into the matrix form introduced by Mateescu [3]

2. Theoretical Model

2.1. *Genetic*

DNA or *deoxyribonucleic acid* is a molecule that carries the genetic information in living cell. Those are *sugar* (deoxyribose sugar), a phosphate, and a base. There are four type of base; *adenine, thymine, cytosine, and guanine*, abbreviated by *A, T, C*, and *G* respectively.

Two single strands of DNA molecule can bind together to form a double stranded DNA molecule. The two strands hold together by (hydrogen) bonds between their bases. These bonds hold between complementarity's bases. That is between *A* in one strand and *T* in the other, and between *C* in one strand and *G* in the other.

2.2. *Restriction enzyme*

Restriction enzymes recognize specific and continuous subpart, called *restriction sites,* of a double stranded DNA molecule and cut it at precisely specifiable positions. Each enzyme has a series of 4-6

IOP Conf. Series: Materials Science and Engineering **567** (2019) 012018 doi:10.1088/1757-899X/567/1/012018

specific base pairs contained in a DNA strand. Each restriction enzyme has a different and specific cutting recognition site especially in the nitrogen base. There are two kinds of restriction endonuclease enzymes which produce blunt ends and sticky or cohesive ends. There are two kinds of sticky ends, first the 5'-end which results in cutting results extending at the 5'-end and 3'-end which results in cutting lengthwise at the 3'-end.

Table 1. Restriction Enzyme Sites

Enzyme	Restriction sites	Enzyme	Restriction sites
*Taq*I	$T\ CGA$	*Hind*III	$A\ AGCTT$
*Rsa*I	$GT\ AC$	*Pvu*I	$CAG\ CTG$
*Hha*I	$GCG\ C$	*Pvu*II	$CAG\ CTG$
*Bsu*RI	$GG\ CC$	*Kpn*I	$GGTAC\ C$
*Eco*RI	$G\ AATTC$	*Bal*I	$TGG\ CCA$
*Eco*RV	$GAT\ ATC$	*Sma*I	$CCC\ GGG$
*Bam*HI	$G\ GATCC$	*Bss*HII	$G\ CGCGC$
*Pci*I	$A\ CATGT$	*Pst*I	$CTGCA\ G$
*Acc*65I	$G\ GTACC$	*Mfe*I	$C\ AATTG$
*Mlu*I	$A\ CGCGT$	*Avr*II	$C\ CTAGG$
*Bmt*I	$GCTAG\ C$	*Bsp*HI	$T\ CATGA$

2.3. *Context-free Grammar*
A context free Grammar G is $G = (V_N, V_T, P, S)$ where V_N is a finite set of variables or nonterminal, S is a special element of V_N called the starting symbol, V_T is a finite set of terminals, and P is a finite set of production. Each production is of the form $a \rightarrow b$ with a in V_N and b in $(V_N \cup V_T)^*$.

A grammar G is a set of rules for constructing a language. The constructed language called the language build by the grammar, is denoted by $L(G)$.

The language on n-cut spliced semi graph is generated by following context-free grammar in the Greiback normal form, $S' \rightarrow S,\ S \rightarrow aSB|\ abB|,\ B \rightarrow b$.

2.4. *Graph*
Definition 2.1: Suppose G is a graph, when splicing G, a new vertices is obtained which is called semi-vertices denoted by V', where $|V'| = p'$.

Semi-Edges

Definition 2.2: Let G be a graph when splicing G, we obtain new edges by decomposition of edges which are called as semi –edges denoted by E', where $|E'| = q'$.

Bipartite Semi graphs

Definition 2.3: G is bipartite if its vertex set V can be partitioned into sets $\{V_1, V_2\}$ such that both V_1 and V_2 are independent.

2.4.1. Euler's Polyhedral formula for Splicing System the Euler's polyhedral formula for DNA spliced semi graph is $P - (q + q') + r = k + 1$, where P is the total number of vertices and semi vertices in the semi graph, q and q' are the number of edges and semi edges, r is the number of faces of semi graph and k is the number of components.

2.4.2. Parikh Matrices Let $\Sigma = \{a_1, a_2, a_3, \ldots, a_k\}$ *with* $a_1 < a_2 < a_3 < \cdots < a_k$ *are regular a alphabet.* $\Psi_k : \Sigma^* \rightarrow M_{k+1}$ *is morphism defined by:* $\Psi_k(\lambda) = I_{k+1}$ *and* $I\ f\ \Psi_k(a_q) =$

$(m_{i,j})_{1<i,j<(k+1)}$, *then for each* $1 \leq i \leq (k+1)$, $m_{i,i} = 1$, $m_{q,q+1} = 1$, *all other elements of the matrix* $\Psi_k(a_q)$ *are 0.*
For a word $w = a_{i_1}a_{i_2}a_{i_3} \ldots a_{i_m}$, $a_{i_j} \in \Sigma$ for $1 \leq j \leq m$ we have

$$\Psi_k(w) = \Psi_k(a_{i_1})\Psi_k(a_{i_2})\Psi_k(a_{i_3}) \ldots \Psi_k(a_{i_m}).$$

2.5. *Norm of Parikh Matrices A function* $\|\cdot\|: \mathbb{R}^{m,n} \rightarrow \mathbb{R}$ *is called a matrix norm on* $\mathbb{R}^{m,n}$ *if for all* $A, B \in \mathbb{R}^{m,n}$ *and all* $\alpha \in \mathbb{R}$

i) $\|A\| \geq 0$ with equality if and only if $A = 0$ (positivity),
ii) $\|\alpha A\| = |\alpha|\|A\|$ (homogeneity),
iii) $\|A + B\| \leq \|A\| + \|B\|$ (subadditivity),
iv) $\|AB\| \leq \|A\|\|B\|$ (submultiplicativity).

Theorem 2.5: For $A \in \mathbb{R}^{m,n}$ we have

i) $\|A\|_1 = \max_{1 \leq j \leq n} \sum_{i=1}^{m} |a_{ij}|$ (max column sum)
ii) $\|A\|_\infty = \max_{1 \leq i \leq m} \sum_{j=1}^{n} |a_{ij}|$ (max row sum)

3. Result

3.1. *2-Cut splicing on DNA molecule with strand 5'-end and 3'-end of sticky ends overhang*
If the double strand DNA molecule's restriction site is 4 bases in length and produce lengthwise cuts in the 5'-end and 3'-end is described in graph form according to the graph splicing scheme that has been given will produce the following graph.

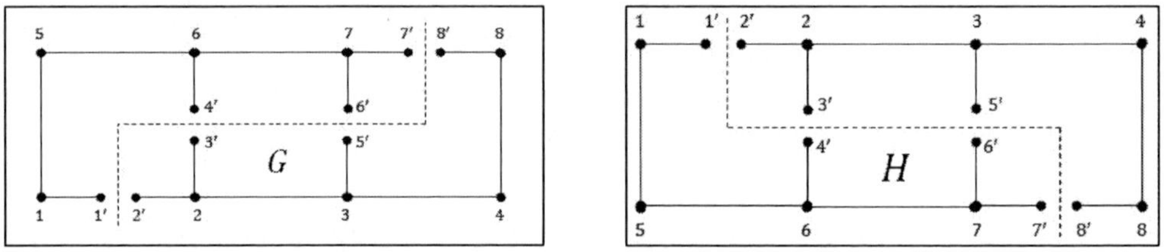

Figure 4.1: Representation *Sticky Ends* Overhang 3'-end (Left) and Overhang 5'-end (Right) with L=4

Because *the Euler's Polyhedral* formula *P- (q + q ')* + r = 16 (6 + 8) + 1 = 2 + 1 = 3 = *k* + *1* is fulfilled, Figure 4.1 is *2-cut splicing*. Based on the *2-cut splicing* in Figure 4.1, two semigraphs are obtained, as follows

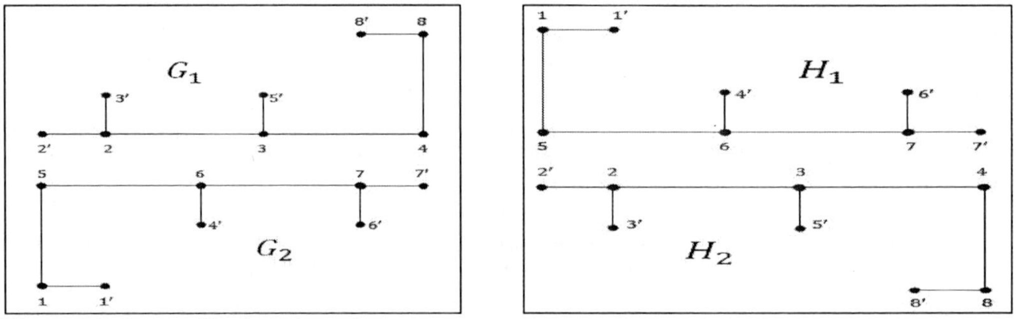

Figure 4.2: 2-cut spliced semi graph

Let the weight of each edge of a spliced semi graph as 'b' then the weight of each semi edge becomes '$\frac{b}{2}$'. After 2–Cut splicing so we have spliced semi graphs in the form of two bipartite semi graphs. Each one has four semi edges and three edges. Therefore let the language be $\left(\frac{b}{2}\right)^4 b^3$. Therefore the language obtained is $L = \left(\frac{b}{2}\right)^4 b^3 = \left\{a^4 b^3 | a = \frac{b}{2}\right\}$.

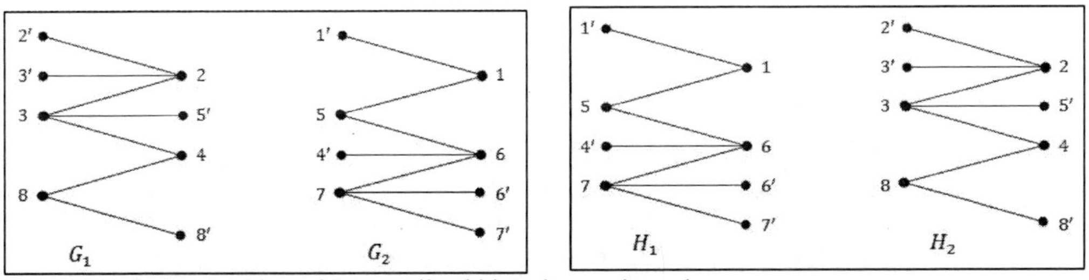

Figure 4.3: Representation of 2-cut spliced bipartite semi graph

Let $G = (V_N, V_T, P, S)$ be the grammar defined on the languages on edges, where $V_N = \{S\}$, $V_T = \{a, b\}$, $P = \{S \to (ASb, Aab),\ A \to a\}$, the set of productions P is given from the language $L(G) = \left\{a^4 b^3 | a = \frac{b}{2}\right\}$.

Since the language generated on the splicing graph such as $L = L(G) = \left\{a^4 b^3 | a = \frac{b}{2}\right\} = \left\{a^{n+2} b^{n+1} | n = 2,\ a = \frac{b}{2}\right\}$, , than L is splicing language. Based on the L, we obtained $V_T = \{a, b\}$ with $a < b$. The whole set of strings formed from the initial string on the L is $V_T{}^*$ as much $\frac{7!}{4!\,3!} = 35$ strings.

The Parikh matrix mapping Ψ_2 is a mapping from $V_T{}^*$ to M_3 with

$$\Psi_2(a) = \begin{bmatrix} 1 & 1 & 0 \\ 0 & 1 & 0 \\ 0 & 0 & 1 \end{bmatrix},\quad \Psi_2(b) = \begin{bmatrix} 1 & 0 & 0 \\ 0 & 1 & 1 \\ 0 & 0 & 1 \end{bmatrix}$$

$\Psi_2(ab) = \Psi_2(a)\Psi_2(b),\ a, b \in V_T{}^*$, where multiplication of matrices is the operation on the right side of this equation.

Table 2. Norm of Parikh Matrices for the 1-Cut splicing system words

Word	$\Psi(w_i)$	$\lVert\Psi(w_i)\rVert_1$	$\lVert\Psi(w_i)\rVert_\infty$
$w_1 = aaaabbb$	$\begin{bmatrix} 1 & 4 & 12 \\ 0 & 1 & 3 \\ 0 & 0 & 1 \end{bmatrix}$	16	17
$w_2 = aaababb$	$\begin{bmatrix} 1 & 4 & 11 \\ 0 & 1 & 3 \\ 0 & 0 & 1 \end{bmatrix}$	15	16
$w_3 = aaabbab$ $w_5 = aabaabb$	$\begin{bmatrix} 1 & 4 & 10 \\ 0 & 1 & 3 \\ 0 & 0 & 1 \end{bmatrix}$	14	15
$w_4 = aaabbba$ $w_6 = aababab$ $w_{11} = abaaabb$	$\begin{bmatrix} 1 & 4 & 9 \\ 0 & 1 & 3 \\ 0 & 0 & 1 \end{bmatrix}$	13	14
$w_7 = aababba$ $w_8 = aabbaab$ $w_{12} = abaabab$ $w_{21} = baaaabb$	$\begin{bmatrix} 1 & 4 & 8 \\ 0 & 1 & 3 \\ 0 & 0 & 1 \end{bmatrix}$	12	13

$w_9 = aabbaba$ $w_{13} = abaabba$ $w_{14} = ababaab$ $w_{22} = baaabab$	$\begin{bmatrix}1 & 4 & 7\\0 & 1 & 3\\0 & 0 & 1\end{bmatrix}$	11	12
$w_{10} = aabbbaa$ $w_{15} = abababa$ $w_{17} = abbaaab$ $w_{23} = baaabba$ $w_{24} = baabaab$	$\begin{bmatrix}1 & 4 & 6\\0 & 1 & 3\\0 & 0 & 1\end{bmatrix}$	10	11
$w_{16} = ababbaa$ $w_{18} = abbaaba$ $w_{25} = baababa$ $w_{27} = babaaab$	$\begin{bmatrix}1 & 4 & 5\\0 & 1 & 3\\0 & 0 & 1\end{bmatrix}$	9	10
$w_{19} = abbabaa$ $w_{26} = baabbaa$ $w_{28} = babaaba$ $w_{31} = bbaaaab$	$\begin{bmatrix}1 & 4 & 4\\0 & 1 & 3\\0 & 0 & 1\end{bmatrix}$	8	9
$w_{20} = abbbaaa$ $w_{29} = bababaa$ $w_{32} = bbaaaba$	$\begin{bmatrix}1 & 4 & 3\\0 & 1 & 3\\0 & 0 & 1\end{bmatrix}$	7	8
$w_{30} = babbaaa$ $w_{33} = bbaabaa$	$\begin{bmatrix}1 & 4 & 2\\0 & 1 & 3\\0 & 0 & 1\end{bmatrix}$	6	7
$w_{34} = bbabaaa$	$\begin{bmatrix}1 & 4 & 1\\0 & 1 & 3\\0 & 0 & 1\end{bmatrix}$	5	6
$w_{35} = bbbaaaa$	$\begin{bmatrix}1 & 4 & 0\\0 & 1 & 3\\0 & 0 & 1\end{bmatrix}$	4	5

Norm of Parikh matrices produced on 2-cut splicing in Table 2 has a range $4 \leq \|\Psi(w)\|_1 \leq 16$ and $5 \leq \|\Psi(w)\|_\infty \leq 17$ and $\|\Psi(w)\|_\infty = \|\Psi(w)\| + 1$

3.2. *4-Cut splicing on DNA molecule with strand 5'–end and 3'-end of sticky ends overhang*
If a 6-base-length, double-stranded DNA restriction site that produces longitudinal pieces at 5 '-end and 3'-end is describe in graph form according to the splicing chart scheme that has been given, it will produce the following graph.

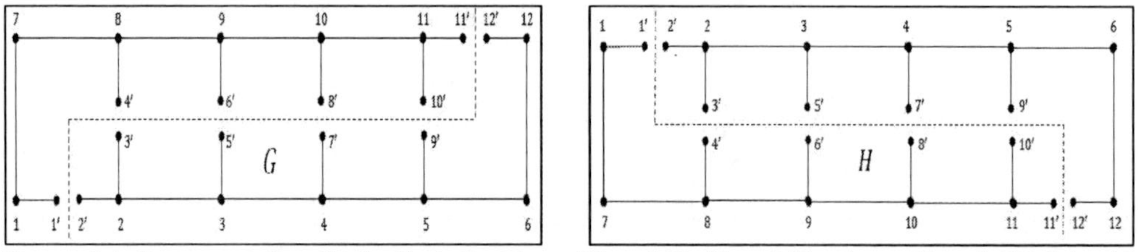

Figure 4.4: Representation *Sticky Ends* Overhang 3'-end (Left) and Overhang 5'-end (Right) with L=6

Consider $L = 6$, the number of set of combinations required for 4-Cut splicing. Here $P = 24$, $q = 10$, $q' = 12$, $r = 1$, $k = 2$. Then $P - (q + q') + r = 24 - (10 + 12) + 1 = 2 + 1 = 3 = k + 1$ therefore is satisfied in 2-Cut splicing, as described below

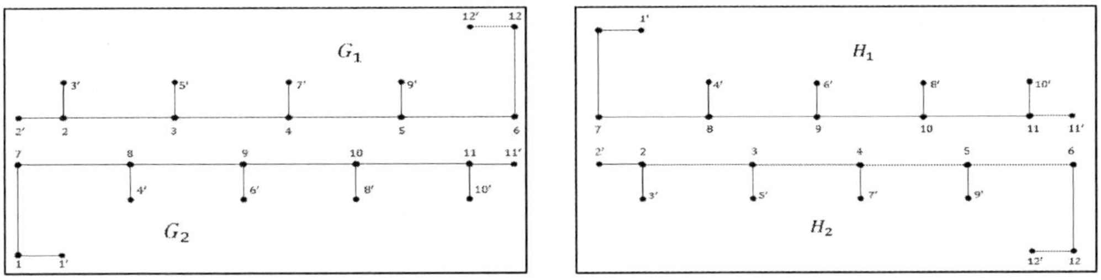

Figure 4.5: 4-cut spliced semi graph

Let the weight of each edge of a spliced semi graph as 'b' then the weight of each semi edge becomes '$\frac{b}{2}$'. After 4–Cut splicing so we have spliced semi graphs in the form of two bipartite semi graphs. Each one has six semi edges and five edges. Therefore let the language be $\left(\frac{b}{2}\right)^6 b^5$. Therefore the language obtained is $L = \left(\frac{b}{2}\right)^6 b^5 = \left\{a^6 b^5 | a = \frac{b}{2}\right\}$.

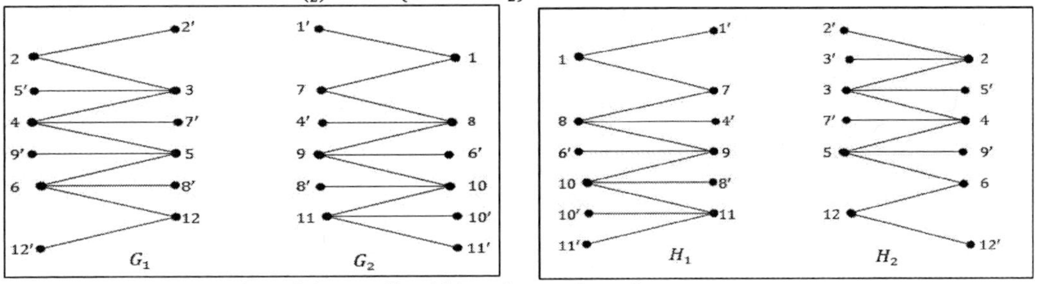

Figure 4.6: Representation of 4-cut spliced bipartite semi graph

The language of the grammar of $G = (V_N, V_T, P, S)$ obtained from the set of result P, where $V_N = \{S\}$, $V_T = \{a, b\}$, $P = \{S \to (ASb, Aab),\ A \to a\}$, is $L(G) = \left\{a^6 b^5 | a = \frac{b}{2}\right\}$.

Since the language generated on the splicing graph such as $L = L(G) = \left\{a^6 b^5 | a = \frac{b}{2}\right\} = \left\{a^{n+2} b^{n+1} | n = 4,\ a = \frac{b}{2}\right\}$, than L is splicing language. Based on the L, we obtained $V_T = \{a, b\}$ with $a < b$. The whole set of strings formed from the initial string on the L is $V_T{}^*$ as much $\frac{7!}{6!\,5!} = 642$ strings.

The Parikh matrix mapping Ψ_2 is a mapping from $V_T{}^*$ to M_3 with

$$\Psi_2(a) = \begin{bmatrix} 1 & 1 & 0 \\ 0 & 1 & 0 \\ 0 & 0 & 1 \end{bmatrix},\ \Psi_2(b) = \begin{bmatrix} 1 & 0 & 0 \\ 0 & 1 & 1 \\ 0 & 0 & 1 \end{bmatrix}$$

$\Psi_2(ab) = \Psi_2(a)\Psi_2(b),\ a, b \in V_T{}^*$ where multiplication of matrices is the operation on the right side of this equation.

Norm of Parikh matrices produced on 4-cut splicing has a range of $6 \leq \|\Psi(w)\|_1 \leq 36$ and $7 \leq \|\Psi(w)\|_\infty \leq 37$ and $\|\Psi(w)\|_\infty = \|\Psi(w)\| + 1$.

Max column sum norm of Parikh matrix obtained by calculating the number of all entries in the third column of each matrix produced by the string where the first entry is the number of ab in the string, the second entry is the number of letters b in the string, and the third entry is 1. And max row sum norm of Parikh matrix obtained by calculating the number of all entries in the first row of each matrix produced by the string where the first entry is 1, the second entry is the number of letters a in the string, and the third entry is the number of ab in the string. The number of a in each string is $n +$

2, the number of b is $n+1$, and the number of substrings ab is $0 \leq |ab| \leq |a||b|$, with n the number of splicing $(n \geq 1)$.

4. Conclusion

DNA molecules hold the characterization of the semi graph at the time of splicing. By using formal language and with the help of Algebra software, we can see the amount of splicing on DNA molecule from the norm value of the Parikh matrix. The result of norm of Parikh matrices is $n+2 \leq \|\Psi(w)\|_1 \leq (n+2)^2$ where n the number of splicing $(n \geq 1)$ and $\|\Psi(w)\|_\infty = \|\Psi(w)\|_1 + 1$.

Acknowledgments

The authors thank to the University of Padjadjaran, due to the funding of this work through to the scheme of Academic Leadership Grand (ALG) with contract number 2297/UN6.D/KS/2018.

References

[1] Head, T. Splicing Representations of Strictly Locally Testable Languages. *Discrete Applied Mathematics vol. 87.* Binghamton, USA: 139-147; 1998.

[2] Freund, R. Splicing Systems on Graphs, Proc. Intelligence in Neural and Biological Systems, IEEE Press, 189-194; 1995.

[3] Lawson, Mark V. 2009. *Finite Automata.* Heriot-Watt University, Edinburgh.

[4] Adleman, L.M. Computing with DNA. *Scientific American.* USA: 54-61; August 1998.

[5] Dinu, L.P. AS garro-A low complexity distance for DNA words fundamental Informatics 73(3),361 372,2006.

[6] Frisco, P. *Theory of Molecular Computing Splicing and Membrane System*. Italy; 1970.

[7] Head, T. Formal Language Theory and DNA: An analysis of the generative capacity of recombinant behaviors. Bulletin of Mathematical Biology, 49, 735-759; 1987.

[8] Jeyabharathi, S., Angayarkanni, M.,Sinthanaiselvi, S., Anusha, R. *Bipartite Semigraph Structure in DNA Splicing System,* International Journal of Computer Application, Vol. VI, E-ISSN 2250-1797; 2012.

[9] Jeyabharathi, S., Sinthanaiselvi, S. *Parikh Representation for Splicing System.* Proceedings of the International Conference on Applied Mathematics and Theoretical Computer Science. 978-93-82338-35-2. 2013.

[10] Jeyabharathi, S., Thiagarajan, K., Sinthanaiselvi, S. *Norms of Parikh Matrices on DNA Splicing System*, International Journal of Advanced Engineering Technology, Vol. VII, E-ISSN 0976-3945; 2016.

[11] Laufer, H.B. *Discrete Mathematics and Applied Modern Algebra.* USA: PWS Pubisher; 1984.

[12] Leon, S.J. *Linear Algebra with Applications,* Ninth Edition. Pearson Education, Inc. United States of America. 2015.

[13] Subramanian, K.G., Huey, A.M., Nagar, A.K. *On Parikh Matrices.* International Journal of Foundations of Computer Science Vol. 20, No. 2, 211-219; 2009.

[14] Available from: https://www.addgene.org/mol-bio-reference/restriction-enzymes/ (cited August 14, 2018).

IOP Conf. Series: Materials Science and Engineering **567** (2019) 012019 doi:10.1088/1757-899X/567/1/012019

Vector space of codons sequence over galois field GF(7^3)

I Aisah[1], N U Sayyidatunnisa[1], B Subartini[1], A Kartiwa[1]

[1]Department of Mathematics, Faculty of Mathematics and Natural Science
Padjadjaran University, Bandung, Indonesia

Corresponding author: isah.aisah@unpad.ac.id

Abstract. DNA and RNA is genetic material that play an important role in living things metabolism process which called protein synthesis. DNA have four nucleic acid, they are adenine (A), guanine (G), cytosine (C), and thymine (T). Protein synthesis process closely related with standard genetic code. Standard genetic code is a set of rules that defines the order of nucleotide bases in DNA or RNA to determine the order of certain amino acids in protein synthesis. The standard genetic code is a combination of three nitrogen bases or triplet's bases. This standard genetic code can mathematically represented by algebraic structure. In this paper we will give that representation using the extended set of 7 elements from the set of four DNA's nucleic acid that are $N = \{D, A, C, O, G, T, P\}$. Then we construct new triplet set from N, called extended triplet set. In the end we analyze the vector space structure of it and find the significant field that correspond with that structure.

1. Introduction

The DNA is which one of nucleic acid that contain genetic information for humans and almost all other organisms. It contains four nucleic acid, that are adenine (A), guanine (G), cytosine (C), and thymine (T). Mostly, RNA have the same structure of DNA. One of the main differences is the RNA's nucleic acid. They are adenine (A), guanine (G), cytosine (C), and uracil (U). RNA is the result of DNA's transcription and RNA is translating into amino acid as protein constituent which our body needs. This whole process involve encoding the triplets or codons. This triplets called standard genetic code and this is our main object.

Recently, many research in analyzing the algebraic structure of DNA sequence was reported [1][2-7]. DNA sequence's analyze over the Galois field of 64 codons [4] and over 125 codons with extended DNA bases into 5 elements [5] have been studied. The question arise when the DNA bases is extended into 7 elements. What kind of algebraic structure it will have and is that extended set can represent the standard genetic code like previous research? Based on that question, in this paper we will extend the DNA bases into 7 elements $N = \{D, A, C, O, G, T, P\}$ and investigate the algebraic structure of it.

2. DNA, Group and Vector Space

Please In this section we give some definitions of DNA, group theory and vector space.

2.1 DNA (Deoxyribonucleic Acid)

DNA is the hereditary material in humans and almost all other organisms, which have four chemical bases that contain the genetic information. The four bases are adenine (A), guanine (G), cytosine (C), and thymine (T), which called the nitrogen bases. Beside, RNA (Ribonucleic Acid) have same structure with DNA but with different nitrogen bases. In RNA we can't find thymine (T) but uracil (U). DNA can replicate and be a source of protein synthesis. Protein synthesis is biological process to form protein particle which essential for living things metabolism. The short process is described by DNA-RNA-Protein which every step is represent by encoding triplet code.

2.2 Group Theory

Definition 2.2.1 Let G is a set that closed under binary operation $*$. G is called group under that operation if satisfy :

1. Associative, that is for every $a, b,$ and c in G, $(a * b) * c = a * (b * c)$ hold.
2. Have the identity element, in other words, there is e in G such that for every a in G, $e * a = a * e = a$ hold.
3. Every element in G, have inverse, that is for every a in G there is a' in G such that $a * a' = a' * a = e$ hold [8].

The example of group is set $N = \{C, U, A, G\}$ with matching of the set $\mathbb{Z}_2 \times \mathbb{Z}_2$, that is $C = (\bar{0}, \bar{0}), U = (\bar{0}, \bar{1})$, $A = (\bar{1}, \bar{0})$, and $G = (\bar{1}, \bar{1})$. It can be show that $(N, +)$ form an Abelian group

2.3 Vector Space

Definition 2.3.1 Set V is called vector space over field F if V is Abelian group under addition and for every $a \in F$ and $v \in V$ there is av in V such that this conditions hold for every a, b in F and u, v in V.

i. $a(u + v) = au + av$
ii. $(a + b)v = av + bv$
iii. $a(bv) = (ab)v$
iv. $1v = v$ [9].

Based on the example of definition 2.2.1, the set NNN form a vector space over $\mathbb{Z}_2$.

3. Vector Space Construction over Galois Field GF(7^3)

In this section, in order to construct the vector space structure of codons sequence, we analyze first the extended triplet set then construct a mapping based on it algebraic structure.

3.1 Extended Triplet Set

The algebra structure of DNA sequence over Galois field with 64 and 125 elements have been studied [8]. In this paper, the DNA nitrogen bases is extend to be $N = \{D, A, C, O, G, T, P\}$. Then we formed $C_{343} = \{X_1, X_2, X_3 \,|X_i \in N, i = 1,2,3\}$ as extended triplet set with cardinality 343 and showed in Figure 1 on the first and third columns.

IOP Conf. Series: Materials Science and Engineering **567** (2019) 012019 doi:10.1088/1757-899X/567/1/012019

	D			A			C			O			G			T			P		
	I	II	III	I	II	III	I	II	III	I	II	III	I	II	III	I	II	III	I	II	III
	0	000	DDD	49	001	DAD	98	002	DCD	147	003	DOD	196	004	DGD	245	005	DTD	294	006	DPD
	1	100	DDA	50	101	DAA	99	102	DCA	148	103	DOA	197	104	DGA	246	105	DTA	295	106	DPA
	2	200	DDC	51	201	DAC	100	202	DCC	149	203	DOC	198	204	DGC	247	205	DTC	296	206	DPC
D	3	300	DDO	52	301	DAO	101	302	DCO	150	303	DOO	199	304	DGO	248	305	DTO	297	306	DPO
	4	400	DDG	53	401	DAG	102	402	DCG	151	403	DOG	200	404	DGG	249	405	DTG	298	406	DPG
	5	500	DDT	54	501	DAT	103	502	DCT	152	503	DOT	201	504	DGT	250	505	DTT	299	506	DPT
	6	600	DDP	55	601	DAP	104	602	DCP	153	603	DOP	202	604	DGP	251	605	DTP	300	606	DPP
	7	010	ADD	56	011	AAD	105	012	ACD	154	013	AOD	203	014	AGD	252	015	ATD	301	016	APD
	8	110	ADA	57	111	AAA	106	112	ACA	155	113	AOA	204	114	AGA	253	115	ATA	302	116	APA
	9	210	ADC	58	211	AAC	107	212	ACC	156	213	AOC	205	214	AGC	254	215	ATC	303	216	APC
A	10	310	ADO	59	311	AAO	108	312	ACO	157	313	AOO	206	314	AGO	255	315	ATO	304	316	APO
	11	410	ADG	60	411	AAG	109	412	ACG	158	413	AOG	207	414	AGG	256	415	ATG	305	416	APG
	12	510	ADT	61	511	AAT	110	512	ACT	159	513	AOT	208	514	AGT	257	515	ATT	306	516	APT
	13	610	ADP	62	611	AAP	111	612	ACP	160	613	AOP	209	614	AGP	258	615	ATP	307	616	APP
	14	020	CDD	63	021	CAD	112	022	CCD	161	023	COD	210	024	CGD	259	025	CDD	308	026	CPD
	15	120	CDA	64	121	CAA	113	122	CCA	162	123	COA	211	124	CGA	260	125	CDA	309	126	CPA
	16	220	CDC	65	221	CAC	114	222	CCC	163	223	COC	212	224	CGC	261	225	CTC	310	226	CPC
C	17	320	CDO	66	321	CAO	115	322	CCO	164	323	COO	213	324	CGO	262	325	CTO	311	326	CPO
	18	420	CDG	67	421	CAG	116	422	CCG	165	423	COG	214	424	CGG	263	425	CTG	312	426	CPG
	19	520	CDT	68	521	CAT	117	522	CCT	166	523	COT	215	524	CGT	264	525	CTT	313	526	CPT
	20	620	CDP	69	621	CAP	118	622	CCP	167	623	COP	216	624	CGP	265	625	CTP	314	626	CPP
	21	030	ODD	70	031	OAD	119	032	OCD	168	033	OOD	217	034	OGD	266	035	OTD	315	036	OPD
	22	130	ODA	71	131	OAA	120	132	OCA	169	133	OOA	218	134	OGA	267	135	OTA	316	136	OPA
	23	230	ODC	72	231	OAC	121	232	OCC	170	233	OOC	219	234	OGC	268	235	OTC	317	236	OPC
O	24	330	ODO	73	331	OAO	122	332	OCO	171	333	OOO	220	334	OGO	269	335	OTO	318	336	OPO
	25	430	ODG	74	431	OAG	123	432	OCG	172	433	OOG	221	434	OGG	270	435	OTG	319	436	OPG
	26	530	ODT	75	531	OAT	124	532	OCT	173	533	OOT	222	534	OGT	271	535	OTT	320	536	OPT
	27	630	ODP	76	631	OAP	125	632	OCP	174	633	OOP	223	634	OGP	272	635	OTP	321	636	OPP
	28	040	GDD	77	041	GAD	126	042	GCD	175	043	GOD	224	044	GGD	273	045	GTD	322	046	GPD
	29	140	GDA	78	141	GAA	127	142	GCA	176	143	GOA	225	144	GGA	274	145	GTA	323	146	GPA
	30	240	GDC	79	241	GAC	128	242	GCC	177	243	GOC	226	244	GGC	275	245	GTC	324	246	GPC
G	31	340	GDO	80	341	GAO	129	342	GCO	178	343	GOO	227	344	GGO	276	345	GTO	325	346	GPO
	32	440	GDG	81	441	GAG	130	442	GCG	179	443	GOG	228	444	GGG	277	445	GTG	326	446	GPG
	33	540	GDT	82	541	GAT	131	542	GCT	180	543	GOT	229	544	GGT	278	545	GTT	327	546	GPT
	34	640	GDP	83	641	GAP	132	642	GCP	181	643	GOP	230	644	GGP	279	645	GTP	328	646	GPP
	35	050	TDD	84	051	TAD	133	052	TCD	182	053	TOD	231	054	TGD	280	055	TTD	329	056	TPD
	36	150	TDA	85	151	TAA	134	152	TCA	183	153	TOA	232	154	TGA	281	155	TTA	330	156	TPA
	37	250	TDC	86	251	TAC	135	252	TCC	184	253	TOC	233	254	TGC	282	255	TTC	331	256	TPC
T	38	350	TDO	87	351	TAO	136	352	TCO	185	353	TOO	234	354	TGO	283	355	TTO	332	356	TPO
	39	450	TDG	88	451	TAG	137	452	TCG	186	453	TOG	235	454	TGG	284	455	TTG	333	456	TPG
	40	550	TDT	89	551	TAT	138	552	TCT	187	553	TOT	236	554	TGT	285	555	TTT	334	556	TPT
	41	650	TDP	90	651	TAP	139	652	TCP	188	653	TOP	237	654	TGP	286	655	TTP	335	656	TPP
	42	060	PDD	91	061	PAD	140	062	PCD	189	063	POD	238	064	PGD	287	065	PTD	336	066	PPD
	43	160	PDA	92	161	PAA	141	162	PCA	190	163	POA	239	164	PGA	288	165	PTA	337	166	PPA
	44	260	PDC	93	261	PAC	142	262	PCC	191	263	POC	240	264	PGC	289	265	PTC	338	266	PPC
P	45	360	PDO	94	361	PAO	143	362	PCO	192	363	POO	241	364	PGO	290	365	PTO	339	366	PPO
	46	460	PDG	95	461	PAG	144	462	PCG	193	463	POG	242	464	PGG	291	465	PTG	340	466	PPG
	47	560	PDT	96	561	PAT	145	562	PCT	194	563	POT	243	564	PGT	292	565	PTT	341	566	PPT
	48	660	PDP	97	661	PAP	146	662	PCP	195	663	POP	244	664	PGP	293	665	PTP	342	666	PPP

Figure 1. Column I is triplet index number, column II the polynomial coefficient, column III the extended triplet

3.2 Mapping The Galois Field with C_{343}

To construct a mapping from Galois field $GF(7^3)$ to C_{343}, defined f as follows.

$$f : GF(7) \to \{D, A, C, O, G, T, P\}$$
$$0 \mapsto D$$
$$1 \mapsto A$$
$$2 \mapsto C$$
$$3 \mapsto O$$
$$4 \mapsto G$$
$$5 \mapsto T$$
$$6 \mapsto P$$

To construct Galois field $GF(p^n)$ we can use $\mathbb{Z}_p[x]$ ring and irredicuble polynomial $p(x)$ n-degrees such that $GF(p^n) = \mathbb{Z}_p[x]/\langle p(x)\rangle$. Then we have $GF(7^3) = \mathbb{Z}_7[x]/\langle p(x)\rangle = \{a_0 + a_1x + a_2x^2 | a_0, a_1, a_2 \in \mathbb{Z}_7\}$ for an irreducible polynomial $p(x)$.

After that, we defined a mapping named φ, from $GF(7^3)$ to C_{343}

$$\varphi : GF(7^3) \to C_{343}$$

with $\varphi(a_0 + a_1x + a_2x^2) = \big(f(a_1)\ f(a_2)\ f(a_0)\big) = (X_1\ X_2\ X_3)$. Consider that the polynomial coefficient a_2 with maximal degree a_2x^2 correspond with bases on second codon/triplet position and polynomial coefficient a_1 correspond with bases on first codon/triplet position and polynomial coefficient a_0 correspond with bases on third codon/triplet position. This order is correspond with error frequency on that three bases from a triplet which analogous as degree of variable on $GF(7^3)$ polynom.

For example, if we have ADA triplet. Based on φ and f that we have defined first, then

$$ADA = \big(f(a_1)\ f(a_2)\ f(a_0)\big) = \big(f(1)\ f(0)\ f(1)\big) = \varphi(1 + 1 \cdot x + 0 \cdot x^2)$$

So we obtained $a_0 = 1, a_1 = 1$ and $a_2 = 0$, the polynomial coefficient which correspond to ADA and can be written

$$110 \leftrightarrow ADA \ldots\ldots (*)$$

Therefore we obtained figure 1 for second column using above mapping which is the polynomial coefficient that correspond with every codons on C_{343}.

According to figure 1, we can see that φ mapping is a bijection and so we can obtained a mapping between the triplet's index number and the triplets as follows.

For ADA, the corresponding index number in figure 1 is 8.

Consider that $8 = 0 \cdot 7^2 + 1 \cdot 7^1 + 1 \cdot 7^0$, so with this result and $(*)$ we can write as follows :

$$8 \leftrightarrow 011 \leftrightarrow 110 \leftrightarrow 1 + x \leftrightarrow ADA$$

According to above result, we can defined a bijection

$$f: S \to GF(7^3)$$

with $S \in \{0,1,2, \ldots, 342\}$ and $f[\alpha] = a_0 + a_1x + a_2x^2$ where a_0, a_1, a_2 is a form of basis 7's number from α. So we have $f[8] = 1 + x$.

3.3 Algebra Structure of C_{343}

Based on definition 2.3.1, in order to analyze the vector space structure of C_{343} we must investigate the group structure first.

3.3.1 Sum operation and Group Structure on C_{343} over Addition To investigate the group structure of C_{343}, first we defined sum operation on C_{343} as follows. For every $X_1Y_1Z_1$ and $X_2Y_2Z_2 \in C_{343}$, define

$$X_1Y_1Z_1 + X_2Y_2Z_2 = \varphi\big[\big(\varphi^{-1}(X_1Y_1Z_1) + \varphi^{-1}(X_2Y_2Z_2)\big) \bmod 7\big]$$

For example, we would find the sum between ADA and OAA. According to figure 1 :

$$ADA \leftrightarrow f[8] = 1 + x$$
$$OAA \leftrightarrow f[71] = 1 + 3x + x^2$$

then

$$\begin{aligned} ADA + OAA &= \varphi(\varphi^{-1}(ADA) + \varphi^{-1}(OAA) \bmod 7) \\ &= \varphi\big((1 + x) + (1 + 3x + x^2) \bmod 7\big) \\ &= \varphi(2 + 4x + x^2) \\ &= GAC \end{aligned}$$

So, $ADA + OAA = GAC$. Then, we can showed that $(C_{343}, +)$ is Abelian group.

3.3.2 Product Operation and Group Structure on $C_{343} - \{OOO\}$ over Multiplication Let $C_{343} - \{OOO\} = C_{343}^*$, and we defined product operation as follows. For all $X_1Y_1Z_1$ and $X_2Y_2Z_2 \in C_{343}^*$, define

$$X_1Y_1Z_1 \cdot X_2Y_2Z_2 = \varphi[\varphi^{-1}(X_1Y_1Z_1)\varphi^{-1}(X_2Y_2Z_2) \bmod g(x)]$$

where $g(x)$ is irreducible polynomial degree 3 with coefficient on $GF(7)$. Based on that definition, we can showed that $(C_{343}^*, \times)$ is Abelian group. For example, we would find the product between ADA and OAA with irreducible polynomial $g(x) = x^3 + 3x^2 + 5x + 4$. Consider that

$$\begin{aligned} ADA \cdot OAA &= \varphi[\varphi^{-1}(ADA)\varphi^{-1}(OAA) \bmod x^3 + 3x^2 + 5x + 4] \\ &= \varphi[(1 + x)(1 + 3x + x^2) \bmod x^3 + 3x^2 + 5x + 4] \\ &= \varphi((1 + 4x + 4x^2 + x^3) \bmod x^3 + 3x^2 + 5x + 4) \\ &= \varphi(1 + 4x + x^2) \\ &= GAA \end{aligned}$$

So, we have $ADA \cdot OAA = GAA$.

3.4 Codon Sequence's Vector Space Construction

After we know that $(C_{343}, +)$ is Abelian group and $(C_{343}, +, \cdot)$ is isomorphic to $GF(7^3)$, we give scalar product construction in order to investigate the vector space structure of C_{343} as follows.

For all $XYZ \in C_{343}$ and $\alpha \in \mathrm{GF}(7^3)$, defined:

$$\times : GF(7^3) \times C_{343} \to C_{343}$$
$$(\alpha, XYZ) \longmapsto \alpha(XYZ)$$

with

$$\alpha(XYZ) = \varphi(\alpha_i \varphi^{-1}(XYZ)\ mod\ 7)$$

Proof

i) Let $\alpha \in GF(5^3)$ and $X_1Y_1Z_1, X_2Y_2Z_2 \in C_{343}$, we will show that $\alpha(X_1Y_1Z_1) \in C_{343}$
Because $X_1Y_1Z_1 \in C_{343}$ then $\varphi^{-1}(X_1Y_1Z_1) \in GF(7^3)$. Because $GF(7^3)$ is field then $GF(7^3)$ closed under multiplication then $\alpha\varphi^{-1}(X_1Y_1Z_1) \in GF(7^3)$.
So $\alpha(X_1Y_1Z_1) = \varphi(\alpha\varphi^{-1}(X_1Y_1Z_1)\ mod\ 7) \in C_{343}$

ii) Let $\alpha \in GF(7^3)$ and $X_1Y_1Z_1, X_2Y_2Z_2 \in C_{343}$, we will show that $\alpha(X_1Y_1Z_1 + X_2Y_2Z_2) = \alpha(X_1Y_1Z_1) + \alpha(X_2Y_2Z_2)$.
According to scalar multiplication definition, we have

$$\begin{aligned} \alpha(X_1Y_1Z_1 + X_2Y_2Z_2) &= \varphi\big(\alpha\, \varphi^{-1}(X_1Y_1Z_1 + X_2Y_2Z_2)\big) \\ &= \varphi\left(\alpha\left(\varphi^{-1}(X_1Y_1Z_1) + \varphi^{-1}(X_2Y_2Z_2)\right)\right) \\ &= \varphi\big(\alpha\, \varphi^{-1}(X_1Y_1Z_1) + \alpha\, \varphi^{-1}(X_2Y_2Z_2)\big) \\ &= \varphi\big(\alpha\, \varphi^{-1}(X_1Y_1Z_1)\big) + \varphi\big(\alpha\, \varphi^{-1}(X_2Y_2Z_2)\big) \\ &= \alpha(X_1Y_1Z_1) + \alpha(X_2Y_2Z_2) \end{aligned}$$

iii) Let $\alpha_1, \alpha_2 \in GF(7^3)$ and $X_1Y_1Z_1, X_2Y_2Z_2 \in C_{343}$, we will show that $(\alpha_1 + \alpha_2)(X_1Y_1Z_1) = \alpha_1(X_1Y_1Z_1) + \alpha_2(X_1Y_1Z_1)$.
According to scalar multiplication definition, we have

$$\begin{aligned}(\alpha_1 + \alpha_2)(X_1Y_1Z_1) &= \varphi\big((\alpha_1 + \alpha_2)\, \varphi^{-1}(X_1Y_1Z_1)\big)\\ &= \varphi\big(\alpha_1\varphi^{-1}(X_1Y_1Z_1) + \alpha_2\varphi^{-1}(X_1Y_1Z_1)\big)\\ &= \varphi\big(\alpha_1(X_1Y_1Z_1)\big) + \varphi\big(\alpha_2(X_1Y_1Z_1)\big)\\ &= \alpha_1(X_1Y_1Z_1) + \alpha_2(X_1Y_1Z_1)\end{aligned}$$

iv) Let $\alpha_1, \alpha_2 \in GF(7^3)$ and $XYZ \in C_{343}$, we will show that $\alpha_1\big(\alpha_2(XYZ)\big) = (\alpha_1\alpha_2)(XYZ)$.
According to scalar multiplication definition, we have

$$\begin{aligned}\alpha_1\big(\alpha_2(XYZ)\big) &= \alpha_1\left(\varphi\big(\alpha_2\varphi^{-1}(XYZ)\big)\right)\\ &= \varphi\left(\alpha_1\varphi^{-1}\left(\varphi\big(\alpha_2\varphi^{-1}(XYZ)\big)\right)\right)\\ &= \varphi\big(\alpha_1\alpha_2\varphi^{-1}(XYZ)\big)\\ &= (\alpha_1\alpha_2)(XYZ)\end{aligned}$$

v) Let $1 \in GF(7^3)$ and $XYZ \in C_{343}$, we will show that $1(XYZ) = XYZ$.
According to scalar multiplication definition, we have

$$\begin{aligned}1(XYZ) &= \varphi\big(1\varphi^{-1}(XYZ)\big)\\ &= \varphi\big(\varphi^{-1}(XYZ)\big)\\ &= XYZ\end{aligned}$$

So we have that C_{343} is one dimensional vector space a vector space over Galois field $GF(7^3)$. with DDA as its canonical base.

Let $S = (C_{343})^N = C_{343} \oplus C_{343} \oplus \dots \oplus C_{343}$ (N-factor) and

$$s + s' = (s_1, \dots, s_N) + (s_1', \dots, s_N') = (s_1 + s_1', \dots, s_N + s_N')$$
$$\alpha s = \alpha(s_1, \dots, s_N) = (\alpha s_1, \dots, \alpha s_N)$$

It can be showed that $(S, +)$ is Abelian group with $(DDD, DDD, \dots, DDD)$ as its identity.

Therefore, $(C_{343})^N$ forms N-dimensional vector space over $GF(7^3)$ with $e_1 = (DDA, DDD, \dots, DDD), \dots e_N = (DDD, DDD, \dots, DDA)$ as their canonic bases. So, we can conclude that for every $s \in S$ there is unique representation as follows

$$s = a_1e_1 + \dots + a_Ne_N$$

or, $(a_1, \dots, a_N)$ is representation coordinate from S over canonical bases e_i.

4. Conclusion

Based on our explanation before, here are some conclusions that we have :

1. The standard genetic code can represented as the elements of extended triplet set (notated by C_{343}) with $X_1X_2X_3$; $X_i \in \{D, A, C, O, G, T, P\}$ as its elements.
2. The C_{343} set form commutative group structure over addition and the set $C_{343} - \{OOO\}$ form commutative group under multiplication with operation that have been defined. Then with scalar product that have been defined, C_{343} form one dimensional vector space structure over Galois field $GF(7^3)$ and the codons sequence is represented as $(a_1, a_2, \dots, a_N) \in (C_{343})^N$.

Acknowledgments

The authors thank to the University of Padjadjaran, due to the funding of this work through to the scheme of Riset Fundamental Unpad (RFU) with contract number 2400/UN6.D/KS/2018.

IOP Conf. Series: Materials Science and Engineering **567** (2019) 012019 doi:10.1088/1757-899X/567/1/012019

References

[1] Hornos J E and Hornos Y M 2006 Algebraic Model for The Evolution of The Genetic Code *Phys. Rev. Lett.* **71** 4401
[2] Sanchez R, Perfetti L A, Grau R and Morgado E 2005 A New DNA Sequences Vector Space on a Genetic Code Galois Field *MATCH Commun. Math. Comput. Chem.* **54** 3
[3] Sanchez R and Grau R 2009 An algebraic hypothesis about the primeval genetic code *Math. Biosci.* **221** 60
[4] Sanchez R, Grau R and Morgado E 2006 A Novel Lie Algebra of the Genetic Code over the Galois Field of Four DNA Bases *Math. Biosci.* **202** 156
[5] Sanchez R and Grau R 2006 A Novel Algebraic Structure of The Genetic Code Over The Galois Field of Four DNA Bases *Acta Biotheor.* **54** 27
[6] Sanchez R, Morgado E and Grau R 2005 Gene Algebra From a Genetic Code Algebraic Structure *J. Math. Biol* **51** 431
[7] Sanchez R, Grau R and Morgado E 2006 A Novel DNA Sequence Vector Space over an extended Genetic Code Galois Field *MATCH Commun. Math. Comput. Chem.* **56** 5
[8] Herstein, I. N. 1996. *Abstract Algebra.* New Jersey: Prentice-Hall
[9] Gallian, Joshep A. 2010. *Contemporary Abstract Algebra.* Belmont: Richard Stratton

IOP Conf. Series: Materials Science and Engineering **567** (2019) 012020 doi:10.1088/1757-899X/567/1/012020

Properties of anti-adjacency matrix of directed cyclic sun graph

M I A Prayitno[1], S Utama[1], S Aminah[1]

[1]Department of Mathematics, Faculty of Mathematics and Natural Sciences, Universitas Indonesia, 16424-Depok, Jawa Barat

Corresponding author: mirfanaps@sci.ui.ac.id, suarsih.utama@sci.ui.ac.id, aminah@sci.ui.ac.id,

Abstract. In this paper we focus on the properties of anti-adjacency matrix of directed cyclic sun graph. Some of these properties are related to the characteristic polynomials and the eigenvalues of the anti-adjacency of its matrix. We will show the general form of characteristic polynomial of the anti-adjacency matrix of directed cyclic sun graph by figuring out the number of the directed induced-cyclic graphs and the directed induced-acyclic graphs. After we find out the general form of the characteristic polynomial, we can find the general form of the eigenvalues of its polynomial by using factorization and Horner methods.

1. Introduction

There are many ways to illustrate a directed graph. Other than making the depiction of the graph, the directed graph can be represented by an adjacency matrix or anti-adjacency matrix [1]. Since the anti-adjacency matrix is an nxn matrix then we can find the characteristic polynomial and the roots / eigen values of its matrix [2].

Some properties of anti-adjacency matrix are related to its characteristic polynomial and its eigenvalues. In Chemical graph theory, the eigenvalues of the anti-adjacency matrix of a graph can be used in accomplishing physical properties of chemical graphs [3]. However, the study about the properties of anti-adjacency matrix of directed cyclic sun graph, especially its eigenvalues or its characteristic polynomials has not been explored yet. Therefore, this adds the importances of the further study about the anti-adjacency matrix of directed cyclic sun graph, which is being discussed in this paper. Hence, we will figure out the general form of the graph's anti-adjacency matrix, its characteristic polynomial, and its eigenvalues. The matrix's coefficients of characteristic polynomial can be obtained by looking for how many directed cyclic-induced subgraphs and directed acyclic-induced subgraphs within the graph [4].

2. Preliminaries

The following section explains the definitions and theorems required to prove the main result.

Definition 2.1. A directed graph $\boldsymbol{G}$ is a finite nonempty set of objects called vertices together with a (possibly empty) set of ordered pairs of distinct vertices of G called arcs or directed edges. The set of vertices of G is denoted by $\boldsymbol{V(G)} = \{\boldsymbol{v_1}, \boldsymbol{v_2}, \boldsymbol{v_3}, \dots, \boldsymbol{v_{n-1}}, \boldsymbol{v_n}\}$ and the set of directed edge denoted by $\boldsymbol{E(G)} = \{\boldsymbol{e_1}, \boldsymbol{e_2}, \boldsymbol{e_3}, \dots, \boldsymbol{v_n}\}$ [5].

Definition 2.2. An adjacency matrix of directed graph G with the set of vertices $V(G) = \{v_1, v_2, v_3, \dots, v_{n-1}, v_n\}$ is an nxn matrix $A = [a_{ij}]$, where $a_{ij} = 1$, if there exists a directed edge from v_i to v_j and $a_{ij} = 0$ otherwise. [1]

Definition 2.3. An anti-adjacency matrix of directed graph G with the set of vertices $V(G) = \{v_1, v_2, v_3, \dots, v_{n-1}, v_n\}$ is an nxn matrix $B = J - A$, where J is a nxn matrix which all the entries are 1 and A is an adjacency matrix of its graph [1]

Definition 2.4. A cycle of order n with m pendant edges attached at each vertex i.e., $C_n \circ \mathrm{m}\, \overline{K_1}$, is called an m-crown with cycle of order n. A 1-crown or a sun graph is a cycle with exactly one pendant edge attached at each vertex of the cycle. We will use $C_n \circ \overline{K_1}$ for this graph. [6]

Definition 2.5. Directed cyclic sun graph is a directed graph obtained by adding a directed edge and a vertex, which are called the outer-vertices, to every vertices in the cycle Graph $\overrightarrow{C_n}$, which are called the inner-vertices. Directed cyclic sun graph is denoted by $\overrightarrow{C_n \circ \overline{K_1}}$. It has a set of vertices $V\left(\overrightarrow{C_n \circ \overline{K_1}}\right) = \{v_1, v_2, v_3, \dots, v_{n-1}, v_n, v_{n+1}, v_{n+2}, \dots, v_{2n}\}$ and a set of edges $E\left(\overrightarrow{C_n \circ \overline{K_1}}\right) = \{\overrightarrow{v_1v_2}, \overrightarrow{v_2v_3}, \dots, \overrightarrow{v_{n-1}v_n}, \overrightarrow{v_nv_1}, \overrightarrow{v_{n+1}v_1}, \overrightarrow{v_{n+2}v_2}, \dots, \overrightarrow{v_{2n}v_n}\}$, where $v_1, v_2, v_3, \dots, v_{n-1}, v_n$ are the inner-vertices and $v_{n+1}, v_{n+2}, \dots, v_{2n}$ are the outer-vertices of it. In this paper we focus on the directed cyclic sun graph which the directed edges of each outer-vertices v_{n+i} always lead to the inner-vertices v_i and the inner-vertices always form a clockwise-cycle graph for $i = 1, 2, \dots, n$. The following is the illustration of it.

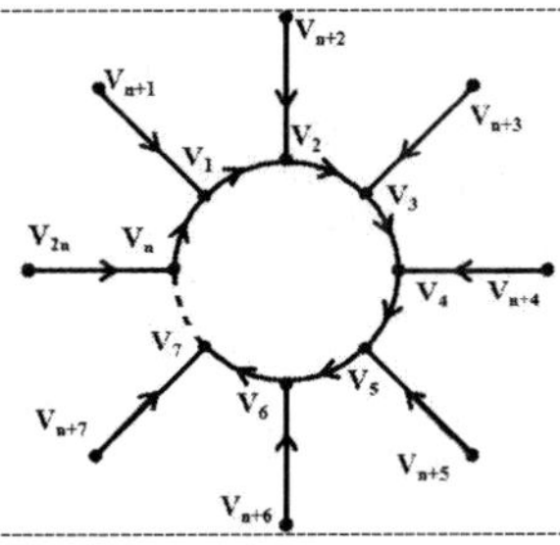

Figure 1. Directed Cyclic Sun Graph $\overrightarrow{C_n \circ \overline{K_1}}$.

The directed cyclic sun graph $\overrightarrow{C_n \circ \overline{K_1}}$ has an anti-adjacency $B\left(\overrightarrow{C_n \circ \overline{K_1}}\right) = [b_{ij}]$, where

$$b_{ij} = \begin{cases} 0, & \{1 \le i \le n-1, j = i+1\}; \{i = n, j = 1\}; \{i = j+n, 1 \le j \le n\} \\ 1, & otherwise \end{cases}$$

Theorem 2.6. Let G be a directed acyclic graph, with $V(G) = \{v_1, v_2, v_3, \dots, v_{n-1}, v_n\}$ and B is the anti-adjacency of G. Then $\mathbf{det}(B) = 1$, if G has a Hamiltonian path and $\mathbf{det}(B) = 0$ otherwise. [2]

Theorem 2.7. Let $\mathbf{P(B(G)) = \lambda^n + b_1\lambda^{n-1} + b_2\lambda^{n-2} + \dots + b_{n-1}\lambda + b_n}$ be a characteristic polynomial of an anti-adjacency matrix $\mathbf{B(G)}$ of a directed graph, let $\left|\mathbf{B(\langle U\rangle_{acyclic})_i^{(j_1)}}\right|$ be the determinant of an anti-adjacency matrix of a directed acyclic induced subgraph with i vertices and $j_1 = 1, 2, \dots, w_1$; where w_1 is the number of a directed acyclic-induced subgraphs $\langle U\rangle_{acyclic}$ with i-vertices of directed cyclic graph $\mathbf{G}$, and let $\left|B(\langle U\rangle_{cyclic})_i^{(j_2)}\right|$ be the determinant the of anti-adjacency matriks of directed cyclic induced subgraph with $\boldsymbol{i}$-vertices and $\boldsymbol{j_2} = \mathbf{1, 2, \dots, w_2}$, where $\boldsymbol{w_2}$ is the number of directed cyclic induced sugraph $\langle\mathbf{U}\rangle_{\mathbf{cyclic}}$ with **i** vertices of directed cyclic graph **G**, then $\boldsymbol{b_i} = (-\mathbf{1})^{\boldsymbol{i}}\left(\sum_{j_1=1}^{w_1}\left|B\left(\langle U\rangle_{acyclic}\right)_i^{(j_1)}\right| + \sum_{j_2=1}^{w_2}\left|B\left(\langle U\rangle_{cyclic}\right)_i^{(j_2)}\right|\right)$, where $\boldsymbol{i} = \mathbf{1, 2, \dots, n}$. [5]

Lemma 2.8. Let G be a directed graph with a set of vertices $\boldsymbol{V} = \{\boldsymbol{v_1, v_2, \dots, v_n}\}$ and B be an anti-adjacency matrix of G. If there exists two vertices which their directed edges go to the same vertex then $\det(\boldsymbol{B}) = \mathbf{0}$.

Lemma 2.9. Let B be an anti-adjacency of directed cycle graph $\overrightarrow{\mathbf{C_n}}$. Then $\det\left(\boldsymbol{B}\left(\overrightarrow{\boldsymbol{C_n}}\right)\right) = \boldsymbol{n} - \mathbf{1}$. [5]

3. Main Result

In the following section we will discuss about the theorems which are related to the main result. In Theorem 3.1, we will discuss about the coefficients b_i; $1 \le i \le 2n$, of the characteristic polynomial of the anti-adjacency matrix of directed cyclic sun graph $\overrightarrow{C_n \circ K_1}$. In Corollary 3.2, we will discuss about the general form of the characteristic polynomial of its anti-adjacency matrix. In Theorem 3.3, we will discuss about the real eigenvalues and complex eigenvalues of its anti-adjacency matrix.

Theorem 3.1. If $P(B\,(\overrightarrow{C_n \circ K_1})) = \lambda^{2n} + b_1\lambda^{2n-1} + b_2\lambda^{2n-2} + \dots + b_{2n-1}\lambda + b_{2n}$ is the characteristic polynomial of the anti-adjacency of directed cylic sun graph $\overrightarrow{C_n \circ K_1}$ then

a) $b_i = (-1)^i(2n)$, for $1 \le i \le n-1$. ... (2)

b) $b_i = (-1)^i(2n-1)$, for $i = n$ ... (3)

c) $b_i = 0$, for $n+1 \le i \le 2n$. ...(4)

Proof:

According to Theorem 2.7, $b_i = (-1)^i\left(\sum_{j_1}^{w_1}\left|B\left(\langle U\rangle_{acyclic}\right)_i^{(j_1)}\right| + \sum_{j_2}^{w_2}\left|B\left(\langle U\rangle_{acyclic}\right)_i^{(j_2)}\right|\right)$.

a) For $1 \le i \le n-1$

- Acyclic Parts

The following is the illustration of its directed acyclic-induced subgraphs.

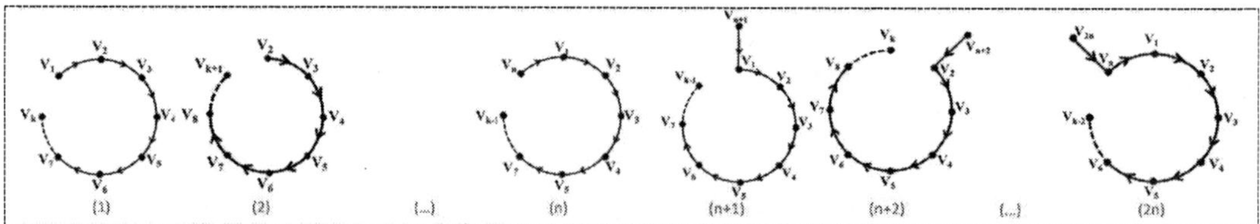

Figure 2. Directed acyclic-induced subgraphs for $i = k$, $1 \le k \le n-1$.

Let B be an anti-adjacency matrix of the directed acylic-induced subgraphs which induce the directed cyclic sun graph $\overrightarrow{C_n \circ K_1}$, for $i = k$, $1 \le k \le n-1$. The set of directed acyclic-induced subgraphs which in the form of Hamiltonian paths is $\{v_1 - v_2 - \dots - v_k,\ v_2 - v_3 - \dots - v_{k+1}, \dots, v_n - v_1 - v_2 - \dots - v_{k-1},\ v_{n+1} - v_1 - v_2 - \dots - v_{k-1},\ v_{n+2} - v_2 - v_3 - \dots - v_k, \dots,\ v_{n+n} - v_n - v_1 - v_2 - \dots - v_{k-2}\}$. From the set of directed cyclic-induced subgraphs, see Fig. 2, we can see that there are $2n$-subgraphs of directed acyclic-induced subgraphs which is in the form of Hamiltonian paths.

Hence, according to Theorem 2.6 and Theorem 2.7; $\det(B) = 1$ and $\sum_{j_1}^{w_1} \left| B\left(\langle U \rangle_{acyclic}\right)_i^{(j_1)} \right| = (2n) \cdot 1 = 2n$.

- Cyclic Parts

Since making directed cyclic-induced subgraph in directed cyclic sun graph needs at least n -vertices, then in the directed cyclic sun graph $\overrightarrow{C_n \circ K_1}$ there does not exist a directed cyclic-induced subgraph which has i-vertices; for $1 \le i \le n-1$. Hence, $\sum_{j_2}^{w_2} \left| B\left(\langle U \rangle_{cyclic}\right)_i^{(j_2)} \right| = 0$.
So we can conclude that
$b_i = (-1)^i(2n + 0) = (-1)^i(2n)$, for $1 \le i \le n-1$. ■

b) For $i = n$
- Acyclic Parts

The following is the illustration of its directed acyclic-induced subgraphs.

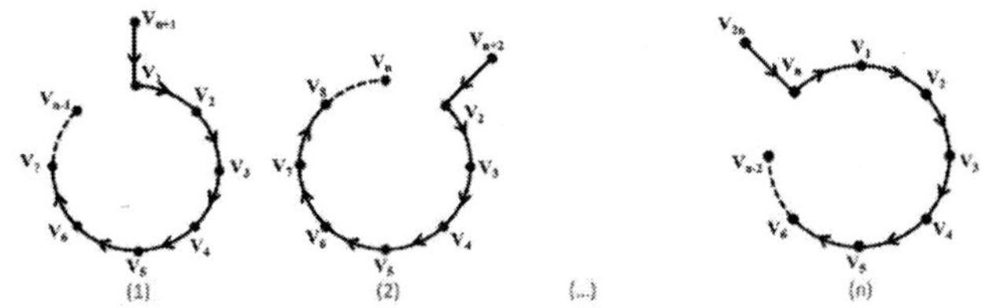

Figure 3. Directed acyclic-induced subgraphs for $i = n$.

Let B be an anti-adjacency matrix of its directed acylic-induced subgraphs which induce the directed cyclic sun graph $\overrightarrow{C_n \circ K_1}$, for $i = n$. The set of directed acyclic-induced subgraphs which in the form of Hamiltonian paths is $\{v_{n+1} - v_1 - v_2 - \cdots - v_{n-1}, v_{n+2} - v_2 - v_3 - \cdots - v_n, \dots, v_{2n} - v_n - v_1 - v_2 - \cdots - v_{n-2}\}$. From the set of directed acyclic-induced subgraphs, see Fig. 3, we can see that there are n-subgraphs of directed acyclic-induced subgraphs which is in the form of Hamiltonian paths. Hence, according to Theorem 2.6 and Theorem 2.7; $\det(B) = 1$ and $\sum_{j_1}^{w_1} \left| B\left(\langle U \rangle_{acyclic}\right)_i^{(j_1)} \right| = (n) \cdot 1 = n$.

- Cyclic Parts

The number of directed cyclic-induced subgraphs which is in the form of directed cycle graph and has n-vertices is one subgraphs. Let B be an anti-adjacency matrix of its subgraph. Hence, according to Lemma 2.9 $\det(B) = n - 1$ and $\sum_{j_2}^{w_2} \left| B\left(\langle U \rangle_{cyclic}\right)_i^{(j_2)} \right| = 1 \cdot (n-1) = n - 1$.
So we can conclude that
$b_i = (-1)^i\left(n + (n-1)\right) = (-1)^i(2n-1)$, for $i = n$. ■

c) For $n + 1 \le i \le 2n$
- Acyclic Parts

Since making directed acyclic-induced subgraph in directed cyclic sun graph needs at most n-vertices then in the directed cyclic sun graph $\overrightarrow{C_n \circ K_1}$ there is not exist a directed acyclic-induced subgraph which has i-vertices; for $n+1 \le i \le 2n$. Hence, $\sum_{j_1}^{w_1} \left| B\left(\langle U\rangle_{acyclic}\right)_i^{(j_1)} \right| = 0$

- Cyclic Parts

The following is the illustration of its the directed cyclic-induced subgraphs.

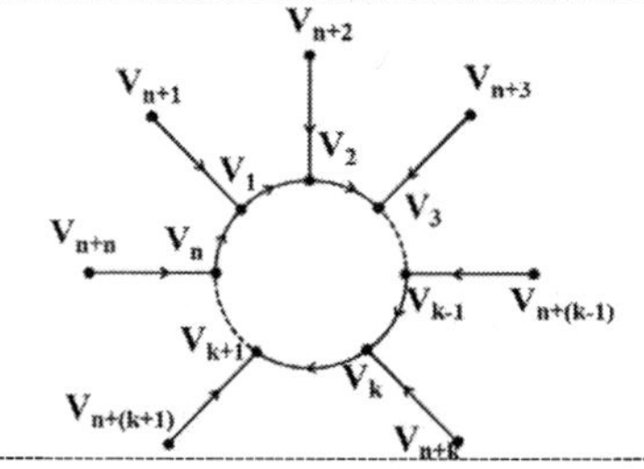

Figure 4. Directed cyclic-induced subgraphs for $n+1 \le i \le 2n$.

Let B be an anti-adjacency matrix of its directed cylic-induced subgraphs for $n+1 \le i \le 2n$. From Fig. 4 we can see that among the directed cyclic-induced subgraphs from directed cyclic sun graph $\overrightarrow{C_n \circ K_1}$ which has $(n+k)$-vertices, there are two vertices which their directed edges go to the same vertex, those are $v_{n+(k+1)} - v_{k+1}$ and $v_k - v_{k+1}$ for $1 < k \le n$, or $v_{n+1} - v_1$ and $v_n - v_1$ for $k = 1$. Then according to Lemma 2.8, det(B) $= 0$ and $\sum_{j_2}^{w_2} \left| B\left(\langle U\rangle_{cyclic}\right)_i^{(j_2)} \right| = 0$.

So we can conclude that

$b_i = (-1)^i(0+0) = 0$, for $n+1 \le i \le 2n$. ■

Corollary 3.2. Let $B\ (\overrightarrow{C_n \circ K_1})$ be the anti-adjacency matrix of directed cyclic sun graph $\overrightarrow{C_n \circ K_1}$, then the characteristic polynomial of its matrix is

$$P\left(B\ (\overrightarrow{C_n \circ K_1})\right) = \lambda^{2n} + \sum_{i=1}^{n-1}(-1)^i(2n)\lambda^{2n-i} + (-1)^n(2n-1)\lambda^n, \quad \text{for } n \ge 3. \quad \ldots (5)$$

Proof:

According to Theorem 3.1, the coefficients of the characteristic polynomial of the anti-adjacency matrix of directed cyclic sun graph $\overrightarrow{C_n \circ K_1}$ are $b_i = (-1)^i(2n)$; for $1 \le i \le n-1$, $b_i = (-1)^i(2n-1)$; for $i = n$, and $b_i = 0$, for $n+1 \le i \le 2n$. Hence, the characteristic polynomial is the equation (5). ■

Theorem 3.3. *Let* $P\left(B\ (\overrightarrow{C_n \circ K_1})\right) = \lambda^{2n} + \sum_{i=1}^{n-1}(-1)^i(2n)\lambda^{2n-i} + (-1)^n(2n-1)\lambda^n;$ *for* $n \ge 3$, is the characteristic polynomial of the anti-adjacency matrix of directed cylic sun graph. Then the eigenvalues of its characteristic polynomial are

a) $\lambda_{1,2,\ldots,2k} = 0$, $\lambda_{2k+1} = (4k-1)$, $\lambda_{2k+2} = 1$, and $\lambda_{2k+3,2k+4,\ldots,4k} = e^{i\left(\frac{m\pi}{k}\right)}$; where $m = 1,2,3,\ldots,2k-2$, for $n = 2k$, $k = 2,3,4,\ldots$ ... (6)

b) $\lambda_{1,2,\ldots,2k+1} = 0$, $\lambda_{2k+2} = (4k+1)$, and $\lambda_{2k+3,2k+4,\ldots,4k} = e^{i\left(\frac{\pi}{2k+1}+\frac{2m\pi}{2k+1}\right)}$, where $m = 1,2,3,\ldots,2k$ for $n = 2k+1$, $k = 1,2,3,\ldots$ ... (7)

Proof:

a) For $n = 2k$, $k = 2,3,4,\ldots$, then the equation (1) is equal to the following: $P\left(\mathrm{B}\ (\overrightarrow{\mathrm{C_n} \circ \mathrm{K_1}})\right) = \lambda^{4k} + \sum_{i=1}^{2k-1}(-1)^i(4k)\lambda^{4k-i} + (-1)^{2k}(4k-1)\lambda^{2k} = 0$

$\Leftrightarrow P\left(\mathrm{B}\left(\overrightarrow{\mathrm{C_n} \circ \mathrm{K_1}}\right)\right) = \lambda^{2k}\left(\lambda^{2k} + \sum_{\mathrm{i=1}}^{\mathrm{2k-1}}(-1)^{\mathrm{i}}(4\mathrm{k})\lambda^{\mathrm{2k-i}} + (4k-1)\right) = 0$

To simplify $\lambda^{2k} + \sum_{\mathrm{i=1}}^{\mathrm{2k-1}}(-1)^{\mathrm{i}}(4\mathrm{k})\lambda^{\mathrm{2k-i}} + (4k-1)$ we use the Horner method in the following:

4k+1	1	-(4k+2)	(4k+2)	-(4k+2)	...	-(4k+2)	(4k+2)	-(4k+2)	(4k-1)
		(4k+1)	-(4k+1)	(4k+1)	...	(4k+1)	-(4k+1)	(4k+1)	-(4k+1)
1	1	-1	1	-1	...	-1	1	-1	0
		1	0	1	...	1	0	1	
	1	0	1	0		0	1	0	

Figure 5. Horner method for $n = 2k$, $k = 2,3,4, \ldots$.

Hence, we will have

$P\left(\mathrm{B}\left(\overrightarrow{\mathrm{C_n} \circ \mathrm{K_1}}\right)\right) = \lambda^{2k}[\lambda - (4k-1)][\lambda - 1]\left[1 + \sum_{i=1}^{k-1}(\lambda^2)^i\right] = 0.$ ... (8)

From the equation (8), we can see the polynomial $\left[1 + \sum_{i=1}^{k-1}(\lambda^2)^i\right]$ is a geometric series with the first term is $a = 1$, the ratio is $r = \lambda^2$, and the number of the terms is k. Hence, we will have $1 + \sum_{i=1}^{k-1}(\lambda^2)^i = \frac{\lambda^{2k}-1}{\lambda^2-1} \Rightarrow P\left(B\left(\overrightarrow{C_n \circ K_1}\right)\right) = \lambda^{2k}[\lambda - (4k-1)][\lambda - 1]\left[\frac{\lambda^{2k}-1}{\lambda^2-1}\right] = 0$ ... (9)

From the equation (9) we will have:

- ❖ $\lambda_{1,2,\ldots,2k} = 0$, $\lambda_{2k+1} = 4k-1$, and $\lambda_{2k+2} = 1$
- ❖ $\frac{\lambda^{2k}-1}{\lambda^2-1} = 0$

  $\Rightarrow \lambda^{2k} - 1 = 0$

  $\Leftrightarrow \lambda^{2k} = 1$ and $\lambda \neq 1, \lambda \neq -1$ ... (10)

  According to [7], the roots from the equation (10) are complex numbers and to solve the equation (10), let $\lambda = r_0 e^{i\theta}$ be the solution of the equation $\lambda^{2k} = 1$. Hence

  $\left(r_0 e^{i\theta}\right)^{2k} = 1 \Leftrightarrow {r_0}^{2k} e^{i\cdot(2k)\cdot\theta} = 1 \Leftrightarrow {r_0}^{2k} e^{i\cdot(2k)\cdot\theta} = 1 \cdot e^{(0+2m\pi)}$ ... (11)

  From the equation (11) we will have $r_0 = 1$ and $\theta = \frac{2m\pi}{2k} = \frac{m\pi}{k}$. ... (12)

  Hence, from the equation (10) and the equation (12) we will have $\lambda_{2k+3,2k+4,\ldots,4k} = e^{i\left(\frac{m\pi}{k}\right)}$, for $m = 1,2,3,\ldots,2k-2$

  ■

b) For $n = 2k+1$; $k = 1,2,3,\ldots$, the equation (1) is equal to the following:

$P\left(B\left(\overrightarrow{C_n \circ K_1}\right)\right) = \lambda^{4k+2} + \sum_{i=1}^{2k}(-1)^i(4\mathrm{k}+2)\lambda^{(4\mathrm{k}+2)-i} + (-1)^{2k+1}(4k+1)\lambda^{2k+1} = 0$

$\Leftrightarrow P\left(B\left(\overrightarrow{C_n \circ K_1}\right)\right) = \lambda^{2k+1}\left(\lambda^{2k+1} + \sum_{i=1}^{2k}(-1)^i(4\mathrm{k}+2)\lambda^{(2\mathrm{k}+1)-i} - (4k+1)\right) = 0$ ... (12)

Similar to a), we use Horner method with the factor equals to $\lambda = (4k+1)$ to simplify the equation (12). Then we will have the following:

$P\left(B\left(\overrightarrow{C_n \circ K_1}\right)\right) = \lambda^{2k+1}[\lambda - (4k+1)]\left[1 + \sum_{i=1}^{2k}(-\lambda)^i\right] = 0.$ ... (13)

From the equation (13), we can see the polynomial $1 + \sum_{i=1}^{2k}(-\lambda)^i$ is a geometric series with the first term is $a = 1$, the ratio is $r = -\lambda$, and the number of the terms is $2k+1$. Hence, we will have

$1 + \sum_{i=1}^{2k}(-\lambda)^i = \frac{-\lambda^{2k+1}-1}{-\lambda-1}$. So we will have the following:

$P\left(B\left(\overrightarrow{C_n \circ K_1}\right)\right) = \lambda^{2k+1}[\lambda - (4k+1)]\left[\frac{-\lambda^{2k+1}-1}{-\lambda-1}\right] = 0$ ... (14)

From the equation (14) we will have

- ❖ $\lambda_{1,2,\ldots,2k+1} = 0$ and $\lambda_{2k+2} = 4k+1$

- $\frac{-\lambda^{2k+1}-1}{-\lambda-1} = 0$
  $\Rightarrow -\lambda^{2k+1} - 1 = 0$
  $\Leftrightarrow \lambda^{2k+1} = -1$ and $\lambda \neq -1$ ... (15)
  According to [7], the roots from the equation (15) are complex numbers and to solve the equation (15), let $\lambda = r_0 e^{i\theta}$ be the solution of $\lambda^{2k+1} = -1$. Hence,
  $(r_0 e^{i\theta})^{2k+1} = -1 \Leftrightarrow r_0{}^{2k+1} e^{i\cdot(2k+1)\cdot\theta} = -1 \Leftrightarrow r_0{}^{2k+1} e^{i\cdot(2k+1)\cdot\theta} = 1 \cdot e^{i(\pi+2m\ )}$ ... (16)
  From the equation (16) we will have $r_0 = 1$ and $\theta = \frac{\pi}{2k+1} + \frac{2m\pi}{2k+1}$. ... (17)
  Hence, from the equation (15) and the equation (17) we will have
  $\lambda_{2k+3,2k+4,\ldots,4k+2} = e^{i\left(\frac{\pi}{2k+1}+\frac{2m\pi}{2k+1}\right)}$, for $m = 1,2,3,\ldots,2k$. ■

4. Conclusion

The general form of the anti-adjacency matrix, the coeficients of its characteristic polynomial, the general form of the characteristic polynomial of its matrix, and the eigenvalues of directed cylic sun graph can be seen in the equation (1), (2), (3), (4), (5), (6), and the following table:

Table 1. The eigenvalues of anti-adjacency matrix of directed cyclic sun graph

The value of n	Separation	
	Real Eigenvalues	Complex Eigenvalues
$n = 2k + 1; k = 1,2,3, \ldots)$	$\lambda_{1,2,\ldots,2k+1} = 0$ and $\lambda_{2k+2} = 4k + 1$	$\lambda_{2k+3,2k+4,\ldots,4k+2} = e^{i\left(\frac{\pi}{2k+1}+\frac{2m\pi}{2k+1}\right)}$, where $m = 1,2,3,\ldots,2k$,
$n = 2k, k = 2,3,4, \ldots)$	$\lambda_{1,2,\ldots,2k} = 0, \lambda_{2k+1} = 4k - 1$, and $\lambda_{2k+2} = 1$	$\lambda_{2k+3,2k+4,\ldots,4k} = e^{i\left(\frac{m\pi}{k}\right)}$; where $m = 1,2,3,\ldots,2k-2$

Acknowledgments

This research is funded by Hibah PITTA UI 2018 (2307/UN2.R3.1/HKP.05.00/2018).

References

[1] Bapat R B 2010 Graphs and Matrices (New Delhi: Springer)
[2] Anton H and Rorres C 2010 Elementary Linear Algebra 9th edition (New York: Wiley)
[3] Wang J, Lu M, Belardo F, and Randic M 2018. The anti-adjacency matrix of a graph: Eccentricity matrix. Discrete Applied Mathematics. 251. 299-309.
[4] Wildan 2013 Polinomial Karakteristik Matriks Anti-adjacency dan Adjacency dari Graf Berarah yang Diberi Orientasi Master Thesis. (Depok: Universitas Indonesia)
[5] Chartrand G, Lesniak L, and Zhang P 2011 *Graphs & Digraphs* fifth edition (Florida: CRC Press)
[6] Irfan M and Semaničová-Feňovčíková A 2016. On total labelings of graphs with prescribed weights. *AKCE International Journal of Graphs and Combinatorics*. 13. 10.1016/j.akcej.2016.06.001.
[7] J. W. Brown and R. V. Churchill, *Complex Variables and Applications* (McGraw-Hill, New York, 2009)

IOP Conf. Series: Materials Science and Engineering **567** (2019) 012021 doi:10.1088/1757-899X/567/1/012021

Inhibition of interleukin-8 production in interleukin-1 stimulated human monocytic THP-1 cells by *N,N* didesmethylgrossularine-1 obtained from an Ascidian *Polycarpa aurata* collected in North Sulawesi

D A Sumilat[1], T Oda[2], D S Wewengkang[3], M Namikoshi[4], H Yamazaki[4]

[1]Faculty of Fisheries and Marine Science, Sam Ratulangi University, Indonesia
[2]Faculty of Pharmacy, Keio University, Minato-ku, Tokyo, Japan
[3]Faculty of Mathematic and Natural Sciences, Sam Ratulangi University
[4]Faculty of Pharmaceutical Sciences, Tohoku Medical and Pharmaceutical University, Sendai, Japan

Email: deiske.sumilat@gmail.com

Abstract. *N,N*-Didesmethylgrossularine-1 (DDMG-1) has a rare β-carboline structure and was isolated from an Indonesian ascidian *Polycarpa aurata* as an active component against tumor necrosis factor (TNF)-α production in lipopolysaccharine (LPS)-stimulated murine macrophage-like RAW 264.7 cells as reported in our previous paper. Further investigation on the inhibitory activity of DDMG-1 against the production of inflammatory cytokines in human monocytic THP-1 cells, we found that DDMG-1 reduced the excess production of IL-8 in LPS-stimulated THP-1 cells through inhibition of the mRNA level of IL-8 IκB-α degradation, and binding of NF-κB to the target DNA site. The THP-1 cells used in this study showed the high production of IL-8 by the stimulation with TNF-α, IL-1β, and 12-O-tetradecanoylphorbol-13-acetate (PMA) as similar to LPS. DDMG-1 inhibited the IL-8 production in IL-1β-stimulated THP-1 cells but did not show an inhibitory activity in the cells stimulated by TNF-α and PMA. Therefore, DDMG-1 was specific to the IL-1β signalling pathway. These results suggested that DDMG-1 could be a useful drug candidate lead compound to control the excess production of IL-8.

1. Introduction

In the previous paper [1], we described the inhibitory activity of *N,N*-didesmethylgrossularine-1 (DDMG-1) on the production of tumor necrosis factor (TNF)-α in lipopolysaccharide (LPS)-stimulated murine macrophage-like RAW 264.7 cells. To compare the difference in the response to DDMG-1 between murine and human immune cells, studies on the effects of DDMG-1 against the production of inflammatory cytokines were performed using human monocytic THP-1 cells, which produced large amounts of interleukin (IL)-8 (abt. 220 ng/ mL) by the stimulation with LPS.

THP-1 cells were derived from an acute monocytic leukemia patient and can be differentiated into macrophage-like cells by for example 12-O-tetradecanoylphorbol-13-acetate (PMA) [2], and, therefore, this cell line is used for the studies on monocyte/macrophage functions, immune modulations [3]. LPS is a component of the Gram-positive bacterial cell wall and frequently used as an inflammatory model because of its property to activate macrophages [4]. Macrophages play important roles in immune reactions, inflammation, and allergy [5]. In response to LPS and other

stimulants, macrophages secrete many kinds of inflammatory cytokines such as IL-1 and IL-6, chemokines such as IL-8, and (TNF)-α via the activation of nuclear factor-kappa B (NF-κB) [6]. IL-8 is a member of the superfamily of C-X-C chemokines and is a chemotactic factor for T cells, neutrophils, and basophils. Expression of IL-8 has been detected in various human cancers and it has been suggested to be a factor in tumor progression and metastasis [7][8][9]. Regulation of IL-8 production is, therefore, an important target for therapeutic research.
In this study, we found that DDMG-1 reduced the IL-8 production in LPS-stimulated THP-1 cells through the inhibition of the messenger RNA (mRNA) level of IL-8, NF-κB inhibitor (IκB)-α degradation, and binding of NF-κB to the target DNA site, which was specific to the IL-1β signaling pathway.

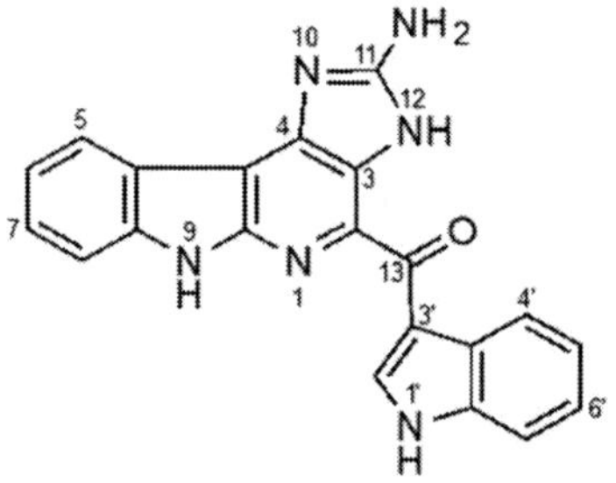

Figure 1. Structure of *N,N*-didesmethylgrossularine-1 (DDMG-1) isolated from *Polycarpa aurata* collected in North Sulawesi.

2. Materials and Methods

2.1. General experimental procedures.

EI mass spectra were obtained by a JEOL JMS-MS 700 mass spectrometer (Tokyo, Japan). 1H and 13C NMR spectra were recorded on a JEOL JNM-AL-400 NMR spectrometer at 400 MHz for ^{1}H and 100 MHz for ^{13}C in CD_3OD (δH 3.30, δC 49.0). Preparative HPLC was carried out with a Hitachi L-6200 system.

2.2. Materials.

Fetal bovine serum (FBS) and other culture materials were purchased from Invitrogen (Carlsbad, CA, USA). Anti-actin was purchased from Santa Cruz Biotechnology (Santa Cruz, CA, USA). All other chemicals, including endotoxin LPS (*Escherichia coli* 055-85) and 3-(4,5-dimethylthiazol-2-yl)-2,5-diphenyltetrazolium bromide (MTT), were purchased from Sigma-Aldrich (St. Louis, Mo., USA).
Isolation of *N,N*-didesmethylgrossularine-1 (DDMG-1).

The solitary ascidian *P. aurata* (978 g, wet weight) was collected by scuba diving in the coral reef at Manado, North Sulawesi, on September 19th 2008, cut into small pieces, and soaked in EtOH on a boat immediately after collection. The EtOH extract was evaporated, and the residue was dissolved in MeOH-H_2O (4:1) and washed with n-hexane. The methanolic solution was concentrated and diluted with water to a MeOH content of 40%, and the solution was extracted with $CHCl_3$. The $CHCl_3$ extract (5.3 g) was separated by a SiO_2 column with $CHCl_3$-MeOH (gradient elution) into six fractions. The 7:3 eluate (3.2 g) was subjected to ODS column chromatography (eluted with 50, 70, 85, and 100% MeOH) to give four fractions. DDMG-1 (5.2 mg) was purified from the second fraction (70% MeOH elution, 0.3 g) by preparative HPLC (ODS, 70% MeOH).
DDMG-1: a yellow pigment; ^{1}H NMR; ^{13}C-NMR EIMS m/z 366 $[M]^+$; 249 $[M - indole]^+$; 221 $[M - indole - CO]^+$. Table 1 [14].

Table 1. NMR Data for DDMG-1 [14]

	^{1}H, δ		^{13}C	
C		δ	mult^{b}	HMBC corr
1				
2		146.6	C	
3		132.4	C	
4		126.5	C	
4a		105.4	C	
4b		119.8	C	
5	8.19 (d, 7.8)	122.7	CH	C-7
6	7.21 (td, 7.8, 1.2)	118.9	CH	C-4b/C-7/C-8
7	7.41 (td, 7.8, 1.2)	127.4	CH	C-6/C-8a
8	7.51 (d, 7.8)	110.6	CH	C-4b
8a		137.5	C	
9	11.6 (s)			C-4a/C-4b
11			159.4	C
12	11.3 (s)			
13		187.0	C	
1’	12.2 (s)			
2’	9.6 (s)	139.7		C-3′/C-7a′
3’		114.5		C
3a’		125.7		C
4’	8.54 (m)	122.3	CH	C-4′/C-9′ C-3a′/C-7a′
4b				
5’	7.26 (m)	121.8^{c}	CH	
6’	7.27 (m)	121.9^{c}	CH	
7’	7.57 (m)	112.2	CH	C-6′/C-7a′
7a’		135.9	C	

2.3. Cell culture.

The culture of THP-1 cells was performed in the RPMI medium containing 10% FBS. THP-1 cells were pretreated with DDMG-1 for 5 min. After addition of each stimulant (LPS, TNF-α, IL-1β, and PMA at the final concentration of 1 μg.mL-1, 20 ng.mL-1, 5 ng.mL-1, and 20 nM, respectively), the cells were cultured for 24 h at 37 °C. The concentration of inflammatory cytokines in each culture supernatant was measured by enzyme-linked immunosorbent assay (ELISA). A control experiment was carried out without the addition of a stimulant.

2.4. Measurement of IL-8 mRNA expression.

THP-1 cells were incubated with 5 or 10 μM DDMG-1 for 1 h, and then 1 μg of LPS was added. After incubation for 6 h, the cells were recovered and washed twice with phosphate-buffered saline (PBS) (–). The total RNA was extracted with an RNeasy Mini Kit (Qiagen Co., Tokyo, Japan). An iScript cDNA synthesis Kit was used to synthesize the first-strand cDNAs from 1 μg of the total RNA. Polymerase chain reaction (PCR) was performed using the CFX96/384 real-time PCR system (BIO-RAD, California, USA). Primers (human IL-8 and GAPDH) for quantitative PCR were purchased from BIO-RAD (USA).

2.5. Western blot analysis.

THP-1 cells were treated with DDMG-1 for 15 min and then stimulated with LPS (1 μg/mL) for 5–30 min. Cytoplasmic extracts of cells were subjected to Western blot analysis with an anti-IκB antibody.

The blots were finally reacted with the ECL reagent (Amersham-Pharmacia, San Francisco, CA, USA) and exposed to X-ray film [10].

2.6. Determination of cell cytotoxicity.
Cell proliferation was evaluated by enumerating viable cells using the MTT formazan production method [11]. THP-1 cells were incubated with various concentrations (0.1, 1, 3, and 10 μM) of DDMG-1 for 24 h. The cells were treated with MTT solution and, after incubation for 3 h, production of formazan was assessed by measuring optical density (OD) at 570 nm.

2.7. Preparation of nuclear extracts and electrophoretic mobility shift assay (EMSA).
Nuclear extracts were prepared from THP-1 cells as described previously [12]. EMSA was conducted on 5% polyacrylamide gels in 1 x Tris-borate/EDTA electrophoresis buffer. The other experimental conditions were the same as described in the previous report [12].

2.8. Statistical analysis.
Each experiment was performed at least three times, and representative data are shown. Means were checked for statistical differences using Student's t-test with error probabilities of $p < 0.1$.

3. Results and Discussion
3.1. Isolation of DDMG-1.
DDMG-1 was obtained as a yellow pigment from *P. aurata* collected in the coral reef at Manado, North Sulawesi. The structure (Fig. 1) was confirmed by the direct comparison of chromatographic and spectroscopic properties with those of the compound isolated and identified previously [13].

DDMG-1 was first isolated from *P. aurata* collected in Chuuk Atoll [14], and we isolated this compound from an Indonesian *P. aurata* and reported the inhibitory activity against the production of TNF-α in LPS-stimulated RAW 264.7 [1]. Three α-carboline alkaloids, grossularines-1 (*N,N*-dimethylamino derivative at C-11 of DDMG-1) and -2 (4-hydroxyphenyl derivative at C-13 of grossularine-1) [15] and DDMG-1 [14], have thus far been isolated from ascidians. Grossularines-1 and -2 showed the weak cytotoxicity against murine leukemia L1210 cells and more potently to human colon (WiDr) and breast (MCF7) tumor cell lines [15][16]. Interestingly, the mechanisms of cytotoxicity by these compounds were suggested to be different [16]. On the other hand, the biological activity of DDMG-1 has not been reported, and we found for the first time the inhibitory activity of DDMG-1 against TNF-α [1].

3.2. Inhibition of LPS-induced IL-8 production by DDMG-1 in THP-1 cells.
We first examined the effect of DDMG-1 on LPS-stimulated THP-1 cells in order to compare the experimental results from the previous study using murine RAW 264.7 cells [1]. In LPS-stimulated THP-1 cells, large amounts of IL-8 (ca 220 ng/mL) were detected in the culture supernatant, but the production of TNF-α was small. Therefore, the effects of DDMG-1 on the inflammatory cytokine production system were observed for the IL-8 production. The results of the quantitative analysis of IL-8 in the culture supernatant detected by ELISA are shown in Fig. 2A, and Fig. 2B shows the effects of DDMG-1 on the IL-8 mRNA expression detected by real-time PCR analysis. These results indicated that DDMG-1 inhibited the IL-8 production in LPS-stimulated THP-1 cells, although the inhibitory effects at lower concentrations of DDMG-1 were smaller than those in RAW cells [1].
Since the transcription factor NF-κB is also greatly involved in the production of IL-8, we examined the effects of DDMG-1 (10 μM) on the induction of NF-κB using Western blot analysis (Fig. 2C) and EMSA (Fig. 2D). These experiments revealed that DDMG-1 suppresses the activation of NF-κB transcription factor. Consequently, DDMG-1 (10 μM) had a very similar inhibitory activity against the TNF-α and IL-8 production in LPS-stimulated RAW 264.7 and THP-1 cells, respectively.

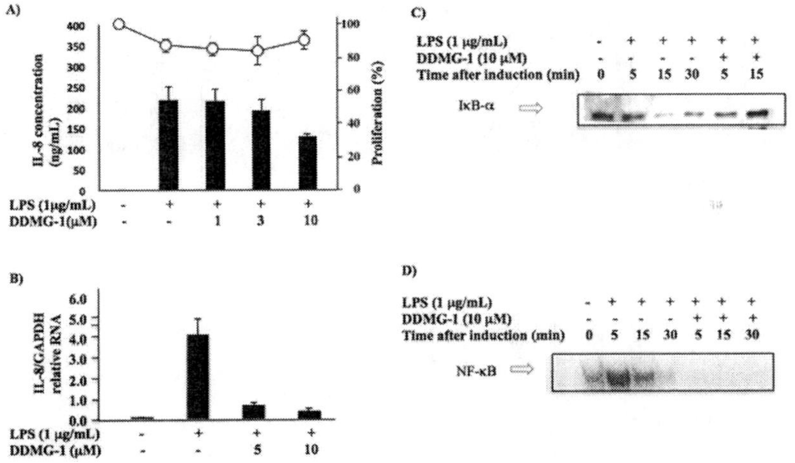

Figure 2. Effects of DDMG-1 on IL-8 production and intracellular signaling molecules in LPS-stimulated THP-1 cells. A): Effects of DDMG-1 on IL-8 production (solid bar) and cell proliferation (open circle). Concentration of IL-8 was measured by ELISA. Statistical difference compared to the control, $p < 0.1$. The effect on cell proliferation was measured by the MTT assay, and data are the means relative (%) to the control value. B): Effects of DDMG-1 on IL-8 mRNA induction. The relative ratio is shown as a percentage, whereby the IL-8 signal was normalized to the GAPDH signal. C): Effects of DDMG-1 on NF-κB transcription by IκB-α degradation. Total proteins were extracted at the indicated times, resolved by 10% SDS-PAGE, blotted, and examined by Western blotting using polyclonal anti-serum against IκB-α. D): Effects of DDMG-1 on DNA 20 binding activities of NF-κB transcription factor. EMSA was performed using 32P-labeled probes that possessed NF-κB binding sites. The NF-κB/probe complex is indicated.

3.3. Effects of DDMG-1 on IL-8 production in THP-1 cells stimulated with IL-1β, TNF-α and PMA

Since the production of IL-8 was also induced in THP-1 cells by IL-1β, PMA, and TNF-α as similar to the stimulation by LPS, we examined the effects of DDMG-1 on the IL-8 production in THP-1 cells stimulated with IL-1α, PMA, and TNF-α (Fig. 3). Interestingly, DDMG-1 showed the different effects on the IL-8 production in THP-1 cells with the different stimulants. DDMG-1 inhibited the IL-8 production in a dose-dependent manner in IL-1β -stimulated THP-1 cells (Fig. 3A). On the other hand, an inhibitory activity of DDMG-1 on the IL-8 production in THP-1 cells stimulated with PMA and TNF- α was not detected (Fig. 3B and C), and increased production was observed in PMA-stimulated THP-1 cells (Fig. 3B). It will be suggested that DDMG-1 enhanced the activation of THP-1 cells by PMA.

These results indicated that the suppressive activity on IL-8 production by DDMG-1 will be specific to the IL-1β signaling pathway.

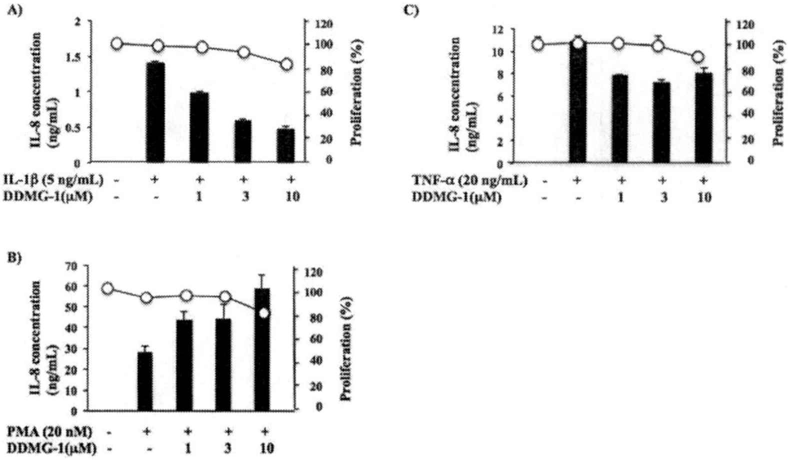

Figure 3. Effects of DDMG-1 on IL-8 production (solid bar) and cell proliferation (open circle) in THP-1 cells stimulated with IL-1β (A), PMA (B), and TNF-α (C). Concentration of IL-8 was measured by ELISA. Statistical difference compared to the control, $p < 0.05$ (A) and 0.1 (B and C). The effect on cell proliferation was measured by the MTT assay, and data are the means relative (%) to the control value.

4. Conclusions.

DDMG-1 (Fig. 1), a rare α-carboline alkaloid, has been obtained from an Indonesian ascidian *P. aurata* as an active component against the production of TNF-α from LPS-stimulated murine RAW 364.7 cells [1]. In this study, we examined the effects of DDMG-1 on the inflammatory cytokine production using human monocytic THP-1 cells to compare the results from RAW 264.7 cells. THP-1 cells used in this study produced large amounts of IL-8 by the stimulation with LPS, IL-1β, PMA, and TNF-α. DDMG-1 inhibited the excess production of IL-8 in LPS-stimulated THP-1 cells by the inhibition of the mRNA level of IL-8, IκB-α degradation, and binding of NF-κB to the target DNA site (Fig. 2). These results were very similar to those obtained with RAW 264.7 cells. Therefore, the effects of DDMG-1 on the production of TNF-α in RAW 264.7 cells and IL-8 in THP-1 cells were ascribable to the resemble mechanism. Furthermore, the inhibitory activity of DDMG-1 was specific to the IL-1β signaling pathway, because DDMG-1 reduced the IL-8 production in THP-1 cells by the stimulation with IL-1β but not with PMA and TNF-α.

Acknowledgements

This work was supported in part by a Grant-in-Aid for Scientific Research (18032033 and 21603012) from the Ministry of Education, Culture, Sports, Science and Technology (MEXT) of Japan. We are grateful to Prof. T. Kasahara (Keio University, Tokyo, Japan) for preparing the IL-8 antibody for ELISA. We also express our thanks to Mr. T. Matsuki and S. Sato of Tohoku Pharmaceutical University for measuring NMR and mass spectra.

References

[1] Oda T, Lee J S, Sato Y, Kabe Y, Sakamoto S, Handa H, Mangindaan R E P and Namikoshi M 2009 Inhibitory effect of *N,N*-didesmethylgrossularine-1 on inflammatory cytokine production in lipopolysaccharide-stimulated RAW 264.7 cells. *Mar. Drugs* 7: 589-599

[2] Bosshart H and Heinzelmann M 2016 THP-1 cells as a model for human monocytes. *Ann. Transl. Med.* 4(21):438-441

[3] Chanput W, Mes J J and Wichers H J 2014 THP-1 cell line: an in vitro cell model for immune modulation approach. *Int. Immunopharmacol.* 23: 37-45

[4] Marlini M, Mabuchi A, Mallard B L, Hairulhisyam N, Akashi-Takamura S, Harper J L and Wheatley A M 2016 Delayed liver regeneration in C3H/HeJ mice: possible involvement of haemodynamic and structural changes in the hepatic microcirculation. *Exp. Physiol.* 101.12: 1492–1505

[5] Ross J A, Auger M J, Burke B, Lewis C E 2002 The biology of the macrophage. In The Macrophage, 2nd ed.; Burke, B., Lewis, C E, Eds. Oxford Medical Publications: Oxford, UK, pp. 1-72

[6] Mendes S S, Candi A, Vansteenbrugge M, Pignon M R, Bult H, Boudjeltia K Z, Munaut C and Raes M 2009 Microarray analysis of the effects of NF-κB or p13K pathway inhibitors on the LPS-induced gene expression profile in RAW264.7 cells Synergistic effects of rapamycin on LPS-induced MMP9-overexpression. *Cell Signaling* 21: 1109-1122

[7] Jin W J, Xu J M, Xu W L, Gu D H and Li P W 2014 Diagnostic value of interleukin-8 in colorectal cancer: A case-control study and meta analysis. *World J. Gastroenterol.* 20 (43): 16334-16342

[8] Albulescu R, Codrici E, Popescu I D, Mihai S, Necula L G, Petrescu D, Teodoru M and Tanase C P 2013 Cytokine Patterns in Brain Tumour Progression *Mediators of Inflammation* Volume 2013, Article ID 979748, 1-7

[9] Konig B, Steinbach F, Janocha B, Dryda A, Stumm M, Philipp C, Alhoff E P and Konig W. 1999 The differential expression of proinflammatory cytokines IL-6, IL-8 and TNF-alpha in renal cell carcinoma. *Anticancer Res.* 19: 1519-1524

[10] Shimizu S, Nomoto M, Naito S, Yamamoto T and Momose K 1998 Stimulation of nitric oxide synthase during oxidative endothelial cell injury. *Biochem. Pharmacol.* 55: 77-83

[11] Carmichael J, DeGraff W G, Gazdar A F, Gazdar A F, Minna J D and Mitchell J B 1987 Evaluation of a tetrazolium-based semiautomated colorimetric assay: assessment of radiosensitivity. *Cancer Res.* 47: 939-942

[12] Oda T, Ueda A, Shimizu N, Handa H and Kasahara T 2002 Suppression of monocyte chemoattractant protein 1, but not IL-8, by alprazolam: effect of alprazolam on c-Re/p65 and c-Rel/p50 binding to the monocyte chemoattractant protein 1 promoter region. *J. Immunol.* 169: 3329-3335

[13] Wang W F, Oda T, Fujita A, Mangindaan R E P, Nakazawa T, Ukai K, Kobayashi H and Namikoshi M 2007 Three new sulfur-containing alkaloids, polycarpaurines A, B, and C, from an Indonesian ascidian *Polycarpa aurata*. Tetrahedron, 63: 409-412

[14] Abas S A, Hossain M B, Helm D, Schmits F J, Laney M, Cabusly R and Schatzman R C 1996 Alkaloids from the Tunicate *Polycarpa aurata* from Chuuk Atoll. *J. Org. Chem.* 61: 2709–2712

[15] Moquin-Pattey C and Guyot M. 1989 Grossularine-1 and grossularine-2, cytotoxic α-carbolines from the tunicate *Dendrodoa grossularia*. *Tetrahedron* 45: 3445-3450

[16] Helbecque N, Moquin C, Bernier J L, Morel E, Guyot M and Henichart J P 1987 Grossularine-1 and grossularine-2, α-carbolines from *Dendrodoa grossularia*, as possible intercalative agents. *Cancer Biochem. Biophys.* 9: 271-279

Blending fishwastes and chicken manure extract as low-cost and stable diet for mass culture of freshwater zooplankton, optimized for aquaculture

E O Ogello[1], S Wullur[2*], A Hagiwara[3]

[1]Department of Fisheries and Natural Resources, Maseno University, P.O.Box Private Bag, Maseno, Kenya
[2]Ins Faculty of Fisheries and Marine Science, Sam Ratulangi University, Manado 95-115, Indonesia
[3]Graduate School of Fisheries and Environmental Sciences, Nagasaki University, 852-8521, Nagasaki, Japan

Corresponding author: stenlywullur@unsrat.ac.id

Abstract: This study investigated the feasibility of fishwastes and chicken manure extract (CME) as cheap diet for mass culture of freshwater zooplankton. CME and fishwastes as well as carbon source were used to make fishwaste diets (FWD). Each diet was triplicated 3 days before inoculation with 5, 2 and 0.4 ind ml^{-1} of rotifers, copepods and cladocerans, respectively in each tank. About 5ml of water was done daily, from which the zooplankters were counted. Harvesting was done at the first exponential growth phase by replacing 50% of the water and FWD. There was a significantly higher density of zooplankton and SGR in FWD_B than FWD_A and control tanks. The zooplankton obtained highest densities on day 7 as follows: rotifers: 100.6±14.8, 146.3±7.0, and 60.0±7.9 ind ml^{-1} in FWD_A, FWD_B and the control tanks, respectively; the copepods: 8.0±11.1, 12.6±13.6 and 4.3±2.1 ind ml^{-1} in FWD_A, FWD_B and control tanks, respectively; the cladocerans: 3.3 ± 6.0, 8.6 ± 8.7 and 3.6±2.5 ind ml^{-1} in FWD_A, FWD_B and control tanks, respectively. The most abundant genera were *Brachionus* sp., *Cyclops* sp. and *Daphnia* sp. for the rotifers, copepods and cladocerans, respectively. This study offers an alternative to expensive on-site microalgal production toward a more cost-effective aquaculture.

1. Introduction

In the recent past, aquaculture activities have increased considerably in the tropical countries, leading to high demand for the zooplankton species such as rotifers, copepods, cladocerans and *Artemia* nauplii as suitable starter foods for fish larviculture [1]. The live foods are highly digestible, more palatable, contain essential nutrients and, preserve water quality unlike inert diets [2]. However, inconsistent supply of the live food resources continues to limit the intensive culture of economically valuable fishes. Naturally, the zooplankton communities form a significant component of aquatic ecosystems, and are the primary prey for most larval fishes and crustaceans. The zooplankters have relatively high reproductive rates, thus making them attractive for use in semi-intensive culture systems, which are popular in the developing countries.

Fresh microalgae are first choice diet for culturing various zooplankton species in hatcheries, for aquaculture [3]. However, the challenges of high density microalgal culture protocols have caused inconsistencies in the production of sufficient quantities of the zooplankton to match the increasing demand for the live foods in aquaculture. Alternatively, cheaper diets such as baker's yeast have been used but culture instabilities due to imbalances of vitamin B_{12} and bacterial flora are common [4]. Other studies have also reported the use of waste-grown bacteria and/or synthetic medium grown-bacteria as rotifer diet [5][6] and dried algae [7]. Indeed, limited zooplankton production capacity often disrupts fish seedling programs in microalgae-based hatcheries. Organic manures especially chicken manure has always been used in zooplankton production. Studies have attributed the effects of the manures on pond productivity to the actions of different sex hormones present in them [8]. For example, egg-laying chicken excrete about 50 and 250 ng g^{-1} dry manure day $^{-1}$ of 17β-estradiol and testosterone respectively [9]. There are evidences that such chemicals are used by zooplankton to regulate reproduction, induce predator defenses, and accomplish selective foraging [10][11][12]. In our previous study, application of 2.0 ml l^{-1} of chicken manure extract (CME) increased the parthenogenetic reproduction of the freshwater rotifer, *Brachionus angularis* [13].

On the other hand, fishwastes contain appreciable amounts of recoverable essential bio-molecules including the *n*-3 series of the long chain polyunsaturated fatty acids (PUFAs) e.g. eicosapentaenoic acid (EPA, 20:5*n*-3) and docosahexaenoic acid (DHA, 22:6*n*-3) [14]. These elements are critical for growth and survival of the fish larvae feeding on the rotifers [15]. The fishwastes also provide excellent substrates for microbial growth, some of which have probiotic properties such as *Bacillus* sp. and lactic acid bacteria (LAB) [16]. Probiotics are known to enhance growth and immunity of rotifers and fish larvae either individually or in combination [17][18][19]. In this study, we hypothesized that a blend of the CME and the FWD can provide positive synergy to enhance growth and reproduction of freshwater zooplankton (i.e. rotifers, copepods and cladocerans), for aquaculture. The zooplankton obtain a substantial proportion of their carbon from detritus pathways via bacteria, and previous studies have reported the role of certain bacterial species in improving population growth and nutrition status of various zooplankton species [20][21].

Composing organic matter yields huge densities of bacterial cells under optimum carbon/nitrogen (C/N) ratio [22]. Bacteria, and by extension the microbial loop are known to play important roles as recycling pathways for C and N in the ecological food webs [23]. Usually, optimum C/N ratio ensures immobilization of inorganic nitrogen into huge bacterial proteins (biomass) and restores good water quality by removing toxic ammonia [24]. Heterotrophic production of bacteria using fish defecates for *in situ* production of zooplankton communities has been exploited in the polyculture of shrimp, catfish and tilapia [25] with adequate carbon source [26]. However, the feasibility of the combined CME and fishwastes for *in situ* zooplankton production is not yet clear. This study explored the applicability of the fishwastes and CME for mass culture of freshwater zooplankton communities in an outdoor tank culture system, optimized for aquaculture production.

2. Materials and Methods

2.1 Study area

This study was conducted at Kenya Marine & Fisheries Research Institute (KMFRI), Kegati Aquaculture Research Centre, located at 00°42'S; 034°47'E. The inoculant zooplankton i.e. rotifers, cladocerans and copepods were obtained from one of the fish production ponds that was previously fertilized with diammonium phosphate (DAP) and urea. One litre of the pond water was randomly sampled and filtered through a kitchen sieve to remove physical debris before determining the zooplankton densities in 1 ml sub-samples from each filtrate using a graduated plate with lugol fixation under stereo microscope, at × 25 magnifications. The densities of the zooplankton were used to determine the quantity of pond water needed to inoculate the experimental tanks. Plankton net of pore size 45 μm was used to filter the pond water to concentrate the zooplankton.

2.2 Chicken manure extract (CME)

One kilogram of fermented chicken manure was boiled in 5 l of pond water for 40 - 50 min and then kept overnight at room temperature. The supernatant liquid was filtered off the sludge using nylon net of pore size 100μm and used as CME.

2.3 Fishwaste diet (FWD)

Barbus altinalis Linnaeus 1758, which is a low-valued bi-catch freshwater fish, were caught from river Mara, Kenya, using an electro-fisher (Smith-Root GPP USA) and transported to the laboratory in a cooler box. The fish were gutted and the fish heads were crushed to make a diet as explained in our patent No. P00201609066 registered in Indonesia. We added 2 ml l^{-1} of CME and 0.2 g l^{-1} of maize flour carbon source into the tanks.

2.4 Experimental design

The experiment was conducted in 9 asbestos tanks each containing 500 l of well water. Three diets i.e. FWD_A (0.5 g l^{-1} of fishwaste + 0.2 g l^{-1} of maize flour), FWD_B (FWD_A + 2 ml l^{-1} of CME) and control (2 ml l^{-1} of CME only) were each triplicated in the culture tanks, 3 days prior to inoculation with a combination of 5, 2 and 0.4 ind ml^{-1} of rotifers, copepods and cladocerans, respectively in each culture tank on day 0. Different plankton nets (mesh sizes 250, 500 and 1000 μm) were used to adjust the density of rotifers, copepods and cladocerans, respectively before inoculation in the experimental tanks. The tanks were covered with mosquito nets to keep off insects and birds. Every day, 5 ml of water was randomly sampled three times at the top 10 cm of each tank from which, the zooplankton were counted with lugol fixation. At the end of the first exponential phase, partial harvesting was done by replacing 50% of all the tank water with new well water and fresh FWD. The experiment lasted for 16 days.

The percent relative abundance (% RA) of each zooplankton species was calculated as [mean abundance (MA)] / total zooplankton taxa (N)] × 100. The population specific growth rate (SGR) (r) was calculated as $r = [\ln N_t - \ln N_o] / t$, where, N_o = initial population density, N_t = population density after the time (t) and t = time (days) and the coefficient of variation (CV) of the mean SGR was calculated as standard deviation / mean SGR, For qualitative composition, the zooplankters were identified up to the genus level according to [27] while the zooplankton diversity was calculated using the Shannon-Weiner diversity index: H' = - SUM [(pi)*ln (pi)], where pi = number of individuals of species i/ total number of samples [28].

2.5 Water quality

Temperature (° C), dissolved oxygen (DO; mg l^{-1}), pH and electrical conductivity (μS cm^{-1}) were measured *in situ* using a multi-probe water checker (U-10 model, Horiba, Tokyo, Japan). The water turbidity (cm) was also measured *in situ* using a secchi disk. The water quality parameter measurements were taken daily at 1200 hours.

2.6 Data analysis

The data was analyzed using R statistical software (version 3.2.1 of the R Foundation for Statistical Computing Platform © 2015). The Bartlett test was used to determine the homogeneity of variances. Two-way analysis of variance (ANOVA) was used to test the effects of the FWD and culture days on the water quality parameters and zooplankton densities. One-way ANOVA was used to test the effect of FWD on the zooplankton mean abundance, SGR and diversity (H'). The Tukey's HSD Post Hoc test was performed to locate any significant differences at $p<0.05$.

3. Results

3.1 Water analysis

Measurements of the water quality parameters are presented in Table 1. The water temperature varied significantly during the culture days (F=2.31, p=0.01) but was not affected by the FWD (F=0.45, p=0.63). Similarly, conductivity varied significantly during the culture days (F=1.76, p=0.04) with no significant effects of FWD (F=0.56, p=0.57). The DO was significantly affected by the FWD (F=4906.74, $p<0.05$) and culture days (F=200.44, $p<0.05$), with higher DO being recorded in the control than in the FWD-treated tanks (p=0.00). FWD_A tanks had significantly higher DO than FWD_B tanks (p=0.00). Turbidity varied significantly within the culture days (F=12.83, $p<0.05$), and between the FWD (F=841.80, $p<0.05$), with higher turbidity being recorded in the FWD tanks than in the control tanks (p=0.00). FWD_B tanks were more turbid compared to FWD_A tanks (p=0.00). The pH was neither affected by the FWD (F=1.18, p=0.31) nor culture days (F=1.07, p=0.38).

Table 1. The mean values of water quality parameters ± SD in the culture tanks for 16 days; Values with different superscript in each row are significantly different at $p<0.05$; Two-way ANOVA; Tukey HSD test, a>b>c, $n=48$.

Water quality parameters	Before experiment	Treatments		
		FWD_A	FWD_B	Control
Temperature (°C)	25.1 ± 1.0	25.6 ± 1.0^a	25.7 ± 1.0^a	25.8 ± 1.2^a
Conductivity (μS cm^{-1})	113.7 ± 2.2	115.9 ± 2.8^a	116.2 ± 2.4^a	115.7 ± 2.1^a
Dissolved oxygen (mgl^{-1})	5.73 ± 0.19	3.94 ± 0.77^b	3.84 ± 0.68^c	5.61 ± 0.21^a
pH	7.51 ± 0.01	6.63 ± 0.03^a	6.65 ± 0.04^a	6.66 ± 0.12^a
Turbidity (cm)	39.6 ± 1.5	35.7 ± 1.3^c	23.4 ± 3.1^a	27.6 ± 3.1^b

3.2 Zooplankton population dynamics

The population density of each zooplankton taxa was significantly affected by the FWD, culture days and the interaction between them ($p<0.05$). There was significantly higher rotifer density in FWD_A than in control tanks from day 4 - 6 (Tukey HSD, $p<0.05$), and higher density in FWD_B than in control tanks from day 3 - 7 and 11 - 12 ($p<0.05$). Meanwhile, there was significantly higher rotifer density in FWD_B than in FWD_A tanks on day 7 and 11 ($p<0.05$). The rotifer densities were 100.6±14.8, 146.3±7.0, and 60.0±7.9 ind ml^{-1} in FWD_A, FWD_B and the control tanks, respectively on day 7 (Figure 1). The population density of copepods was significantly higher in FWD_A than in control tanks on day 5 and 8 ($p<0.05$), and higher in FWD_B than in control tanks from day 5 - 8 and 12 ($p<0.05$). There was no significant difference between the FWD tanks ($p>0.05$). The copepods' densities were 7.8±2.5, 12.1±2.7 and 3.6±0.4 ind ml^{-1} in FWD_A, FWD_B and control tanks, respectively on day 7 (Figure 1). For the cladocerans, there was no significant difference in the population density between FWD_A and control tanks ($p=0.08$). However, there was significantly higher density in FWD_B than in control tanks from day 6 - 8 and 11 - 13 ($p<0.05$), and higher density in FWD_B than in FWD_A tanks on day 7 and 11-13 ($p<0.05$). The densities of the cladocerans were 3.1±1.2, 7.7±1.7 and 2.3±0.5 ind ml^{-1} in FWD_A, FWD_B and control tanks, respectively on day 7 (Figure 1). The summary of the composition of zooplankton taxa, mean and percent abundance is presented in Table 2. The most abundant genera were *Brachionus* sp., *Cyclops* sp. and *Daphnia* sp. for rotifers, copepods and cladocerans, respectively.

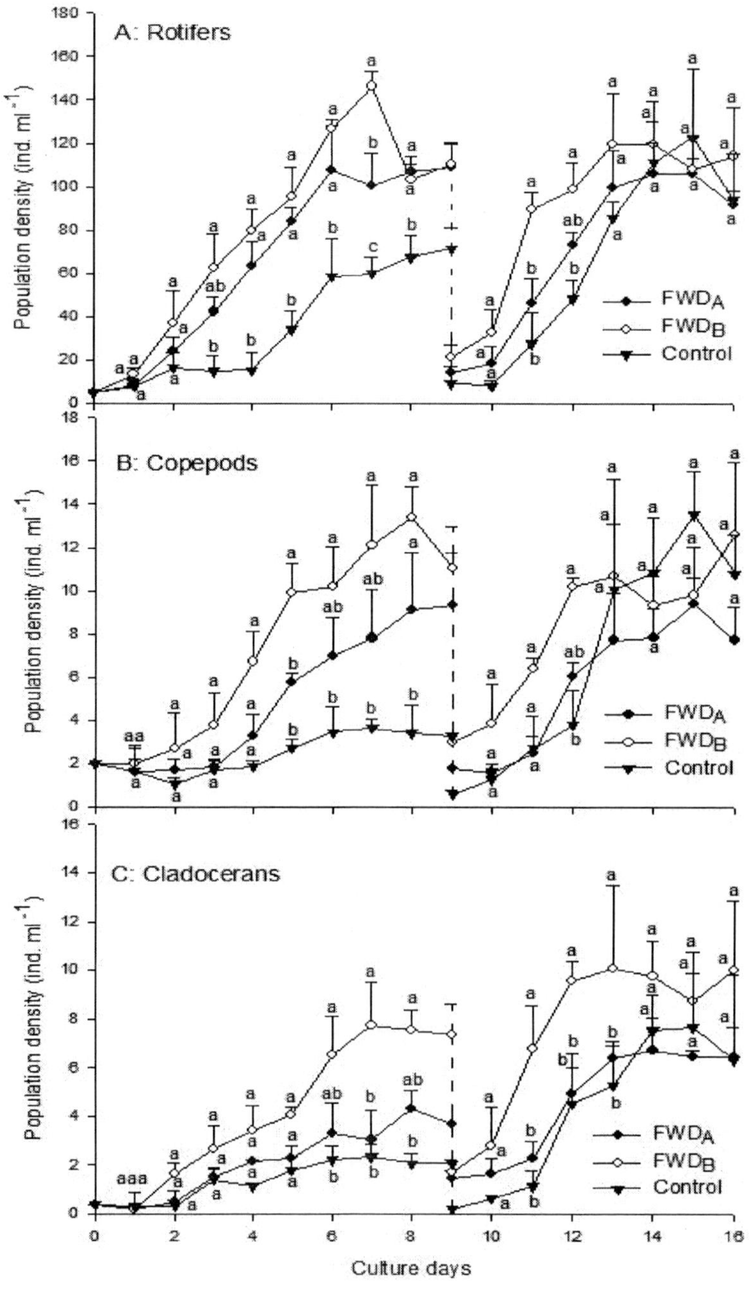

Figure 1: The population density curves of each zooplankton taxa. Each plot and vertical bar represents mean ± SD. Half of the culture water was replaced with new media on day 9 in each treatment as shown by the dotted lines. Different superscripts on each day represent significantly different mean population densities at $p<0.05$; Two-way ANOVA, Tukey HSD test; $n=3$, a>b>c

The SGR of each zooplankton taxa was significantly affected by the FWD (One-way ANOVA, $p<0.05$) with higher SGR for the rotifers in FWD_B than in FWD_A ($p=0.05$) and control tanks ($p=0.00$), and higher SGR in FWD_A than in control tanks ($p=0.00$) (Table 3). The SGR for the copepods was significantly higher in the two FWD than in the control tanks ($p<0.05$), but not significantly different between the FWD tanks ($p=0.11$). There was significantly higher SGR for the cladocerans in FWD_B than in FWD_A ($p=0.02$) and control tanks ($p<0.05$), but not significantly different between FWD_A and control tanks ($p=0.66$) (Table 5-2.4). The FWD did not affect the diversity (*H'*) of the rotifers (One-way ANOVA; $F=0.06$, $p=0.93$), copepods (One-way ANOVA; $F=0.09$, $p=0.91$) and cladocerans (Kruskal-Wallis test; $\chi^2=1.67$, df=2, $p=0.43$) (Table 3). Similarly, there was no significant difference in coefficient of variation (CV) among the treatments (One-way ANOVA, $F=1.97$, $p=0.22$) (Table 3).

Table 2: Zooplankton taxa composition, mean abundance (MA) and percent relative abundance (% RA) of the zooplankton species in each treatment at day 16; the values are mean ind/ml ± SD. Different superscripts in a row denote significant differences for each zooplankton species at $p<0.05$, One-way ANOVA, Tukey HSD test, $n=6$, a>b>c; -, absent.

Zooplankton taxa (ind/ml)	Treatments					
	FWD_A		FWD_B		Control	
	MA	% RA	MA	% RA	MA	% RA
Rotifers						
Brachionus sp	38.6 ± 8.6^{ab}	25.3	51.8 ± 12.1^{a}	25.5	33.6 ± 3.7^{b}	23.7
Filinia sp	-	-	9.6 ± 1.2	4.9	-	-
Lecanea sp	13.8 ± 4.6^{a}	9.1	10.6 ± 6.1^{a}	4.6	12.0 ± 3.2^{a}	8.4
Keratella sp	17.3 ± 3.0^{a}	11.0	12.2 ± 2.9^{b}	5.4	7.5 ± 2.8^{c}	5.6
Asplanchna sp	19.0 ± 2.2^{b}	12.3	25.5 ± 6.2^{a}	12.7	24.3 ± 2.1^{ab}	16.7
Others	4.6 ± 3.0	3.2	5.5 ± 2.2	2.9	3.3 ± 2.1	2.1
Copepods						
Cyclops sp	16.5 ± 2.6^{b}	11.0	21.3 ± 2.8^{a}	10.3	18.3 ± 1.6^{ab}	12.6
Diaptomus sp	10.8 ± 3.4^{b}	7.1	20.1 ± 5.7^{a}	9.8	14.5 ± 2.0^{ab}	10.5
Cladocerans						
Diaphanosoma sp	9.3 ± 2.2	5.8	16.3 ± 3.2	7.1	-	-
Daphnia sp	15.6 ± 3.7^{a}	10.4	16.0 ± 3.1^{a}	7.8	15.0 ± 5.1^{a}	10.5
Moina sp	7.6 ± 1.8^{a}	5.2	10.0 ± 7.1^{a}	4.9	13.6 ± 3.7^{a}	9.8
Ceriodapnia sp	-	-	4.5 ± 2.8	2.5	-	-
Total (*N*)	**154**		**204**		**143**	

Table 3: Specific growth rate (SGR) and diversity (*H'*) of each zooplankton taxa in every treatment, and the coefficient of variation (CV) of each treatment. The values are mean ± SD. One-way ANOVA, Tukey HSD test, different letters in each row represent significantly differences at $p<0.05$; a>b>c

		Treatments		
Parameters	Zooplankton taxa	FWD_A	FWD_B	Control
SGR (day $^{-1}$)	Rotifera	0.42 ± 0.02^b	0.48 ± 0.01^a	0.35 ± 0.02^c
(*n*=3)	Copepoda	0.42 ± 0.04^a	0.48 ± 0.03^a	0.32 ± 0.02^b
	cladocera	0.51 ± 0.06^b	0.65 ± 0.03^a	0.48 ± 0.03^b
Diversity (*H'*)	Rotifera	1.44 ± 0.14^a	1.51 ± 0.08^a	1.36 ± 0.14^a
(*n*=6)	Copepoda	0.67 ± 0.05^a	0.69 ± 0.01^a	0.68 ± 0.02^a
	cladocera	1.05 ± 0.17^a	1.30 ± 0.06^a	0.69 ± 0.20^a
CV (*n*=3)		0.09 ± 0.03^a	0.05 ± 0.02^a	0.06 ± 0.01^a

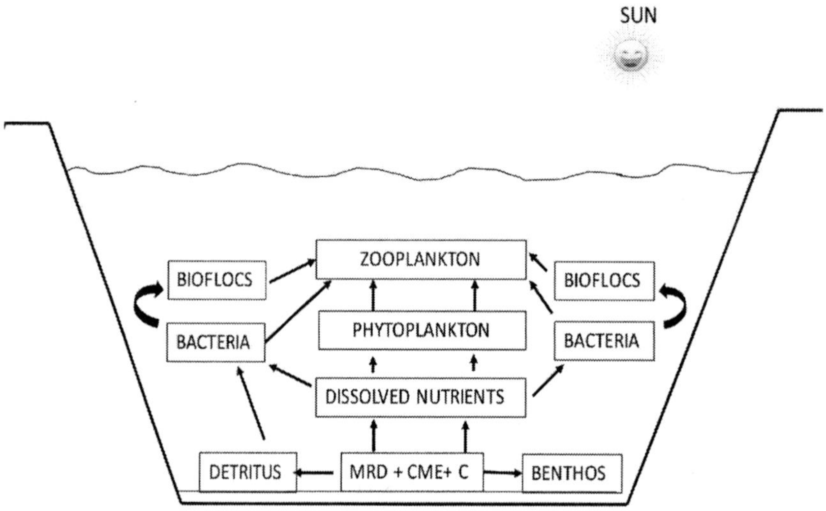

Figure 2: Schematic flow of nutrients from FWD, CME, detritus and carbon source (C) to biotic communities in the tank culture system. FWD = fishwaste diet; CME = chicken manure extract

4. Discussion

In larviculture, management of zooplankton forage base is a critical phase for successful transition of fish larvae to fingerling stage. The dynamic characteristics of zooplankton populations have led researchers to apply different techniques to produce high densities of desired zooplankton species until fish are either harvested, or able to consume inert feeds. Despite significant progress in zooplankton cultivation techniques, cost-effective and stable protocols for producing sufficient quantities of the desired zooplankton species without algae are scanty. High density algal production is expensive and laborious, thus becoming a burden for most hatcheries. This study demonstrated the feasibility of FWD made by blending CME and fishwastes for culturing zooplankton species in an outdoor mass culture system.

Despite changes in DO, conductivity and pH during the experimentation period (Table 1), the values remained within the acceptable limits for freshwater pond aquaculture [29]. More than half of zooplankton species in every sample consisted of rotifers (Table 2). Rotifers, especially the *Brachionus* spp. are normally the first zooplankters to reach large numbers in newly colonized habitats, thus taking competitive advantage [30,31]. With the shortest lifespan of 5 - 12 days, rotifers reach peak reproductive levels much earlier (about 3.5 days) than other zooplankters [32]. Copepods and cladocerans have similar life span (about 50 days) but have different peak reproductive periods which take about 24 and 15 days to reach, for copepods and cladocerans, respectively [33]. The copepods only reproduce sexually and therefore require longer periods to increase their population [32]. In addition, overcrowding decreases fecundity of the calanoid copepods and the reproduction of the harpacticoid copepod, e.g. *Tigriopus japonicus* is more successful when mixed-cultured with rotifers, perhaps due to inter- and intra-specific interactions [34].

Addition of CME to the fish wastes caused significant increase in SGR and population densities of the zooplankton species and therefore, appears as an improvement to the FWD. Decomposing fish tissues are known to provide substrates for the proliferation of diverse microbial flora, some of which have probiotic properties [16]. The CME probably facilitated phytoplankton growth in the tanks, thus expanding forage base (i.e. bacteria and phytoplankton) for the zooplankton growth and reproduction. This demonstrated the importance of the synergy of FWD and CME for zooplankton growth and reproduction. Chicken manure is also an excellent substrate for probiotic bacteria [12,35], and contains growth promoting compounds e.g. 17α and β-estradiol that can positively influence zooplankton reproduction history [9] either individually or in combination [18]. The efficacy of CME in enhancing the production of *Diaphanosoma celebensis* and *Tigriopus japonicus* has been reported where up to 14 ind ml^{-1} was produced within 12 days [36]. In our previous study, 2 ml l^{-1} of CME enhanced parthenogenetic reproduction of *B. angularis* significantly [13]. It is hypothesized that omnivorous zooplankton species feed on detritus-bacteria complex in the absence of live prey. Therefore, fungi and bacteria associated with decaying organic substances could supplement phytoplankton foraging with essential proteins, lipids and vitamins to cause high growth effects for zooplankton. For this reason, phytoplankton alone do not necessarily increase zooplankton populations, thus explaining the relatively low zooplankton density in the control experiment. In this study, the zooplankton population densities increased to new peaks comparable to the previous ones after the first harvest, suggesting that the system can be self-sustaining, perhaps with frequent harvesting to reduce organic loads. Generally, the lower CV of the treatments suggests stability of the FWD technology. With application of FWD_B, it is possible to obtain about 150, 12, and 8 ind ml^{-1} of rotifers, copepods and cladocerans, respectively, on weekly basis. Comparatively, [37] obtained about 76 rotifers ml^{-1} within 5 days using conventional chicken manure methods in ponds and tanks, respectively. The results of the present study demonstrate the superiority of FWD technique compared to the normal traditional live food culture methods using chicken manure.

The FWD technology embodies a complex microbial control system that leads to degradation of waste materials, facilitates the recycling of nutrients and proliferation of microflora that flourishes under optimum C/N ratio. The FWD technology represents a biofloc microcosm where the decomposing fish tissues produce diverse microbial flora, which also form zooplankton diet. The system also attracts opportunistic benthic community that feed on the detritus and bioflocs. In addition, the dissolved nutrients from FWD, in presence of sunlight, facilitate primary production of phytoplanton that is grazed upon by the zooplankters. The complex food web that exists in FWD aided culture system is summarized in Figure 2.

Initiatives geared towards developing nutrition strategies such as bioflocs and periphyton that maximizes the contribution of natural and supplemental feeds in culture facilities would expand aquaculture production. Even though FWD can be used as cheap and stable diet for zooplankton production, the success of the FWD technology could be limited by presence of pathogenic microbes, which may affect the cultured animals. Copepods produced extensively may cause mass mortalities in cultured fishes through transmission of viruses and parasites.

In terms of production costs, successful microalgal culture in conventional larviculture systems basically depends on availability of optimal parameters such as nutrients, light, pH range, aeration, temperature and salinity. In addition, high ratio of algal biomass to target species (usually about 5-10:1) elevates the cost of microalgae production [38]. On average, mono-specific algal cultures cost up to USD 120-200 kg^{-1} DW, where labour accounts for 50-85%, pumping; 4-24%, nutrients; 4-20% and mixing; 5-8% of the total production costs [38]. Consequently, high microalgae production costs influence the zooplankton production costs. For example, the estimated cost of rotifer mass production using large scale batches is USD 4.5/ million rotifers where feeds, live algae and yeast account for 72, 50 and 22% of the production costs, respectively [38]. Comparatively, the FWD is significantly cheaper because fishwaste, which is the main ingredient in the diet, can be obtained for free. In addition, the efficacy of the FWD can be improved by addition of cheap carbon source. In general, the estimated cost of rotifer production in 500 l culture tanks using FWD is approximately USD 0.01/ million rotifers.

FWD medium can be used as an enrichment emulsion for low-quality live food diets such as Baker's yeast. Other home-made emulsion products have been produced using fish oil and egg yolk as cheap ways of enriching rotifer cultures, but these products are prone to oxidation and have short shelf life that limits their application in aquaculture. The bacteria-held PUFAs (in FWD) are more protected against oxidation, and provide a variety of other natural nutrients that meet the species-specific nutritional requirements of the rotifers, and their predators [39]. The FWD technology is a universal innovation that can be applied for both marine and freshwater live food production. Bioflocs are important in aquaculture, and has been used to improve growth rate of rotifers [40] and cultivable fishes [41]. FWD appears to be a major leap toward making pre-planning of fish seedling production in aquaculture facilities feasible.

5. Conclusion

To eliminate microalgae from aquaculture production chain, FWD is a promising alternative. FWD combined with CME produced about 150, 12, and 8 ind ml^{-1} of rotifers, copepods and cladocerans, respectively, on weekly basis in outdoor culture tanks. FWD production protocol is cheap and costs about USD 0.01/ million rotifers. Therefore, FWD is convenient for profitable aquaculture production particularly in the regions without sophisticated infrastructural investments for high dense microalgal production. However, further studies should document a comprehensive account of the microbial flora associated with the FWD, including their successive colonization overtime. There is need to test the suitability of other environmental wastes e.g. from livestock abattoirs for live food production and the suitability of the FWD-fed zooplankton species for larviculture of the local fish species.

The FWD offers an opportunity to 1) reduce environmental pollution sources by reusing poorly discarded fishwastes; 2) reduce or eliminate direct dependence on the immediately cultured or the expensive on-site microalgae production and, 3) to lower the cost of zooplankton enrichment, thus making it convenient for aquaculture production, especially in the less developed countries, where malnutrition is prevalent.

Acknowledgements

This study was partly supported by JSPS KAKENHI grant No. JP17H03862 and JSPS, Asia-Africa No. 24380108 to Atsushi Hagiwara and by The Ministry of Research, Technology and Higher Education of Indonesia (Kemenristekdikti) under PPUPT research grant to SW. Erick Ochieng Ogello is grateful to the Ministry of Education, Culture, Sports, Science and Technology for the MEXT PhD. scholarship. We thank our anonymous reviewers for strengthening the scientific merit of our manuscript.

References

[1] Conceicao L E, Yufera M, Makridis P, Morais S and Teresa-Dinis M 2010. Live feeds for early stages of fish rearing. *Aquacult. Res.* 41, 613–640.

[2] Kitajima C and Fujita S 1983. Nutritional values of live organisms used in Japan for mass propagation of fish. A review. *Aquaculture* 34, 115–143.

[3] Maruyama I, Nakao T, Shigeno I, Ando Y and Hitayama K 1997. Application of unicellular algae Chlorella vulgaris for the mass culture of marine rotifer Brachionus. *Hydrobiol.* 358, 133–138.

[4] Hirayama K and Funamoto H 1983. Supplementary effect of several nutrients on nutritive

deficiency of beaker yeast for population growth of the Brachionus plicatilis. *Nipp. Suis. Gakka*. 49, 505–510.

[5] Hirata H, Murata O, Yamada S, Ishitani H and Wachi M 1998. Probiotic culture of the rotifer Brachionus plicatilis. *Hydrobiol*. 387, 495–498

[6] Loo P L, Chong V C, Vikineswary S and Ibrahim S 2016. Waste-grown phototrophic bacterium supports culture of the rotifer *Brachionus rotundiformis*. *Aquacult. Res*. 47, 3029–3041

[7] Ogello E O, Sakakura Y and Hagiwara A. 2017. Culturing *Brachionus rotundiformis* Tschugunoff (rotifera) using dried foods: application of Gamma-Aminobutyric Acid (GABA). *Hydrobiol*. 796, 99-110

[8] Hakk H, Millner P and Larsen G 2005. Decrease in water-soluble 17β-estradiol and testosterone in composted poultry manure with time. *J. of Env. Qual*. 34, 943–950.

[9] Shore L S, Harel-Markowitz E, Gurevich M and Shemesh M 1993. Factors affecting the concentration of testosterone in poultry litter. Journal of Environmental Sciences and Health, part A: *Env. Sc. Eng. and Toxicol*. 28, 1737–1749.

[10] Gilbert J J, 1966. Rotifer ecology and embryological induction. *Science* 151, 1234–1237.

[11] Preston B L, Snell T W, Robertson T L and Dingmann B J 2000. Use of freshwater rotifer Brachionus calyciflorus in screening assay for potential endocrine disruptors. *Env. Toxicol. and Chem*. 19, 2923–2928.

[12] Rapatsa M M, Moyo N A 2013. Performance evaluation of chicken, cow and pig manure in the production of natural fish food in aqua dams stocked with Oreochromis mossambicus. *Phy. and Chem. of the Earth* 66, 68 – 74.

[13] Ogello E O and Hagiwara A 2015. Effects of chicken manure extract on the population growth, mixis induction and body size of the freshwater rotifer *Brachionus angularis* Gosse 1851. *Asian Fis. Sci*. 28, 174–185.

[14] Cho S, Kim S-Y, Yoon M and Kim S-B 2014. Physicochemical profiles of Chub Mackerel *Scomber japonicus* bones as a food resource. *Fish. and Aqua Sci*. 17(2), 175–180.

[15] Ogello E O, Wullur S, Sakakura Y and Hagiwara A. 2018. Composting fishwastes as low-cost and stable diet for culturing *Brachionus rotundiformis* Tschugunoff (Rotifera): influence on water quality and microbiota. *Aquaculture*, 486, 232–239

[16] Balcazar J L, Venderll D, de-Blas I, Ruiz-Zarzuela I, Muzquiz J L and Girones O 2008. Characterization of probiotic properties of lactic acid bacteria isolated from intestinal microbiota of fish. *Aquaculture* 278, 188–191.

[17] Gatesoupe F J 1991. The effect of three strains of lactic bacteria on the production rate of rotifers, Brachionus plicatilis, and their dietary value for larval turbot, *Scophthalmus maximus*. *Aquaculture* 96, 335–342.

[18] Verschuere L, Rombaut G, Sorgeloos P and Verstraete W 2000. Probiotic bacteria as biological control agents in aquaculture. *Microbiol. and Mol. Biol. Rev*. 64(4), 655–671.

[19] Nayak S K 2010. Probiotics and immunity: A fish perspective. *Fish & Shell. Immunol*. 29, 2–14.

[20] Pace M L, Porter K G, Feig Y S 1983. Species- and age-specific differences in bacterial resource utilization by two co-occurring cladocerans. *Ecol*. 64, 1145–1156.

[21] Hagiwara A, Hamada K, Hori S and Hirayama K 1994. Increased sexual reproduction in *Brachionus plicatilis* (Rotifera) with the addition of bacteria and rotifer extract. *J. of Exp. M. Biol. and Ecol*. 181, 1–8.

[22] Avnimelech Y 2003. Control of microbial activity in aquaculture systems: active suspension ponds. *World Aquacult*. 34(4), 19–21.

[23] Azam F, Fenchel T, Field J G, Gray J S, Meyer-Reil L A and Thingstad F 1983. The ecological role of water-column microbes in the sea. *Mar. Ecol. Progr. Series* 10, 257–263.

[24] Azim M E and Little D C 2008. The biofloc technology (BFT) in indoor tanks: Water quality, biofloc composition, and growth and welfare of Nile tilapia (*Oreochromis niloticus*). *Aquaculture* 283, 29–35.

[25] Bouvy M, Arfi R, Guiral D, Pagano M and Saint-Jean L 1994. Role of bacteria as food for zooplankton in a eutrophic tropical pond (Ivory Coast). *The Nerthelands Journal of Aquatic*

Ecology 28(2), 167–174.
[26] De Schryver P, Crab R, Defoirdt T, Boon N and Verstraete W 2008. The basics of bio-flocs technology: the added value for aquaculture. *Aquaculture* 277, 125–137.
[27] Shiel R J 1995. A guide to identification of rotifers, cladocerans and copepods from Australian Indian waters. Co-operative Research Centre for Freshwater Ecology, Murray-Darling. Albury. p 67.
[28] Shannon C E and Weaver W 1948. A mathematical theory of communication. *The Bell System Technical Journal*, 27, 379–423
[29] Boyd C E 1982. Water quality management for pond fish culture. Elseiver Publishers, 45pp
[30] Shandhu J S, Toor H S and Sehgal H S 1985. Zooplankton community in manured fish tanks. *Ann. of Biol*. 1, 94–97.
[31] Louette G, De Meester L and Declerck S 2008. Assembly of zooplankton communities in newly created ponds. *Freshwater Biol.* 53, 2309–2320.
[32] Allan J D 1976. Life History patterns in zooplankton. American Naturalist 110, 165–180.
[33] Filenko O F, Isakova E F and Gershkovich D M 2011. The lifespan of the cladoceran Ceriodaphnia affinis Lilljeborg in a laboratory culture. *Inland Water Biology* 4(3), 283–286.
[34] Hagiwara A, Jung M M, Sato T and Hirayama K 1995. Interspecific relations between marine rotifer Brachionus rotundiformis and zooplankton species contaminating in the rotifer mass culture tank. *Fisheries Science* 61, 623–627.
[35] Elsaidy N, Abouelenien F G, Kirrella A K 2015. Impact of using raw or fermented manure as fish feed on microbial quality of water and fish. *Egyp. J. of Aq. Res.* 41, 93–100.
[36] Hagiwara A, Kim H-J, Matsumoto H, Ohta Y, Morita T, Hatanaka A, Ishizuka R and Sakakura Y 2016. Production and use of two marine zooplanktons, *Tigriopus japonicus* and *Diaphanosoma celebensis*, as live food for red sea bream Pagrus major larvae. *Fisheries Science,* 82, 799–80
[37] Kyewalyanga M and Mwandya A W 2002. Effect of frequency of fertilization on abundance of rotifers and protozoa in flooded ponds and simulation tanks. *Western Indian Ocean Journal of Marine Science* 1(1), 11–18.
[38] Treece G D 1995. The production of live-food organisms for fishes. *World animal sciences*, 21, 369-382.
[39] Harel M, Koven W, Lein I, Bar Y, Behrens P, Stubblefield J, Zohar Y and Place A R 2002. Advanced DHA, EPA and ArA enrichment materials for marine aquaculture using single cell heterotrophs. *Aquaculture* 213, 347–362.
[40] Nesta K, Ogello E O, Sakakura Y, Hagiwara A 2018. Fish processing wastes as an alternative diet for culturing the minute rotifer, *Proales similis* de Beauchamp. *Aquaculture Research* DOI: 10.1111/are.13707.
[41] Ogello E O, Safina M, Aura C M, Abwao J O and Munguti J M 2014. A critical appraisal of feasibility of tilapia production in earthen ponds using biofloc technology, a review. *International Journal of Aquatic Sciences*, **5** (1): 21-39.

IOP Conf. Series: Materials Science and Engineering **567** (2019) 012023 doi:10.1088/1757-899X/567/1/012023

Antihyperglicemic capacity of basil (*Ocimum basilicum* L.) leaves extracts coated with the marine fish scales derived nanochitosan

I F M Rumengan[1], L Mandey[2], G Citraningtiyas[3], A H Luntungan[4]

*[1]Faculty of Fisheries and Marine Science, Sam Ratulangi University
*[2]Faculty of Agriculture and Food Technology, Sam Ratulangi University
*[3]Faculty of Math and Science, Sam Ratulangi University
*[4]Postgraduate School, Sam Ratulangi University

Corresponding email: innekerumengan@unsrat.ac.id

Abstract. Basil (*Ocimum basillicum* L.) is well-known as a medicinal plant and culinary herb because of its phytochemical contents. Crude extracts of basil leaves are of particular interest for health benefits, including in reducing blood glucose level. Nanochitosan is potentially applied to increase the efficacy of phytopharmaceuticals due to its low viscosity that easily absorbed by the tissues. This study was conducted to evaluate the antihyperglicemic capacity of basil leaf extracts when coated with nanochitosan. Nanochitosan was prepared from the parrot fish *(Scarus* sp) scales derived Chitosan by using the gelatin ionic method. Both chitosan and nanochitosan were characterized by Fourier Transform Infrared Spectroscopy (FTIR), SEM, particle sizing and heavy metals analyses. Extraction of basil leaves were performed by a maceration method using ethanol 96% and the extract was evaporated. Nanochitosan was then applied as coatings of the extracts, and the nanochitosan coated extract was separated from the solution by centrifugation. In vivo assay was applied using 24 male rats, which divided into extract treated groups and the control groups. All the animals were fasted for 12 h, but were allowed free access to water, before commencement of the experiments. Hyperglycemia condition was induced by a single intraperitoneal dose of 120 mg/kg. The rats were fasted for 12 h and blood was taken from the tail artery of the rats. The extract of basil leaves at the dose of 400, 800, 1000 mg/kgBW significantly lowered blood glucose level *($P<0,05$)* of rat in hyperglycemic after 5 hours of extract administration. The efficacy of antihypeglicemic dose of basil leaves extracts coated with nanochitosan was found to be 400 mg/kgBW. Based on the result of the study, it is suggested that nanochitosan could increase the efficacy of phytopharmaceuticals.

1. Introduction

Diabetes mellitus is one of the most common metabolic disorders [1]. It is mainly characterized by a loss of glucose homeostasis by an interruption in carbohydrate, fat and protein metabolism and defects in insulin secretion, insulin action, or both [2]. Deficiency of insulin can cause some part in body fall short in using glucose from the blood circulation. This results in elevation of blood glucose level, which is known as hyperglycemia [3]. Hyperglycemia is directly related to increased hepatic glucose production. The ensuing β-cell dysfunction promotes decreased insulin synthesis and secretion, further perpetuating the associated hyperglycemia. As for the vascular endothelial, chronic hyperglycemia is strongly correlated with many Diabetes melitus related microvascular complications, including retinopathy, nephropathy, and neuropathy [4]. Glycemic management of diabetes is needed to avoid complications of the disease that will occur. Studies have conclusively determined that reducing blood sugars decreases the onset and progression of microvascular complications [5]

A possible antyhyperglicemic capacity in *O. basilicum* has been reported in many studies [6][7] *O. basilicum* is a common herb that is known for its ornamental and therapeutic importance. The chemical constituents which have been isolated from the plant include terpenoids, alkaloids, flavonoids, tannins, saponin glycosides and ascorbic acid [8]. These compounds have been related to reduce blood glucose level. However, the active compounds that provide pharmacological effects such as flavanoid compounds and polyphenols have weaknesses due to its unstable against influence temperature and high light intensity that make easily oxidized [9]. The solution to protect these compounds is to use coating method with nanoparticle. Nanoparticle is defined as a solid particle with size range of 1-1000 nm [10]. The superiority of chitosan such as biocompatible, biodegradable, and non-toxic can be developed as a coating for an extract.

Chitosan is obtained from the deacetylation of chitin, a naturally occurring and abundantly available (in marine crustaceans) biocompatible polysaccharide [11]. In this study, chitosan was extracted from parrot fish scales (*Scarus* sp) because it refers to a previous study which has successfully tested the active group of chitin in parrot fish scales that can be developed into chitosan [12].

Modification basic structure of chitosan to nanochitosan as a coating of an extract provide more benefits caused by its low viscosity properties so that it is more easily absorbed by the body and able to deliver nutrients to the target cell [13]. Nanochitosan coated on basil leaf extract also was gave increased the total phenolic compound [14]. This study aims to analyze the antihyperglicemic capacity of basil (*O. basillicum* L.) leaves extracts coated with nanochitosan derrived the marine fish scales.

2. Materials And Methods

2.1. Plant Material

Basil plants varieties Minahasa were collected from market local in Minahasa. The basil leaf extraction process refers to the modified Sangi and Katja (2011) methods [15]. Wet basil (*O. basillicum*) leaves are sorted, sliced, dried. The dried simplicia is smoothed with a blender. Basil leaf powder was extracted by maceration with 96% ethanol. A total of 200 g of basil leaf powder was macerated in 2000 ml of 96% alcohol for 24 hours. The filtrate from maceration is then evaporated using a vacuum evaporator to remove the ethanol solvent to obtain a thick extract. The thick extract of basil leaves was heated in an oven at 400C for 24 hours.

2.2. Extraction of chitosan

The fish scales of parrotfish (*Scarus* sp.) was collected from local markets in North Sulawesi. Preparation of chitosan fish scales refers to the modified Suptijah *et al* (1992) method [16], and starts with extraction chitin as reported by Rumengan *et al* (2017) [12]. Chitosan is obtained with deacetylation of chitin with add 40% NaOH (1:10), then heated at a temperature of 100-110°C for 1hour. The solids obtained are washed with distilled water until neutral pH beforedried in the sun.The chitosan obtained was weighed and stored at room temperature.

2.3. Preparation of nanochitosan and coating method

Preparation of nanochitosan using the ionic glass method [17]. Tween 80 as much as 0.1% added as homogenizer and tripoliposphate (TPP) 0.1% as stabilizer. Coating of leaf extract basil with nanochitosan refers to Kurniasari and Atun (2017) method which has been modified [18]. The coating process is carried out with add basil leaf extract coated with nanochitosan at extract ratio Chitosan: extract was 2 : 1. Then homogenized using a magnetic stirrer for 30 minutes. Solution of the basil leaf extract coated with nanochitosan were centrifuged.

2.4. Experimental design

In this experiment, a total of 24 adult male Wistar rats was used. All the animals were fasted for 12 h, but were allowed free access to water, before commencement of the experiments.The rats were divided into 8 groups of 3 rats each. The animal of group 1 served as negative control was given accolades; group 2 served as a positive control group was given glibenclamid at dose 0,45 mg/kgBw; group 3 to 5 were given extract basil leaf at dose 400 mg/kgBw (BLE 1) ; 800 mg/kgBw (BLE 2) ; and 1000 mg/kgBw (BLE 3) respectively. Group 6 to 8 were given extracts basil leaf coated with nanochitosan at dose 400 mg/kgBw (BLECN 1), 800 mg/kgBw (BLECN 2), and 1000 mg/kgBw (BLECN 3) respectively.

Diabetes mellitus was induced in overnight-fasted rats by single intraperitoneal injection of freshly prepared Alloxan monohydrate dissolved in sterile normal saline at a dose 150 mg/kg body weight. Diabetes mellitus was confirmed by measuring fasting blood glucose levels after 3 days. Rats with fasting blood glucose of more than 126 mg /dL were considered as diabetics and used for further experiment [19]. Treatment was given immediately after the determination of diabetes in rats. Measurement of fasting blood sugar was carried out every 1 hour after treatment was given. All blood samples were collected from the tail artery of the rats at the interval.

Determination of the blood glucose levels was done using a GlucoDr TM instrument. Blood glucose levels for each group were expressed in mg/dl as mean ± SD. The data were statistically analyzed using ANOVA with multiply comparisons versus control group and *t* test to see optimal dose of extract coated with nanochitosan. The values of $p<0.05$ were considered as significant [20].

3. Results and Discussion

Table 1. Fasting blood glucose level before and after administrated alloxan

Group	Fasting blood glucose pra alloxan	Fasting blood glucose post alloxan	Δ(%)
Control (-)	75 ±5.000	140 ± 9.165	46
Control (+)	64 ±1.528	136 ± 1,.32	53
BLE 1	86 ±7.211	137 ± 15.716	37
BLE 2	69 ±3.215	136 ± 8.505	49
BLE 3	67 ±15.275	139 ± 11.533	52
BLECN 1	65 ±5.000	139 ±35.726	53
BLECN 2	68 ±3.055	136 ± 15.177	50
BLECN 3	68 ±6.429	136 ± 6.506	51

Control (-) : aquades; Control (+) : glibenclamid at dose 0.45 mg/kgBW; BLE 1 : basil leaf extract at dose 400 mg/kgBW; BLE 2 : basil leaf extract at dose 800 mg/kgBW; BLE 3: basil leaf extract at dose 1000 mg/kgBW. BLECN 1 : basil leaf extract coated nanochitosan at dose 400 mg/kgBW; BLECN 2 : basil leaf extract coated nanochitosan at dose 800 mg/kgBW; BLECN 3 : basil leaf extract coated nanochitosan at dose 1000 mg/kgBW

The average of fasting blood glucose level in all groups before administration of alloxan according to Table 1 were in the range of 64-86 mg / dL. The highest fasting blood glucose level was found to be in BLE 1 group which was 86 mg / dL and the lowest was in the positive control group which was 64 mg / dL. After administration of alloxan, the rats' fasting blood glucose level in all groups were increased. The highest fasting blood glucose level was increased in the positive control group and BLECN 1 with an increase of 53%. The lowest increase was in the negative control group with a percentage increase of 46%.

Normal fasting blood glucose levels in rats was ≤100 mg / dL [21]. It means that the level of fasting blood glucose level before administration of alloxan was in the normal category. In pharmacology / bioactivity testing of experimental animals, the state of diabetes mellitus can be induced by pancreactomy and the administration of chemicals. In this study was using alloxan as a chemical to make diabetic rats characterized by a hyperglycemic state. Alloxan is a hydrophilic compound and unstable. As diabetogenic, alloxan can be used intravenously, intraperitoneally and subcutaneously. This study was used an intrapertitoneal injection method because it is considered more effective.

Table 2. Change in fasting blood glucose level after treatment for five hours

Group	Fasting blood glucose post alloxan	Time (Hours)				
		1	2	3	4	5
Control (-)	140 ± 9.165	139 ± 19.553	148 ± 9.018	152 ± 6.110	153 ± 6.658	158 ± 26.633
Control (+)	136 ± 1.732	128 ± 6.807	119 ± 5.568	93 ± 2.887	90 ± 0.577	71 ± 8.544
BLE 1	137 ± 15.716	136 ± 14.526	137 ± 13.50	126 ± 2.039	102 ± 4.509	94 ± 3.786
BLE 2	136 ± 8.505	129 ± 9.018	104 ± 7.371	84 ± 5.132	91 ± 19.757	79 ± 1.000
BLE 3	139 ± 11.533	137 ± 6.928	101 ± 17.03	98 ± 9.000	84 ± 6.028	115 ± 15.011
BLECN 1	139 ±35.726	128 ± 32.047	100 ± 10.44	89 ± 9.644	80 ± 2.000	73 ± 6.506
BLECN 2	136 ± 15.177	125 ± 18.028	142 ± 35.53	165 ± 44.44	130 ± 20.55	107 ± 12.014
BLECN 3	136 ± 6.506	132 ± 8.185	100 ± 9.609	87 ± 8.888	79 ± 8.327	72 ± 9.713

Control (-) : aquades; Control (+) : glibenclamid at dose 0.45 mg/kgBW; BLE 1 : basil leaf extract at dose 400 mg/kgBW; BLE 2 : basil leaf extract at dose 800 mg/kgBW; BLE 3: basil leaf extract at dose 1000 mg/kgBW. BLECN 1 : basil leaf extract coated nanochitosan at dose 400 mg/kgBW; BLECN 2 : basil leaf extract coated nanochitosan at dose 800 mg/kgBW; BLECN 3 : basil leaf extract coated nanochitosan at dose 1000 mg/kgBW

According to Table 2, there were a changed of fasting blood glucose (FBG) level to all group after being treated with basil leaf extract with or without coated by nanochitosan and glibenclamide drug for five hours of observation. In the negative control group there was an increase in FBG level every hour during 5 hours of observation, while in the other groups had a decreased even though not all groups had a stable decreased. At the first hour after the treatment all groups didn't show a significant change in FBG. It was changed occur at the second hour until the fifth hour after treatment.

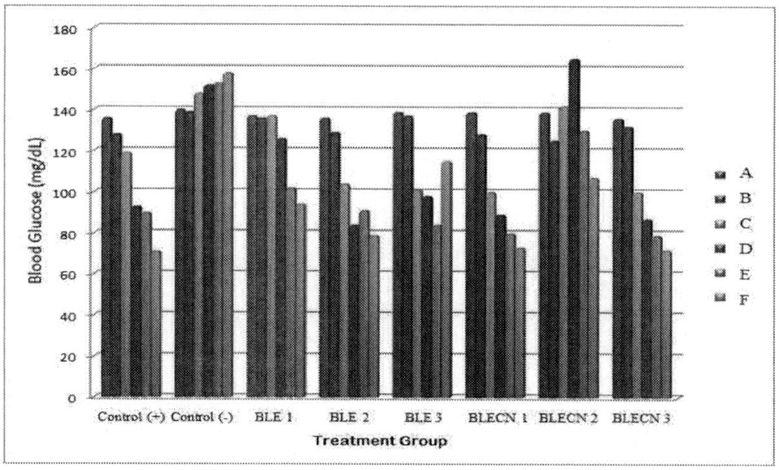

A: early blood glucose post aloxan C : blood glucose at 2nd hour E : blood glucose at 4th hour
B: blood glucose at 1st hour D : blood glucose at 3rd hour F : blood glucose at 5th hour

Figure 1. Graph of changes in blood glucose level during 5 hours of observation

In Figure 1 it is clear that the positive control group, BLE 1, BLECN 1 and BLECN 3 have a stable graph of FBG level reduction compared to the other groups for 5 hours of observation. In the negative control group there was a steady increase in FBG level. In the fifth hour most of the rats were in the normal FBG category in the range of 70-100 mg / dL except in BLE 3 group (115 mg / dL) and BLECN 2 (107 mg / dL) which were still in the category of impaired fasting blood glucose level which defined in the range of 100 - 125 mg / dL [22]. Based on One Way Anova test, indicate that there was an effect of giving basil leaf extract and basil leaf extract coated with nanochitosan on blood glucose of hyperglycemic mice, with a significance number = 0.004 (<0.05). This value shows that among treatment groups had different mean values significantly. Before testing ANOVA, the data has been tested for normality and homogeneity.

The comparison of each treatment group with the positive control group showed no significant difference. That means the administration of glibenclamide drug which is a commercial drug for diabetics was almost the same as giving basil leaf extract in its function to reduce high blood glucose levels (hyperglycemia). Glibenclamide is a second generation sulfonyluea drug which has the potential as an antihyperglycemic agent. The mechanism of action stimulates insulin secretion from beta langherhans cells, usually used for damage to langherans with less severe cells [23].

The results of this study are in line with the research of Mohammed *et al* (2007) who used basil species in Africa (*Ocimmum gratissium*) to test the antidiabetic effects of basil leaves in streptozocin-induced wistar rats [7]. The results of the study stated that at the 2nd hour of observation, there was a decrease in the group given the extract, but the increased in blood glucose level occurred after 4th hour of treatment. The group was given basil leaf extract at a dose of 500 mg /kgBW was reported to have the most effective effect with the presentation of glycemic changes of 81.3% for 8 hours of observation. While at doses of 250 mg /kgBW and 1000 mg / kg BW did not show a significant changes. Basil leaf extract at a dose of 400 mg / kgBW can increase antidiabetic effects and inhibit hepatic glucose mobilization and metabolism of carbohydrate enzymes for 2 hours of treatment [24] . In addition, basil leaf extract can also increase insulin secretion from the pancreas in 70 minutes of observation [25].

The effectiveness of nanochitosan in coating basil leaf extract was seen in the BLECN group at a dose of 400 mg /KgBW when compared to the BLE group with the same dosage (p = 0.023). The results of this study indicate that nanochitosan can increase the effectiveness of basil leaf extract as antihyperglycemic. A similar study was conducted by Malapermal *et al* (2017) using silver nanoparticles to increase the antidiabetic effect of basil leaves (*O. basillicum*) and successfully reduce blood glucose level by 79.74% [26].

Active compounds such as flavonoids have an activity in increasing insulin secretion by increasing the entry of Ca^{2+} ions through Ca channels [27]. Saponin compounds work by reducing absorption in the intestine by decreasing glucose absorption and modifying carbohydrate metabolism, increasing glucose utilization in peripheral tissues, glycogen storage and increasing sensitivity of insulin receptors in tissues. Tannin also acts as an antihyperglycemic with a mechanism to inhibit intenstinal glucose uptake and inhibits adipogenesis and antioxidants beside its ability to regenerate damaged pancreatic β cells [28]. Basil leaf extract has antihyperglycemic activity caused by its phenolic components contained in it which play a role in lowering blood glucose levels [29].

4. Conclusion

The administrate of basil leaf extract in hyperglycemic diabetic rats was shown to reduce blood glucose levels in all basil leaf extract group.The effectiveness of nanochitosan coating on basil leaf extract was shown at BLECN 1 at dose of 400 mg/kgBW *(p = 0.023)*. It is indicated that the basil leaf extract has a antihyperglicemic capacity and its phytopharmaceuticals increased caused by coated with nanochitosan.

Acknowledgment
This work was a part of the research project sponsored by University of Sam Ratulangi Manado with a research scheme RTUU for fiscal year 2018. The authors thank the Directorate Research and Community Service staff and the involved board of the ministry for financial support.

References

[1] Wild S, Roglic G, Green A, Sicree R, King H. Global prevalence of diabetes: estimates for the year 2000 and projections for 2030. Diabetes Care 2004; 27(5): 1047-53.
[2] Katzung BG, Masters SB, Trevor AJ. Basic & clinical pharmacology. New York: McGraw-Hill Medical; 2011.
[3] Jarald E, Joshi SB, Jain DC. Diabetes vs herbal medicines. Iran J Pharmacol Ther 2008; 7(1): 97-106.
[4] Brownlee M. 2005. The pathobiology of diabetic complications: A unifying mechanism. Diabetes 54: 1615–1625.
[5] Stratton IM, Adler AI, Neil HA. Association of glycaemia with macrovascular and microvascular complications of type 2 diabetes (UKPDS 35): prospective observational study. BMJ 2000; 321:405-412.
[6] El-Beshbishy HA, Bahashwan SA. Hypoglycemic effect of basil (*O. basillicum*) aqueous extract is mediated through inhibition of a glucosidase and a-amylase activities: an in vitro study. Toxicol Ind Health 2012; 28(1): 42-50.
[7] Mohammed, A., Y. Mohammed, M.A.Okasha, R.A Magaji, and A.H. Yaro. 2007. Effects of aqueous leaves extract of *Ocinum gratissimum* on blood glucose levels of streptozocin-induced diabetic wistar rats. *African Journal of Biotechnology*.6(18) : 2087-2090
[8] Bariyah, S K. 1D. Ahmed and 2M. Ikram.2012. *O. basillicum*: A Review on Phytochemical and Pharmacological Studies. Pak. J. Chem. 2(2):78-85, 2012. ISSN (Online): 2222-307X
[9] Anlysn, E and D Dougherty. 2006. *Modern Physical Organic chemistry*. California: University Science Books Sausalito
[10] Tiyaboonchai, W. 2003. Chitosan nanoparticles: A promising system for drug delivery. *Naresuan University Journal*. 11 (3): 51–66.
[11] Tajik H, M Moradi, S M R Rohani, A M Erfani, and F S S Jalali. 2008. Preparation of Chitosan from Brine Shrimp (Artemia urmiana) Cyst Shells and Effects of Different Chemical Processing Sequences on the Physicochemical and Functional Properties of the Product. *Molecules*. 13: 1263-1274.
[12] Rumengan, I F M, P Suptijah, S Wullur, and A Talumepa. 2017. Characterization of chitin extracted from fish scales of marine fish species purchased from local markets in North Sulawesi, Indonesia. *IOP Conf. Series: Earth and Environmental Science*. do i :10.1088/1755-1315/89/1/012028
[13] Tao Y, Z Hongliang, G Bing, G Jiao, H Yinming, and S Zhengquan. 2011. Water soluble chitosan nanoparticls inhibit hypercholesterolemia induced by feeding a high- fat diet in male Sprague-dawley rats. *Nanomaterials*. 1 (1)
[14] Luntungan,A H, L C Mandey, and I F M Rumengan. The effects of nanochitosan coating to phenolic compounds of basil leaf *Ocimum basillicum* L. Extracts. J. Ilmu dan teknologi pangan. Vol.5(2)
[15] Sangi M S dan D G Katja. 2011. Aktivitas antioksidan pada beberapa rempah-rempah masakan khas Minahasa. *Chemical Program*. 4(2) :66-74
[16] Suptijah P, H Salamah, S Sumaryanto, J Purwaningsih, dan Santoso. 1992. Pengaruh Berbagai Isolasi Khitin Kulit Udang Terhadap Mutunya. *Laporan Penelitian Jurusan Pengolahan Hasil Perikanan*. Institut Pertanian Bogor : Bogor.
[17] Suptijah P, M J Agoes, dan R Desie. 2011. Karakterisasi nanokitosan cangkang udang vannamei (Litopenaeus vannamei) dengan metode gelas ionik. *Jurnal Pengolahan Hasil Perikanan Indonesia*.14(2)

[18] Kurniasari D dan S Atun. 2017. Pembuatan dan Karakterisasi Nanopartikel Ekstrak Etanol Temu Kunci (*Boesenbergia pandurata*) Pada Berbagai Variasi Komposisi Kitosan. *Jurnal Sains Dasar.*6(1) :31-35

[19] American Diabetes Association. 2015. Standards of medical care in diabetes. *Diabetes Care.* 38(1) :1-94

[20] Duncan R C, Knapp RG, Miller MC (1977). Test of hypothesis in population means. In: Introductory Biostatistics for the health sciences. John Wiley and Sons Inc. NY pp. 71-96.

[21] Wulandari O dan S Martini. 2010. The difference in incidence of type II diabetes mellitus complications according to random blood sugar. *Jurnal Berkala Epidemiologi.* 1(2)

[22] Ministry of Health. 2014. Indonesian Ministry of Health's Data and Information Center. Jakarta: Ministry of Health RI

[23] Syarif A. 2007. *Farmakologi dan Terapi edisi 5.* Fakutas Kedokteran : Universitas Indonesia : Jakarta

[24] Ezeani C, I Ezenyi, T Okoye and C Okoli. 2017.*Ocimum basillicum* extract exhibits antidiabetic effects via inhibition of hepatic glucose mobilization and carbohydrate metabolizing enzymes. *Journal of Intercultural Ethnopharmacology.* 6 (1): 22-28

[25] Hannan J M A, L M Ali, B Rokeya, P R Flatt, and Y H A Wahab. 2006. *Ocimum sanctum* leaf extracts stimulate insulin secretion from perfused pancreas, isolated islets and clonal pancreatic–cells. *Journal of Endocrinology.*189 :127–136

[26] Malapermal V, I Botha, S Babu, N Krishna, and J N Mbatha. 2017. Enhancing antidiabetic and antimicrobialperformance of Ocimum basillicum, and Ocimumsanctum (L.) using silver nanoparticles.*Saudi Journal of Biological Sciences* 24: 1294–1305

[27] Suarsana I N 2009. Activity of hypoglicemic and antioxidative extract methanol of trivial on diabetic rats. *Thesis.* Post graduate of Institute Agliculture Bogor.

[28] Kumari, M and S Jain. 2012. Tannins : an nutrient with positive effect to manage diabetes. *Research journal of recent science.* 1(12)

[29] Suanarunsawat T 2014 Anti-Oxidative, Anti-Hyperglycemic and Lipid-Lowering Effects of Aqueous Extracts of *Ocimum sanctum* L. Leaves in Diabetic Rats. *Food and Nutrition Sciences.* 5 : 801811.

Relationship of primary productivity and phytoplankton abundance in Muara Kuala Raja, Bireuen district, Aceh

N Nurfadillah[1,2], S Tarina[1], E Miswar[2,3], I Dewiyanti[1,2], S Agustina[1,2]

[1]Marine Science Department, Faculty of Marine and Fisheries, Syiah Kuala University, 23111 Indonesia
[2]Center for Marine and Fishery Studies, Syiah Kuala University, Banda Aceh, 23111, Indonesia
[3]Utilization of Fisheries Resources Department, Marine and Fisheries Faculty, Syiah Kuala University, Indonesia

Corresponding Author: nurfadillah@unsyiah.ac.id

Abstract. Phytoplankton is a major producer in oceans which require the availability of nutrients, pH, temperature and light intensity in order to increase the productivity. This study aims to analyse the relationship between the primary productivity and the abundance of phytoplankton in Muara Kuala Raja Bireuen, Aceh. This study was conducted on May 2017. Location sampling have 3 station points. The measure of primary productivity was examined by a light-dark bottle oxygen method. The relationship between the primary productivity and the abundance of phytoplankton was analysed by using Principal Component Analysis (PCA). The results showed that the highest average of gross and net primary productivity were found at station 1 of 209,12 mgC/m^3/h and 172,00 mgC/m^3/h. The highest abundance of phytoplankton was found at station 1 of 11477,76 ind/l. The result of PCA showed that each of the parameter spread between the two axes indicates the data diversity value. The data diversity value on the 1- axis was 37% and the 2-axis was 59%, so the total of data diversity between 1-axis and 2-axis was 96%. Based on the PCA analysis, it can be concluded that primary productivity has a strong relationship with the abundance of phytoplankton.

1. Introduction

Kuala Raja is one of the beaches located in Bireuen City. There are many activities in this area such as fishing activities, aquaculture, and marine tourism. Various activities can affect phytoplankton. The aquaculture area contributes nitrate and phosphate contents that can increase the growth of phytoplankton. Phytoplankton is a major producer in the waters, which require the availability of nutrients, pH, temperature and light intensity to increase productivity [1][2]. In otherwise, poor water quality and unbalanced nutrient availability will result in decreased productivity in phytoplankton. Although nutrient plays important role in photosynthesis of phytoplankton, the excess nutrient may not directly indicate high primary productivity [3].

Primary productivity is the total amount of organic matter by the activity of plant photosynthetic. Primary productivity is a food supply for heterotrophic organisms such as bacteria, fungi and animals. The value of primary productivity in waters is more influenced by light intensity compared to the presence of nutrients [4]. Light intensity will affect the presence of phytoplankton. Therefore, it is necessary to study the presence of phytoplankton and its relationship with the primary productivity of Muara Kuala Raja Bireuen, Aceh. The purpose of this study is to describe the value of primary productivity, the value of phytoplankton abundance and analyse the relationship of primary productivity and abundance of phytoplankton in the waters of Muara Kuala Raja Bireuen, Aceh.

IOP Conf. Series: Materials Science and Engineering **567** (2019) 012024 doi:10.1088/1757-899X/567/1/012024

2. Research Methods

2.1. Data collection

This research was conducted in Muara Kuala Raja in May 2017. The research area was divided into 3 observation stations, namely station 1 is a domestic and pond area, Station 2 is an estuary area and station 3 is a coastal area. Identification of sample was done at Marine Biology Laboratory of Syiah Kuala University and BARISTAND laboratory Banda Aceh. The location of the research are described in Figure 1.

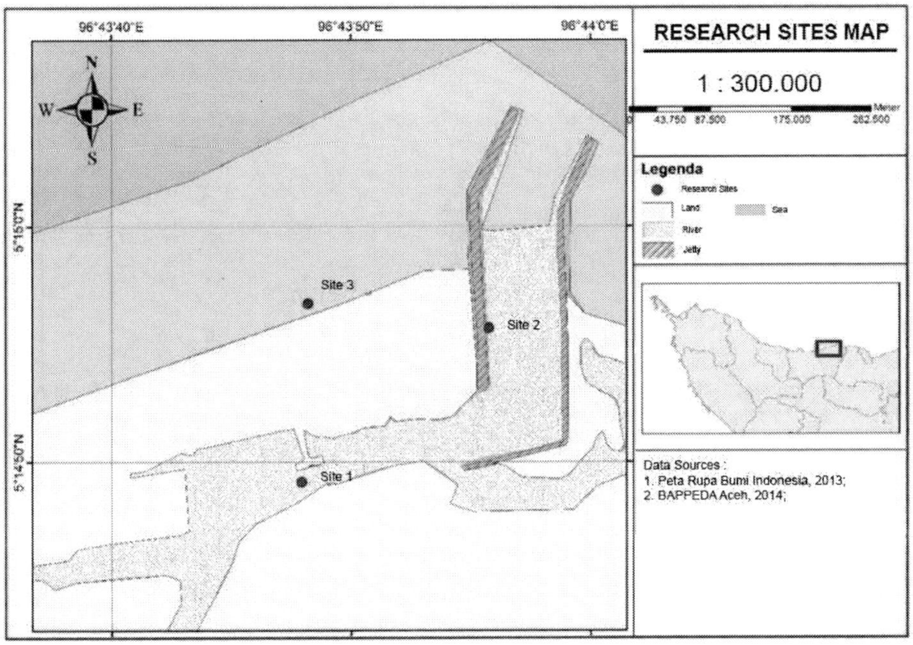

Figure 1. The Map Location Research

2.2. Measurement of primary productivity

Measurement of primary productivity using oxygen method by taking water samples at each research location using light bottles, dark bottles and initial bottles as initial oxygen (DO_o). Light bottles and dark bottles filled with water samples were incubated for 6 hours starting at 8.30 am - 2.30 pm. Next, measurement of oxygen values for each of the incubated bottles was carried out.

2.3. Sampling of phytoplankton

Phytoplankton sampling was carried out by filtering 100 litres of water samples cumulatively using planktonnet mesh size 40μm. Filtered water samples (30 ml) were included in a collection bottle labeled and then preserved with 2% lugol 8-10 drops. Furthermore, the water samples were observed by a microscope, and identified using the book references [5].

2.4. Measurement of aquatic parameter

Observation of water quality at each station includes current speed, visibility, temperature, pH, DO and salinity. Nitrate and phosphate measurements were carried out in the BARISTAND laboratory using a spectrophotometer method.

IOP Conf. Series: Materials Science and Engineering **567** (2019) 012024 doi:10.1088/1757-899X/567/1/012024

2.5. Data analysis
There are some calculations for data analysis in this study. The formula used to calculate primary productivity in waters is as follows [6]:

$$\text{Gross Primary Productivity (mgC/m\^{}3/hr)} = \frac{(O2\ LB\ ) - (\ O2\ DB)\ x\ 1000}{(PQ)(t)} x0.375$$

$$\text{Net Primary Productivity (mgC/m\^{}3/hr)} = \frac{(O2\ LB\ ) - (\ O2\ IB)\ x\ 1000}{(PQ)(t)} x0.375$$

Where O_2LB is concentration of oxygen in the light bottle; O_2DB is concentration of oxygen in the dark bottle; PQ is coefficient of photosynthetic=1.2; t is time of incubation (6 hours); 0.375 is ratio of carbon and oxygen atomic mass

Calculation of phytoplankton abundance in this study uses field view method by using the formula [7]:

$$\text{N (ind/l)} = n\ (ind) \frac{Vt\ (ml)}{Vcg\ (ml)}\ x\ \frac{Acg\ (mm)^2}{Aa\ (mm)^2}\ x\ \frac{1}{Vd\ (l)}$$

Data analysis about the relationship between phytoplankton abundance and primary productivity and several other variables using the Principal Component Analysis (PCA) method.

3. Results and Discussion
3.1. Primary productivity
The value of primary productivity at each research station shows differences, this is caused by the differences in the activity of aquatic organisms in using oxygen both for respiration and decomposition of bacterial activity. Gross primary productivity is total photosynthesis which includes organic material used for respiration during measurement. While the net primary productivity is the storage of organic materials during measurement. The results of gross and net primary productivity can be seen in Table 1.

Table 1. Value of Primary Productivity in Muara Kuala Raja

No	Primary Productivity (mgC/m^3/hour)	Station 1 mgC/m^3/hour	Station 2 mgC/m^3/hour	Station 3 mgC/m^3/hour
1	Gross Primary Productivity	209.12	57.29	61.65
2	Net Primary Productivity	172.00	46.87	47.04

In table 1, the highest average gross and net productivity values are found at station 1, that is 209.12 mgC/m ^3/hour and 172.00 mgC/ m ^ 3/hour. The high value of primary productivity at station 1 is thought to be due to the influence of high brightness values on the station so that it affects the intensity of light entering the waters. Increased light intensity will always be followed by increasing the value of primary productivity to the optimum point [4]. The intensity of exceeding the optimum light is the inhibiting light and under optimum light is also a limiting light [8]. Increased light intensity is followed by an increase in temperature so that it affects the presence of phytoplankton and the value of primary productivity [9]. In addition to light intensity, the high value of primary productivity at station 1 is due to the high nitrate content (1.6 mg / l) and phosphate (0.11 mg / l) derived from household activities and aquaculture areas. The presence of nutrients in the waters affects the presence of phytoplankton. Phytoplankton utilize nutrients for their growth and breeding rate. Changes in environmental parameters including light intensity, DO, salinity, temperature, nitrate, silica and phosphate affect variations in the distribution and abundance of phytoplankton in the waters [10]. The high nutrient values and primary productivity of phytoplankton in the waters showed the natural ability of the aquatic environment to absorb the effects of enrichment of nutrients from land such as nitrate, silica and phosphate [11].

The lowest primary productivity is found in station 2. The low primary productivity in this region is thought to be due to the low contribution of nutrients from the land and the low water temperature compared to other stations (26.3°C). Temperature plays an important role in controlling the condition of aquatic ecosystems. Aquatic organisms have a certain temperature range in their lives. The difference in the intensity of light entering the water column gives a difference in temperature distribution at each depth. Temperature effects on primary productivity depended on light intensity, under light limited conditions, temperature had a negative effect on primary productivity, because warming strongly increases grazing activity of phytoplankton consumers and community respiration, whereas the carbon incorporation process is limited by light [9].

3.2. Phytoplankton Abundance

In this study, the composition of phytoplankton was obtained in 2 classes, namely Bacillariophyceae and Dinophyceae. Table 2 shows that the percentage level of Bacillariophyceae class is 81% higher than the Dynophyceae class at 19%. This is consistent with research conducted by Aryawati and Thoha in the Berau waters of East Kalimantan that found as many as 28 genera of phytoplankton which were divided into 2 classes namely Bacillariophyceae as many as 24 genera and Dynophyceae as many as 4 genera [12]. This is suspected of Bacillariophyceae or diatoms have high productivity compared to Dynophyceae. Bacillariophyceae is commonly found because it is easy to adapt to the environment, and a wide distribution of sea water, fresh water [13][14]. The existence of phytoplankton is very dependent on the condition of the waters that can affect and produce different species communities in each water column. According Onyewa [15] that phytoplankton composition is not always evenly distributed in a waters.

The highest abundance of phytoplankton is found at station 1 at 11477.76 ind / L and the lowest abundance value is at station 3 of 3487.68 ind / L. Dytilum sol is an abundant species, it is suspected that nitrate phosphate levels in the sampling area are in accordance with the environment of phytoplankton and suitable for the habitat of this species. Tomas [16] states that *Ditylum sol* live in warm waters. The temperature at station 1 is 31°C which is classified as warm water. While the value of the lowest abundance of phytoplankton species is the type of *Nitzschia* sp and *Synedra* sp. It is suspected that this species is at a temperature not available with the habitat of his life. *Synedra* sp lives in cold waters [16]. The types of *Nitzschia* sp and *Synedra* sp abundant in Lake Laut Tawar with low temperature criteria (24°C) [2]. Temperature increases have been reported to have a positive effect on the abundance of species in the sea and in some terrestrial ecosystems [17][18][19].

Tabel 2. Abundance of Phytoplankton

NO	Genus	LOCATION		
		Abundance Site 1 (ind/l)	Abundance Site 2 (ind/l)	Abundance Site 3 (ind/l)
	Bacillariophyceae			
1	*Asterionella* sp.	639,36	0	0
2	*Biddulphia* sp.	351,36	0	31,68
3	*Chaetoceros affinis*	799,68	0	0
4	*Chaetoceros antlanticus*	0	0	384
5	*Chaetoceros* sp.	543,36	0	0
6	*Coscinodiscus nitidus*	543,36	0	0
7	*Ditylum sol*	3360	864	1152
8	*Hyalodiscus* sp.	31,68	447,36	0
9	*Navicula* sp.	63,36	511,68	0
10	*Nitzschia longissima*	96	0	0
11	*Coscinodiscus* sp.	607,68	0	0
12	*Pleurosigma* sp.	543,36	351,36	0
13	*Rhizosolenia hebetata*	799,68	415,68	864
14	*Rhizosolenia cal-caravis*	511,68	31,68	384
15	*Rhizosolenia* sp.	543,36	63,36	0
16	*Synedra* sp.	0	384	0
17	*Thalasiothrix* sp.	31,68	0	0
18	*Triceratium favus*	511,68	0	0
	Dynophyceae	0	0	0
19	*Ceratium furca*	988,8	480	672
20	*Ceratium cariense*	0	351,36	0
21	*Dynophysis homunculus*	511,68	31,68	0
22	*Streptotheca indica*	0	384	0
	Total	11477,76	4316,16	3487,68
	Total Abundance of Phytoplankton (ind/l)		19281,6	

3.3. Relationship of primary productivity to phytoplankton abundance

The relationship of primary productivity with several parameters is done by the Principal Component Analysis (PCA) method. Based on PCA analysis shows a very close relationship with factor analysis value of 96%. Analysis can be seen in Figure 2.

Based on Figure 2 showed that some parameters have a positive correlation to primary productivity including abundance of phytoplankton, nitrate, phosphate, visibility, pH, temperature. On the negative axis are salinity and current. Parameters that are close to the positive axis have an influence on primary productivity and are different from salinity and currents that are in the negative axis, which means that they do not have a major influence on primary productivity. Parameters included in the positive correlation (phytoplankton abundance, nitrate, phosphate, visibility, pH, temperature) appear to have a significant effect on primary productivity. This is consistent with the research of Yulianto *et al.*, in Pulau Panjang Jepara waters that the value of primary productivity at 181.25 mgC/m ^ 3/ hour, and obtaining abundance of phytoplankton at the same station was 56.559 ind / l which means in this study primary productivity is affected by phytoplankton abundance [20].

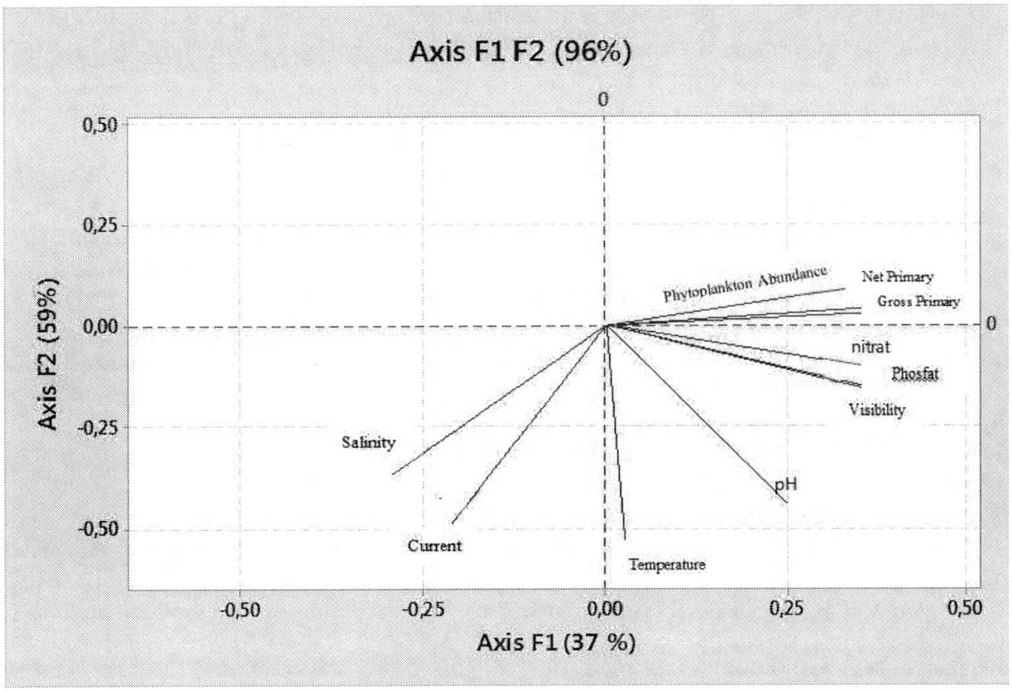

Figure 2. PCA Analysis

Temperature did not affect phytoplankton diversity, but species richness directly increased primary productivity, probably as result of the strong selection effect [9]. But, The influence of abiotic factors, such as light, temperature, and hydrology is more evident on the temporal fluctuations of productivity rates, when compared to that of biotic factors [1]. However, primary productivity and producer biomass are separate ecosystem functions, with productivity measuring carbon flux and biomass measuring carbon accumulation [21].

4. Conclusion

The highest value of gross and net primary productivity at station 1 is 209.12 mgC / m ^ 3 / hour and 172.00 mgC / m ^ 3 / hour while the highest phytoplankton abundance value at station 1 is 11477.76 ind / l and the lowest in Station 3 is 3487.68 ind / l. Based on PCA analysis shows a close relationship between primary productivity and phytoplankton abundance of 96%. Phytoplankton abundance, nitrate, phosphate, visibility, pH, temperature to have a significant effect on primary productivity.

Acknowledgment

We acknowledge the collaboration of the crews of marine biology laboratory to sampling and thanks to dean Marine and Fisheries Faculty, Syiah Kuala University Banda Aceh.

References

[1] Feresin E G, M S Arcifa, L H S Silva, A L H Esguícero *J. Acta Limnologica Brasiliensia* **22**(4): 384-396.
[2] N Nurfadillah, E M Adiwilaga, A Damar 2012 *J. Depik,* **1**(2): 93-98.
[3] Edwards E M, T Platta, S Sathyendranatha 2004 *J. Ecological Modelling* **171**(1–2): 103-125.
[4] Alianto, E M Adiwilaga, A Damar 2008 *J. Ilmu-Ilmu Perairan dan Perikanan Indonesia* **15**(1) : 21-26.
[5] Hoppenrath M, M Elbrächter, G Drebes 2009 *Marine phytoplankton* Schweitzer Vertagebuch Handlung Stuttgart Publisher ISBN-10: 3510613929.264 p.
[6] Vijay K K, M M Ranga 2015 *Primary productivity and status of Anasagar Lake, Ajmer (Rajasthan)* LAP LAMBERT Academic Publishing ISBN-10: 3659791229. 112 p.
[7] [APHA] American Public Health Association 2005 *Standard methods for the examination of water and wastewater* 21edition Washington: APHA, AWWA (American Waters Works Association) and WPCF (Water Pollution Control Federation). Hal 3 – 42.
[8] Miller C B 2004 *Biological oceanography* Blackwell Science Ltd, USA.
[9] Lewandowska A M, P Breithaupt, H Hillebrand, H G Hoppe, K Jürgens, U Sommer 2012 *J.of Sea Research* **72**: 87–93.
[10] Chang H 2008 *Water Res* **42** (13), 3285—3304.
[11] Marlin N, A Damar, Effendi 2015 *J. Ilmu Perikanan Indonesia* **20**(3) : 272-279
[12] Aryawati R, H Thoha 2011 *J. Maspari* **2** : 89-94.
[13] Abida I W 2010 *J. Kelautan* **3**(1): 36-41
[14] Damar A 2003 *Effects of enrichment on nutrien dynamics, phytoplankton dynamics and primary production in Indonesian tropical waters : a comparison between Jakarta Bay, Lampung Bay and Semangka Bay* Forschung-und Technologiezentrum Westkueste Publ. Ser No. 199: 196 p.
[15] Onyewa I C 2007 *J. Turkish of Fisheries and Aquatic Sciences* **7**: 89-96
[16] Tomas C R 1997 *Identifying Marine Phytoplankton* Acedemic Press Harcount Brace & Company London.
[17] Beaugrand G., M Edwards, L Legandre 2010 *Proceedings of the National Academy of Sciences of the United States of America* 1–5.
[18] Tittensor D P, C Mora, W Jetz, H K Lotze, D Ricard, E V Berghe, B Worm 2010 *Nature* **466**, 1–4.
[19] Menéndez R, A G Megías, J K Hill, B Braschler, S G Willis, Y Collingham, R Fox, D B Roy, C D Thomas 2006 *Proceedings of the Royal Society* **273** (1593), 1465–1470.
[20] Yulianto D, M R Muskananfola, P W Purnomo 2014 *Diponegoro Journal Of Maquwares* **3**(4) : 195-200.
[21] Stachowicz J J, J F Bruno, J E Duffy 2007 *Understanding the effects of marine biodiversity on communities and ecosystems* Annual Review of Ecology, Evolution, and Systematics 38, 739–766.

Litter production and decomposition of mangrove in the Northern Coast of Aceh Besar district, Aceh province

I Dewiyanti[1,3], N Nurfadillah[2,3], T Setiawati[1], F Yanti[1], S A Elrahimi[1]
[1]Marine Science Department, Faculty of Marine and Fisheries, Syiah Kuala University. Darussalam, Banda Aceh, Indonesia
[2] Aquaculture Department, Faculty of Marine and Fisheries, Syiah Kuala University. Darussalam, Banda Aceh, Indonesia
[3]Center for Marine and Fishery Studies, Syiah Kuala University, Banda Aceh, 23111, Indonesia

Corresponding Author: irma_alfian@yahoo.com / irmadewiyanti@unsyiah.ac.id

Absract. Mangrove vegetation plays an important role and supplier of organic matter derived from litter fall and litter decomposition The purposes of this research was to obtain the productivity of mangrove vegetation by analyzing the litter fall, decomposition coefficient, and percentage of litter decomposed in the northern coast of Aceh Besar District. The research used purpossive sampling method to determine the 6 stations devided into 2 substations (landward and seaward). This research was conducted in July to September 2016. Each substation was put one of the litter trap. The litter trap method was used to know the mangrove litter production. Litter decomposition focused on *R. mucronata* leaves were studied by using litterbag technique. The litter bags were made of synthetic nylon which had size 20cm x 30cm and mesh size was 1 x 1.25 mm^2. There were 3 types of mangrove found, litter production of *Rhizophora mucronata* as much as 79.21 $g/m^2/day$, *Avicennia marina* as much as 0.70 $g/m^2/day$ and *Sonneratia alba* was 17.46 $g/m^2/day$. Based on the substation influenced by tidal, mangrove litters found in the seaward higher than in the landward, the average litter fall was 52.05 $g/m^2/day$ and 43.32 $g/m^2/day$. Result of mangrove decomposition showed that the percentage of decomposition for the 49th day in the seaward substation was 48.58 % and the landward was 41.53 %. Mangrove litter decomposition increased during the time at both substation. Decomposition rate of litter for *Rhizopora mucronata* species classified very quickly where the substation on the seaward was 0.012 (d^{-1}) and landward was 0.010 (d^{-1}). The half-time took in the leaf litter of mangroves for the landward was 69 days and seaward was 57 days, *R. mucronata* leaves litter decomposited in seaward substation was faster than landward.

1. Introduction

Ecologically, mangrove acts as a habitat for feeding, nursering and spawning ground of various aquatic biota [1]. Mangrove also serves as a supplier of organic materials derived from litter so that it can provide nutrients for organisms that live in the surrounding water. Litter is the dead material located above ground and it will experience decomposition and mineralization process. Castillo et al [2] explained that litter of mangrove represented that leaf litter was the greatest percentage of total annual litter production (70%), followed by flower litter (15%), twigs (10%), and propagules (5%).

Litter decomposition is a process in which organic matter produced by mangrove forests is transferred to the sediment influenced by dry and rainy seasons [3, 4]. Generally, four basic steps that decomposition of litter transferred to their habitat: (1) removal of leaves material compound by water, (2) microorganisms colonization, (3) consumption by herbivores, and (4) action of physical environmental factors [5, 6]. Measuring primary productivity in mangrove ecosystems are intricate but litter production and litter decomposition has been widely used as a measure of productivity, particularly to predict contribution of nutrient to the estuarine ecosystem [7]. The productivity is generally dependent on nutrients derived from the litter present in the water. Factors that contribute to litter decomposition are litter type, temperature, pH, and microbial activity [8].

Aceh Province has potential marine biota resources such as crustaceans (shrimps, crabs) and mollusks (shells, snails). This high potential is because Aceh area has some coastal ecosystem that supports the primary productivity in the estuary and sea. For coastal areas, the existence of mangrove forests primarily as a green line along the coast is very important for the supply of fish and shrimp as well as maintaining the quality of the fishery ecosystem and also protecting the settlements from abrasion. Some of communities in Aceh Besar and Banda Aceh entrust their life as fishermen and they commonly use mangrove area for fishpond in order to increase their income. According to DKP [9], Aceh Besar has coral cover area of 1.155 ha and mangrove ecosystem area is 133.94 ha. The decomposition of mangrove leaf litter has been studied in tropical and subtropical regions on the word (e.g Southwest India [10], Peninsular Malaysia [11], Gazi Bay Kenya [12], Sabang Indonesia [5]. The decomposition of mangrove leaf litter has been studied in previous research explained that the percentage of leaves litter decomposed (*Rhizophora stylosa)* was 56.68% ± 0.67 in site next to the land and it was 59.30% ± 1.81 in site next to the sea [5].

Gampong Leungah, Beurenut, Lamreh, Lampanah, Ladong are some villages located on the north coast of Aceh which has a large community of mangrove forest. *Rhizophora* sp. and *Avicennia* sp. are the most dominant mangrove species in this mangrove area. According to Dewiyanti [13] *Rhizophora apiculata*, *R. stylosa,* and *Avicennia marina* were the dominant species of mangrove pre tsunami in Ulee Lheue coastal, Banda Aceh city. The mangrove ecosystem is decreasing in the northern coastal area of Aceh Besar with increasing the amount of pond area and the excessive exploitation. Those make the role of mangrove as feeding and spawning area for macro fauna is also decreasing and it impacts low coastal productivity. Considering the importance of mangrove litter and litter decomposition to support the production of aquatic fauna in coastal areas, it is necessary to know the amount of litter production available on the mangrove forest floor and its decomposition rate. Therefore, the present study aimed (1) to measure the productivity of mangrove vegetation by analyzing the litter fall and also the factors that affected the mangrove litter, (2) decomposition coefficient, and percentage of litter decomposed in the northern coast of Aceh Besar District, Aceh Province.

2. Methodology

2.1 Time and Research Location

This research was conducted for 3 months starting from July to September 2016. The method used in this research was survey method by taking sampling directly in mangrove ecosystem. The determination of research station was done by using purposive sampling method based on the desired goal. According to field assessment, there were 6 observation stations, namely Alue Naga village (station 1), Ladong village (station 2), Beurenut village (station 3 and 4), Lampanah village (station 5), and Leungah village (station 6). At each station divided into 2 substations (landward and seaward), three transects as a plot (10m x 10m) were established in each substation for replication. The location of the research can be seen in Figure 1.

2.2 Measurement of Litter Production

Litter fall production used litter trap technique with transect size 1m x 1m [14], and the net mesh size for litter trap was 1cm x 1cm. Litter trap was placed with a height of 1.5 m above ground level, so it can avoid from the tidal effect. Measurement procedure for litter fall production was taken every 8 days (weekly) for two months by taking all litter fall accommodated in the litter trap. Litter samples were separated based on litter components including leaves, branches, twigs and reproductive organs (flowers and fruits). The litter sample were oven dried at 60°C for 48 hours and weighed until constant and final dry weight was recorded with a precision of 0.01 g.

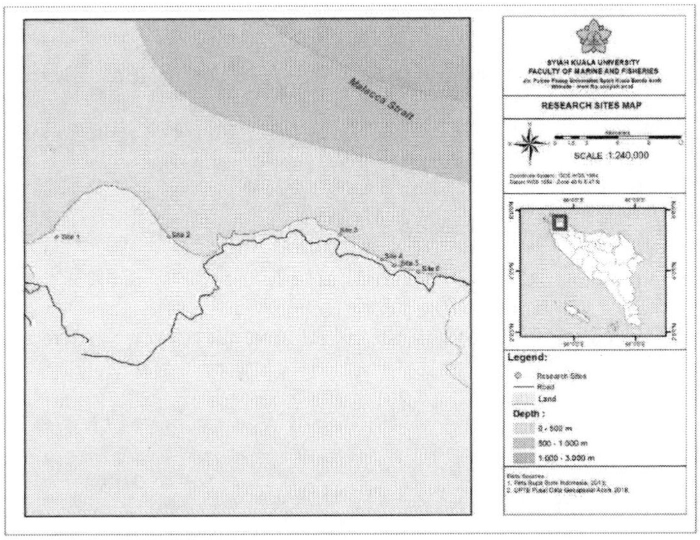

Figure 1. The map of study sites located at Aceh Besar districts.

2.3 Measurement of Litter Decomposition

Litter samples taken for this study were *Rhizophora mucronata* leaves because they were easily obtained and abundant in the mangrove area. Criteria of the leaves taken were senescent leaves, and the leaves were still intact located on the floor of mangrove forest. Leaf litter decomposition is commonly measured by using the litter bag technique [11]. Commonly, litter bag method is used for the decomposition research [15]. The litter bags were made of synthetic material and had size 20cm x 30cm [16] with mesh size 1.5 x 1.5 mm^2.

Leaves litter that has been collected from the field was taken to the laboratory and oven dried at 60°C for 48 hours or until the litter is completely dry. Furthermore, 12 g of leaves after oven dried was placed in each litter bag and every bag was tied in the mangrove prop root in order not to drift or disappear during the tide. Litter bags were placed as many as 6 bags on each substation with 3 replications. Sampling time was daily, weekly and biweekly, i.e the 3rd day, 7th day, 14th day, 21st day, 35th day, and 49th day. Sampling time was determined based on the results of several studies that have been found the litter decomposition process started between 7-15 days. Monitoring of the litter bag was carried out at certain intervals to estimate the litter decomposition rate [17]. Totally, 36 litterbags were needed in each observation and for 6 time observations in 10 weeks, dry leaves needed were 2592 g and litter bags were 216 bags. After collecting sample, the bags were returned to laboratory where they were rinsed in a sieve to remove sediment, and then continue with oven dried to have a constant mass. The samples were oven dried at 60ºC for 2 days and weighed until constant and then final dry mass was recorded.

2.4 Statistical Analysis

2.4.1 Production of Mangrove Litter. Litter production analysis was conducted using the equation below [18]

$$Dry\ Weight = \frac{g}{m^2/day}$$

Where g is gram of dry weight; m^2/day is squared meter per day

2.4.2 Decomposition of Mangrove Litter. Litter which was already decomposed and percentage of leaves decomposed were calculated by using formula [11]

$$D\ (\%) = \frac{B1 - B2}{B1}\ x\ 100\%$$

Where D is litter decomposed; B1 is dry weight before decomposition; B2 is dry weight after decomposition.

The relationship between percentage dry mass remaining in litter bags and sampling time at all sites best fitted in a negative single exponential model [11]. The formula was :

$$Xt = Xoe^{-Kt}$$

Xt is percentage of initial material; Xo is remaining after time t (days) and K is a decay coefficient (d^{-1}).

The times required for decomposition of half the initial material (t50) were determined by using the formula:

$$t_{50} = \frac{ln2}{K}$$

3. Results and Disscussion

3.1. Litter Fall Mangrove Production

The result showed that there were three mangrove species found namely, *Rhizophora mucronata, Avicennia marina* and *Sonneratia alba*. In the observation area, the existing mangrove vegetation was dominated by *R. mucronata,* The dry weight average of litter fall from *R. mucronata* was 37.44 g/m^2/ day in seaward and in landward substation was 41.77 g/m^2/day, totally 79.21 g/m^2/day leaves litter contributed by *R. mucronata*. *Avicennia marina* had weight average 0.70 g /m^2 /day in seaward and *Sonneratia alba* was 3.55 g/ m^2/day in landward and 13.90 g /m^2 /day in seaward substation (Table 1). According to Andrianto et al, [19] the shape and size of leaves of *R. mucronata* larger than the other species causing the leaves to fall easily when the wind blows and *R. mucronata* often appears in every station, it has a high density of 20 individuals /100 m^2. The results of research conducted by Andrianto et al, [19] in Durian Village and Batu Manyan Village, Padang showed that the productivity of *R. mucronata* mangrove litter obtained as much as 34.8 g/ m^2 /day with a density of 10 individuals/10 m^2, while *R. stylosa* was 15.74 g/ m^2/day with density of 3 individuals/10 m^2. These information indicated that the litter mangrove productivity on the northern coastal area of Aceh Besar was higher than previous study.

In average, mangrove leaves litter production obtained in seaward was 52.05 g/m^2/day, and in landward was 45.32 g/ m^2/day. The productivity of mangrove litter found in seaward higher than landward due to the mangrove density in seaward was 30 individuals/100 m^2 and in landward was as many as 25 individuals/100 m^2. In additional, the litter fall higher in seaward than landward assumed because of environmental condition such as salinity, temperature and dissolved oxygen. Salinity is one of aquatic parameters which is influence the litter production. Lestarina [20] mentioned that the higher the salinity, the higher the productivity of the mangrove litter. In additional, high contribution of leaves to litter productivity is related adaptation mangrove to reduce water loss in order to survive in high salinity conditions. The average of salinity in field study was 33.5 ppt in seaward and 28 ppt in landward. Mangrove vegetation that inhabit at a higher salt content will produce a high mangrove litter as well, it is related the ability to accumulate excess salt in the leaf components and then it fall as a litter [20].

Leaf is the largest contributor of litter compared to other plant components (Table 1). There was a correlation between peaks in litter fall and peaks in leaf fall whereas leaf fall contributed 67% of total litter fall [21]. The litter composition was primarily structured of leaves which accounted for 65–75% of total litter production in our study sites [22]. The result showed that leaves litter production in study area was 67.07 g/m^2/day. The leaf component as a mangrove litter was more than other component as much as 17.69 g/m^2/day [23]. This was thought due to leaf shape, size wide and thin, moreover the characteristics make leaves easy to fall. It was also caused by the physiological properties of the leaves, where the leaves play an important role in the process of photosynthesis. The old leaves will fall and be replaced by relatively young leaves. Mangrove is generally located in the forest tightly, the old leaves located at the bottom or inside of the canopy get less light and they can't photosynthesis perfectly

IOP Conf. Series: Materials Science and Engineering **567** (2019) 012025 doi:10.1088/1757-899X/567/1/012025

affected leaves fall faster [24]. In the energy flow of mangrove forests, leaves litter as a source of nutrients for organisms in surrounding area [25]. On the other hand, litter of flowers and fruits were 22.69 $g/m^2/day$. Lack of fruit and flower production due to the observation time wasn't in flowering season. In additional, they were produced by some individuals' mangrove species (*R. mucronata*) while leaves were produced by all individual observed of mangrove species.

Table 1. Average of Mangrove Litter Dry Weight in Study Area

Species	Litter component	Average Litter dry weight ($g/m^2/day$)		
		Landward	Seaward	**Total**
Rhizophora mucronata	Leaf	27.9	26.17	54.07
	Fruit	7.05	5.17	12.22
	Flower	2.55	2.77	5.32
	Twig	4.27	3.33	7.60
Total		**41.77**	**37.44**	**79.21**
Avicennia marina	Leaf	-	0.70	0.70
	Fruit	-	-	-
	Flower	-	-	-
	Twig	-	-	-
Total		**-**	**0.70**	**0.70**
Sonneratia alba	Leaf	2.79	9.51	12.3
	Fruit	0.76	4.39	5.15
	Flower	-	-	-
	Twig	-	-	-
Total		**3.55**	**13.90**	**17.45**
Total		**45.32**	**52.05**	**97.37**

3.2. Decomposition of Mangrove Litter

Observation of litter decomposition was conducted for 49 days. *R. mucronata* leaves observed because of dominance species in study area. Decomposition of mangrove leaves litter from the 3rd day to 49th day showed that percentage of litter decomposed increased with increasing time of decomposition period (Figure 1). The dry weight of the mangrove litter during the decomposition process decreased from the initial dry weight. Dry weight was required to know the percentage of decomposition rate during observation. Where the initial dry weight of the litter was 12 gr each bag. Dry weight of mangrove litter began to decline on the 7th day and continued to decrease with increasing incubation time. The dry weight of leaves litter has an average 7.02 g in landward and 6.17 g in seaward at the last observation period on the 49th. Leaves decomposed were high in the beginning and at the end of decomposition process and they were fluctuated during experiment period. Decomposition process is initially rapid in the beginning than slow, after one week showed a rapid initial mass loss indicated by fresh leaves had lost 31.4% and dried leaves had lost 38.3% because of leaching in a mangrove of Southwest India [10]. Furthermore, this condition was suspected by the existence of influence in quality and quantity of the litter element content that as food for the soil microbe. The abundance of bacteria, microorganisms and invertebrates enormously accelerates the decomposition process of the organic material produced by mangrove forest [26].

Furthermore, the percentage of litter decomposed in seaward was higher than landward during the observation. On the 49th day, the percentage of decomposed litter was 41.53% (landward) and 48.58% (seaward) (Figure 2). The mangrove location next to the sea decayed faster than the location next to the land, the comparison between both sites at the end of observation on the 68th day showed that the percentage of decomposed litter next to the sea was 59.30% ± 1.81 and percentage of decomposed litter in the to the land was 56.68% ± 0.67 [5]. In toward the sea, the decomposition process is influenced by physical factors such as the movement of tidal currents and inundation by longer seawater. The mechanism of loss of soluble materials from litter is caused by rain or water flow [26]. Increasing decomposition rates because of water soaking caused leaching of labile material and high inundation caused more litter will be decomposed rapidly [11].

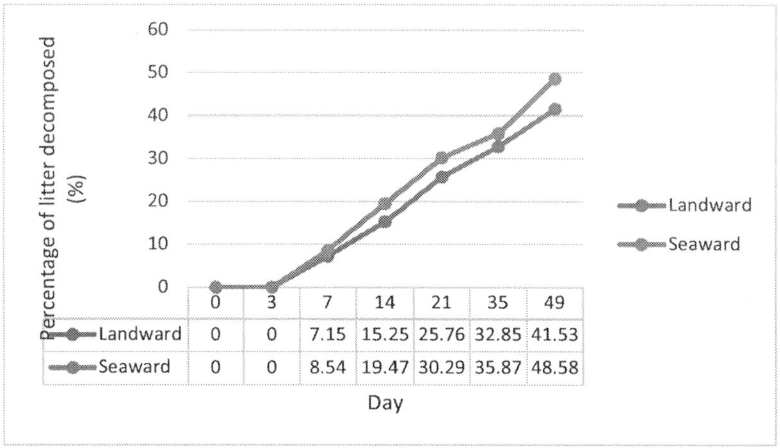

Figure 2. Percentage of Litter Decomposed During the Observation Period

The decomposition of the litter was also caused by the erosion of litter by the movement of waves. The moist substrate conditions of the water as compared to the mainland also play a role in litter decomposition, an environment that is always alkaline and moist causing the litter decomposition process faster. The compounds contained in the litter also affects the ability of a microbe to decompose complex compounds contained in the litter [8]. The average of temperature in seaward and in landward ware 39.5^0C and 37.2^0C. The high temperature causes a faster decomposition process due to high temperatures can increase the level of chemical weathering. Climate plays an important role in the decomposition rate of organic matter, increasing temperature and humidity causes the decomposition process quickly, on the other side low temperatures may impair the decomposer communities [27]. Furthermore, there was statistically correlation between temperature of water and rate of decomposition ($P<0.05$); a higher the temperature the less remaining litter [5]. The breakdown of the leave is defined as weight loss due to some physical-chemical properties caused by environmental conditions such as high temperature and decomposition process increases in accordance with temperature [28, 29]. Other studies have shown that temperature to be a necessary element for mangrove decomposition, increasing a few degrees of temperature could be adequate to increase the decomposer activity whereas consequently accelerating the process [30]. The tree abiotic factors which are significantly liable for the leaf breakdown rate involved dissolved oxygen, electrical conductivity in the winter and temperature in the summer [31].

3.3. Decomposition/Decay coefficient (K) and half-time (t50)

K (d^{-1}) is decay coefficient that can determine the rate of decomposition. Average of K in landward was 0.010 d^{-1} and in seaward was 0.012 d^{-1} (Table 2). The exponential value of K can be classified into three classes i.e. very fast ($K> 0.01$), medium ($K = 0.005$-0.01) and slow ($K <0.005$) [10]. The decay coefficient in study area was categorized fast and medium. The decay coefficient in seaward was faster than landward, it assumed due to the influence of the tides. According to the result found by Dewiyanti [5] the average value of K at the substation next to the land was 0.010 d^{-1} and next to the sea was 0.011 d^{-1} but the K value wasn't significantly different between both substations. Seaward substation received greater influence from the physical tide directly impacting the fragmentation of leaves. Increasing the level of decomposition due to the water immersion leads to washing of unstable materials and high inundation causes waste to decompose, the decomposition rate is influenced by environmental factors such as temperature, nutrient content, leaf density and soil moisture. [11, 32] Litter mass loss was greater on the tidal side of the floodgates than landward and lower decomposition rates at the landward side are due to tides not reaching the landward side [33]. Moreover, decay coefficient found in study area was similar to those reported for other species of *Rhizophora*, i.e. in Malaysia (K = 0.01 day^{-1} [11]; Trinidad and Tobago (K = 0.01 day^{-1} [33].

Table 2. Value of decay coefficient (K) and t_{50} of *Rhizopora mucronata*

Species	Sub Station	Decay Coefficient/ K (d^{-1})	t_{50} (days)
Rhizophora mucronata	Landward	0.010	69
	Seaward	0.012	57

Time required for the decomposition of half the initial material (t_{50}) was 69 days in landward substation and 57 days in Seaward substation. The t_{50} is the time needed for the litter to break down 50% (half) of the initial amount before the litter is perfectly decomposed. The t_{50} of *R. mucronata* can be seen in Table 2. The time required for the decomposition of half the leaves litter (t_{50}) was 67 days and 63 days for next to the land and site next to the sea [5]. The t_{50} indicated that the time to decompose litter from half of the starting material in seaward decomposed faster than landward. Ashton et al. [11] found that the required t_{50} of *Rhizophora apiculata* litter was 76 days and 43 days in the Malaysian Peninsula mangrove forest. Several factors are suspected to influence the high leaves decomposed in the field study such as factor of inundation, environment and decomposer. Overall leaf mass may be lost by leaching up to 33% during the initial phase from several days to several weeks [34]. Furthermore, in previous study showed that week showed a faster initial mass loss indicated by residual leaves that had been lost 38.3% due to immersion in the Southwest Indian mangrove forests after one week [10].

4. Conclusion

The highest mangrove litter production was *Rhizophora mucronata* with the average 79.21 $g/m^2/day$ compared *Avicennia marina* and *Sonneratia alba*. The productivity of mangrove litter found in seaward higher (52.05 $g/m^2/day$) than landward (45.32 $g/m^2/day$) due to the mangrove density and environment condition. Based on the mangrove litter components, leaves was the largest contributor of litter compared to other plant components within leaves litter was 67.07 g/ m^2/day. Some factors was influencing mangrove litter productivity i.e. temperature, salinity and mangrove density. The decomposition percentage of *R. mucronata* leaves increased with increasing period observation time, and the percentage of litter decomposed in seaward was higher than landward were 48.58% and 41.53% at 49th day, respectively. The coefficient of decomposition showed very fast and medium category. Time required for the decomposition of half the initial material (t_{50}) was 69 days in landward and 57 days in seaward, the values showed that the leaves litter incubated in seaward decomposed faster than landward substation.

Acknowledgment

We would like to express our gratitude to the Directorate General of Higher Education through Syiah Kuala University for providing research grant.We are grateful to all who helped out in any way during Fieldwork. Our deepest thanks to the local people in study area for their supporting and very good cooperation. Thanks to Dila, Afdal, Titin and Yanti for their field work during the research study.

References

[1] Sjöling S, Salim M, Thomas J, Jasson J K 2005 *Journal Estuarine Coastal and Shelf Science* **63**:397-406.
[2] Castillo J A, Gabriela V, Jorge L P 2006 *Hydrobiologia* **559**:101–111.
[3] Margarita L Naranjo, Jeffrey A, Jorge C 2018 *Journal of Coastal Research* 00 (0): 0000.
[4] Polidoro B A, Carpenter K E, Collins L, Duke N C, Ellison A M, Ellison J C, Farnsworth E J, Fernando E S, Kathiresan K, Koedam N E, Livingstone S R, Miyagi T, Moore GE, Nam V N, Ong J E, Primavera J H, Salmo SG, Sanciangco J C, Sukardjo S, Wang Y, Yong J W H 2010 *PLoS ONE* **5**(4): e10095.
[5] Dewiyanti I 2010 *Jurnal Biodiversitas* **11** (3): 139-144.
[6] Kathiresan K, Bingham B L 2001 *Advances in Marine Biology* **40**: 81–251.
[7] Morrisey D, Beard C, Morrison M, Craggs R, Lowe M 2007 *Auckland Regional Council Technical Publication Number* **325**.
[8] Osono T, Takeda H 2006 *Mycologia* **98**: 172- 179.
[9] Dinas Kelautan dan Perikanan Provinsi Aceh 2012 *Laporan Kegiatan Assessment lapangan ecosystem perairan Aceh Besar.*
[10] Ananda K, Sridhar K R, Raviraja N S, Baerlocher F 2007 *Wetlands Ecol Manage* **112**: 73-81
[11] Ashton E C, Hogarth PJ, Ormond R 1999 *Hydrobiologia* **413**: 77-88
[12] Bosire J O, Dahdouh-Guebas F, Kairo J G, Kazungu J, Dehairs F, Koedam N 2005 *Biological Conservation* **126**: 287–295.
[13] Dewiyanti I 2005 *Jurnal Natural* **5**(2):12-16.
[14] Ribeiro C, Madeira M, Araujo M C 2002 *Forest Ecology and Management* **171**: 31-41.
[15] Mackay A P, Smail G 1995 *Hydrobiologia* **332**: 93-98.
[17] Haraguchi A K, Hisaya C, Hesegawa, Takahashi T I 2002 *Europ Jurnal Soil Biology* **38**: 89-95.
[18] Hamidy R, Soelaksono S, Adianto, Taufikurrahman 2002 Biologi **2** (13):755-768.
[19] Andrianto F, Bintoro A,Yuwono S B 2015 *Jurnal Kehutanan dan Pertanian* **1** (3): 13-15.
[20] Lestarina M P 2011 *Produksi dan laju dekomposisi serasah mangrove dan potensi kontribusi unsur hara di perairan mangrove Pulau Panjang Banten* Tesis Institut Pertanian Bogor.
[21] Juman R 2005 Revista de Biolog´ıa Tropical 53(1); 207–217.
[22] Mchega I S S, Ali A I 2017 *Journal of East African Natural History* **106**(1): 5–18.
[23] Aida G R, Yusli W, Fahrudin A, Mohammad M K 2014 *Jurnal Ilmu Pertanian Indonesia* **19** (2) : 91-97.
[24] Munir A M 2004 *Pendugaan Produktivitas Serasah Hutan Mangrove di Pulau Gili Sulat, Nusa Tenggara Barat* Tesis Institut Pertanian Bogor.
[25] Bunyavejchewin S, Nuyim T 2001 *Silvicultural Research Report* 28-38.
[26] Sa'ban M, Ramli W N 2013 *Jurnal Mina Laut Indonesia* **3** (12): 132-146.
[27] Mfilinge P L, Tsuchiya M 2008 *Journal of Sea Research* **59** (1-2): 94-102.
[28] Graça M A S, Baerlocher F, Gessner M O 2005 *A practical guide; methods to study litter decomposition* Springer Netherlands.
[29] Kuers K, Simmons J 2006 *Leaf litter decomposition* CAWS Litter decomposition study.
[30] Kristensen E, Bouillon S, Dittmar T, Marchand C 2008 *A review. Aquatic Botany* **89** (2): 201-219.
[31] Rezende R S, Marcelo de O P, José F G, Mauricio M P 2013 *Acta Limnol Bras* **25** (2): 158-168.
[32] Robertson A I 1988 *J Exp Mar Biol Eco*l **116**: 235-247.
[33] Todd M D, Olusegun O 2000 *Aquatic Botany* **68**: 273–280.
[34] Davis SE, Corronado M C, Childers D L, Day J W Jr 2003 *Aquatic Bot* (75): 199-215.

IOP Conf. Series: Materials Science and Engineering **567** (2019) 012026 doi:10.1088/1757-899X/567/1/012026

Study on existence of the fisheries resources abundance by using environmental deoxyribonucleic acid (e-DNA) approach at fishing grounds in the Sulawesi Sea, Indonesia

Kawilarang W A Masengi[1, 2], I F Mandagi[1,2], L Manu[1], F Silooy [1], I L Labaro, A W R Masengi[1], N Sebua[1], E I K G Masengi[3], Benny Pinontoan[3], Y Hutabarat[4], F Hukom[5], M Iwata[6], Y Abe[6], Y Sato[7], R Kimura[8] and K Yamahira[2]

[1]Faculty of Fisheries and Marine Sciences, Sam Ratulangi University, Indonesia
[2]Tropical Biosphere Research Center, University of The Ryukyus, Japan
[3]Faculty of Mathematics and Natural Sciences, Sam Ratulangi University, Indonesia
[4]Faculty of Fisheries and Marine Sciences, Diponegoro University, Indonesia
[5]Center of Oceanography, Indonesian Institute of Sciences, Indonesia
[6]Aquamarine Fukushima and Marine Sciences Museum, Iwaki, Japan
[7]Center for Strategic Research Project, University of The Ryukyus, Japan
[8]Faculty of Medicine, University of The Ryukyus, Japan

Corresponding author: alex_masengi@unsrat.ac.id

Abstract. Here, we report the results of our preliminary study on deep sea eDNA at fishing ground to approach the fisheries resources abundance at Sulawesi sea by using deep-sea water sampling collected from 10 sites ranging from 110m-200m in depth at front side of the International Coelacanth Research Center and Museum Base at Lolak Waters and Manado Bay North of Sulawesi using Nansen Bottle Sampler (1500 cc). The collected waters were filtered using Power Water Sterivex DNA Isolation Kits and preserved with the DNAiso Reagent then transported to Center for Strategy Research Project, University of the Ryukyus, Okinawa Japan where eDNA analyses were conducted. Our results revealed that the concentrations of eDNA has a good quality were measured with a NanoDrop Lite spectrophotometer, indicating eDNA was successfully extracted. Therefore, by using universal primers for eDNA, MiFish-U-F/R for the 1st-PCR (mt-12S amplification) and 2nd-PCR (tag-indexing) for library preparation to accommodate sequence variations and show that intense signal of MiFish eDNA amplification. Using a high-throughput Illumina MiSeq platform for sequencing analyses, we detected eDNA from 40 fish's species with dominantly by *Caranx sexfasciatus, Encrasicholina punctifer*.

1. Introduction

Indonesia is an Archipelago country are consisting of the 18.110 islands with a coastline length about 99,000 Km and consisting of about 9,900,000 tons of marine resources potency per year. This potency is distributing around the surface, middle and bottom waters layers. To monitor these marine resources existence abundance, the lots of the budget and times are needed.

IOP Conf. Series: Materials Science and Engineering **567** (2019) 012026 doi:10.1088/1757-899X/567/1/012026

A new technology approach can be overcoming of this problem were available can be implemented at around Indonesian marine waters by using the environmental DNA approach. Environmental DNA (eDNA) in aquatic environments refers to genetic material found in the water column. In the case of multicellular organisms, eDNA originates from various sources, such as metabolic waste, damaged tissue or sloughed skin cells [1]. Ficetola et al., was the first study demonstrating the use of eDNA for detecting an aquatic vertebrate species (invasive American bullfrog) from controlled environments and natural wetland, published in 2008 [2]. Here, we report the results of our preliminary study on deep sea eDNA at fishing ground to approach the fisheries resources abundance at Sulawesi sea.

2. Materials and Methods

Deep-sea water sampling was collected from 10 sites ranging from 110m-200m in depth at front side of the International Coelacanth Research Center and Museum Base at Lolak Waters and Manado Bay North of Sulawesi using Nansen Bottle Sampler (1500 cc) as shown in Figure 1. The positions were follows the discovered of coelacanth by Green Eye Project on 2007-2015 [3].

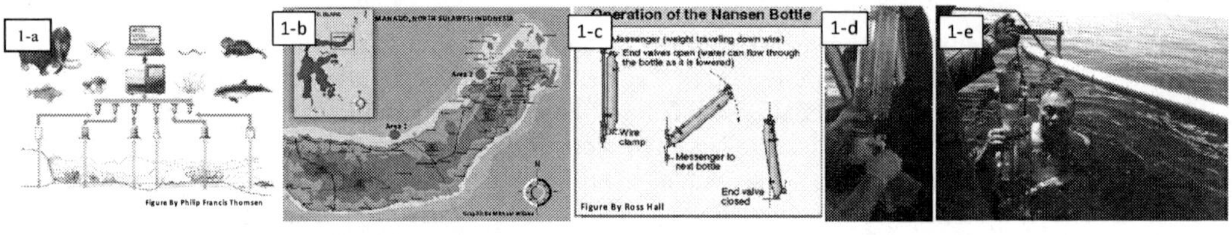

Figure 1. The concept of preliminary study of eDNA and the collections sites of deep seawaters Sampling.

The collected waters were filtered using Power Water Sterivex DNA Isolation Kits and preserved with the DNAiso Reagent and kept in a deep freezer -25^0 C at Faculty of Fisheries and Marine Science, Sam Ratulangi University, until they were transported to Center for Strategy Research Project, University of the Ryukyus, Okinawa, Japan (Figure 2.) where eDNA analyses were conducted following MiFish protocol [4]. Therefore, e-DNA extraction was conducted at Center for Strategy Research Project at University of the Ryukyus Okinawa Japan by using PowerWater Sterivex DNA Isolation Kit Samples by followed its protocol as follow [5].

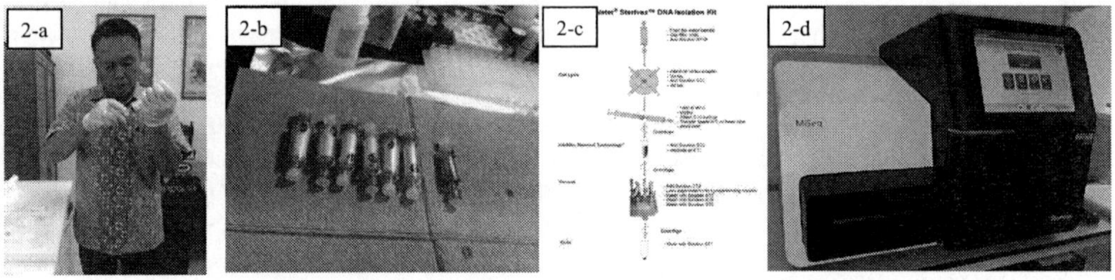

Figure 2. Laboratory Works on Deep-sea Environmental DNA Research at Faculty of Fisheries and Marine Science Sam Ratulangi University and University of the Ryukyus.

3. Results and Discussion

Our Results of eDNA Water sampler quality After Extraction and Amplification of Ten Collection sites at Manado Bay and Lolak Waters, Sulawesi Sea, Indonesia was used to analysis the fisheries resources existence abundance after check the quality of DNA as shown in Table 1. Some number of the species have been analyses at The Center the Strategic research Projects of University of the Ryukyus. The eDNA water sampler was collected on 2018.06.17 and kept under -25^0 C, then was extracted on 2018.07.18. According to this table, the quality of DNA from 10 locations at the sea and one water sample site from the freshwater river as positive control shows that the DNA quality ranging from 1.8-12.6 ng/mL.

Table 1. Results of eDNA Water sampler quality After Extraction and Amplification of Ten Collection sites at Manado Bay and Lolak Waters, Sulawesi Sea, Indonesia

Sample ID	Date of Collection	Date of Extraction	Collection Sites	Quality of DNA (ng/mL)	Nanodrop Consentration (ng/mL)
478	2018.06.17	2018.07.18	Tanjung Ompu 1	1.8	2.4
479	2018.06.17	2018.07.18	Tanjung Ompu 2	2.3	2.22
480	2018.06.17	2018.07.18	Lolak River	3.5	2.7
481	2018.06.17	2018.07.18	Molosing Island 1	3.1	2.28
482	2018.06.17	2018.07.19	Molosing Island 2	1.9	1.63
483	2018.07.03	2018.07.19	Big Fish Manado 1	2.9	2.12
484	2018.07.03	2018.07.19	Big Fish Manado 2	12.6	2
485	2018.07.03	2018.07.19	Minanga Manado 3	4.8	2.2
486	2018.07.03	2018.07.19	Recla. Manado 4	3.3	3.39
487	2018.07.03	2018.07.19	Trad. F.G. Manado 5	5	2.13
488	2018.04.20	2018.07.19	Sariouw River (*)	10.7	1.79

(*) Water sampler from Sariouw River as a positive control

Suggesting that the quality of the DNA is good enough to conduct for sequencing analysis of the next process. That argument are strong support by the results are showing at Figure 3a. The blue line are the quality DNA (ng/ml) of the seawater samples and the orange color indicated the standard of the the quality DNA as a control. From this figure shows, all of the seawater samples are laid at the upper sites of the orange colors. We conducted electrophoresis for each part 1st-PCR and 2nd-PCR

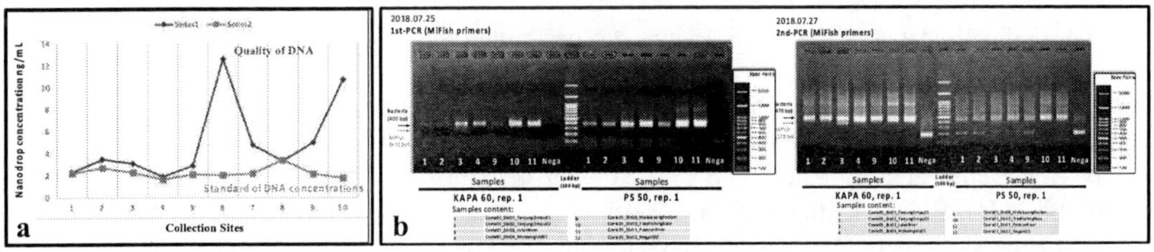

Figure 3. a. The comparison between Nanodrop concentration and Standard Concentration of water samples (ng/mL) of each collection site using in this study, **b**. The 1st-PCR and 2nd-PCR Shows Intense

Signal of MiFish eDNA by Using Universal Primers MiFish-U-F/R

amplification products and the results show as River samples as positive control exhibited intense signal of MiFish eDNA amplification and negative controls (DW) showed no clear bands. It's clearly showed (Figure 3b) that the water samples was collected from those two areas of Sulawesi waters are so cleared and feasible to be continued to next of the genome work. Then it can be explained (table 2) from all 10 collection sites we found that the number of fish species have been obtained based on fish base database. By using a high-throughput Illumina MiSeq platform for sequencing analyses, we detected eDNA from 40 fish's species and the dominantly are *Caranx sexfasciatus* and *Encrasicholina punctifer* fishes.

Table 2. Results of collections site, satellite positions, sea`s depth and the number of fish's species was detected at Lolak Waters and Manado Bay North of Sulawesi.

Stations	Positions (Degrees)		Sea's Depth (Meter)	Species (Number)
	Latitude (N)	Longitude (E)		
1	0.9298	124.0351	115	8
2	0.9338	124.0066	150	6
3	0.9339	123.9961	196	14
4	0.9217	124.9889	170	2
5	0.9128	123.9869	110	11
6	1.4699	124.8199	95	3
7	1.4707	124.8179	120	2
8	1.4686	124.8153	135	8
9	1.4659	124.8112	120	6
10	1.4661	124.8156	90	4

According to the fisherman around the surveyed areas that about 2 weeks before those locations were plenty of the *Encrasicholina spp.* were caught there. In addition, to eDNA, this metabarcoding approach is applicable for species identification of the marine fisheries resources' existence abundance in Sulawesi sea. Then this bioinformatics data was compared to the fish's distributions and ecological as reported from The Green eye project 2007-2015 where the site also known as fishing ground of those target fishes. Therefore, the results of collections site, satellite positions, sea`s depth and the number of fish's species was detected at Lolak Waters and Manado Bay North of Sulawesi, explained that the collection site at Lolak Waters relatively higher fisheries resources existence abundance compared to other collection site in Manado Bay (table 2).

The relationships between the collection sites number and sea`s depth as shown at Figure 4-a by radar chart approach and the number of the fishes' species detected (Figure 4-b) was analyzed by using the balloons pattern. Therefore, Figure 4-c performed the relationships between two variables, seas` depth and the number of the fish species have been detected by environmental DNA tools approach and examined by simple regression analysis applied. Based of this results we found that the relationships between variable sea`s depth and the number of the fish species are had any relation even relatively low indicated by value of $r = 0.257$ with the equation is Number Fish Spec. = 1.309 + 0.0401 Sea`s Depth (M). Comparing to these three graph, there are strong supported each other. On the other hand, in these figures explained that the surveyed locations have been taken the water sample at Manado bay are relatively shallower the sea depth than them at Lolak water of Bolaang Mongondow regency.

IOP Conf. Series: Materials Science and Engineering **567** (2019) 012026 doi:10.1088/1757-899X/567/1/012026

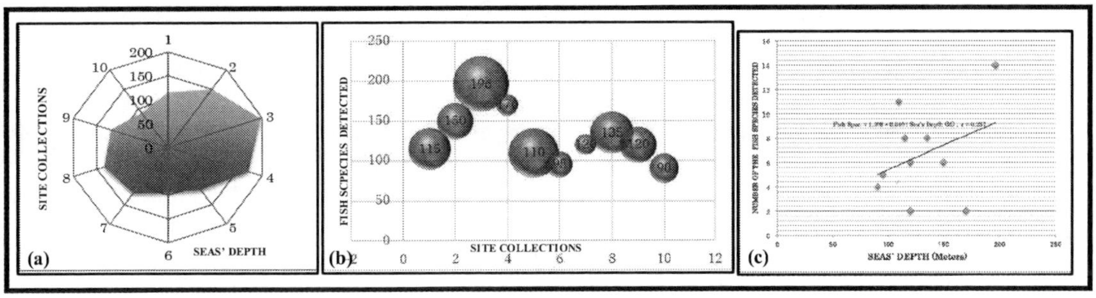

Figure 4. The Relationships between Collection site, Sea`s Depth and the Number of Fish Species detected.

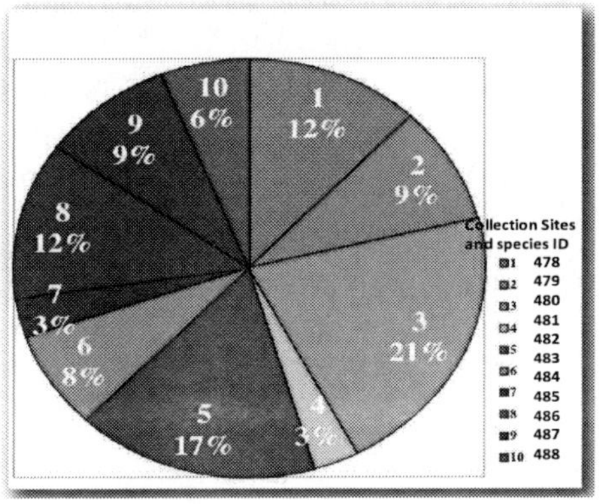

Figure 5. Comparison of the number of fish's species detected in percentage at each surveyed sample locations (from 1-10 sites), where each color represented each sample ID.

In Figure 5, explained that the number of the fish's species detected in percentage at each surveyed sample locations. From this figure shows that the location number 3 were the biggest fish's species in the number are located at Lolak waters as indicated at number 1-5 than them at Manado Bay of locations number of 6-10 in accumulations of 64 fish's species. From this information could be said that the fisheries abundance of Lolak waters were detected 41 species (64.06%) and are bigger than that at Manado bay of 23 fish's species (35.94%). This indicated the deep-sea marine protected area of The International Coelacanth Research Center and Marine Museum base site of Lolak waters are relatively higher of the fisheries resources existence abundance than at the Manado bay.

4. Concluding Remarks

Based on our results analysis to developed universal primer MiFish in a metabarcoding approach to fish eDNA we confirmed that the Lolak Waters are having relatively higher the fisheries resources existence abundance comparing of that on Manado bay. In implementation of the deep-sea environmental DNA research, the un-contamination aspect during fields work is absolutely necessary, therefore for effectiveness and efficiencies research of the marine fisheries resource's existence abundance point of view, the environmental DNA technology approach is suitable to be applied.

Finally, this marine environmental DNA technology could be got fruitful if can be implemented to Indonesian marine and freshwater due to Indonesia country have very wider territorial.

Acknowledgments
We would like to thank first to Japanese Government through the Ministry of Education, Culture, Sports, Science and Technology (MEXT) which gave opportunity to us for doing this project due to the very rare moment, to the Dean of Faculty of Fisheries and Marine Science, Sam Ratulangi University, who giving support to the team during surveyed and gave us the letter of the material transferred agreement (MTA) of deep-sea water samples and brought to Japan. Then to the Tropical Biosphere Research Center and The Center for Strategic Research Project University of the Ryukyus which have been funding and very eager to support this project and allowed us to conducted the eDNA laboratory works.

References
[1] Kelly RP, Port JA, Yamahara KM, Martone RG, Lowell N, Thomsen PF, Mach ME, Bennett M, Prahler E, Caldwell MR. 2014 Harnessing DNA to improve environmental management. *Science* **344**, 1455–1456. (doi:10.1126/science. 1251156)

[2] Ficetola GF, Miaud C, Pompanon F, Taberlet P. 2008 Species detection using environmental DNA from water samples. *Biol. Lett.* **4**, 423–425. (doi:10.1098/ rsbl.2008.0118)

[3] Masamitsu Iwata., K.W. A Masengi., I.F.Mandagi., J.Seba., Biological Survey on The Indonesian coelacanth. Research Report 2007-2015

[4] Miya M *et al.* 2015 MiFish, a set of universal PCR primers for metabarcoding environmental DNA from shes: detection of more than 230 subtropical marine species. *R. Soc. open sci.* **2**: 150088. http://dx.doi.org/10.1098/rsos.150088

[5] Anonymous (1993) DNA Isolation Kit Sample (For isolation of genomic DNA from Sterivex TM filter units, Millipore catalog# SVGPL10RC), (Catalog No. 14600-S) USA.

IOP Conf. Series: Materials Science and Engineering **567** (2019) 012027 doi:10.1088/1757-899X/567/1/012027

Lenght-weight relationship and condition factor of huluu fish (*Giuris margaritacea*) in Limboto Lake

N Auliyah[1]

[1]Departement of Fisheries, Gorontalo University, Gorontalo, Indonesia

Corresponding author : nurulauliyah05@gmail.com

Abstract. *Giuris margaritacea* (Valenciennes), the snakehead gudgeon is a species of fish in the family Eleotridae, live in Limboto Lake. Decline in the quality of the lake water, sedimentation lakes and eutrophication give bad conditions resulting to Limboto Lake. Diversity resources fishing lake has degradation. Population the fish original and important has decline, one of that is huluu fish, increasingly difficult to found. The purpose of this study is to find out lenght- weight relationship and condition factor Huluu fish (*Giuris margaritacea*) in Limboto Lake. The research was done in april until june in Limboto Lake with 7 points of sampling in Limboto Lake. Sampling was carried out by purposive sampling, Data analysis used to this research is *Model Allometric Linear* (LAM) used to calculate parameter a and b through the measurement of changes in lenght-weightt and condition factors. The result of this research known that fish huluu derived from Limboto Lake generally have model growth allometric positive (b>3) with the condition factors range between 0.12 - 1,91 whereby the meaning of the amount is fish categorized as flat fish.

1. Introduction

Giuris margaritacea (Valenciennes), the snakehead gudgeon is a species of fish in the family Eleotridae. In the world, distribution area spreading of the Giuris across Asia and Africa[1]. In Gorontalo, this fish are called Huluu and can be found in Limboto Lake. Limboto Lake is located in Limboto District, Gorontalo Regency, where has a strategic roles include the resources for fisheries (aquaculture and catch). Now the condition of Limboto Lake increasingly alarming and indicated sedimentation occur. In 1930, Limboto Lake depth was 30 meters with an area of 8000 Ha, then in 2007 the depth and area were decrease to 2.5 meters with an area of 3000 Ha[2].

Community activities that produce domestic waste and aquaculture activities carried out in the lake cause a decrease to water quality of lake, lake sedimentation and eutrophication. In addition, the native people also catches fish using poison (potas), fish bombs and large-scale fishing equipment causes a decrease in the genetic diversity of fish and water quality. Impact of the conditions that occur in Limboto Lake, the diversity of fisheries resources in the Lake has been degraded. The population of important native and economical fishes has decreased including Huluu (*Giuris margaritacea)* [3].

In 2016 the Fish Quarantine Center of Gorontalo Province issued a list of types of fish that were included in the protected species, barred and invasive (JADDI), and status of Huluu Fish was included in the JADDI fish list so it needed to be protected [4]. This statement is supported by The IUNC red List of Threatened Species in 2017 that the status of *Giuris margaritacea* is last concern [5]. Based on that knowledge was we need to do this research. Research objectives that is to know relations lenght-weight and condition factors of Huluu fish (*Giuris margaritacea*) in Limboto Lake. Lenght-weight relationship give information about growth pattern and condition factor give information about how the fish as related to its environment [6].

2. Research Method

Research was done in april until june in Limboto Lake with 7 points of sampling, that is Village Tabumela, Hutuo, Dembe, Lupoyo, Lekobalo, Kayu bulan and Hunggulawa. The points of sampling on map in Figure 1. Sampling was carried out by purposive sampling, which was based on fishing activities in Limboto Lake especially related to Hulu'u (*Giuris margaritacea*). In this study the process of fishing was carried out by using the type of trap fishing called "Bunggo". The time of fishing was carried out at 09.00-16.00 WITA. Furthermore, the fish caught was measured in length and weight.

Data analysis used to research is *Model Allometric Linear* (LAM) used to calculate parameter a and b through the measurement of changes in the weight and length

$$W = aL^b \ldots\ldots (1)$$

Where; W is the weight of the fish (gram), L is the length of the fish (cm), a is the linear regression intercept, b is the regression coefficient. The value of b from this calculation can illustrate the pattern of fish growth. If the value of b = 3, then the growth pattern is isometric or weight gain equals to the length of the fish growth and if the value of b ≠ 3, then the growth pattern is allometric. Allometric growth patterns are divided into two, namely allometric positive and allometric negative. If the b value below 3 is called allometric negative (the length increase is faster than the weight gain), and if the b value above 3 is called allometric positive (weight gain is faster than the length increase).

The condition factor is obtained through the formula:

$$Kn = \frac{W}{aL^b} \ldots\ldots\ldots (2)$$

Where; Kn is the condition factor, W is the weight of the fish (gram), L is the length of the fish (cm), a is the linear regression intercept, b is the regression coefficient.

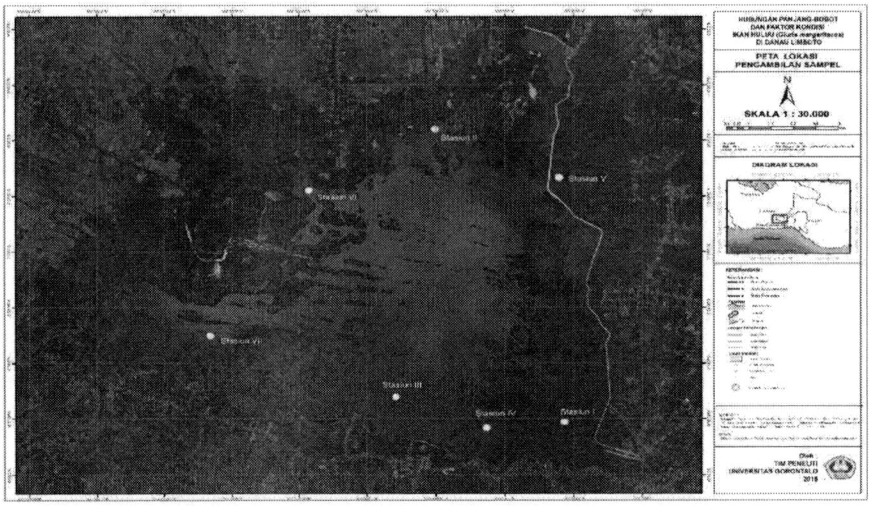

Figure 1. Map station for collecting sample of Hulu fish in Limboto Lake

3. Result and Discussions

Based on the results of research conducted in lake limboto on 7 point of sampling there are Tabumela village , Hutuo , Dembe , Lupoyo , Lekobalo , Kayu bulan and Hunggulawa .As for point the sample collection can be seen in Figure 1. Fish huluu who caught at the station I is 66 of fish, station II is 55 of fish , station III is 55 of fish 1 , station IV is 35 of fish , station V is 46 of fish , station VI is 52 of fish and station VII is 54 of fish. The quantity of fish caught every month fluctuating.In april is the highes catches about 141 of huluu fish t, in May that is 110 of fish and June that is 112 of fish.

Tabel 1. Relationship between the lenght - heavy and factors condition of Huluu (*Giuris margaritacea*) in Limboto Lake

Station	a	B	R^2	L	W	Growth of pattern	Condition Factor
I	-5,70	3,35	0,93	14,58	37,7	Alometric positive	0,14
II	-5,48	3,23	0,82	15,2	39,7	Alometric positive	0,15
III	-5,27	3,13	0,96	15,3	41,5	Alometric positive	0,16
IV	-5,51	3,25	0,76	15,6	42,2	Alometric positive	0,15
V	-5,79	3,36	0,96	15,2	36,6	Alometric positive	0,12
VI	-5,58	3,25	0,79	15,6	38,4	Alometric positive	0,14
VII	1,1	1,21	0,74	14,6	37,3	Alometric negative	1,91

Based on analysis of the was done with lenght and weight that show fish huluu who caught at the station I until VI having a pattern growth alometric positive which means growth the weight of the faster than the length of the body. Fish huluu who caught at the station VII having a pattern growth different namely alometric negative means the development the length of the body faster than the growth weight. Fish are not always have a similar growth, relations lenght and weigh is differently affected environmental water conditions s [7].

The results of the analysis condition factors were made Huluu fish discovered that the the range of condition factor between 0,12 - 1,91 whereby the meaning of the value a fish is categorized as fish flattened. Flattened of fish is result from the environment thick in the waters that is no longer support for fish to growth [8]. Factors that have brought about flattened of fish shaped in environmental factor thick in the waters, the availability of food and diseases [9].

4. Conclusion

The results of the study of the length and weight of Hulu'u fish (*Giuris margaritacea*) in Limboto Lake, showed that the Hulu'u fish growth pattern at all I-VI stations was positive allometric ($b > 3$) and at station VII Hulu'u fish growth pattern is a negative allometric ($b < 3$) meaning that the increase in body length is faster than the growth of body weight. The condition factor analyzed for Huluu fish, it is known that the range of condition factors obtained is between 0.12 - 1.91 where the meaning of the value is the fish included in the category of flat fish.

Acknowledgments

Thanks to Rector of Gorontalo University who always support in full activities in the context of the development of the quality of Lecturer .Thanks to KEMENRISTEK DIKTI which gave a chance to an author to receive research grants

References

[1] *https://www.fishbase.in/summary/Giuris-margaritacea.html (di akses 12/6/2017,pukul 15.18)*

[2] Akuba R dan Rugaya Biki 2007. *Master Plan Pengelolaan Danau Limboto.* Kerjasama PSL Universitas Negeri Gorontalo, Balitbagpedalda ProvinsiGorontalo, dengan Kementerian Negara Lingkungan Hidup Republik Indonesia.

[3] Krismono A, A Warsa, A Suryandari, A Nurfiarini, A S Nastiti 2007. *Rehabilitasi Sumber Daya Perikanan di danau Limboto*. Balai Riset Kelautan dan Perikanan. Departemen Perikanan dan Kelautan.
[4] Balai Karantina Ikan dan Pengendalian Mutu (BKIPM) 2016. *Pemetaan Sebaran Jenis Ikan /Agen Hayati Yang Dilindungi dan Bersifat Invasif (JADDI)*. Laporan Akhir. Provinsi Gorontalo.
[5] Larson H & Sparks J S 2017 *Giuris margaritacea. The IUCN Red List of Threatened Species* 2017: e.T196317A58335977. http://dx.doi.org/10.2305/IUCN.UK.20173.RLTS.T196317A58335977.en.
[6] Olurin K B And O A Aderibigde 2006 *Lenght-Weight Relationship and Condition Factor of Pond Reared Juvenille Oreochromis niloticus.* World Journal of Zoology 1 (2) : 82-85.
[7] A DeeRobert and K Wiliam 2008 *Weight-Legth Relationship In Fisheries Studies: The Standard Allometric Model Should Be Applied With Caution.* (American Fisheries Society **137**,2008): 707-719.
[8] R. H. Moore and Q . Bone 2008 *Biology of Fishes. Third Edition. Taylor and Francis Group*, NewYork. 478 p.
[9] Y P Pamungkas dan Burnawi 2015 *A long weights tilapia (Oreochromis niloticus) in lake kerinci* (BTL Vol. 13 No. 2. Jambi. Desember 2015): 67 – 70.

IOP Conf. Series: Materials Science and Engineering **567** (2019) 012028 doi:10.1088/1757-899X/567/1/012028

Analysis of chemical elements of the rocks on the coastal areas in the Eastern Bolaang-Mongondow District, North Sulawesi Indonesia

P A Angmalisang[1], L Manu[2], H F Sangian[3]

[1]Department of Aquatic Resource Management, Sam Ratulangi University, INDONESIA.
[2]Department of Fishery Product Processing, Sam Ratulangi University, INDONESIA
[3]Department of Physics, Sam Ratulangi University, INDONESIA.

Corresponding author: hannysangian@yahoo.co.id

Abstract. The chemical elements making up the rocks located on the coastal areas in the Eastern Bolaang Mongondow were successfully mapped and investigated by employing the XRF method. The rocks collected from the Central Buyat, South Buyat, Motongkat and Jiko coastal areas of the sub-districts were milled until the size was close to 120 mesh. The samples were characterized and measured to obtain the chemical elements deposited inside the rocks. It was found that the elements which make up the rocks are relatively different depending on the coastal locations. The silicon (Si) was the highest percentage of all elements found and followed by iron (Fe), calcium (Ca) and aluminum (Al). The composition of Si deposited in the rocks from Buyat coastal area was of 42.4% which the biggest percentage compared the rest of the samples. The Iron (Fe) element had a chemical composition recorded at 28.9% that was the highest figure in which rocks were obtained from Jiko coast. The aluminum (Al) composition was relatively similar for all samples collected from that location and observed at 10%.

1. Introduction

The chemical elements, such as *the rare earth elements* (REE) which have an atomic number from 57 until 71, are very essential for technological applications used by people. Those materials have a magnetic property and can be used as alloy substances as energy storage. The REE are also useful for renewable energy technology, an electrical car, LED lamps, *wind power* as well as are important of weapon industry. They are deposited inside stones, sands, and crust expanding from zero to ten kilometers below the ground. The spread of those elements is not uniform depending on the geological age and volcanic activity [1].

The Indonesia country has the longest beach line in the world after Canada. It is giving the unique landscape in which their appearances, such as colors and environments are significantly different each other. It indicates, that chemical elements, compositions, and material structure are unique each coastal line. The coastal line latitude measured from the surface of seawater and beach profile influence the chemical elements compositions [2][3][4].

The Eastern Bolaang Mongondow District (BOLTIM) is consists of sub-districts which have coastal lines, they are *Kotabunan, Tutuyan, Motongkat, and Nuangan*. If they are watched physically materials, or minerals deposited in coastal lines not only uniqueness and interesting to carry out a study but also explored for supporting a regional development.

The study of the contents of the precious elements and minerals inside stones has been conducted for hundreds of years. The gold and cuprum materials were metal substances that the most explored and treaded for years. The materials, however, were mixed with other elements, such as *iron oxide copper-gold (IOCG)* as published previously [5]. The iron ores were found inside stones and to analyze them using *La*ser-Induced Breakdown Spectroscopy method [6]. The tungsten deposited inside rocks was characterized by utilizing radioactivity of an isotope [7]. Meanwhile, silicon material widely used for semiconductor industry was discovered on crusts mainly in sands and soils called a *silicate*. The method to measure the Silicon content also employs Si isotope between silicate and iron [8] while, cuprum element was analyzed using methods INAA, ICP-MS, and LA-MC-ICP-MS [9].

The study and exploration of the minerals content on the coastal area are still limited in Indonesia because of environmental considerations. However, some coastal areas have been investigated of the minerals deposited inside rocks and sands. Authors [10] were successful to characterize of the chemical compositions on the coastal zones employing gamma-ray methods. It was found that the calcium element deposited inside calcium carbonate, or calcite dominated of the materials on the beach rocks. The method of *computer Controlled Scanning Electron Microscopy (CCSEM)* was applied to measure the chemical data and mineral ores of the coastal sendiments [11]. While the *rare earth elements* have been discovered on the sands of a desert and beach using a spatial distribution whereby the results found that there was a correlation with inland and beach materials [12]. The radiometric technique was used in analyzing a sediment motion on the beach by detecting a gamma rays emmite [13].

The minerals deposited inside sands have been well mapped and classified [14]. Authors classified simplicity of the minerals in sands with ratio SiO_2/Al_2O_3, Fe_2O_3/K_2O and Ca content. The mineral investigation inside sands and sediment using a portable XRF (*X-ray fluorescence*) has been reported by authors previously [15]. The XRF was powerful to detect the chemical elements, K, Ca, Ti, Cr, Mn, Fe, Co, Cu, Zn, Rb, Sr, Zr, Pb, U, S, As, Br, Mo and Hg, but it was not accurate for P, V, Cr, Ni, Se, Ag, Cd, Sn, Sb, I, Ba and Bi. Meanwhile, the combination methods, XRD, XRF, TG, and SEM were successful to measure minerals of titanium, quartz, and iron-rich structures and alloy iron–titanium [16].

As previously mentioned, it was found that the methods can be employed in analyzing the chemical elements deposited inside rocks, sands, and soils. This study was to map and to investigate the chemical elements deposited in rocks in some regions located on the coastal areas in the Eastern Bolaang Mongondow District, North Sulawesi Province, Indonesia by using XRF. The procedures conducted as follows: Collecting the rocks taken from the coastal area; milling the rocks until their size close to 120 mesh; and characterizing the samples.

2. Methods

2.1. Material Preparations

The rocks were obtained from four the coastal areas of the Motongkat, Central Buyat, South Buyat and Jiko sub-districts in Eastern Bolaang Mongondow District, North Sulawesi Province. Then the rocks were crushed using a heavy tool and they were milled and screened passing the particles with size 120 mesh.

2.2. Measurement

The particles obtained were shipped to the Central Lab belonged by The State Malang University for a measurement. The characterization of samples using XRF (Merk : PANalytical, Type : Minipal 4) followed by methods previously reported [17][18]. The XRF employed was to know the element composition that made up the rocks as previous investigations [19][20][21]. The procedures as follows: The samples prepared were put inside the XRF instrument (The MiniPal 4) and then were rotated and the measurement of elemental compositions was conducted for 1200 seconds. The standard each element was available to find the sample content (in concentration) and the measured intensities and the data were obtained.

3. Results and Discussion

The data obtained are the composition of the element which makes up of the rocks in some coastal areas in Eastern Bolaang Mongondow as shown in Table 1. The chemical composition of elements is significantly different from each other. There are some rare elements appear in the table but in the low percentage for all locations.

Table 1. The element compositions deposited in rocks located in Motongkat, Central Buyat, Buyat and Jiko coastal areas

No	Element	Symbol	Composition (%)			
			Motongkat	Central Buyat	South Buyat	Jiko
1	Aluminum	Al	9.7	10.0	9.8	10.0
2	Silicon	Si	39.3	39.6	42.4	36.3
3	Phosphorus	P	0.57	0.68	-	-
4	Kalium	K	2.42	3.21	3.04	3.17
5	Calcium	Ca	21.4	18.4	17.3	18.1
6	Titanium	Ti	1.33	1.10	1.33	1.47
7	Vanadium	V	0.073	0.075	0.080	0.10
8	Chromium	Cr	0.072	0.068	0.071	0.078
9	Manganase	Mn	1.68	0.59	0.68	0.81
10	Iron	Fe	22.0	24.8	23.8	28.9
11	Copper	Cu	0.15	0.13	0.13	0.16
12	Zinc	Zn	0.03	-	0.01	-
13	Strontium	Sr	0.71	0.69	0.68	0.55
14	Barium	Ba	-	-	-	0.09
15	Europium	Eu	0.3	0.4	0.3	0.3
16	Rhenium	Re	0.3	-	0.2	-

The strontium element, a soft silver was discovered in all coastal regions whose percentages were 0.71% (Matongkat), 0.69% (Central Buyat), 0.68% (South Buyat) and 0.55% (Jiko). The Sr that is prepared with barium in the form of an oxide is widely used in the high technology devices, such as color television cathode ray tubes which it resists the X-ray emission and it is added on the fireworks for a red color [22].

The barium element was only discovered in South Buyat sub-district coast observed at 0.2%. This metal generally alloyed with aluminum, is employed as a cleaner of unwanted gases in vacuum columns and tubes in electronic apparatuses. Since it is easily condensed, barium is suitable for gases cleaner and is attracted to O_2, N_2, CO_2, and H_2O [23]. Fig. 1 shows the XRF spectra of a chemical element discovered on the coast in the Matongkat sub-district and is plotted as an intensity with respect to energy. To find the element appearing in the spectrum should be compared with the data which are available. The manganese element was found in quite a high percentage in this regional coast recorded at 1.68%. The metal is very applicable in the modern technology and life as well as is widely used in an alloy with iron [24].

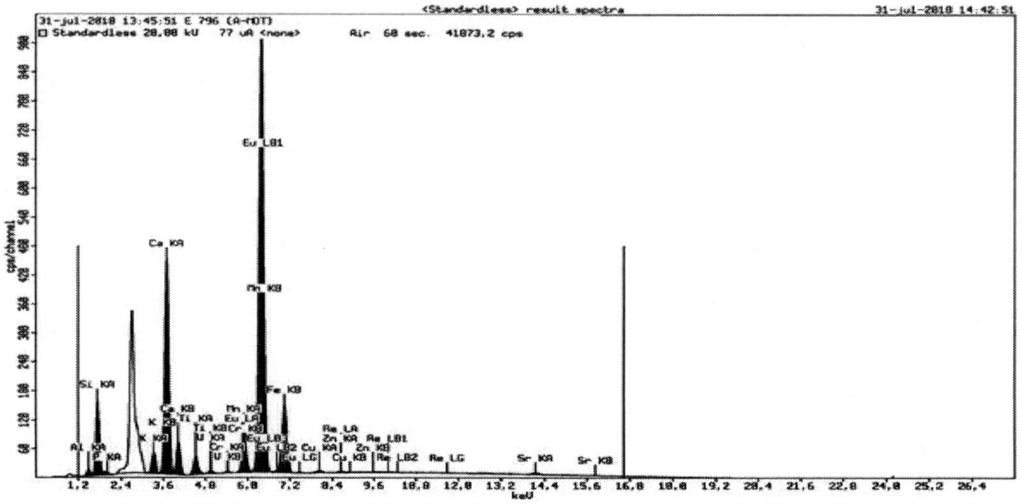

Figure 1. The XRF spectrum of the chemical elements appearing inside rocks obtained from Matongkat coastal region.

The other rare materials found were europium and rhenium whose composition is similar at 0.3% while, iron, copper, and zinc was of 22.0, 0.15 and 0.03%, respectively. Iron metal was also in high composition discovered in all regional coasts in order to it is perspective of an exploration for industrial needs.

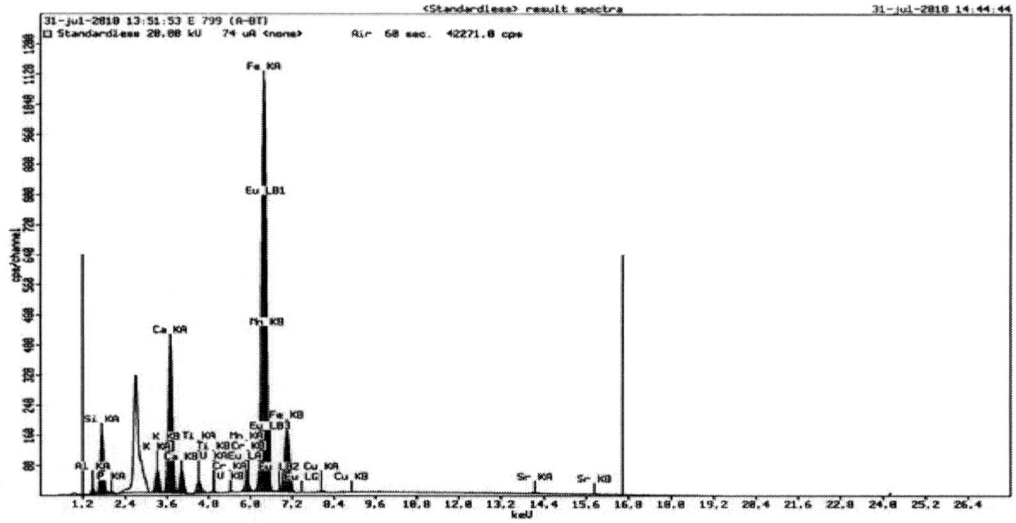

Figure 2. The element composition inside rocks obtained from Central Buyat coast.

The vanadium and chromium measured were recorded at 0.073 and 0.072% which were similar to other places as described below. Kalium, calcium, and titanium are relatively high proportion discovered on the coastal area in Motongkat sub-district whereby percentages are 2.42, 21.4 and 1.33, respectively. Three elements are also exploited economically for their compositions are significantly big. These materials are very used in the chemical industry; kalium and calcium are widely used in fertilizer and food industries, and titanium is applied in a semiconductor-based device. The composition of a chemical element found in the coastal area in Central Buyat is relatively different if compared to Motongkat sub-district coast as shown in Fig. 2. The zinc, barium, and rhenium elements did not appear in the spectrum but the iron was significantly increased to 24.8%. The aluminum, silicon, and calcium are the slightly different compared with the previous region in which their compositions were of 10.0, 39.6 and 18.4%.

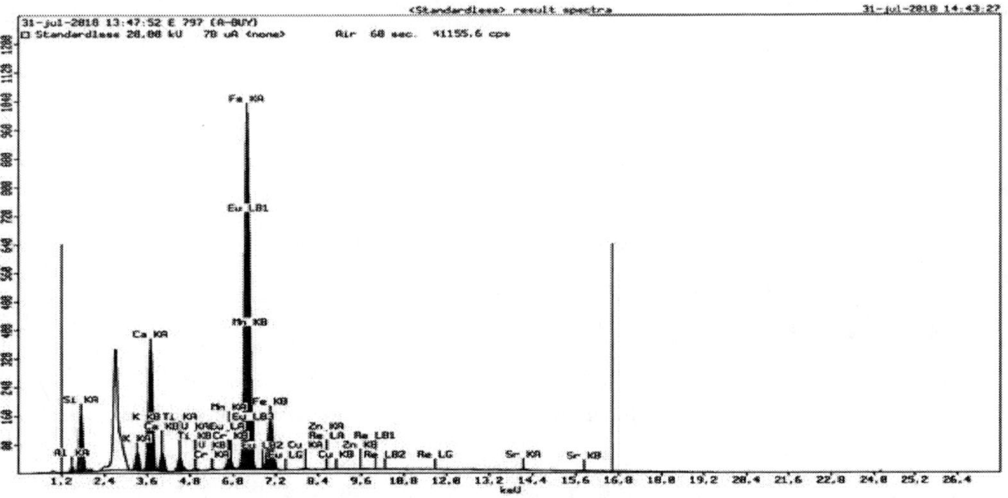

Figure 3. The element composition inside rocks obtained from South Buyat coast.

The silicone composition was the biggest discovered inside rocks of the coastal region in South Buyat Sub-district and was observed at 42.4% as drawn in Fig. 3. The phosphorous and barium did not appear in the XRF spectrum. The aluminum element was recorded at 9.8% which was relatively same with another sub-district zone. The kalium which was produced for agricultural-based products was observed at 3.04% that was similar to other coastal regions.

Calcium that has been produced for building development utility has a big portion in South Buyat coastal region appearing at 17.3%. The titanium which was the most applicable in semiconductor devices was discovered at 1.33 that was similar to that of Motongkat sub-district and was only slightly different with the rest of coastal regions.

The vanadium, chromium and manganese compositions were recorded at low portions of 0.080, 0.071 and 0.68% and were comparable with that of other areas. The iron element gave a higher percentage of composition that was a number 23.8%. If compared with other coastal areas, iron is possible to conduct an exploration in Eastern Bolaang Mongondow District. The copper, zinc, and strontium were the low portions that were of 0.13, 0.01 and 0.68%. Meanwhile, the europium and rhenium which were included in the rare material were observed at 0.3 and 0.2%, respectively.

The coastal Jiko rocks were not found the three elements as follows phosphorus, zinc, and rhenium as presented in Fig. 4. The highest chemical element composition belonged to silicon followed by iron, calcium, and aluminum whose their percentages were of 36.3, 28.9, 18.1 and 10.0%, respectively. The middle compositions were kalium, titanium, manganese, strontium, and copper. The titanium was the biggest percentage compared with other locations which were observed at 1.47%. While, the lower elements were vanadium, chromium, barium, and europium were noted at 0.10, 0.078, 0.09 and 0.3%. The chemical and physical properties of the heavy materials found in coastal areas could be referred to the previous work [25]. While the characteristics and properties of the Rare Earth Element were investigated in the Western Indonesia [26].

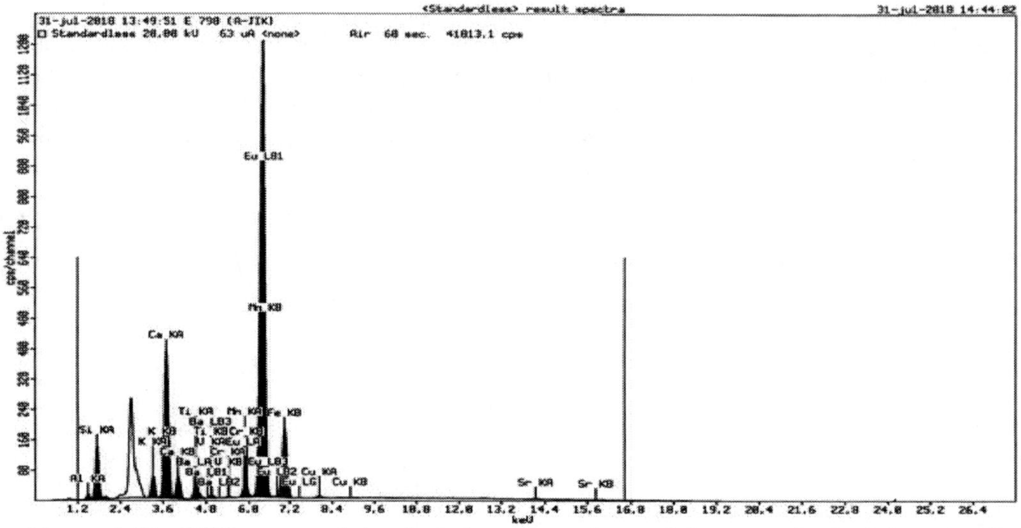

Figure 4. The XRF spectra of chemical elements made up of the rocks obtained from the Jiko coast.

4. Conclusions

The compositions of the chemical elements which were contained in the rocks of the four coastal regions in Eastern Bolaang Mongondow District were successfully determined to employ an XRF measurement. It was found that the iron, aluminum, and silicon were the highest portions of the rocks discovered of all coastal locations. The kalium, calcium, and titanium were in the middle portions spread from 1.10 to 21.4% which were possible to be carried out an exploration. The rare materials, such as vanadium, chromium, manganese, strontium, europium, barium and rhenium were also discovered in all coastal areas but the gold deposit was not discovered. The exploration of the materials was possible to be conducted since some chemical elements were in big percentages.

Acknowledgments

Authors would like to express our very great appreciation to the Ministry of Research, Technology and Higher Education of Indonesia which has funded this research. We would also like to thank the Rector of Sam Ratulangi University, Head of Service and Research Institute and Dean of Fishery and Marine Science Faculty of Sam Ratulangi University for their constructive contribution and approval of the program proposed.

References

[1] Long K R Van Gosen B S Foley N K Cordier Daniel 2010 U.S. Geological Survey Scientific Investigations Report 2010–5220 96, Available at http://pubs.usgs.gov/sir/2010/5220/.
[2] Delgado J Nieto J M Boski T. 2010 Estuarine, Coastal and Shelf Science 88, 71-83.
[3] Darby D A and Yu Tsang W 1987 Journal of Sedimentary Petrology 57 (5), 831-838.
[4] Razjigaeva N G and Naumova V V 1992 Journal Of Sedimentary Petrology 62 (5), 802—809.
[5] Groves D I Bierlein F P Meinert L D Hitzman M W 2010 Economic Geology 105, 641–654.
[6] Grant K J Paul G L O'neill J A 1991 Applied Spectroscopy 45 (4), 701-705.
[7] Landis G P Rye R O 1979 Economic Geology 69, 1025-1059.

[8] Shahar A Ziegler K Young E D Ricolleau A Schauble E A Y Fei Y 2009 Earth and Planetary Science Letters 288, 228–234.
[9] Cooper H K Duke M J M Simonetti A Chen G C 2008 Journal of Archaeological Science 35, 668-680.
[10] Ravisankar R Rajalakshmi A Eswaran P Meenakashisundram V Gajendiran V Manikandan E Magudapathy P Panigrahi B K Nair K G M 2007 International Journal of Pixe 17(3 & 4, 193−203.
[11] Bernstein S Frei D McLimans R K Knudsen C Vasudev VN 2008 Journal of Geochemical Exploration 96, 25–42.
[12] Kasper-Zubillaga J J Acevedo-Vargas B Bermea O M Zamora G O 2008 Chemie der Erde, 68, 45–59.
[13] Carvalho C Anjos RM Veiga R Macario K 2011 Journal of Environmental Radioactivity 102, 185-192.
[14] Herron M M 1998 Journal of Sendimentary Petrology 58(5), 820-829.
[15] Kenna T C Nitsche F O Herron M M Mailloux B J Peteet D Sritrairat S Sands E Baumgarten J 2011 J. Anal. At. Spectrom 26, 395–405.
[16] Kaminsky H A W Etsell T H Ivey D G Omotoso O 2008 Minerals Engineering 21 (4), 264–271.
[17] Worasith N Goodman B A Neampan J Jeyachoke N Thiravetyan P 2011 Clay Minerals 46, 539–559.
[18] Nayak P S Singh B K 2007 Bull. Mater. Sci. 30 (3), 235–238.
[19] Chia C H Gong B Joseph S D Marjo C E Munroe P Rich A M 2012 Vibrational Spectroscopy 62, 248–257.
[20] Eisazadeh A Kassim K A Nur H 2012 Applied Clay Science 67–68, 5–10.
[21] Knappett C Pirrie D Power M R Nikolakopoulou I Hilditch J Rollinson G K 2010 Journal of Archaeological Science 38, 219-232.
[22] Widory D Liu X Dong S Atmospheric Environment 2010 44(30), 3679-3687.
[23] Tsydenova O Bengtsson M 2011 Waste Management 31 (1) 45-58.
[24] Rekha K Nirmala M Manjula Nair G Anukaliani A 2010 Physica B: Condensed Matter 405(15), 3180-3185.
[25] Routray S Vanamu J Swain R 2018 IOP Conf. Series: Materials Science and Engineering 338, 012012 1-6.
[26] Handoko A D Sanjaya E 2018 IOP Conf. Series: Earth and Environmental Science 118 (2018) 012077 1-6.

IOP Conf. Series: Materials Science and Engineering **567** (2019) 012029 doi:10.1088/1757-899X/567/1/012029

Trophic status of Limboto lake in Gorontalo Province

S Y Noor[1]

[1]Department of Fisheries, Gorontalo University, Gorontalo, Indonesia

Corresponding author: yuyun.noor@gmail.com

Abstract. Limboto Lake were on the worried condition now. It happened because of the process depreciation and sedimentation which caused by sedimentation and waste ground that threaten the existence of ecosystem in the future. The purpose of the research is to determine trophic status of Limboto Lake in Gorontalo Province. Measurements of several water quality parameters were carried out insitu at 8 (eight) stations conducted by sampling and laboratory analysis of several parameters of water quality carried out at the Manado Research and Standardization Reasearch Centre. To determine the trophic status of Limboto Lake waters calculated using the *Trophic State Index* (TSI) of Carlson. The results showed that the average value *Trophic State Index* (TSI) of Limboto Lake range from 50.3071 – 61.9297. This shows that the trophic status of Limboto Lake waters is classified as eutrophic.

1. Introduction

Limboto Lake is one of the natural resources assets of Gorontalo Province. The area of Limboto Lake covers 920 km^2 and there are around 23 rivers and tributaries that enter the lake. Limboto Lake functions as providers clean water, plant and animal habitat, regulator of hydrological functions, natural disaster prevention, stabilization of systems and natural processes, producer biological natural resources, transportation, recreation and sports, fisheries resources (agriculture and capture fisheries), source of income, flood control, research and education.

Limboto Lake is in a very worrying condition at this time because it experiences shrinkage and siltation due to sedimentation, which threatens its existence in the future. The decreasing area of lake waters has caused the decline in the function of the lake as a water storage area, which has the potential to cause flooding around the lake area even outside the Limboto Lake area.

In 1932 the depth average of Limboto Lake was 30 meters with an area of 7,000 hectare, and in 1961 the depth average of Limboto Lake was reduced to 10 meters and the large area became 4,250 hectare. While the years 1990 - 2004 the depth average of the Limboto Lake is 2.5 meters with the large area of 3,000 hectare. The siltation of Limboto Lake was mainly caused by erosion and sedimentation due to agricultural efforts that did not heed the conservation of land and forest logging activities (illegal logging) in the upstream area (water catchment), especially in the Limboto watershed. Because of the siltation of Limboto Lake, most of the areas that used to be lakes turned into residential settlements or converted into agricultural businesses [1].

Eutrophication and water pollution are environmental problems that affect the waters of the lake in general where the consequences will affect the survival of human life. Therefore, the problem around Limboto Lake currently is the uncontrolled growth of water hyacinth, poison fishing (*Potasium Clorida*), fish bombs and large-scale fishing devices that have caused a decrease in the genetic diversity of fish and other aquatic organism, and decreased lake water quality.

The conditions that occur at this time are inseparable from the activities of the surrounding communities such as shifting cultivation activities that have been going on for a long time. As a result, environmental damage was marked by erosion, flooding during the rainy season, and drought in the dry season in the Gorontalo Province in the last 2 years. The direct effect that occurs in waters of Limboto Lake is silting and eutrophication because of increasing nutrients and pollutants into the bodies of the lake waters. The fast process of eutrofication and sedimentation in Limboto Lake has

IOP Conf. Series: Materials Science and Engineering **567** (2019) 012029 doi:10.1088/1757-899X/567/1/012029

caused the main function of the lake to decrease, such as a flood damper in the rainy season and a provider of water in the dry season, as well as a habitat for several species of fish. Based on these problems, the research was carried out with the title trophic status of Limboto Lake in Gorontalo Province

2. Method

2.1. Research Locations

This research was conducted at Limboto Lake with take measurements of some parameters the quality of water in 8 (eight) station and testing the other parameters of water quality carried out in Laboratory Manado Research and Standardization Reasearch Centre.

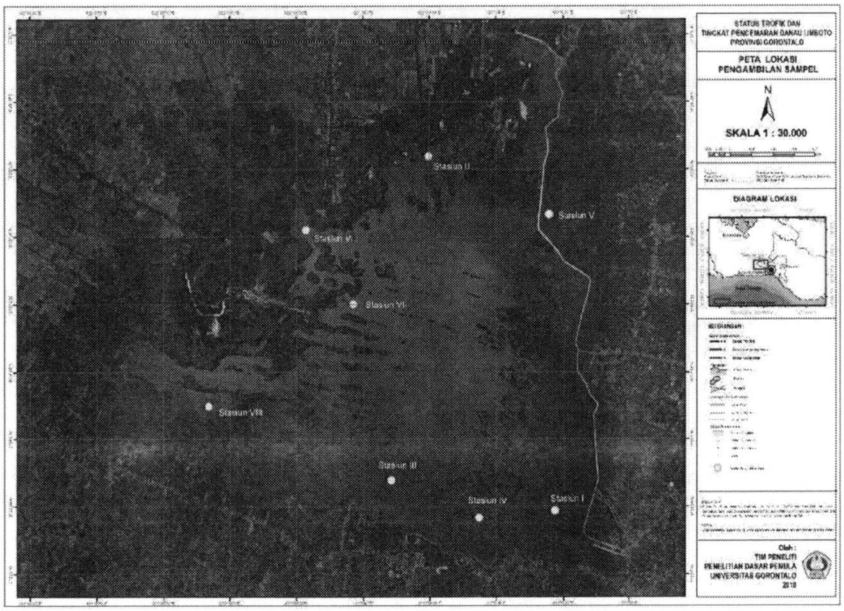

Figure 1. Map of sampling location

IOP Conf. Series: Materials Science and Engineering **567** (2019) 012029 doi:10.1088/1757-899X/567/1/012029

Table 1. The Condition of Water Sampling

Station	Location	Position	Description
1	Tabumela Village	N 0° 36' 25,10" E 123° 02' 33,59"	Outlet Limboto Lake
2	Hutuo Village	N 0° 36' 12,09" E 122° 59' 39,13"	Agriculture Area
3	Dembe Village	N 0° 33' 22,62" E 122° 59' 38,67"	Central Cage and Fisheries rumpon activity
4	Lekobalo Village	N 0° 33' 32,81" E 123° 00' 01,07"	Housing Area
5	Lupoyo Village	N 0° 36' 02,43" E 123° 00' 24,09"	Water Hyacinth Area
6	Kayu Bulan Village	N 0° 35' 54,74" E 122° 59' 18,70"	Housing Area
7	Hunggaluwa Village	N 0° 35' 35,10" E 122° 58' 33,59"	Estuary Biyonga River
8	Hunggaluwa Village	N 0° 36' 25,10" E 123° 02' 33,52"	Water Hyacinth Area

2.2. Sample Collection And Measurement Water Quality

Water quality parameters measured insitu consist of temperature, pH, brightness and dissolved oxygen (DO). Taking water samples is done using a point sampler at a depth of 5 meters from the surface water. Water samples that have entered the point sampler are transferred to the sample bottles which have been labeled according to the sampling station. The sample bottle containing the water sample is wrapped in aluminum foil and stored in the coolbox so that the water sample is protected from bacterial contamination and the sample water quality conditions do not change the sample.

2.3. Measurement Water Quality in the Laboratory

Parameters of water quality measured in the laboratory consist of phosphate, nitrate, ammonia, chlorophyll-a, and Biological Oxigen Demand. The water from 8 (eight) station sampling analysed to each parameters.

a. Phosphate Analysis

   Testing the level of phosphate in water using spectrophotometry was carried out with a method of ascorbic acid.

b. Nitrate Analysis

   The sample was put into a tube reaction as much as 5.0 ml. and Bruchine solution was added as much as 0.5 ml. After that, a concentrated Sulfuric Acid (H_2SO_4) solution was added as much as 5 ml, then the nitrate content was measured using a spectrophotometry with a wavelength of 420 nm.

c. Ammonia Analysis

   Ammonia analyzed by using the method of the spectrophotometry fenat at wavelengths 640 nm.

d. Chlorophyll-a Analysis

   Measurement of chlorophyll-a were carried out by filtering water sample using whatman paper and than checking absorbance with a spectrophotometry at wavelengths 750 and 665 nm.

IOP Conf. Series: Materials Science and Engineering **567** (2019) 012029 doi:10.1088/1757-899X/567/1/012029

2.4. Data Analysis

To determine the trophic status of waters in Limboto Lake are calculated using the *Trophic State Index* (TSI) of Carlsons. Based on the absorption of radiation energy by different atoms at the basic power level. The formula of average *Trophic State Index* (TSI)

TSI-P = 14,42 x Ln[TP] + 4,15 (μg/l)

TSI-Cla = 30,6 + 9,81 x Ln[Chlor-a] (μg/l)

TSI-SD = 60 – 14,41 x Ln[Secchi] (meters)

$$Average\ TSI = \frac{(\mathrm{TSI-P\ +\ TSI-Cla\ +\ TSI-SD})}{3}$$

Based on the TSI value obtained, trophic status of water classified into the ultra oligotrophic (<30), oligotrophic (30-40), mesotrophic (40-50), eutrophic (50-60), heavy eutrophic (60-70), hypereutrophic (70-80) and algae scum (>80).

3. Results And Discussion

3.1. Trophic Status of Limboto Lake

The results of the study show that the average value of the Limboto Lake *Trophic State Index* (TSI) of range from 50.3071 – 61.9297. This shows that the trophic status of waters in Limboto Lake is classified as eutrophic, that is, lake water contains high levels of nutrients, this status indicates that water has been polluted by increasing levels of nitrogen and phosphorus. This is in accordance with the opinion of Horne and Goldman [2] that waters are categorized as eutrophic if they have high nutrient depth of less than 10 meters. The high nutrient content causes abundant aquatic organisms, marked by algae blooms. Florida Lakewatch [3] which states that an eutrophic waters containing chlorophyll 25 mg/m^3 can interfere with the lives of other aquatic organisms which can further disrupt the designation of the water body.

Table 2. TSI value of phosphorus, chlorophyll-a, and Secchi Disc at Limboto Lake

Station	**TSI TP (μg/l)**	**TSI chlorophyll (μg/l)**	**TSI SD (m)**	**TSI Carlson**
1	29.9872	53.5731	67.3610	50.3071
2	60.8469	58.9545	65.9876	61.9297
3	41.1366	55.6864	69.9883	55.6037
4	43.2001	55.5331	67.3610	55.3647
5	35.8340	59.9881	71.1898	55.6706
6	45.8292	58.1010	70.8799	58.2700
7	42.2052	56.5590	70.2794	56.3479
8	39.9824	62.6371	71.8303	58.1499

3.2. *Water Quality*

Table 3. The Condition of Water Quality Limboto Lake

Parameters	Unit	Station								Quality Standard
		1	2	3	4	5	6	7	8	
Temperature	^{0}C	31	34	30	30	31	32	33	32	20-30
pH		8.1	8.5	8.4	8.3	8.5	8	8	8.1	6-9
DO	ppm	14.4	19.8	14.6	9.8	18.7	10.3	11.8	12.3	4
BOD	ppm	13	13	7	7	7	26	40	50	3
Nitrate	ppm	3.52	4.87	3.51	3.33	3.76	4.59	4.6	4.45	10
Ammonia	ppm	0.060	0.028	0.024	0.0011	0.021	0.031	0.057	0.017	0.02

The result measurement of water quality (temperature, pH , and Disolved Oxygen) in Limboto Lake ranged from 30-34 °C , 8-8.5 and 9.8-19.8 ppm. The pH conditions of Limboto Lake waters still in the normal range according to quality standards of PPRI No. 82 in 2001 for class II for freshwater aquaculture while temperature and DO have passed the standard quality standards of PPRI No. 82 of 2001 [4].

The result measurement of water quality (Biological Oxygen Demand, nitrate and ammonia) in the laboratory ranged from 7-50 ppm, 3.33-4.87 ppm, and 0.0011-0.06 ppm. The condition of biological Oxygen Demand in waters Limboto Lake had passed the quality standard of PPRI No. 82 in 2001, while nitrate and ammonia will continue to be around normal of quality standard by PPRI No. 82 in 2001.

The wáter quality in Limnoto Lake is very important to know, because Limboto Lake is one of waters potential to experience environmental changes due to the disposal of agricultural waste, industrial waste and household waste. Water quality plays an important role for the organism's ability to maintain its life [5].

Conclusion

Trophic State Index (TSI) of a Limboto Lake range 50.3071 – 61.9297 so trophic status waters of Limboto Lake is classified as eutrophic.

Acknowledgments
Thanks to Kemenristekdikti and Gorontalo University who have supported research so that this research can be done.

References

[1] Dinas Kelautan Perikanan. 2009. *Laporan tahunan. Kabupaten Gorontalo*. Gorontalo.Hal 25-40
[2] Horne A. J. and Golman, C. R, 1994. *Limnologi. Mc Graw Hill International Book Company*. New York.
[3] Florida lakewatch 2000, *A Beginner'sGuide To Water Management Nutrients*. Departemen Of FisheriesAnd Aquatic Sciences, Intitute Of Food And Agricultural Sciences University Of Florida.
[4] Peraturan Presiden Republik Indonesia. 2001. *Pengelolaan Kualitas Air dan Pengendalian Pencemaran.*
[5] Wahyuni Hasti, Setia Budi, Dwi P. 2013. *Kandungan Logam Berat pada Air, Sedimen dan Plankton di Daerah Penambangan Masyarakat Desa Batu Belubang Kabupaten Bangka Tengah*. Prosiding Seminar.

Growth response of rotifers on a bacterial-based diet made from fishwastes

S Wullur[1], E L Ginting[1], V Waraow[1], I F M Rumengan[1], E O Ogello[2], A Hagiwara[3]

[1]Faculty of Fisheries and Marine Science, Sam Ratulangi University, Manado, Indonesia.
[2]Department of Fisheries and Natural Resources, Maseno University, P.O.Box Private Bag, Maseno, Kenya
[3]Graduate School of Fisheries and Environmental Sciences, Nagasaki University, 852-8521, Nagasaki, Japan

Corresponding author: stenlywullur@unsrat.ac.id

Abstract. Fishwastes are potential low-cost diet sources for replacing microalgae, which is the popular diet for rotifers. Growth responses of rotifer were tested under different weights of the diet (i.e. 0.05, 0.1, 0.2 and 0.4 g/L), different culture containers (i.e. using small tall (ST), small-short (SS), medium-tall (MT) and large-short (LS) containers), different initial densities of the rotifer (i.e. 10, 50 and 200 rotifers/mL) and mass cultured trials for 15 and 30 days. Population densities of rotifers increased significantly (ANOVA, p=0.001) when providing 0.1 to 0.4 gr/L than those fed 0.05 gr/L of the diet (Tukey HSD test, p<0.05). The densities also were significantly higher (ANOVA, p=0.001) when cultured using larger containers (LS and MT) than using smaller containers (ST, SS) (Tukey HSD test, p<0.05), but it was not significant when inoculated at different initial densities (ANOVA, p>0.05). The rotifers were successfully mass cultured for 30 days and were successfully harvested for 7 times at densities of about 2500 to 3300 ind./mL.

1. Introduction

Aquaculture growth is on the rise almost all over the world, and there is increased demand for live food biomass i.e. rotifers, as a suitable starter diet for larval fish rearing in hatcheries [1][2][3][4]. Indeed, studies have shown that much smaller rotifers are valuable in marine fish larviculture [5] Despite the current success in ultrahigh-density cultures technique for rotifers [6][7][8], the technique still requires a continuous supply of high dense microalgae as food for the rotifers.

So far, microalgae are the first choice for feeding rotifers [9], thanks to their excellent influence on rotifer growth rate [7][8][10][11], rotifer nutrition [12][13] and for the nutrition of fish larvae that feeds on the rotifers [14]. However, culturing microalgae require high investment and running expenses [15], with difficulties in production [16][1], harvesting [17], and storage / preservation [18]. [19] introduced bakers's yeast as a cost-effective method to replace microalgae for rotifer production. However, baker's yeast is deficient of essential nutrition [20][21], for proper growth and development of the rotifers and their predators [12][22]. Thus, requires further enrichment with expensive chemical

IOP Conf. Series: Materials Science and Engineering **567** (2019) 012030 doi:10.1088/1757-899X/567/1/012030

emulsions. On the other hand, there are evidences that rotifers, as top predators in the microbial web [23][24] can utilize soluble compounds of bacterial excrete that are found in the culture media as food [25]. Indeed, previous studies have reported that presence of bacteria in rotifer cultures provide better effects on growth and nutritional value of the rotifers and fish larvae that feeds on them [26][27]. Nonetheless, information on effects of pathogenic bacteria on rotifer growth is scanty, even though addition of lactic acid bacteria to rotifer cultures increases resistance of fish larvae against pathogenic bacteria e.g. of the vibrio genera [28].

In this study, we hypothesized that a fishwastes-based diet can provide an excellent substrate for proliferation of billions of bacterial cells, which are directly ingested by rotifers, as their food source. To date, the use of bacteria as food source for rotifers is limited to experimental scale and mainly added as supplementary to microalgae or baker's yeast [29]. In the present study, we investigated the possibility of culturing rotifers using the fishwastes in the same culture facility with the rotifers.

2. Materials and methods

2.1. Materials

Two strains of rotifer *Brachionus rotundiformis* Perth and Poigar strains were used in this study. The rotifer *B. rotundiformis* Perth strain was obtained from a clone culture at laboratory of Aquaculture Biology, Nagasaki University, while *B. rotundiformis* Poigar strain was isolated from estuary around Poigar, North Sulawesi, Indonesia. A diet based on fishwastes was used as dietary source for the rotifers for several generations. The fishery industrial waste was prepared following the protocols described in a patent registered in Indonesian with number: P00201609066 and was kept in freezer (-20^0C) prior of using in this study.

2.2. Methods

Growth responses of the rotifers on the diet was investigated by culturing the rotifers at different diet weights, different culture containers and different initial density of rotifers using the diet as food source. Experiment on different weight of the diet (i.e. 0.05, 0.1, 0.2, 0.4 g/L in triplicate) was performed in 15 small-tall flasks (volume 0.075L: length: 65mm and width: 30mm). The flasks were filled with 75 mL filtered (Glass microfiber filters, Whatman) and sterilized (autoclaved at 121^0C for 30 minute) seawater at salinity 22 ppt. The Perth strain rotifers were transferred into each flask at density of 10 ind./ml following the addition of respective treatments of diet weight in triplicates. The diet was added once at the beginning and no other addition afterwards till end of experiment. As control, microalgae *Nannochloropsis oculata* was prepared at density of $1.2x10^7$ cell/mL. *N. oculata* density was maintained daily by adjusting the density similar to initial density. Observation was made once a day by counting the numbers of rotifer in 200 μl sample taken from three different layers (surface, middle and bottom) of each flask. A one-way ANOVA and Tukey-Kramer's test was performed to compare population densities of rotifer among the MRD weights.

Experiment on different culture containers was performed in four different shapes and volumes of culture containers in triplicate i.e. ST (small tall glass flasks: vol: 25 mL, length: 65mm and width: 30mm), SS (small-short glass beakers: vol: 75ml, length: 95mm and width: 55mm), MT, (medium-tall glass beakers, volume: 750ml, length: 181mm and width: 95mm) and LS (large-short polycarbonate tanks; volume: 30 l, length: 320cm and width: 350 cm). All flasks and tanks were firstly sterilized and were filled with sterilized seawater at salinity of 22 ppt. The Perth strain rotifers were introduced to each flasks and tanks at density of 10 ind./ml following the addition of the diet at 0.2 g/L. All flasks were placed in an incubator at constant temperature (28^0C) while tanks were placed outdoor in a water bath at temperature of about 28^0C. Observation on population density of rotifer was made once a day by counting the numbers of rotifer in 200 – 1000 μl sample taken from three different layers (surface, middle and bottom) of each container depends on volume of culture. A one-way ANOVA and Tukey-Kramer's test was performed to compare population densities of rotifer among the culture containers

Experiment on different initial rotifer density was performed in 30-L transparent polycarbonate cylindrical tanks. Three different initial densities of rotifer *B. rotundiformis* Perth strain at 10, 50 and 200 ind./mL were introduced to nine 30-L transparent polycarbonate cylindrical tanks. All tanks were filled with artificial seawater (Marine Art Hi, Tomita Pharmaceutical, Japan) at salinity 20-22 ppt and were placed inside water baths at constant temperature at 27-29^{0}C. The diet (0.2 g/L) was added to the tanks soon after inoculating rotifers at respective densities. Observation on population density of rotifer was made once a day by counting the numbers of rotifers in 1000 μl sample taken from three different layers (surface, middle and bottom) of each tank. A one-way ANOVA and Tukey-Kramer's test was performed to compare population density of rotifer among the initial density treatments
Mass culture trials on the Poigar strain rotifer using the fishwastes as exclusive dietary source for the rotifer was conducted for 15 and 30 days. The rotifers were semi-continuously mass cultured in two fibre tanks (volume 750 L, length 125 cm, width 81 cm and height 75 cm). All tanks were filled with 750 litres of artificial seawater (salinity 20 ppt) and were placed outdoor. About 10 and 400 ind./ml of rotifers were initially introduced to each tanks following the addition of the diet at 0.27 gr/L . The cultures were harvested every 2 to 5 days depending on density of the rotifer following the addition of new diet to the cultures. No water exchange and no aeration were provided to the cultures during the experiment. Observation was made every 6 hours by counting numbers of rotifer in 1000 μl samples taken from three different layers (surface, middle, bottom) of culture.

3. Results and discussions
Population growth curves of rotifer S-type *B.rotundiformis* revealed that the fishwastes-based diet used in this study has significant effect on growth of rotifers. The rotifers are capable of utilizing organic materials produced by the microbial decomposing processes on the diet as their nutritional source for development and reproduction [30, 31]. The rotifer showed a significant response in growth when cultured under different weight of the diet (Anova *test*, F =8.13, P = 0.00), culture duration (F = 30.71, P = 0.00) and the interaction between them (F = 6.33, P = 0.00) on rotifer population density (Table 1).

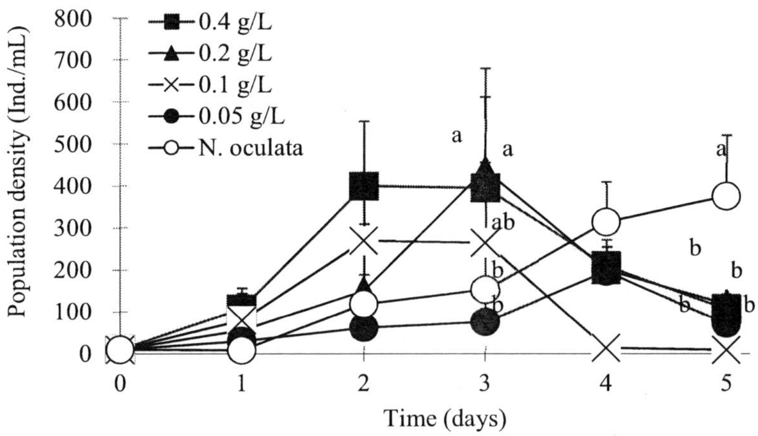

Figure 1. Population density curves of rotifer batch cultured under different MRD weight (0.4 (■), 0.2 (▲), 0.1 (X) and 0.05 g/L (●) and control (1.2×10^7 *Nannochloropsis oculata* cells/mL(O). The values and error bars represent mean ± SD of 3 replicates each. Different alphabetical letters indicate significant differences at p<0.05, a>b

Higher population density was observed at diet weight 0.2 and 0.4 g/L compared to control experiment on day 3, before experiencing culture crash. The rotifer density in the control experiment increased significantly on day 4 and 5 (Figure1). Population densities of rotifers cultured with 0.4 and 0.2 g/L of the diet were significantly higher than those cultured with 0.1 and 0.05 g/L (TukeyHSD test, $p<0.05$). Mean maximum densities of 400±155 and 438±174 ind./mL were obtained at 0.4 and 0.2 g/L on 3, while 264±193 ind./mL were obtained in the control. Results of the present study indicated that the organic materials produced from microbial decomposing processes on the diet could support nutritional requirement of rotifer for certain culture period, beyond this period, the system become exhausted and required a new diet resource. It is suggested that the diet weights between 0.2-0.4 g/L in this study supported nutritional requirement of rotifer until day 3. Unlike in the control treatment, population density of rotifer increased until the end of experiment due to daily addition of *N. oculata* to the rotifer.

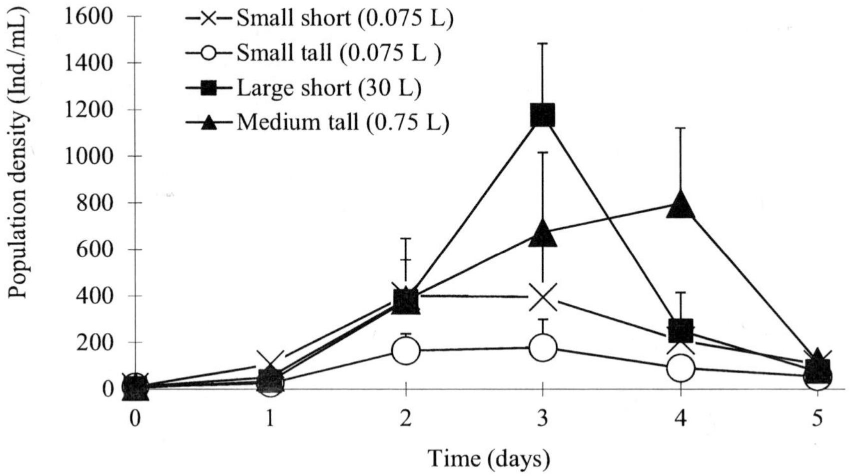

Figure 2. Population density curves of rotifer batch cultured under different volumes and shapes of culture containers i.e. ST ((O) small tall), SS ((X) small-short), MT ((▲) medium-tall) and LS (■) large-short polycarbonate tanks). The values and error bars represent mean ± SD of 3 replicates each. Different alphabetical letters indicate significant differences at $p<0.05$, a>b.

Growth response of rotifer on the fishwaste-based diet was significant to culture container ($F = 10.47$, $P = 0.00$), culture duration ($F = 26.89$, $P = 0.00$) and their interaction ($F = 5.59$, $P = 0.00$) (Table 1). Rotifer cultured using the diet showed higher densities in 30-L tanks and 750 mL flasks than those in 25 mL and 75 mL flasks on day 3 (Tukey's HSD *test*, $p<0.05$). Maximum rotifer densities of 1170 ± 306 were obtained in 30-L, on day 3 (Figure 2). In general, rotifer cultured in small containers is mostly for maintaining rotifer stock, while those cultured in larger containers are for mass production. Result of the present study shows that the fishwastes-based diet can be used for maintaining stock and also for mass production purpose. High population density in the LS tanks (30L tanks) indicating a positive effect of the diet on growth of rotifer in larger containers. In regard to the current high-cost mass production of rotifer based on microalgae, result of the present study can be a breakthrough in mass production of rotifers without microalgae. The fishwastes-based diet is made from cheap sources thus offers opportunity to reduce the cost of rotifer production for profitable aquaculture production.

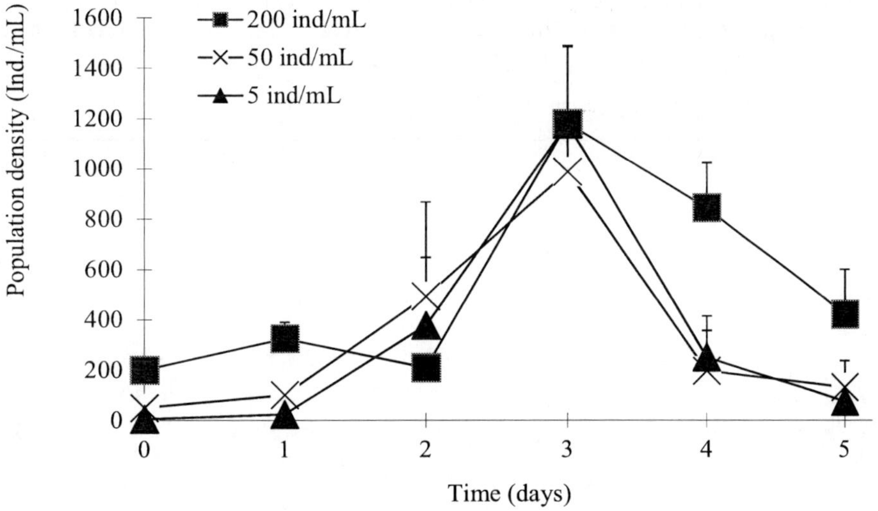

Figure 3. Population density curves of the rotifer batch cultured under different rotifer starter densities of 200 ind./mL (■), 50 ind./mL (X) and 5 ind./mL (▲). The values and error bars represent mean ± SD of 3 replicates each. Different alphabetical letters indicate significant differences at $p<0.05$, a>b

Initial density is closely related to competition between individual rotifer for food which, has strong influence on growth and reproduction performance of rotifers [32]. Response in growth of rotifers cultured using the fishwastes-based diet at different rotifer initial densities is shown in Figure 3. Initial density of rotifer had significant effect on population density of rotifer during 5 days culture (ANOVA test, $p=0.01$). The density of rotifer at higher initial density treatment (200 ind./mL) was significantly higher than those at lower initial density (5 and 50 ind./mL) on day 4 and 5 of culture (TukeyHSD test, $p<0.05$). Maximum densities of rotifer in all treatments were attained on day 3 at about 1170 ind./mL and the densities decreased sharply afterwards. The rotifers at initial densities of 200 ind./mL showed a little decrease on day 2 then sharply increased until day 3 of culture. It is suggested that until day 2, competition on food source had occurred on higher initial density treatment (200 ind./mL) indicated bydecrease in population density, which was not seen in the lower initial density treatments (5 and 50 ind./mL). The fast growth pattern of rotifers till day 3 and crash afterwards, support our previous assumption that nutritious materials produced by microbial decomposing processes on the diet become exhausted for certain period. Lubzens reported a similar growth pattern by using the diet as food source for the rotifer [29].

Table 1. One-way ANOVA output of diet weight, culture container and rotifer starter density on population density of rotifer in all the treatments. df: degrees of freedom, SS: Sum of squares, MS: Mean square, *F*: F-ratio, *P*: level of significance (*denote significant difference at $p < 0.05$)

	df	SS	MS	*F*	*P*
Diet weight					
Day	5	16.38	3.28	30.71	0.00***
Diet weight	4	3.47	0.87	8.13	0.00***
Day × Diet weight	20	13.52	0.68	6.33	0.00***
Residuals	60	6.40	0.11		
Culture container					
Day	5	3141948	628390	26.89	0.00***
Culture container	3	733988	244663	10.47	0.00***
Day × culture container	15	1959188	130613	5.59	0.00***
Residuals	48	1121588	23366		
Rotifer initial density					
Day	5	6400008	1280002	30.38	0.00***
Rotifer starter density	2	516363	258182	6.13	0.00**
Day × rotifer starter density	10	870373	87037	2.07	0.05
Residuals	36	1516940	42137		

Unlike the batch culture system discussed above, semi-continuous culture system mainly performed for long term production to supply food requirement of fish larvae. In the present study, we successfully used the fishwastes-based diet to mass produce the rotifers using outdoor semi-continuous culture system without any addition of microalgae as the current mass production protocols.

In regards to the current problems on the cost of microalgae production for feeding rotifer in larviculture industry [30], the fishwastes-based diet can be a promising alternative for a profitable aquaculture production by reducing or eliminating the production cost of microalgae. Although the use of fishwastes-based diet for rotifer is just in the trial phase and still require optimizations, it had a strong positive influence on growth of rotifer in this culture system. The maximum density achieved (around 1357 – 3305 ind./mL), the long duration of culture (30 days culture) and the yield obtained (7 times harvest) during the trial can be a good starting point for future improvement.

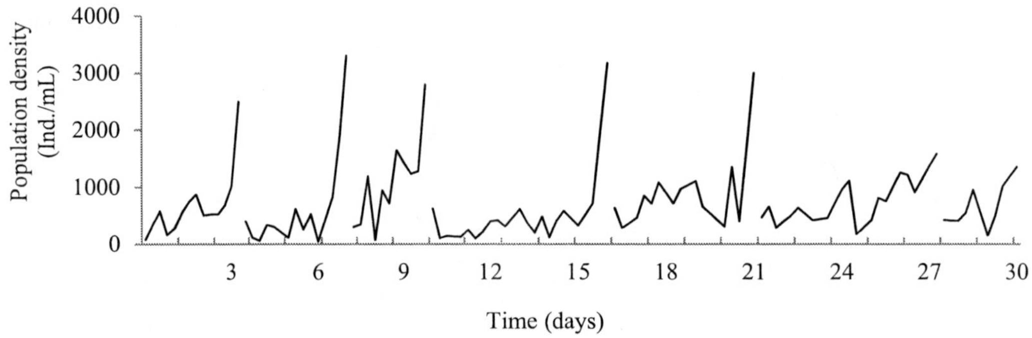

Figure 4. Population density curve of rotifer semi-continuously mass cultured in a 750 L outdoor culture tanks for 30 days.

Acknowledgments
This study was supported by the Ministry of Research, Technology and Higher Education of Indonesia (Kemenristekdikti) under KLN research grant awarded to SW.

References

[1] Lee, Y-K, 2001. Microalgal mass culture systems and methods: Their limitation and potential. J. Appl. Phycol., 13; 307-315.

[2] Wullur, S., Y. Sakakura, A. Hagiwara. 2009. The minute monogonont rotifer *Proales similis* de Beauchamp: culture and feeding to small mouth marine fish larvae. Aquaculture 293, 62-67.

[3] Wullur S, Sakakura Y, and Hagiwara A, 2011. Application of the minute monogonont rotifer *Proales similis* de Beauchamp in larval rearing of seven-band grouper *Epinephelus septemfasciatus*. *Aquaculture*, *315*(3-4), 355-360.

[4] Hagiwara A, Wullur S, Marcial H S, Hirai N, and Sakakura Y, 2014. Euryhaline rotifer *Proales similis* as initial live food for rearing fish with small mouth. *Aquaculture*, *432*, 470-474.

[5] Wullur S, Kumagai S, Sakakura Y, Hagiwara A, 2018. Assessment of different minute zooplankton in the larval rearing of rusty angelfish *Centropyge ferrugata*. *AACL Bioflux, 11,*1495-1501.

[6] Suantika G, Dhert P, Nurhudah M, and Sorgeloos P. (2000). High-density production of the rotifer *Brachionus plicatilis* in a recirculation system: consideration of water quality, zootechnical and nutritional aspects. *Aquacultural engineering*, *21*(3), 201-213.

[7] Yoshimura K, Tanaka K, and Yoshimatsu T. 2003. A novel culture system for the ultra-high-density production of the rotifer, *Brachionus rotundiformis* a preliminary report. *Aquaculture*, *227*(1-4), 165-172.

[8] Yoshimatsu, T., and M A.Hossain, 2014. Recent advances in the high-density rotifer culture in Japan. Aquaculture International 22, 1587-1603.

[9] Maruyama I, Nakao T, Shigeno I, Ando Y, and Hirayama K. 1997. Application of unicellular algae *Chlorella vulgaris* for the mass-culture of marine rotifer Brachionus. In *Live Food in Aquaculture* (pp. 133-138). Springer, Dordrecht.

[10] Hagiwara, A., M.D. Balompapueng, N. Munuswamy, and K. Hirayama. 1997. Mass production and preservation of the resting eggs of the euryhaline rotifer *Brachionus plicatilis* and *B. rotundiformis*. Aquaculture, 155(1), 223-230.

[11] Dhert P, G. Rombaut, G. Sauntika, P. Sorgeloos. 2001. Advancement of rotifer culture and manipulation techniques in Europe. Aquaculture 200:129-146

[12] Watanabe, T., C. Kitajima, S. Fujita. 1983. Nutritional values of live organisms used in Japan for mass propagation of fish: A review. Aquaculture 34, 115–143.

[13] Brown, M.R., 2002. Nutritional value and use of microalgae in aquaculture. Avances en Nutrición Acuícola VI. Memorias del VI Simposium Internacional de Nutrición Acuícola, 3, pp.281-292.

[14] Reitan K I, Rainuzzo J R, Oie G, Olsen Y, 1997. A review of the nutritional effects of algae in marine fish larvae. Aquaculture, 155; 207-221.

[15] Borowitzka M.A., 1997. Microalgae for aquaculture; opportunities and constraints. J Appl Phycol 9, 393-401

[16] Acién F G, Fernández J M, Magán J J, and Molina E. 2012. Production cost of a real microalgae production plant and strategies to reduce it. *Biotechnology advances*, *30*(6), 1344-1353.

[17] Barros A.I., A.L. Gonçalves, M. Simões, and J.C. Pires. 2015. Harvesting techniques applied to microalgae: A review. Renewable and Sustainable Energy Reviews 41:1489-1500.

[18] Camacho-rodriguez, J., M.C. Ceron-Garcia, M.D. Macias-sanches, J.M. Fernandez-Sevilla, L. Lopez-Rosales, E. Molina-Grima. 2015. Long-term preservation of concentrated *Nannochloropsis gaditana* cultures for use in aquaculture. J. Appl. Phycol, 28, 299-312.

[19] Hirata H and Mori Y. 1967. Cultivation of the rotifer *Brachionus plicatilis*, fed on a mixed diet of marine Chlorella and baker"s yeast. Saibai-gyogyo 5: 36–40.

[20] Hirayama K and Funamoto H. 1983. Supplementary effect of several nutrients on nutritive deficiency of baker's yeast for population growth of the rotifer *Brachionus plicatilis* Bull. Jap. Soc. Sci. Fish. 49: 505-510.

[21] Hirayama K, and Satuito C G, 1991. The nutritional improvement of baker's yeast for the growth of the rotifer, *Brachionus plicatilis*. In *Proceedings of a US-Asia Workshop. Hawaii* (pp. 151-162).

[22] Lubzens E, Tandler A, and Minkoff G, 1989. Rotifers as food in aquaculture. *Hydrobiologia*, *186*(1), 387-400.

[23] Starkweather P L, Gilbert J J, and Frost T M, 1979. Bacterial feeding by the rotifer *Brachionus calyciflorus*: clearance and ingestion rates, behavior and population dynamics. *Oecologia*, *44*(1), 26-30.

[24] Yasuda K, and Taga N, 1980. A mass culture method for Artemia salina using bacteria as food. *Mer*, *18*(53), 62.

[25] Sakamoto, H., Hirayama, K., 1983. Dietary effect of Thiocapsa roseopersicina Ž . photosynthetic bacteria on the rotifer *Brachionus plicatilis*. Bull. Fac. Fish., Nagasaki Univ. 54, 13–20.

[26] Gatesoupe F J, Arakawa T, and Watanabe T, 1989. The effect of bacterial additives on the production rate and dietary value of rotifers as food for Japanese flounder, *Paralichthys olivaceus*. *Aquaculture*, *83*(1-2), 39-44.

[27] Gatesoupe F J, 1990. The continuous feeding of turbot larvae, *Scophthalmus maximus*, and control of the bacterial environment of rotifers. *Aquaculture*, *89*(2), 139-148.

[28] Gatesoupe F J, 1991. The effect of three strains of lactic bacteria on the production rate of rotifers, *Brachionus plicatilis*, and their dietary value for larval turbot, *Scophthalmus maximus*. *Aquaculture*, *96*(3-4), 335-342.

[29] Lubzens E, 1987. Raising rotifers for use in aquaculture. In *Rotifer Symposium IV* (pp. 245-255). Springer, Dordrecht.

[30] Ogello E O, Wullur S, Sakakura Y and Hagiwara A, 2017. Composting fishwastes as low-cost and stable diet for culturing *Brachionus rotundiformis* Tschugunof (Rotifera): influence on water quality and microba. Aquaculture 486:232-239.

[31] Nesta K, Ogello E O, Sakakura Y, Hagiwara A 2018. Fish processing wastes as an alternative diet for culturing the minute rotifer, *Proales similis* de Beauchamp. *Aquaculture Research* DOI: 10.1111/are.13707.

[32] Ciroz-Perez, J., Gomez, A., Serra, M., 2001. On the taxonomy of three sympatric sibling species of the *Brachionus plicatilis* (Rotifera) complex from Spain, with the description of *B. ibericus* n. sp. J. Plankt. Res. 23, 1311–1328.

The convergence properties of new hybrid conjugate gradient method

Y Salih[1], M A Hamoda[2], Sukono[3], M Mamat[1]

[1] Department of Computer Science and Mathematics, Faculty of Informatics and Computing, Universiti Sultan Zainal Abidin, 22200 Terengganu, Malaysia
[2]Department of Mathematics, Faculty of science, AlAsmarya Islamic University, Libya
[3]Universitas Padjadjaran, Bandung, Indonesia

Corresponding author: Si2023@putra.unisza.edu.my

Abstract. In this paper, a new Hybrid Conjugate Gradient Methods is presented, which produced sufficient descent search direction at every iteration and global Convergence Properties for solving large-scale nonlinear optimization problem under exact line search. The numerical experiments show that a hybrid method has the best efficiency for the test problems.

1. Introduction

Let us consider the nonlinear unconstrained optimization problem

$$\min\{f(x) : x\epsilon R^n\}, \tag{1}$$

Where f is smooth and its gradient g is available. Conjugate gradient methods are very efficient for solving (1), especially when the dimension n is large. The iterates of (CG) method for solving (1) are obtained by

$$x_{k+1} = x_k + \alpha_k d_k \ , \ k = 0,1,2,\ldots \tag{2}$$

Where α_k is the current iterate, α_k is a positive scalar and called the step length which is determined by some line search, and d_k is the search direction defined by

$$d_k = \begin{cases} -g_k, & if\ k = 0 \\ -g_k + \beta_k d_{k-1}, & if\ k \geq 1 \end{cases} \tag{3}$$

Where $g_k = \nabla f(x_k)$ and β_k is the conjugate gradient coefficient. Well-known formulae for β_k include the Hestenes-Steifel (HS) method [1], the Fletcher-Reeves (FR) method [2], the Polak-Ribiere-polyak (PRP)method [3][4], the Conjugate Descent (CD) method [5], Liu-Storey (LS) method [6], and the Dai-Yuan (DY) method [7], which are given, respectively, by

$$\beta_k^{HS} = \frac{g_k^T(g_k - g_{k-1})}{d_k^T(g_k - g_{k-1})} \beta_k^{FR} = \frac{\|g_k\|^2}{\|g_{k-1}\|^2} \beta_k^{PRP} = \frac{g_k^T(g_k - g_{k-1})}{\beta_{k-1}^T g_{k-1}}$$

$$\beta_k^{CD} = \frac{\|g_k\|^2}{d_{k-1}^T g_{k-1}} \beta_k^{LS} = \frac{g_k^T(g_k - g_{k-1})}{-d_{k-1}^T g_{k-1}} \beta_k^{DY} = \frac{\|g_k{}^2\|}{d_{k-1}^T(g_k - g_{k-1})}$$

Many studies have been achieved to explain the poor performanceof some methods for some problems, and to establish enough descent property of conjugate gradient techniques [8-17].

2. Hybrid Conjugate Gradient Method and Algorithm

In this section, we present a new Hybrid conjugate gradient method, namely YHM2, where YHM2 denotes Yasir, Hamoda and Mamat. The hybridization manner process happened while two or more classical CG algorithms are blended. The combination of classical CG techniques Used in this paper is referred to PRP and DPRP methods respectively [3][18],

$$\beta_k^{PRP} = \frac{g_k^T(g_k - g_{k-1})}{\|g_{k-1}\|^2} \tag{4}$$

$$\beta_k^{DPRP} = \frac{\|g_k\|^2 - \frac{\|g_k\|}{\|g_{k-1}\|}\left|g_k^T g_{k-1}\right|}{\mu\left|g_k^T d_{k-1}\right| + \|g_{k-1}\|^2}, \quad \mu > 1 \tag{5}$$

Using the above formula, we proposedβ_k^{YHM2} is defined by

$$\beta_k^{YHM2} = \begin{cases} \dfrac{g_k^T(g_k - g_{k-1})}{\|g_{k-1}\|^2 + \left|g_k^T d_{k-1}\right|} if \|g_k\|^2 \geq \left|g_k^T g_{k-1}\right| \\ \beta_k^{DPRP} \text{otherwise} \end{cases} \tag{6}$$

The Algorithm is given as follows:

Step1: Initialization

Given$x_0 \epsilon R^n$,$\varepsilon > 0$, set$d_0 = -g_0$, $k = 0$.

Step2: Computing conjugate gradient coefficientβ_k, by formula (6).

Step3: Compute $d_k$based on (3) if $g_k = 0$ then stop.

Step4: Computing step size by

$$\alpha_k = \arg min_{\alpha \geq 0} f(x_k + \alpha d_k),$$

Step5: Update new point base on (2).

Step6: Set $k = k + 1$ goes to Step 2.

3. Global convergence Analysis

Firstly we need to simplify theβ_k^{YHM2}, in order that the proof may be easier. From (6), we know that,

$$\beta_k^{YHM2} = \begin{cases} \dfrac{g_k^T(g_k - g_{k-1})}{\|g_{k-1}\|^2 + \left|g_k^T d_{k-1}\right|} if \|g_k\|^2 \geq \left|g_k^T g_{k-1}\right| \\ \beta_k^{DPRP} \text{otherwise} \end{cases}$$

Now, if $\|g_k\|^2 \geq \left|g_k^T g_{k-1}\right|$then,

$$\frac{g_k^T(g_k - g_{k-1})}{\|g_{k-1}\|^2 + \left|g_k^T d_{k-1}\right|} = \frac{\|g_k\|^2 - g_k^T g_{k-1}}{\|g_{k-1}\|^2 + \left|g_k^T d_{k-1}\right|} \leq \frac{\|g_k\|^2}{\|g_{k-1}\|^2}$$

Also,

$$\frac{g_k^T(g_k - g_{k-1})}{\|g_{k-1}\|^2 + \left|g_k^T d_{k-1}\right|} \geq \frac{\|g_k\|^2 - \left|g_k^T g_{k-1}\right|}{\|g_{k-1}\|^2 + \left|g_k^T d_{k-1}\right|} \geq 0$$

From [18] we know that $0 \leq \beta_k^{DPRP} \leq \frac{\|g_k\|^2}{\|g_{k-1}\|^2}$

Hence we obtain,

$$0 \leq \beta_k^{YHM2} \leq \frac{\|g_k\|^2}{\|g_{k-1}\|^2}$$

3.1. Sufficient descent condition

For the sufficient descent condition to hold,

$$g_k^T d_k \leq -c\|g_k\|^2 \ \forall k \geq 0, c > 0 \tag{7}$$

The following theorem shows that our new formula with exact line searches possess the sufficient descent condition.

Theorem 1. Let $\{x_k\}$ and $\{d_k\}$ be sequence generated by the method of the form (2), (3) and the above Algorithm, where the step size α_k is determined by the exact line search, then (7) holds for all $k \geq 0$.

Proof

We proof by induction, that if $k = 0$ then $g_0^T d_0 = -c\|g_0\|^2$

Hence, the condition holds true, now we need to prove that:

$g_k^T d_k \leq -c\|g_k\|^2$ For $k \geq 1$

From (3) we have $d_{k+1} = -g_{k+1} + \beta_{k+1} d_k$

Multiply both sides by g_{k+1}^T

$$g_{k+1}^T d_{k+1} = g_{k+1}^T(-g_{k+1} + \beta_{k+1} d_k) = -\|g_{k+1}\|^2 + \beta_{k+1} g_{k+1}^T d_k \tag{8}$$

For exact line search, we know that$g_{k+1}^T d_k = 0$.Thus $g_{k+1}^T d_{k+1} = -\|g_{k+1}\|^2$

Hence this condition holds true for$k + 1$. Therefore, the sufficient descent condition holds. □

3.2. Global convergence Properties

In this subsection, we will prove the global convergence properties of the YHM2 under the following assumption.

Assumption1

i) The level set $\Omega = \{x \in R^n,\ f(x) \leq f(x_0)\}$ is bounded.

ii) Function $f: \mathbb{R}^n \to \mathbb{R}$ is continuously differentiable with Lipschitz continuous gradient in$N$, there exists a constant$L > 0$, such that,

$$\|g(x) - g(y)\| \leq L\|x - y\|, \qquad \forall\, x, y \in N$$

The following lemma is often used to prove global convergence which was provided by Zoutendijk[19].

Lemma 1.

Let x_k be generated by Algorithm for which Assumption1 hold. Consider any method in form 1 and 2 and $d_k$satisfies $g_k^T d_k < 0$ for all k andα_k is obtained by$\alpha_k = \arg min_{\alpha \geq 0} f(x_k + \alpha d_k)$, then,

$$\sum_{k=0}^{\infty} \frac{(g_k^T d_k)^2}{\|d_k\|^2} < \infty \tag{9}$$

The following theorem is based on Lemma 1.

Theorem 2

Suppose that the conditions in Assumption 1 hold. Consider the YHM2 method and step sizeα_k is obtained by the exact line search then,

$$\liminf_{k \to \infty} \|g_k\| = 0 \tag{10}$$

Proof. Case 1:

If$\|g_k\|^2 \geq |g_k^T g_{k-1}|$ then

$$\beta_k^{YHM2} = \frac{g_k^T(g_k - g_{k-1})}{\|g_{k-1}\|^2 + |g_k^T d_{k-1}|}$$

Toprove this theorem, we use contradiction. If Theorem 2 is not true, then a constant $c > 0$ exists, such that

$$\|g_k\| \geq c \tag{11}$$

From equation (3) we have $d_k = -g_k + \beta_k^{YH}\ d_{k-1}$and squaring both sides of the equation, we get

$$\|d_k\|^2 = \|g_k\|^2 - 2\beta_k^{YHM}\ g_k d_{k-1} + (\beta_k^{YH}\ )^2\|d_{k-1}\|^2 \tag{12}$$

Dividing both sides by $(g_k^T d_k)^2$, then,

$$\frac{\|d_k\|^2}{(g_k^T d_k)^2} = \frac{\left(\beta_k^{YHM2}\right)^2 \|d_{k-1}\|^2}{(g_k^T d_k)^2} - \frac{2}{g_k^T d_k} - \frac{\|g_k\|^2}{(g_k^T d_k)^2}$$
$$= \frac{\left(\beta_k^{YHM2}\right)^2 \|d_{k-1}\|^2}{(g_k^T d_k)^2} - \left[\frac{1}{\|g_k\|} + \frac{\|g_k\|}{g_k^T d_k}\right]^2 + \frac{1}{\|g_k\|^2}$$
$$\leq \frac{\left(\beta_k^{YHM2}\right)^2 \|d_{k-1}\|^2}{(g_k^T d_k)^2} + \frac{1}{\|g_k\|^2} \leq \frac{1}{\|g_k\|^2}$$

Because of an exact line search $g_k^T d_{k-1} = 0$. Hence,

$$\frac{\|d_k\|^2}{(g_k^T d_k)^2} \leq \sum_{i=0}^{k} \frac{1}{\|g_i\|}$$

Therefore,

$$\frac{(g_k^T d_k)^2}{\|d_k\|^2} \geq \frac{c^2}{k}$$

Thiscontradicts the Zoutendijk condition in $\sum_{k=0}^{\infty} \frac{(g_k^T d_k)^2}{\|d_k\|^2} < \infty$

The proof is completed.

Case 2:

When $\beta_k^{YHM2} = \frac{\|g_k\|^2 - \frac{\|g_k\|}{\|g_{k-1}\|}|g_k^T g_{k-1}|}{\mu|g_k^T d_{k-1}| + \|g_{k-1}\|^2}, \quad \mu > 1$

the proof of this theorem can be seen in[18].

4. Numerical experiments and discussions

In this section, we selected some preliminary numerical experiments considered in Andrei [20]. The performance of this method s compared with other coefficients such as FR [2],PRP [3][4], WYL [17], and DPRP [18] methods we considered $\varepsilon = 10^{-6}$ and the gradient value as the stopping criteria as suggested by Hillstrom [21]. For every test function used, there are four or five initial points has been tested randomly from it. A list of problem function and the initial points used are shown in table1. All problems mentioned are solved by MATLAB version 8.3.0.532 (R2014a), We used the exact line search to compute the step size, the CPU processor used was Intel(R) Cor™ i5-M520 (2.40GHz), with RAM 4GB.

Figure 1 and Figure 2 list the performance of the above methods relative to the number of iterations and CPU time respectively, using a performance profile introduced by Dolan and More [22].

Table 1. List of test functions

NO	Function	Dimension	Initial points
1	Six Hump Camel	2	-10, -8, 8, 10
2	Three Hump Function	2	-10, 10, 20, 25
3	Booth	2	10, 25, 50, 100
4	Treccani	2	5, 10, 20, 50
5	Zettl	2	5, 10, 20, 30
6	Ex – Rosenbrock	2,4,10,100,500,1000,10000	13, 25, 30, 50
7	Diagonal 4	2,4,10,100,500,1000	1, 3, 6, 12
8	Shalow	2,4,10,100,500,1000,10000	10, 25, 50, 70
9	Ex - Tridiagonal1	2,4,10,100,500,1000,10000	12, 17, 20,30
10	Ex- white and Holst	2,4,10,100,500,1000,10000	3, 5, 7, 10
11	Perturbed Quadratic Function	2,4,10,100,500,1000	1, 3, 5, 10
12	Ex- Denschnb	2,4,10,100,500,1000,10000	8, 13, 30, 50
13	Ex- Beale	2,4,10,100,500,1000,10000	-1, 3, 7, 10
14	Ex – Himmelblau	10,100,500,1000,10000	50, 70, 100, 125
15	Generalized Quartic	2,4,10,100,500,1000,10000	1, 2, 5, 7
16	Hager	2,4,10,100,500,1000,10000	5, 10, 15, 20
17	Ex – Penalty	2,4,10,100	80, 100. 111, 120
18	Quadrtic QF2	2,4,10,100,500,1000	5, 20, 50, 100
19	Ex - Quadratic Penalty qp2	2,4,10,100,500,1000	10, 20, 30, 50
20	Diagonal 2	2,4,10,100,500,1000	1, 5, 10, 15
21	Raydan1 Function	2,4,10,100	1, 3, 5, 7
22	Sum Squares Function	2,4,10,100,500,1000	1, 3, 7,10
23	Generalized Tridiagonal 1	2,4,10,100	7, 10, 13, 20
24	Fletcher	4,10,100,500,1000,10000	3, 5, 8, 9
25	Quadratic QF1	2,4,10,100,500,1000	3, 5, 8, 10
26	Generlized Tridiagonal 2	2,4,10,100	15, 18, 20, 22
27	Leon Function	2	2, 5, 8, 10
28	Ex-Wood	4	3, 5, 20,30
29	Quartic Function	4	5, 10, 15, 20
30	Matyas Function	2	1, 5, 10, 15
31	Colville Function	4	2, 4, 7, 10

IOP Conf. Series: Materials Science and Engineering **567** (2019) 012031 doi:10.1088/1757-899X/567/1/012031

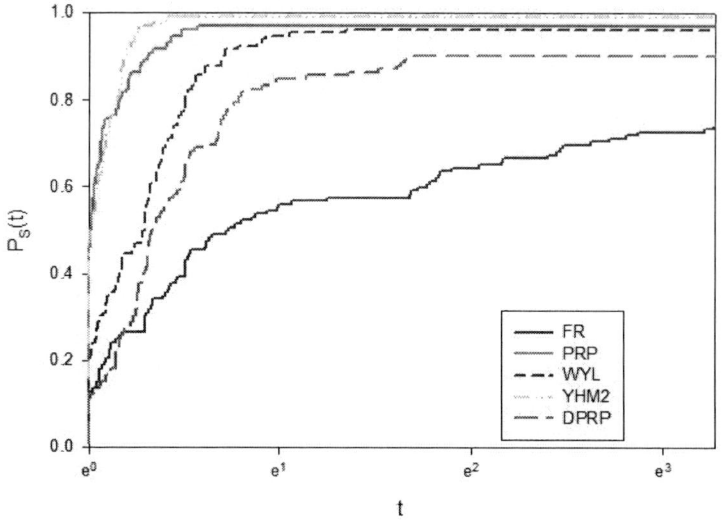

Figure 1.Performance profile relative to the number of iterations

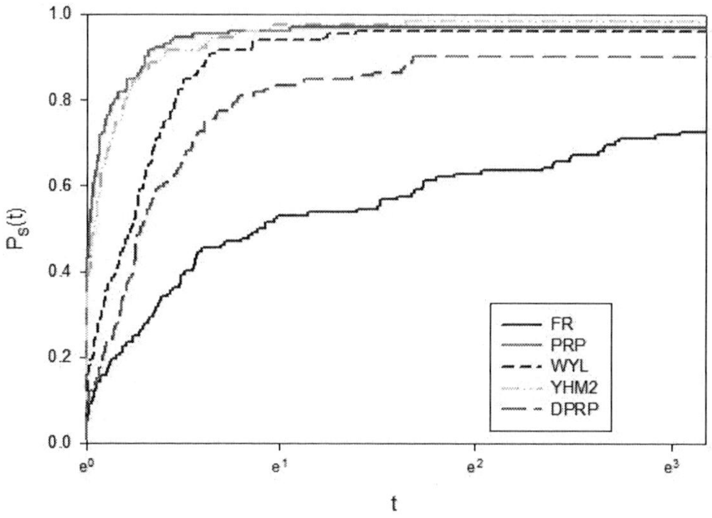

Figure 2. Performance profile relative to the CPU Time

From the above Figures, we can see the YHM2 coefficient better than other coefficients (FR, PRP, DPRP and WYL) because YHM2 solves all of the test problems and achieve 100%.

5. Conclusion

The new hybrid conjugate gradient methodshave been proven that satisfies the sufficient descent condition and global convergence prosperities under exact line search. The numerical results for this method are more efficient when as compared to other CG methods.

Acknowledgements
The authors would like to thank the Ministry of Higher Education of Malaysia for the funding of this research under the Fundamental Research Grant Scheme (Grant no.FRGS/1/2017/STG06/UniSZA /01/1) and also the Government of Sudan for the funding of this study.

Reference

[1] Hestenes MR and Stiefel E, Methods of conjugate gradients for solving linear systems. Vol. 49. 1952: NBS.
[2] Fletcher R and Reeves C M, Function minimization by conjugate gradients. The computer journal, 1964. 7(2): p. 149-154.
[3] Polak E and Ribiere G, Note sur la convergence de méthodes de directions conjuguées. Revue française d'informatique et de recherche opérationnelle. Série rouge, 1969. 3(16): p. 35-43.
[4] Polyak BT, The conjugate gradient method in extremal problems. USSR Computational Mathematics and Mathematical Physics, 1969. 9(4): p. 94-112.
[5] Fletcher R, Practical methods of optimization. Vol. 1, Unconstrained Optimization. 1980.
[6] Liu Y and Storey C, Efficient generalized conjugate gradient algorithms, part 1: theory. Journal of optimization theory and applications, 1991. 69(1): p. 129-137.
[7] Dai YH. and Yuan Y, A nonlinear conjugate gradient method with a strong global convergence property. SIAM Journal on optimization, 1999. 10(1): p. 177-182.
[8] Zhang L and Zhou W, Two descent hybrid conjugate gradient methods for optimization. Journal of Computational and Applied Mathematics, 2008. 216(1): p. 251-264.
[9] Andrei N, A hybrid conjugate gradient algorithm for unconstrained optimization as a convex combination of Hestenes-Stiefel and Dai-Yuan. Studies in Informatics and Control, 2008. 17(1): p. 57.
[10] Andrei N, Hybrid conjugate gradient algorithm for unconstrained optimization. Journal of Optimization Theory and Applications, 2009. 141(2): p. 249-264.
[11] Al-Bayati AY and Al-Kawaz RZ, A new hybrid WC-FR conjugate gradient-algorithm with modified secant condition for unconstrained optimization. Journal of Mathematical and Computational Science, 2012. 2(4): p. 937-966.
[12] Mamat M, Rivaie M, Mohd I, and Fauzi M, A new conjugate gradient coefficient for unconstrained optimization. Int. J. Contemp. Math. Sciences, 2010. 5(29): p. 1429-1437.
[13] Rivaie M, Abashar A, Mamat M, The convergence properties of a new type of conjugate gradient methods. 2016.
[14] Hamoda M, Rivaie M and Mamat M, "A New Nonlinear Conjugate Gradient Method With Exact Line Search For Unconstrained Optimization," *Journal of Humanities and Applied Science (JHAS)*, pp. 1-16, 2017.
[15] Aini N, Hajar N, Mamat M, Zull N and Rivaie M, Hybrid quasi-newton and conjugate gradient method for solving unconstrained optimization problems. 2017: Journal of engineering and applied sciences. .
[16] Powell M J D, Buhmann M D, and Iserles A, "Approximation theory and optimization: tributes to M. J. D. Powell," ed: Cambridge University Press, 1997, pp. 1-10.
[17] Wei Z, Yao S, and Liu L, The convergence properties of some new conjugate gradient methods. Applied Mathematics and Computation, 2006. 183(2): p. 1341-1350.
[18] Dai Z and Wen F, Another improved Wei–Yao–Liu nonlinear conjugate gradient method with sufficient descent property. Applied Mathematics and Computation, 2012. 218(14): p. 7421-7430.
[19] Zoutendijk G, Nonlinear programming, computational methods. Integer and nonlinear programming, 1970. 143(1): p. 37-86.
[20] Andrei N, An unconstrained optimization test functions collection. Adv. Model. Optim, 2008. 10(1): p. 147-161.
[21] Hillstrom KE, A simulation test approach to the evaluation of nonlinear optimization

algorithms. ACM Transactions on Mathematical Software (TOMS), 1977. 3(4): p. 305-315.
[22] Dolan E.D and Moré JJ, Benchmarking optimization software with performance profiles. Mathematical programming, 2002. 91(2): p. 201-213.

Optimization of impeller design for stirred tank using computational fluid dynamics

K Priyadi[1], Chin-Tu Lu[2], H Sutanto[1]

[1]Department of Mechanical Engineering, Atma Jaya Catholic University, Jakarta 12930, Indonesia
[2]Department of Mechanical Engineering, Southern Taiwan University of Science and Technology, Tainan 710, Taiwan ROC

Corresponding author: hadi.sutanto@atmajaya.ac.id

Abstract. Stirred tanks are widely used for mixing to pseudo-plastic fluids in chemical, food and many processing industries. Impeller stirred tanks play a very important role throughout process industries. Design of such systems and the effective control of the relevant processes require an understanding and prediction of the flow characteristics. One technique to develop high efficiency of this fluid production such as bioethanol is to reduce the stagnant or dead zones formation phenomena inside the stirred tank using computational fluid dynamics. The results have shown that the new impeller design as dependent variables were able to eliminate the dead zone formation related with the independent variables such as power number P_0, pumping number N_q, wall shear stress τ, mixing time τ, and average effective viscosity μ.

1. Introduction

Stirred tanks are among the most common process equipment used in chemical and other process industries since their good mixing ability and the characteristics for scale-up. Stirred tanks can be classified in different ways. Firstly, they can be sorted by the way of operation of the units, such as batch, semi batch or continuous. Secondly, by thermal operation as isotherm or adiabatic. Thirdly, by the type of applied impeller, such as turbine, jet, blade, etc. Normally in a stirred tank with the centrally positioned impellers, a rotating motion with a pair of vortices behind each blade, one above and one below the disk is generated. The flow generated in the stirred tanks are predominantly turbulent due to high impeller rotation speeds to achieve necessary process conditions [1].

The most commonly used type of stirred tanks is a fluid stirring reactor type. The problem that often arises in the fluid stirrer reactor is the emergence of stagnant or dead zone or cavern formation phenomenon as shown in Figure 1. In the non-Newtonian fluidizing process, the viscosity value of the fluids are vary depending on the shear stress occurring in the fluid. The higher the shear stress occurs, the lower the viscosity of the fluid as related with the shear-thinning in pseudo-plastic property. Hence, the fluid located close to the slats tend to have lower viscosity, while the fluid away from the blades has a higher viscosity value resulting in a stagnant zone or cavern formation.

The stirred units have many parts, therefore experimental information or data during the production phase are difficult to obtain. Hence, to build an accurate and validated model of the units can require much time which cannot be afford nowadays. Experiments in laboratory scale can support the modeling process but the scale-up of the tanks to industrial scale can be a problem because the developed flow pattern of scale-up levels can be significantly different. Using CFD or Computational Fluid Dynamics, the operation of equipment can be studied in detail without disrupting the production process. The CFD approach has the possibility to model the entire geometries in three dimensions even multiphase systems [2]. In addition, CFD simulations applicable not only in development [3], safe operation [4] and modeling the dynamic behavior of the experimental system, but also after integration of the mathematical description of considered parameters it can be describe mass and energy with related to transport process [5]. CFD can lead to a better understanding of the behavior of stirred vessel and macro mixing phenomenon in three-dimensional turbulent flow [6].

The purpose of this study is to provide a solution to the stagnant or dead zone or cavern formation phenomenon that occurs in the fermentation stirring process in the fluid stirrer reactor to produce bio-ethanol that caused by the weak propulsion of the blade. A new blade design for a pseudo-plastic type of fluid behavior has to produce a higher propulsion or pumping rate and also maintaining the stability of the power required to rotate the blades, using CFD software with ANSYS Fluent.

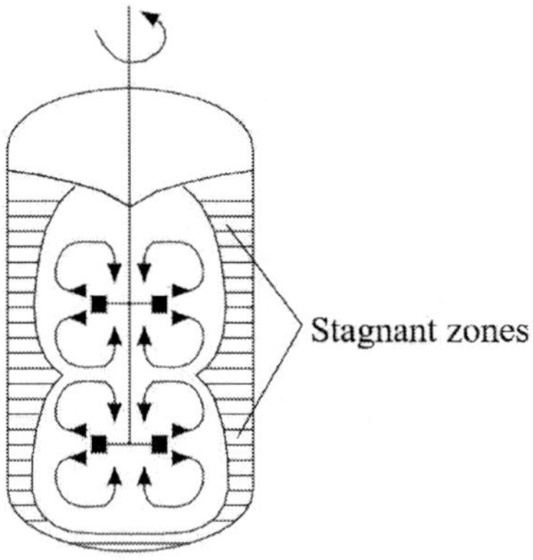

Figure 1. Stagnant or dead zones [7]

2. Stirred Tank Reactor

Stirred tank as biological reactor or bioreactor like a fermenter, which is producing a product, using the work of micro-organisms or enzymes studied by the industrial microbiology domain. Bioreactors always set the condition of their tanks in a controlled environment therefore the micro-organisms can live and work. Hence, there is a classification of bioreactors based on the needs of the desired results. The classification of the bioreactor according to the moisture content contained in the tank can be divided into submerged-culture fermentation and solid-state fermentation. For stirred tank reactor it is categorized into submerged-culture fermentation.

The power number *Po* to turn the wheel of fluid is:

$$P_0 = \frac{P}{(\rho x N^3 x D^5)} \text{ where } P \propto N^3 x D^5$$

and N is the impeller angular velocity and D is the impeller diameter. This equation becomes the basis for the power or newton number (*Po*) equation, which goes into the list of dimensionless parameter groups [8].

The pumping number *Nq* is obtained based on the fluid flowing from the blade to the outer blade of the region at any time, which can be represented by the amount of flow flowing through the surface area. The mass flow out through the surface area can be calculated by the following equation,

$$Q = \frac{\dot{m}_{axial} + \dot{m}_{radial}}{\rho}$$

where, $\dot{m}$ is the mass flow rate. Hence the pumping number is then obtained by using the following equation,

$$Nq = \frac{Q}{NxD^3}$$

The wall shear stress τ is obtained related with viscosity that can be modelled with fluids consisting of several layers following both the plates on the top and bottom without slippage between them. The expression of shear stress for Newtonian fluids with the variable of shear rate is [8]

$$\tau = \mu x \dot{\gamma}$$

In the fluid stirrer, the mean specific energy dissipation rate can be calculated by the equation, where all energy occurring in the fluid, as a result of the impulse movement in the vessel, changed or dissipated into heat energy,

$$\overline{\varepsilon_T} = \frac{P}{\rho xV} = \frac{P_0 xN^3 xD^5}{V}$$

The predictable mixing time or the time required of a mixture to reach the level of homogenization, can be calculated with the following equation,

$$\tau_n = \left(\frac{\vartheta}{\overline{\varepsilon_T}}\right)^{1/2}$$

where, ν is the kinematic viscosity of the fluid and V is the total volume of liquid inside the stirred tank. Mixing time has been used as a key parameter for assessing the performance of a mixing system. For Newtonian fluid, this phenomenon can occur due to the vertical stress drop σ_v occurring in the fluid or the horizontal fluid stress increase σ_h in the stagnant zones area.

The flow density as the flow decreases, causing the increase of the horizontal stress at a point from the flow to the stagnant zone. Due to the velocity difference at that point the formation of wall shear stress and the vertical stress value in the stagnant area increases that result to a periodical slip. On the same principle for non-Newtonian fluids, that there is a velocity difference between flowing fluids and stagnant zones or cavern formation. In the fluid stirrer reactor, the velocity difference due to the increased concentration of shear stress at the center of the blade region will result for the viscosity difference between the fluid that flows and at the stagnant zones.

3. Research Methodology

The stirred tank geometry represents a typical "standard" configuration where the tank diameter T is 0.8 m and other dimensions may be related to T as follows,

Fluid height $H_L = T$
Impeller diameter D = 0.4 T
Impeller clearance C = $T/3$
Blade thickness = 0.001 m
Tank bottom type = ASME 6% bottom
Shaft diameter = 0.03 m
Angular velocity = 60 rpm

This study uses data validation for power number and pumping number from experimental results by Lane and Koh [9, 10] as Po = 4.67 and Nq = 0.73. Data for effective viscosity μ = 0.3 kg/ms and specific gravity 1.21 from reference [8]. For shear thinning non Newtonian fluids, where the viscosity decrease with increasing shear rate, the data obtained from the study of Fraiha et.al [8].

The standards used to design the geometry of the new blade design is that it should increase the pumping number and shear rate generated by the blade since they allow for a solution to reduce the stagnant or dead zone. The new design also has additional effect to stabilize power consumption of the stirred tank. Numerical simulation set up based on the new design of impeller was specified for turbulence using ***k-ε*** model with standard wall function and the simulation run as a transient problem with standard initial condition of absolute reference frame zero velocity at all grid nodes [10][11].

4. Results and Discussions

The study on development of stirred tank reactor did an analysis of causes and effects of the phenomenon from the mechanical approach of blade geometry using numerical analysis. Results from numerical simulation have shown the percentage of error for power number *Po* and pumping number *Nq* are small, as seen in Table 1.

Table 1. Validation data for *Po* and *Nq*

VARIABLES	SYMBOL	EXPERIMENTAL RESULT	NUMERICAL RESULT	ERROR [%]
Power Number	*Po*	4.67	4.35	7.40
Pumping Number	*Nq*	0.73	0.70	3.94

After analyzing a correlation between the angle of attack to the performance of the blade, the 30^0 angle of blade is selected since the angle of blade has the ratio of *Nq* to *Po* to the second highest while maintaining the stable wall shear stress. The design of the new blade has a basic principle that the larger surface area of blade to drive fluids, the greater the resulting *Nq*, but the *Po* value increases. Hence the number of blades was reduced to 3 pieces or AM 30 (Axial-flow Mix-solidity) with addition to two wings or flaps on the top and bottom of blades (Figure 2). The shape of the blade resembles hydrofoil in order to lower *Po* and increase the number of surfaces that push the fluid to increase *Nq*. Figure 3 shows the relationship between impeller type and the ratio of *Nq/Po* and average effective viscosity.

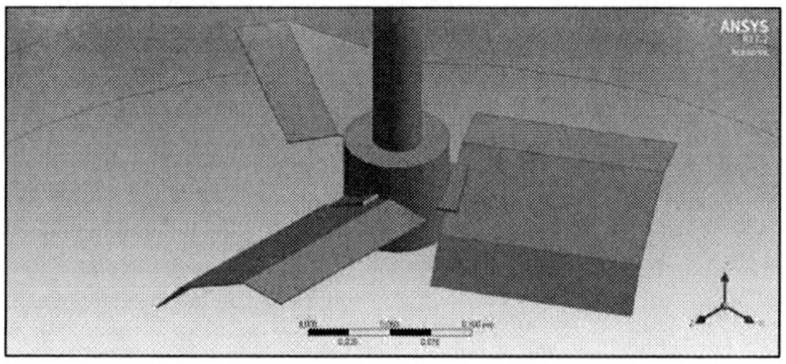

Figure 2. ISO view of new blade design 3 AM 30

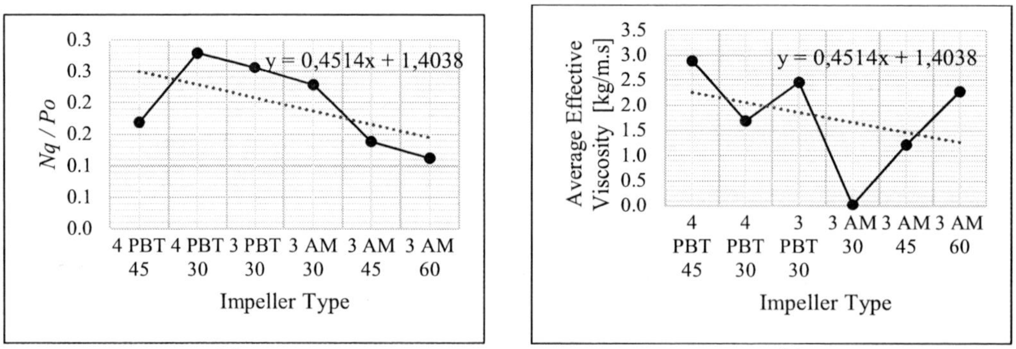

(a) (b)

Figure 3. Graph of impeller type vs **(a)** *Nq/Po* and **(b)** average effective viscosity

IOP Conf. Series: Materials Science and Engineering **567** (2019) 012032 doi:10.1088/1757-899X/567/1/012032

The new blade design 3 AM 30 has the ratio of *Nq/Po* is 0,229 but it has the lowest effective viscosity which is useful in reducing the stagnant or dead zone phenomenon in stirred tank. This phenomenon is possible to occur only in a fluid stirrer reactor for non-Newtonian fluids aggregation processes. The fluid viscosity may change depend on the amount of shear stress that occurs in the fluids, as seen in Figure 4.

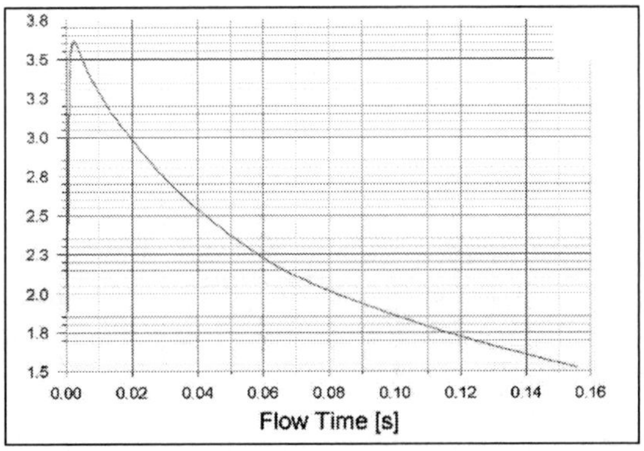

Figure 4. Average effective viscosity vs flow time

The stagnant or dead zone phenomenon has a negative impact on the fermentation process in the stirred tank reactor as it decreases the homogeneity level of the fermentation mixture.

A comparison between two contours for the tendency to form a cavern formation is shown in figure 5, where the blades 3 AM 30 has areas of less effective viscosity than stirring results using blades 4 PBT 45 type. The average effective viscosity of blade design 3 AM 30 is 0.026 kg/ms which is smaller than the average effective viscosity of blade 4 PBT 45 as 2.889 kg/ms. The maximum effective viscosity of blade design 3 AM 30 is 43.173 which is smaller than the maximum effective viscosity of blade 4 PBT 45 as 75.838 kg/ms.

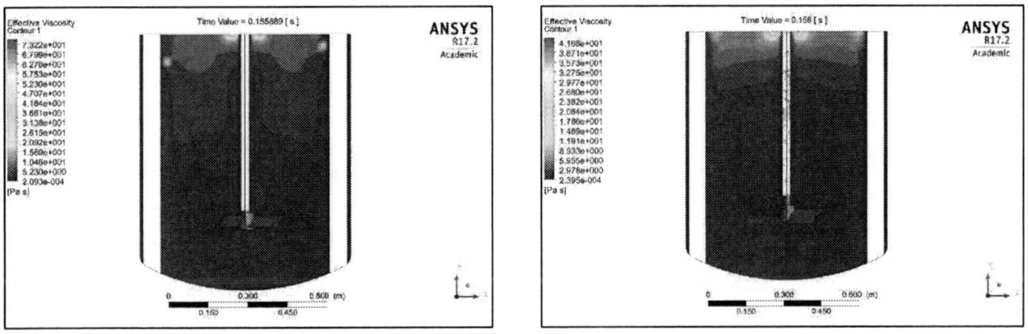

(a) **(b)**

Figure 5. Contour of effective viscosity in cross section of stirred tank for impeller type **(a)** 4 PBT 45 and **(b)** 3 AM 30

5. Conclusions

CFD analysis on the optimization of high-efficiency stirred tank reactor using mechanical approach from blades can minimize or eliminate stagnant or dead one or cavern formation using a new design blades. Independent variables such as power number, pumping number and wall shear stress are also compatible variables to develop high efficiency stirred tank. Some dependent variables such as impeller design and the angle of attack have a correlation with high efficiency stirred tank reactor. The new design of the impeller to produce high efficiency stirred tank can be developed to produce high efficiency stirred tank can be developed by reducing the number of blades, increasing the surface area of the impeller blade and make it similar to hydrofoil.

The new impeller design as 3 AM 30 is able to reduce the power consumption but maintain the pumping number at the similar value as 4 PBT 45. Optimization of impeller design to 3 AM 30 can prevent the phenomenon of stagnant or dead zone in stirred tank reactor with "standard" configuration of T equal to 0.8 m.

References

[1] Joshi JB, Nere NK, Rane CV, Murthy BN, Mathpati CS, Patwardhan AW and Ranade VL 2011 CFD simulation of stirred tanks: Comparison of turbulence models Part I : Radial flow impellers *Can. J. Chem Eng* **89 (1)** pp 23-82

[2] Tamburini A, Cipollina A, Micale G, Brucato A and Ciofalo M 2013 CFD prediction of solid particle distribution in baffled stirred vessels under partial to complete suspension conditions *Chemical Eng Transaction* **32** pp 1447-1452

[3] Wei H 2010 Computer-aided design and scale-up of crystallization processes: integrating approaches and case studies *Chemical Eng Research and Design* **88 (10)** pp 1377-1380

[4] Rudniak L, Machniewski PM, Milewska A and Molga E 2004 CFD modelling of stirred tank chemical reactors_ homogenuous and heterogeneous reaction systems *Chemical Engineering Science* **59 (22-23)** pp 5233-5239

[5] Sommerfield M and Decker S 2004 State of the art and future trends in CFD simulation of stirred vessels hydrodynamics *Chem. Eng. Technology* **27**, pp 215-224

[6] Ranade VV, Tayalia Y and Krishnan H 2015 CFD prediction of flow near impeller blades in baffled stirred vessels: Assessment of computational shapshot approach *Chemical Engineering Communications* **189:7** pp 895-922

[7] Edward PL, Atiemo-Obeng VA and Kresta SM 2004 *Handbook of Industrial Mixing-Science and Practice* John Wiley and Sons, NJ

[8] Fraiha M, Biagi JD and Ferraz CO 2011 Rheological behavior of corn and soy mix as feed ingredients, *Ciencia e Technologia de Alimentos* **31 (1)**, Campinas pp 129-134

[9] Lane GL and Koh PLT 1997 CFD simulation of Rushton turbine in a baffled tank *Int. Conf. on CFD in Mineral & Metal Processing and Power Generations, CSIRO Division of Minerals*, Clayton South, pp 377-386

[10] Lane GL, Schwarz MP and Evans GM 2015 Comparison of CFD methods for modeling of stirred tanks in *Proceedings of 10th European Conference on Mixing*, Elsevier, pp 273-280

[11] ANSYS Fluent 2013 *Theory Guide 15* Ansys, Inc., Canonsburg, PA

[12] Gerasimov A 2006 *Modeling Turbulent Flows with FLUENT* Ansys, Inc., PA.

IOP Conf. Series: Materials Science and Engineering **567** (2019) 012033 doi:10.1088/1757-899X/567/1/012033

Container terminal landside operation analysis and discrete event simulation in container terminal in port: a case study of Terminal 3 Ocean-going PT Pelabuhan Tanjung Priok

A R Nasution[1], Arviansyah[1]

[1]Department of Management, Faculty of Economics and Business, Universitas Indonesia, INDONESIA.

Corresponding Author: aditya.ridwan@ui.ac.id, arviansyah@ui.ac.id

Abstract. This study discusses the landside operation and presents the application of Truck Appointment System (TAS) and lane segmentation strategy through discrete event simulation in Terminal 3 Ocean-going PT Pelabuhan Tanjung Priok. This study aims to analyze the operation planning system in Terminal 3 Ocean-going and identify problems that exist on terminal 3. This study also seeks to examine the landside operation process and offer alternative scenarios to solve the congestion problem in the terminal. Alternative scenarios provided are a combination of Truck Appointment System (TAS) and lane segmentation strategy. Truck Appointment Strategy (TAS) with 100% utilization is proved to reduce average total time in the system by 76% and reduce average total time in peak days by 88%, from 151.5 minutes to 38.4 minutes per truck. Truck Appointment System (TAS) is also proved to reduce queue length by 97%, from 162 trucks to 5 trucks.

1. Introduction

Shipping lines and seaport holds an essential part in global trade. According to UNCTAD (2017) [1], more than 80% of global trade volume and more than 70% of global trade value are transported through maritime transportation. In 2016, global trade through maritime transport increases by 2.6%, improved from 1.8% growth in 2015. Total volumes reached 10.3 billion tons in 2016, including the increase of more than 260 million tons of cargo. The total volume consists of 3.1 billion tons of tanker trade (oil and gas), and 7.2 billion tons of dry cargo. Containerized trade contributes 23.8% of dry cargo shipments, convert to 1.72 billion tons of containerized shipments. According to UNCTAD (2017) [1] in 2016, containerized trade has 3.1% growth with a total shipment of 140 million 20-foot equivalent unit (TEUs). Based on the throughput volume in the container port, Asia dominated global trade with 64% of global trade volume. 64% of global trade volume is distributed, mainly from countries in East Asian and South East Asia. The rest of the global trade are divided into Europe with 16%, North America with 8%, Developing America with 6%, Africa with 4%, and Oceania with 2%.

Seaports are strategically crucial for industrial activity, trade, global production, and economic growth. Indonesia's most important ports rank much lower than the neighboring countries such as Singapore and Malaysia [2][3]. Port of Singapore has a throughput of 30.9 million TEUs, becomes the second biggest port in the world based on throughput. Malaysia's biggest port and its second biggest port, which is Port Kelang and Port Tanjung Pelepas, have a throughput of 13.1 million TEUs and 8 million TEUs and are ranked 11th and 19th. Indonesia biggest port, Port Tanjung Priok has a throughput of only 5.5 million TEUs, which make it the 26th biggest port based on throughput [1].

This study aims to reveal the problems at Terminal 3 Ocean-going and examine the current situation for viable improvement based on the low throughput compared to other ports. Based on the problems analysis, this study demonstrates the implementation of a truck appointment system (TAS) as a recommendation to improve the performance of the landside operation in Terminal 3 Ocean-going. For this purpose, this study determines the activities and process required in landside operation in Terminal 3 Ocean-going. Those activities and process are translated into a discrete-event simulation using Arena simulation software. The simulation is used to assess the current situation at Terminal 3 Ocean-going landside operation and to simulate the implementation of alternative scenarios.

2. Literature review

We review the literature based on the interlinked subjects; these are terminal operations in each area, strategies to reduce congestion at the terminal gate and discrete event simulations. Operation areas of container terminal consist of four areas, namely berth area, transport area, yard area, and hinterland area. Operation process on hinterland area is the transport process outside of container terminal, consist of transporting container from the hinterland to terminal and from the terminal to the hinterland. Transportation between hinterland and terminal were done with external transportation such as trucks or trains [4][5].

2.1. Operation areas: berth, transport, and yard

The berth area is a place to vessel berthing and container handling process between vessel and terminal. Operation of the container handling uses quay cranes. Quay cranes are used to move import container from vessel to internal truck and move the export container from internal truck to vessel according to the operation plan. Operation on the transport area consists of two operations, which is transport between berth area and yard area and transport between yard area and gate area and to the hinterland. The first one uses internal trucks to move between the berth and yard area. Transportation between the yard area and gate area uses external trucks to transport containers between these areas. Yard areas are used as temporary storage for containers before being sent to their destinations. A yard area is a buffer area to connect between different or same mode of transportation, for example between water and land transportation. A yard area is divided between several blocks according to its use, e.g., export or import containers. These blocks are marked with lines to determine places where containers can be positioned within those blocks. There are some spaces between blocks dedicated to truck lanes and for handling operation between trucks and yard cranes [6][7].

2.2. Strategy to solve congestion at the port container terminal

The port of California implements 24 hours of operation at its container terminal to reduce congestion during peak hours [8]. The purpose of this strategy is to disperse volumes during peak hours to off-peak hours such as early morning or midnight when the terminal is still open for operation. Another strategy applied at other container terminal is to implement a time-varying toll charge [9]. This strategy applies a toll charge for every truck entering the terminal. This strategy is utilized to force trucks switching their schedules to off-peak hours with low toll charge or even free of charge [10]. The next strategy is to implement a truck appointment system (TAS) [11][12]. With this strategy, the terminal implements a booking system for every truck to use. Every truck entering the terminal must use the booking system and booking available time slots in the system. Every time slot has a one-hour interval. Trucks must arrive at the terminal within those interval time. There will be an additional charge for trucks passing the time interval; they cannot enter the terminal when longer than 15 minutes behind schedule and must book another time slot. There will be limited capacity for each time slots based on the terminal's volume during those hours. This strategy is used to disperse truck volumes to available time slots which mainly during off-peak hours.

TAS implementation is usually is combined with another strategy such as lanes segmentation. Lanes segmentation divides existing terminal gate based on types of containers such as general containers or reefer containers. This strategy can also be used to divide terminal gates into the gate for trucks using TAS and the gate for trucks not using TAS. This strategy offers another incentive for trucks to use TAS [8].

3. Research methodology

This study utilized exploratory study to identify and analyze problems occurs in the Terminal 3 Ocean-going and descriptive study to assess the benefit of implementing a truck appointment system and lanes segmentation to Terminal 3 Ocean-going [13]. Simulations in this study adopt the research model from Li, Wu, & Goh (2015) [14] to examine problems occurs at container terminals based on operation areas and the research model from Gracia et al. (2017) [8] to examine truck appointment systems and lanes segmentation implementation in Terminal 3 Ocean-going.

This study is conducted from January to May 2018 starting from primary and secondary data collection followed by the data analysis and discussions. This study utilized in-depth interviews with control and planning division of PT Pelabuhan Tanjung Priok and representatives of Terminal 3 operator division. Data related to the organization and business processes or activities are collected as well, such as truck arrival and distribution. The duration of each activity in the landside operation is collected using observation method in the terminal area.

The obtained data related to operation problems occur at Terminal 3 Ocean-going are analyzed qualitatively. Moreover, we examine the data to construct discrete event simulations using Arena 14 to simulate the implementation of a truck appointment system and lanes segmentation at Terminal 3 Ocean-going [15] [16]. The simulation model focuses on operation process simulation of container trucks in the terminal regardless of their activities in the terminal (carrying or acquiring containers).

4. Results

Based on in-depth interviews and observations at the terminal area, we identify and analyze several prevailing problems at the Terminal 3 Ocean-going according to every operation planning area in the terminal. These problems arise due to many causes, such as undisciplined operators at the terminal, lack of coordination between area and between terminal operator and its stakeholders, and problems from its stakeholders' requirements.

Table 1 shows the problems identified in this study. Arguably, we determine two critical problems requiring solutions drawn from interviews and observation analysis. The first problem is the limited yard space. When the yard area cannot support the berth area, then the berth area has limited productivity percentage depending on how much yard area can support. When the terminal operator predicts that yard area will be over capacity when the next ship arrives, some of the containers need to be moved to another area called Lini 2. This movement requires time for coordination between division and equipment mainly yard cranes and trucks which could be used for something more productive.

The second problem is the congestion in the terminal when there is a massive volume surge. This problem usually occurs every Friday to Sunday when the volume decreases again. This sudden increase is caused by the weekly cycle that occurs in this terminal. Friday and Saturday are the time when most of the ship arrives at the terminal, so most container owners will transport their containers to the terminal when the ship arrives on Friday or Saturday. The congestion makes landside operations in the terminal a lot longer than usual. This congestion also disrupts access to terminal 2 since the queue line are stretched to terminal 2 gates.

Furthermore, this study focuses on analyzing the implementation of a truck appointment system and lanes segmentation to reduce and solve the congestion problem occurs in Terminal 3 Ocean-going. By implementing these strategies, the terminal operator can disperse truck volumes on peak hours to off-peak hours and from peak days (Friday-Sunday) to off-peak days (Monday-Wednesday).

The landside operation process for external container trucks at Terminal 3 Ocean-going is divided into three parts, namely the customs gate, terminal gate, and yard area. Before trucks can enter the terminal, they need to prepare the required documents based on their containers, shipping export containers or take import containers. Those documents will be checked at the customs gate, and the terminal gate, and both gates will update the container status on their system. After the truck has passed the arrival gate, trucks will start their container handling processes at the yard area. Finally, trucks will need to pass through the terminal and customs gates again to document the transactions and update the status on the terminal system once again.

Table 1. Identified Problems in Terminal 3 Ocean-going and Alternative Solutions

Problem	Problem Description	Alternative Solution
Terminal Area		
Terminal Design are not compatible	**Initially, Terminal 3 designed as a multifunction terminal, so the terminal did not have enough yard area to support its berth area**	**Add more yard area in Lini 2**
Congestion in Terminal Gate	**Congestion happened because of truck volume increase exponentially**	**Implementing a truck appointment system and lanes segmentation**
Limited crane operator	Crane operator did not provide enough to support when there is a high volume in the terminal. Terminal need time to add more operator to help its operation	Implement detailed operation schedule for every operator according to volume trends
Operator in the terminal gate is not discipline	Gate operator is not disciplined in changing the container status in the terminal's system	Implement a system to change container status in the terminal automatically
The equipment is broken	There are two quay cranes and one yard-crane broken	Fix or replace the equipment
Berth Allocation Area		
Ship delay enter to the terminal	Delay caused by unfinished document or insufficient export container in the terminal	Remind the shipping lines to prepare all documents required
Ship delay when departure from terminal	Delay caused by unfinished container operation in the ship	Monitoring closely in the operation process
Unscheduled ship arrival	Majority of the ship will be berthing in Terminal 3 Ocean-going unscheduled	Make a partnership with shipping lines to make berthing scheduled date
Yard Planner Area		
Uneven yard area	There are uneven floors in some parts of the yard	Fix uneven floors when occurred
Ship Planner Area		
Planning process requires much time and depends on the planner	Long planning time cause the whole planning to become long too and the time and quality of the plan heavily depends on the planner	Implement a system to assist or to make the plan needed for terminal operation
The difference in EDI and CVIA file	EDI and CVIA document received did not have the same number of containers	Remind the shipping lines to check the documents before sending it to the terminal operator

Next, the simulation model has several assumptions to produce a precise output. These assumptions are as follows:

a. The container truck arrival data examined in this study consist of truck arrival data at Terminal 3 Ocean-going from 23-29 April 2018, i.e., truck arrival data for one week.
b. The process duration for each activity is collected through observation at Terminal 3 Ocean-going

c. The simulation period on every scenario is run for a week or 168 hours based on the arrival data being used.
d. Each scenario is replicated ten times.
e. The α value being used for the validation test is 5%.
f. All lanes are assumed to be empty at the beginning of the week.
g. Terminal 3 Ocean-going operates for 24 hours and is divided into three shifts.
h. All equipment is only used for the landside operation process of the truck at the terminal.
i. The operation activity at the terminal is stopped at 04:00-05:00, 07:00-08:00, 12:00-13:00, and 18:00-19:00 each day.

Table 2. The Output of the Validation Test on the Base Case Scenario

	System Output (Truck)
Mean	5,659
Standard Deviation	76.53
$T_{9;0,25}$	2.262
Confidence Interval	95%
Upper Tail	5,825
Lower Tail	5,492
Real System Output	5,666
Validity	**Valid**

After running the simulation, the output is then validated using a test with a specific confidence interval to compare the simulation model output and the real system output. The validation test uses a 95% confidence level and an α value of 5%. The test output is shown in Table 2. The parameter used in this test is the truck output from the system for a week. A simulation model is constructed using this output which replicates a real system with an output of 5,666 within the model simulation confidence interval of 5,492-5,825 trucks. Furthermore, the simulation model can be used to simulate the implementation of strategies in the system.

In this part, we illustrate the TAS and segmentation implementation through five alternative scenarios. Scenario 1 will implement a 15% utilization of TAS, and another 85% truck still use the existing system. In Scenario 2, it applies 15% utilization of TAS and lanes segmentation in the form of adding one arrival terminal gate exclusive for truck using TAS booking system. For Scenario 3, it increases TAS utilization to 40%. Scenario 4 is similar to Scenario 2, besides 40% utilization of TAS; it also adds one more arrival terminal gate. In Scenario 5, the utilization of TAS become 100% utilization.

Table 3. The Output of Paired-T Test between the Base Case and Alternative Scenarios (Hours)

	Base Case	**Scenario 1**	**Scenario 2 Non TAS**	**Scenario 2 TAS**	**Scenario 3**	**Scenario 4 Non TAS**	**Scenario 4 TAS**	**Scenario 5**
Mean	10.98	7.54	8.05	7.3	4.19	4.82	4.16	1.85
Std Dev	1.58	0.77	1.24	1.37	1.66	0.51	1.39	0.13
Upper Tail		4.28	4.19	5.04	7.82	7.15	7.81	10.19
Lower Tail		2.59	1.66	2.32	4.88	5.15	5.82	8.05
		Statistically Significant						

IOP Conf. Series: Materials Science and Engineering **567** (2019) 012033 doi:10.1088/1757-899X/567/1/012033

Each alternative scenario is tested with Paired-T test to examine its differences to the base case scenario. Table 3 shows the output of Paired-T test on alternative scenarios. Every alternative scenario is significantly better than the base case scenario. In Scenario 2 and 4, the output differentiates trucks using TAS and trucks not using TAS. The parameter used on this test is the average total time needed in the system.

5. Discussion

By running simulation, we can conclude that TAS and lanes segmentation can reduce congestion at Terminal 3 Ocean-going significantly based on the Paired-T test. The parameters used in this study are the average total time needed in the system and the average truck queue length in the system. The average total time in the base case scenario can be reduced in each alternative scenario; Scenario 5 generates the most significant reduction. Scenario 1 can reduce the average total time by 34% from the base case scenario. Scenario 2 is proved to provide incentives to trucks using TAS with a decrease in average total time of 40% from base case scenario compared to trucks not using TAS with a decrease of 29% from the base case scenario.

Furthermore, Scenario 3 can reduce the average total time by 58% from the base case scenario. Scenario 4 is also proved to provide incentives to trucks using TAS with an average reduction in total time of 65% compared to trucks not using TAS with a decrease of 60%. Scenario 5 shows that if the TAS is 100% used by all trucks entering the container terminal, the average total time reduction is 76%.

The reduction of total time on congestion in the terminal will be better if we see from the average total time from Friday to Sunday because it focuses on peak days at the terminal. Scenario 1 can reduce the average total time by 38%, and Scenario 2 reduces the total time by 43% for trucks using TAS and 32% for trucks not using TAS. Furthermore, Scenario 3 can reduce the average total time in all three days by 65% while Scenario 4 can reduce the average total time by 72% for trucks using TAS and 67% for trucks not using TAS. Finally, Scenario 5 can reduce the average total time for three days by 88% from the base case scenario.

The average reduction in total time is due to the transfer of truck volume during other days such as Monday to Thursday when there is no truck buildup in the terminal. Figure 1 depicts a comparison between the base case scenario and the alternative scenarios; the average total time for Friday, Saturday and Sunday have decreased in each alternative scenario while the average total time for Monday, Tuesday, Wednesday, and Thursday has experienced an increase in some scenarios, especially in Scenario 5. This result occurs as the result of the transfer of truck volume in the base case scenario from Friday to Sunday; in the alternative scenario, it is switched to Monday to Thursday using TAS with an amount based on each scenario each is 15%, 40%, and 100% use of TAS.

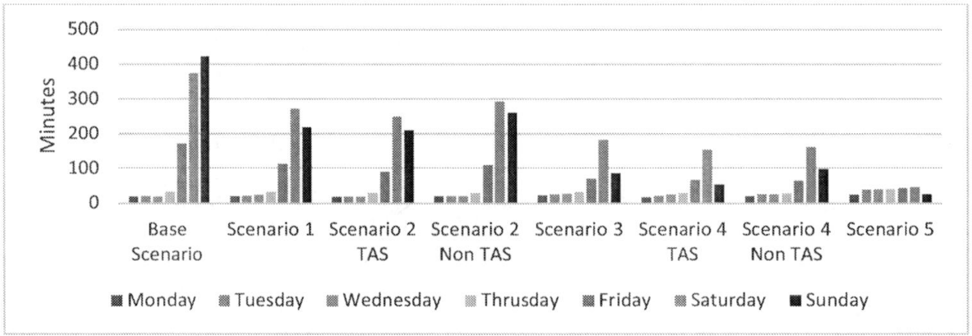

Figure 1. Average Total Time for every Scenario

IOP Conf. Series: Materials Science and Engineering **567** (2019) 012033 doi:10.1088/1757-899X/567/1/012033

In the figure above, the use of 100% TAS can provide the most significant time reduction compared to other scenarios with a decrease in the total time of the week by 76% and a decrease in the average total time for Friday to Sunday of 88% from base case scenario.

Lanes segmentation strategy in the form of adding lanes specifically for trucks using TAS is proven to provide additional incentives for trucks using TAS. In Scenario 2, trucks using TAS have an average total time for Friday to Sunday of 184.4 minutes, decrease by 17% from trucks not using TAS where the average total time for Friday to Sunday reaches 221.2 minutes. In Scenario 4, the differences between trucks using TAS and those not using TAS decreased to 15%. Trucks using TAS have an average total time for Friday to Sunday of 91.8 minutes compared to trucks that do not use TAS for 108 minutes.

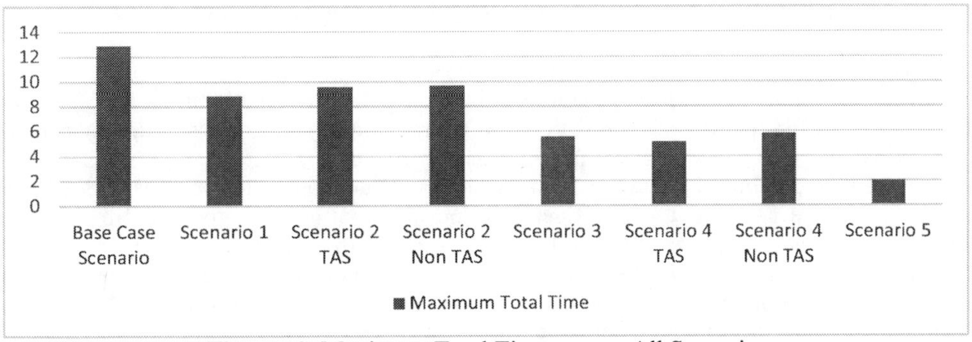

Figures 2. Maximum Total Time among All Scenarios

In Figures 2, the maximum total time in the landside operation process at Terminal 3 Ocean-going for base case scenario and alternative scenarios. In the base case scenario, the highest total time is 12.92 hours for one truck. This amount can be reduced by alternative scenarios that have been made. Scenario 1 can reduce the highest total time by 31% from the base case scenario. In Scenario 2, the maximum total time can be reduced by 26% for trucks using TAS and 25% for trucks not using TAS.

Furthermore, Scenario 3 can reduce the maximum total time by 57% from the base case scenario. While Scenario 4 can reduce the maximum total time by 60% for trucks using TAS and 55% for trucks that do not use TAS. Scenario 5 can reduce the maximum total time by 84%, from 12.92 hours in the base case scenario to 2.02 hours.

Lanes segmentation strategy does not give a significant difference between trucks using TAS and those not using TAS in Scenario 2. Trucks that use TAS have a maximum total time faster by 1% from trucks that do not use TAS. Different results are shown in Scenario 4. The maximum time difference between trucks using TAS and trucks that do not use TAS is 11%. Trucks not using TAS have a maximum time of 5.81 hours compared to trucks using TAS which have a maximum time of 5.15 hours.

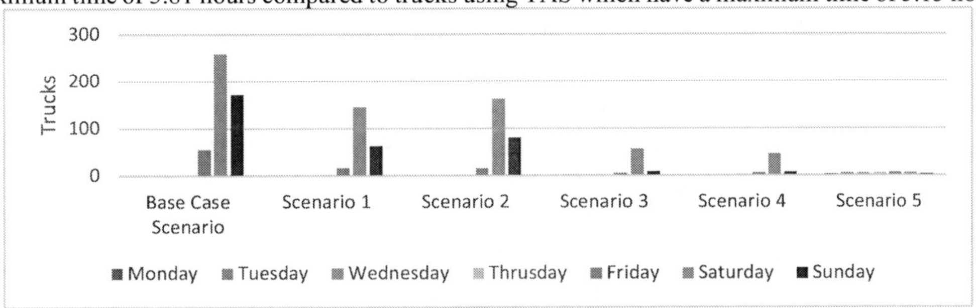

Figures 3. Average Queue Length for Each Scenario

The average queue length shows a very significant decrease, comparing the base case scenario and alternative scenarios. In Figures 3, the longest average queue is from Friday to Sunday where the volume of trucks coming is much larger than other days. The decrease in queue length occurs from Friday to Sunday. Each alternative scenario can reduce the queue length on these three days significantly. Scenario 1 can reduce the average queue length for the three days by 54%, from 162 trucks to 75 trucks and Scenario 2 can reduce the average queue length by 47%. Furthermore, Scenario 3 and 4 can reduce the average queue length for the three days by 86% and 88%. Scenario 5 can reduce the average queue length for the three days by 97% compared to the base case scenario (the average queue length of 162 trucks) to only five trucks on average queue length at Terminal 3 Ocean-going.

Scenario 5 shows that the implementation of TAS can reduce and almost eliminate all the queues at Terminal 3 Ocean-going, from 162 trucks in the base case scenario to 5 trucks when using TAS with 100% utilization when all container trucks are entering the Ocean-going terminal.

6. Conclusion

This study reveals specific problems at Terminal 3 Ocean-going in every planning area. The biggest challenge faced by the operator is a limited yard area for temporary container storage and congestion at the terminal entrance gate due to volume increase every week.

The landside operation for container trucks includes (1) document transactions at both customs and terminal arrival gate, (2) the container handling process, and (3) document transactions at customs and terminal departure gate. Problems arise when there is a surge volume of the truck during peak days leading to congestion at Terminal 3 Ocean-going entrance, i.e., the customs gate. During peak days, each truck needs 11 hours to complete its activity at Terminal 3 Ocean-going with around 10 hours spent in the queue line.

This study proffers the truck appointment system (TAS) as a viable solution for Terminal 3 Ocean-going and lanes segmentation. Base on the results of Scenario 1 through 5, we conclude that a truck appointment system and lanes segmentation can reduce congestion at the terminal significantly even if the utilization is just 15% in Scenario 1. In Scenario 2 and 4, we show scenarios to use an additional terminal gate as incentives for trucks using TAS based on the difference in improvement between trucks using TAS and regular trucks. Furthermore, this study demonstrates the full utilization of TAS, simulated in Scenario 5, reduces the total time in the system significantly with 83.2% decrease from the base case scenario.

Acknowledgements

The presentation of this paper is funded by Universitas Indonesia under International Publication of Students Thesis Grant (Hibah PITTA) number: 1732/UN2.R3.1/HKP.05.00/2018

References

[1] UNCTAD. (2017). *Review of Maritime Transport 2017*

[2] PT Pelabuhan Tanjung Priok. (2016). Laporan Tahunan 2016 - Strengthening Operation, Delivering Value, 195–202. https://doi.org/10.1016/B978-0-12-378591-6.00010-2

[3] De Langen, P. W. (2015). Port Management 3. *United Nations Conference on Trade and Development*, *15*(9), 46. https://doi.org/10.1057/9781137475770

[4] Li, W., Wu, Y., & Goh, M. (2015a). Maritime Terminal Operational Problems. In *Planning and Scheduling for Maritime Container Yards* (pp. 5–29). https://doi.org/10.1007/978-3-319-17025-1

[5] Alderton, P. M. (Patrick M. (2008). *Port Management and Operations. Igarss 2014*. https://doi.org/10.1017/CBO9781107415324.004

[6] Roa, I., Peña, Y., Amante, B., & Goretti, M. (2013). Ports: Definition and study of types, sizes, and business models. *Journal of Industrial Engineering and Management*, *6*(4), 1055–1064. https://doi.org/10.3926/jiem.770

[7] Arvis, J., Mustra, M., Panzer, J., Ojala, L., & Naula, T. (2016). Connecting to Compete: Trade Logistics in the Global Economy. *World Bank*, 1–76. Retrieved from http://lpi.worldbank.org/international/global

[8] Gracia, M. D., González-Ramírez, R. G., & Mar-Ortiz, J. (2017). The impact of lanes segmentation and booking levels on a container terminal gate congestion. *Flexible Services and Manufacturing Journal*, *29*(3–4), 403–432. https://doi.org/10.1007/s10696-016-9256-4
[9] Chen, G., & Jiang, L. (2016). Managing customer arrivals with time windows: a case of truck arrivals at a congested container terminal. *Annals of Operations Research*, *244*(2), 349–365. https://doi.org/10.1007/s10479-016-2150-3
[10] Chen, X., Zhou, X., & List, G. F. (2011). Using time-varying tolls to optimize truck arrivals at ports. *Transportation Research Part E: Logistics and Transportation Review*, *47*(6), 965–982. https://doi.org/10.1016/j.tre.2011.04.001
[11] Zehendner, E., & Feillet, D. (2014). Benefits of a truck appointment system on the service quality of inland transport modes at a multimodal container terminal. *European Journal of Operational Research*, *235*(2), 461–469. https://doi.org/10.1016/j.ejor.2013.07.005
[12] Saroni, U. (2015). Managing optimum workload through terminal appointment system (TAS) : Case of Jakarta International Container Terminal.
[13] Cooper, D. R., & Schindler, P. S. (2014). *Business Research Methods. Business Research Methods.*
[14] Li, W., Wu, Y., & Goh, M. (2015b). Planning and scheduling for maritime container yards: Supporting and facilitating the global supply network. *Planning and Scheduling for Maritime Container Yards: Supporting and Facilitating the Global Supply Network*, (Stromberg 2015), 1–110. https://doi.org/10.1007/978-3-319-17025-1
[15] Altiok, T., & Melamed, B. (2007). *Simulation Modeling and Analysis with ARENA*.
[16] Law, A. M. (2015). Simulation Modelling and Analysis.pdf. McGraw-Hill Education.

IOP Conf. Series: Materials Science and Engineering **567** (2019) 012034 doi:10.1088/1757-899X/567/1/012034

Analysis of supply chain management in ABC poultry using system dynamics approach

J W Jufiardi[1], **F A Pinagara**[1]

[1]Department of Management Science, Faculty of Economics and Business, Universitas Indonesia.

Corresponding author: jose.jufiardi@gmail.com

Abstract. This study discusses supply chain management to produce broilers in ABC Poultry using System Dynamics approach. Small poultry farm in Indonesia must make a decision to sustain itself in the midst of needs to increase its capacity and facing issue with dependency to day-old chick producers. The purpose of this study is to analyse the supply chain management to increase profitability of the poultry in between two options of decision, buying the day-old chicks from third-party supplier or doing vertical integration to produce its own day-old chicks. This study found that doing vertical integration is the best option for the poultry because it can increase the profitability of the poultry and reduce the sensitivity of day-old chick price to the poultry.

1. Introduction

Poultry sector is one of an important sector in Indonesia's food industries. Poultry is one of the cheapest protein source, and it is consumed by majority of Indonesian people that has low income. Poultry farming in Indonesia is dominated by broiler farm. Broiler meat is 66% of the national meat consumption. Broiler farming industry in Indonesia is dominated by multinational corporation about 80% of the market. These multinational corporations have technology, resource, and skilled workers. The Indonesian Commission for the Supervision of Business Competition (KPPU) in 2016 suspected that 12 companies had involved in chicken cartel practices in day-old chick prices. Individual broiler farmers face the problems to acquire day-old chick supply with competitive prices and to produce more broiler in their farm.

This study discusses the impact of supply chain management decision in individual poultry farm in Indonesia using system dynamics in 5, 10 and 15 years. The study use ABC Poultry Farm in Bogor to be the case study material. There are four scenarios available to choose for the ABC poultry to increase its profitability:

1. Buy day-old chicks from 3rd party suppliers and increase their farm capacity to rear more chicken with traditional type of farm (Opened-house)
2. Buy day-old chicks from 3rd party suppliers and increase their farm capacity to rear more chicken with modern-type of farm (closed-house)
3. Do vertical integration to produce its day-old chicks and increase their farm capacity without any probability to buy day-old chicks from 3rd party suppliers.
4. Do vertical integration to produce its day-old chicks and increase their farm capacity with probability to buy day-old chicks from 3rd party supplier if it is necessary.

These scenarios were simulated and calculated using system dynamics approach with Powersim application and the references of this paper is primary data of the ABC Poultry and secondary data from previous studies. The simulation was made based on several assumptions.

2. Literature Review

2.1. Supply Chain Management
Supply chain is the business link that need to fulfill customer request of particular goods or services [1]. Supply chain needs to be managed to maximize value of goods or services (supply chain surplus) and the success of managing supply chain depends on decisions-making related to flow of information, product and cash.

2.2. Vertical Integration
Vertical integration is one of supply chain management strategy that can be implemented by an organization to enhance ability to replace third party supplier (backward integration) or third party distributor (forward integration) in supply chain. The advantages of vertical integration:
1) Decreasing transaction cost.
2) Increasing in certainty of supplies.
3) Increasing coordination. [2]

2.3. System Dynamics
System Dynamics is a method to enhance learning in complex system that was developed by Jay W Forrester [3]. System Dynamics, formerly known as Industrial Dynamics, finds the feedback and the characteristic of the feedback that is generated by industrial activity or agent activity in the system. The decision that is taken by one agent in the system will affect positively or negatively to other agent on the system.

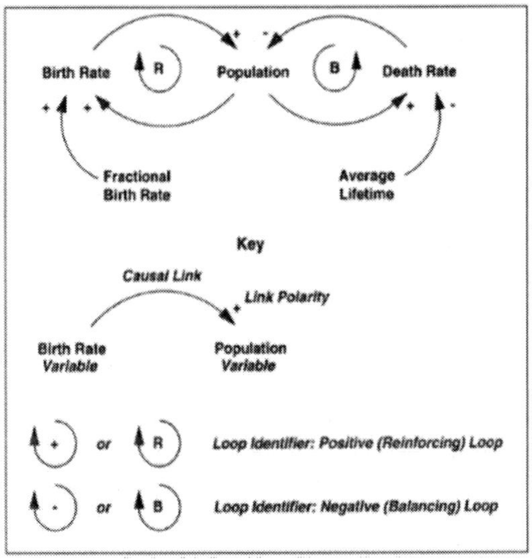

Figure 1. Causal Loop Diagram's Notation

Causal Loop Diagram is a one of tools in System Dynamics that is usually used to learn about the system and to picture the feedback in the system. Stock and Flow Diagram is another tool that is used to learn about the feedback effect in numbers and mathematical model.

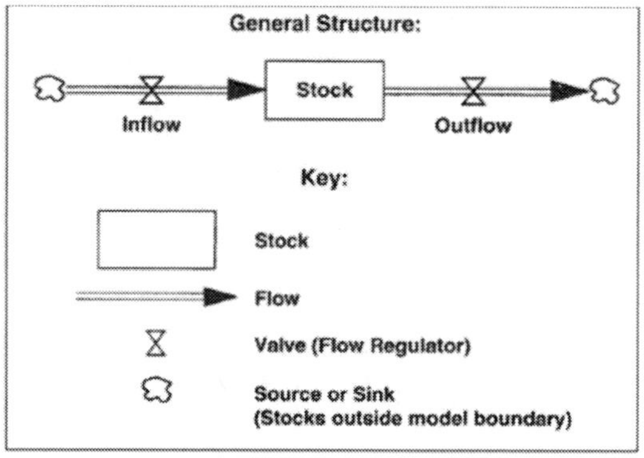

Figure 2. Stock and Flow Diagram's Notation

3. Research Methodology

This study made to help decision-making in ABC Poultry Farm. This study used interview and observation to collect primary data in ABC's farm that located in Bogor, West Java. The secondary data was using the farm historical report in two periods of producing. All of the data have been collected for six months.

After data were collected, then Causal Loop Diagram was made based on the collected data. The Causal Loop Diagram helped to find the main loops and the secondary loops in the system. Next step, Stock and Flow Diagram was made based on the Causal Loop Diagram and the collected data. After the Stock and Flow Diagram was made, The SFD was verified using Barlas's steps of model validation. Then, the simulation was done and the simulation record was analysed [4]. The analysis result was used to choose the best scenario and explain the constraint.

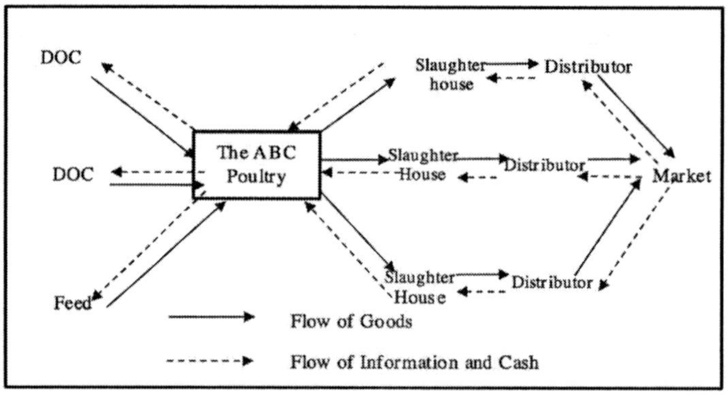

Figure 3. The Supply Chain of ABC Poultry

4. Result

The ABC Poultry focuses on rearing of broiler chicken. This study is focused on a farm that located on Gadog, Bogor with 6,500 chickens for one farm capacity. Figure 3 showed the supply chain of The ABC Poultry that consists of day-old chick suppliers, feed suppliers, slaughterhouses, distributors and market. The ABC Poultry orders day-old chicks from suppliers once the farm is empty and the ABC Poultry orders feed every two weeks. The slaughterhouses buy adult chicken directly from the farm and chicken that has been slaughtered are brought to the market by distributors.

The Simplified Causal Loop Diagram, shown on Figure 4, made based on interview with the owner of The ABC Poultry, observation, literature review and Integrated Supply Chain Model for Sustainable Poultry Production in Bangladesh: A System Dynamics Approach by Shamsuddoha in 2014 [5]. The CLD consists of 1 reinforcing loop, shown by indicator "R", and 7 balancing loops, shown by indicator "B".

The Stock and Flow Diagram made based on historical data and the CLD. The SFD shows the system of production chain in ABC Poultry. The cash on hand generator in this model is chicken for sale inventory and the costs of this system consist of feed total cost, day-old chick cost, Farm build cost and maintenance cost. The stock and flow diagram simulated in 5, 10 and 15 years and using weeks as a timestep.

The Stock and Flow Diagram of 1st and 2nd Scenarios, shown on Figure 5, was made based on a real condition on a farm. This SFD model used a traditional farm type (Opened-house) with each farm capacity equals to 6,500 chickens and 20,000 chickens with a build cost equals to Rp 150,000,000/farm. In thess scenarios, ABC Poultry bought day-old chicks from third party suppliers with a price equals to Rp 6,000/chick.

The Stock and Flow Diagram of 3rd Scenario, shown on Figure 6, was modified from the model of 1st scenario. This SFD model used a modern farm type (Closed-house) with each farm capacity equals to 20,000 chicken with a build cost equals to Rp 2,500,000,000/farm, in this scenario, ABC Poultry also bought day-old chicks from third party suppliers with a price equals to Rp 6,000/chick.

The Stock and Flow Diagram of 4th Scenario, shown on Figure 7, was modified from the model of 2nd scenario. This SFD model also used modern farm type (Closed-house) with farm capacity and build cost same as the 2nd scenario model. In this scenario, ABC Poultry could produce and sell its own day-old chicks with a price equals to Rp 5,000/chick, but couldn't buy day-old chicks from third party suppliers.

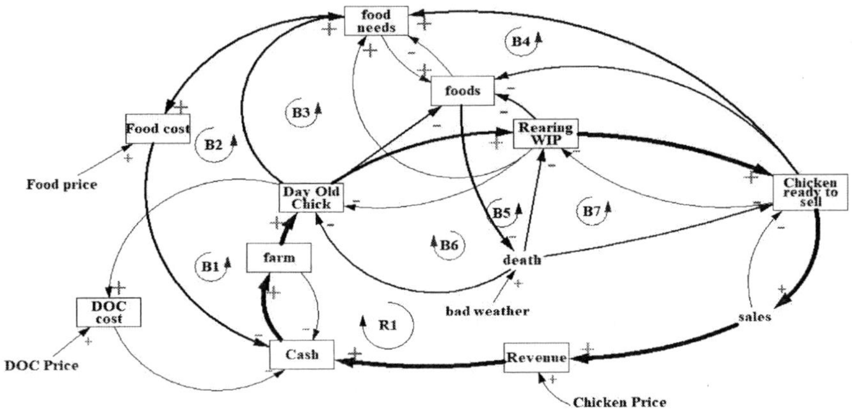

Figure 4. The Causal Loop Diagram of ABC Poultry

Input for each variable of the Stock and Flow Diagram:

1) Farm Capacity: 6,500 chickens (Scenario 1), 20.000 chickens (Scenarios 2, 3 and 4)
2) Initial Cash: Rp 1,000,000,000.
3) Day-Old-Chick Price: Rp 6,000/chick
4) Chicken Price: Rp 18,500/kg
5) Food Price: Rp 7,100/kg
6) Building Cost: Rp150,000,000/farm (Scenario 1), Rp2,500,000,000/farm (Scenarios 2, 3 and 4)
7) Farm Depreciation: 10%

IOP Conf. Series: Materials Science and Engineering **567** (2019) 012034 doi:10.1088/1757-899X/567/1/012034

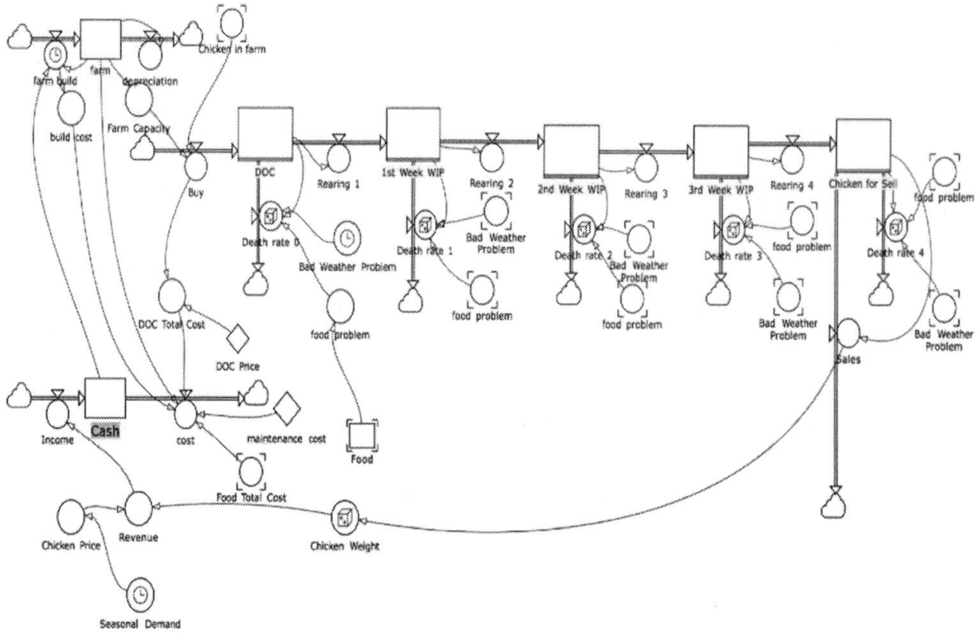

Figure 5. The Stock and Flow Diagram of ABC Poultry for 1st and 2nd Scenarios

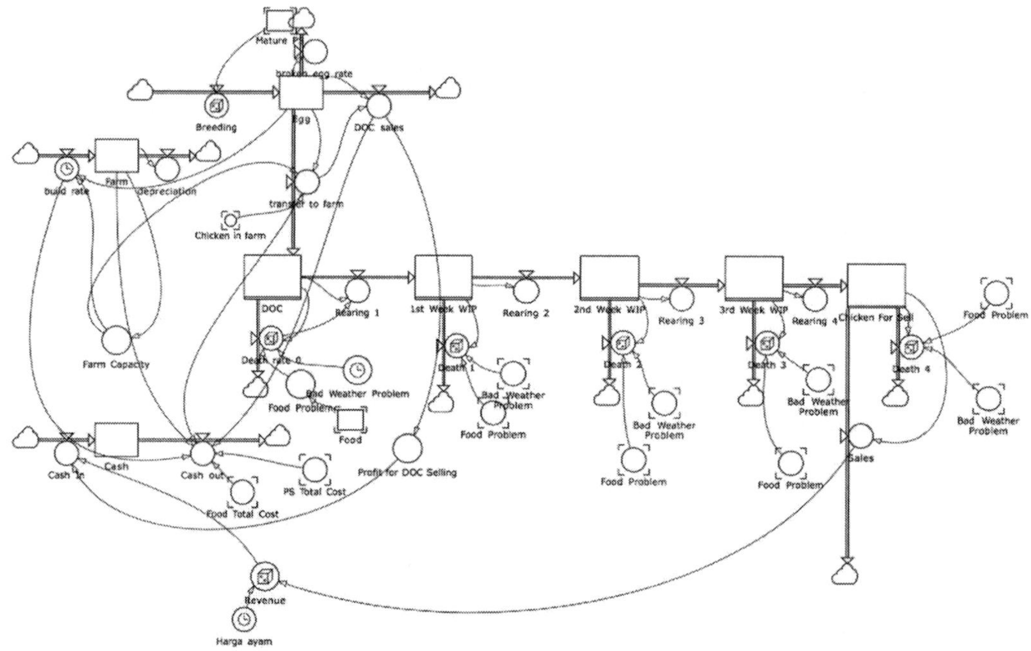

Figure 6. The Stock and Flow Diagram of ABC Poultry for 3rd Scenario

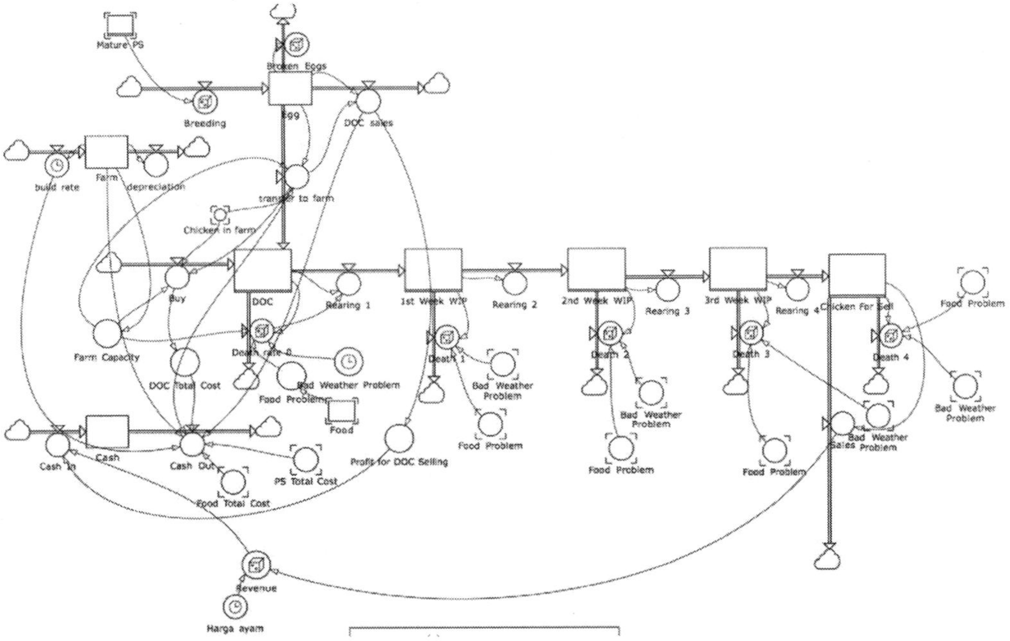

Figure 7. The Stock and Flow Diagram of ABC Poultry for 4th Scenario

The result of all scenarios in 5 years is shown on Table 1, The NPV result was calculated with rate of 4.75% (Bank Indonesia 7 days Repo Rate). The result showed that the 4th scenario generates highest NPV of cash on hand in the amount of Rp 26,036,541,756 with production capacity equal to 160,000 chickens in one cycle of production. Followed by 3rd Scenario with cash on hand in amount of Rp 19,360,590,397 and 100,000 chickens for production capacity. The least preferable scenario is 1st Scenario with cash on hand equals to Rp 262,937,439 and product capacity equals to 24,000 chickens.

Table 1. The Simulation Result (5 years)

Scenario	NPV (Normal Condition) (Rp)	Production Capacity (year 5th) (chicken)
1st Scenario	262,937,439	24,000
2nd Scenario	1,005,656,132	160,000
3rd Scenario	19.360.590.397	100,000
4th Scenario	26,036,541,756	160,000

The result of all scenarios in 10 years is shown on Table 2, The NPV result was calculated with rate of 4.75% (Bank Indonesia 7 days Repo Rate). The result showed that the 4th scenario generates highest NPV of cash on hand in the amount of Rp 55,903,825,170 with production capacity equal to 240,000 chickens in one cycle of production. Followed by 3rd Scenario with cash on hand in amount of Rp 31,436,846,956 and 120,000 chickens for production capacity. The least preferable scenario is 1st Scenario with cash on hand equals to Rp 1,939,664,849 and production capacity equals to 40,000 chicken.

Table 2. The Simulation Result (10 years)

Scenario	NPV (Normal Condition) (Rp)	Production Capacity (year 10th) (chicken)
1st Scenario	1,939,664,849	40,000
2nd Scenario	6,596,184,815	240,000
3rd Scenario	31,436,846,956	120,000
4th Scenario	55,903,825,170	240,000

The result of all scenarios in 15 years is shown on Table 3, The NPV result was calculated with rate of 4.75% (Bank Indonesia 7 days Repo Rate). The result showed that the 4th scenario generates highest NPV of cash on hand in the amount of Rp 89,629,854,208 with production capacity equals to 300,000 chicken in one cycle of production. Followed by 3rd Scenario with cash on hand in amount of Rp 44,554,368,851 and 120,000 chickens for production capacity. The least preferable scenario is 1st Scenario with cash on hand equals to Rp 4,946,468,051 and production capacity equals to 50,000 chickens.

Table 3. The Simulation Result (15 years)

Scenario	NPV (Normal Condition) (Rp)	Production Capacity (year 15th) (chicken)
1st Scenario	4,946,468,051	50,000
2nd Scenario	17,311,688,979	300,000
3rd Scenario	44,554,368,851	120,000
4th Scenario	89,629,854,208	300,000

5. Conclusion

The simulation result of this study showed that the most preferable scenario to choose by poultry is 4th scenario. The 4th scenario in 5 years generates NPV of cash on hands equal to Rp 26,036,541,756, in 10 years equals to Rp 55,903,825,170 and in 15 years equals to Rp 89,629,854,208. The most accurate result of simulation is the five-year simulation, because the longer period of simulation has lesser accuracy.

The model of this study showed that the best decision to the ABC Poultry is the 4th Scenario or do the vertical integration with buying from 3rd party supplier and using closed-house farm type. This 4th scenario generates highest cash on hand and decrease the cost of day-old-chick transaction. The second of the best decision is the 3rd scenario that is vertical integration without buying from third party supplier.

The implications of choose the 4th scenario are:

1) The poultry needs to provide initial cash on hand in the amount of Rp 1,000,000,000 to afford operational expenses.
2) The poultry needs to administer the permits to produce and sell its own day-old-chick.
3) The poultry needs to provide available land lots to build ten farms and provide Rp 2,500,000,000 to build farm each year.
4) The poultry needs to provide vehicles to transporting day-old-chick.

Acknowledgements
The presentation of this paper is funded by Universitas Indonesia under International Publications of Student Thesis Grant (Hibah PITTA) number: 1732/UN2.R3.1/HKP.05.00/2018.

IOP Conf. Series: Materials Science and Engineering **567** (2019) 012034 doi:10.1088/1757-899X/567/1/012034

Reference

[1] Chopra S and Meindl P 2016 *Supply Chain Management: Strategy, Planning and Operation*. Harlow: Pearson Education Limited.

[2] Buzzel R D 1983 *Is Vertical Integration Profitable?* Boston, MA: Harvard Business School Publishing.

[3] Forrester J W 1961 *Industrial Dynamics*. Portland: Productivity Press.

[4] Barlas Y. 1994 *Model Validation in System Dynamics*: Methodological Issues

[5] Shamsuddoha M 2014 *Integrated Supply Chain Model for Sustainable Poultry Production in Bangladesh*: A System Dynamics Approach. p 1834.

[6] Heizer J and Render B. 2014 Operations Management Global Edition 11th Edition. Harlow, Essex: Pearson Education Limited.

[7] Sterman J D 2000 *Business Dynamics-System Thinking and Modeling for a complex world*. New York: McGraw-Hill.

IOP Conf. Series: Materials Science and Engineering **567** (2019) 012035 doi:10.1088/1757-899X/567/1/012035

Supplier evaluation and order allocation using fuzzy analytical hierarchy process and augmented epsilon constraint methods

Farizal[1], R M Dewi[1], D S Gabriel[1]

[1]Department of Industrial Engineering, Faculty of Engineering, Universitas Indonesia, Kampus Baru UI Depok 16424, Indonesia

Corresponding author: farizal@eng.ui.ac.id

Abstract. Environmental issues nowadays affect the way to run business. These issues push firms to have effective and efficient green supply chain management. One of critical aspect in green supply chain management is green supplier selection. Choosing suitable supplier is an important part in procurement activity. Almost 70% of the total production cost is derived from raw material purchasing cost. This research proposes two phase meta-model for supplier selection and order allocation that takes into consideration environmental criteria besides traditional criteria such as quality, cost and delivery. For the purpose, fuzzy set and analytical hierarchical process (AHP) were combined. AHP was used to allow uncertainties and vagueness due to human decision making and subjective criteria. For order allocation phase, multi-objective mathematical programming method (MOMP), the augmented ε-constraint (AUGMECON) method was used to find Pareto optimal solutions for multiple sourcing. These proposed methods were tested in one of tire manufacturing company in Indonesia. The results show that the methods gave a 0.16% of the total cost lower than the existing one as addition to fulfilling green criteria.

1. Introduction

Nowadays, enterprises realize that effective and efficient supply chain management practices can affect both direct and indirect profit [1]. It is because enterprise must focus on reducing operational cost and enlarging overall profit to keep competitive in this globalization era [2]. However, due to environmental awareness issues, enterprises have to consider environmental aspect in managing their operations and supply chains. This consideration is often called as green supply chain management [2]. Global competition push firms to have effective and efficient green supply chain management.

Accoring to [2], green supply chain management is the management of funds, information and product flow along all supply chain stages to find the right balance between environmental and economic aspects. One of critical aspect in supply chain management is supplier selection [3]. Choosing suitable supplier is important responsibility for any organization in procurement activity [4]. It can help manufacturer to have better performance such as increasing customer satisfaction and

IOP Conf. Series: Materials Science and Engineering **567** (2019) 012035 doi:10.1088/1757-899X/567/1/012035

reducing purchasing cost because almost 70% of the total production cost is derived from raw material purchasing cost [5].

There are two types of supplier selection; i.e. single-sourcing and multiple-sourcing. In single-sourcing, a single supplier can fully satisfy all demand of a company and the company only needs to decide the selected supplier. On the other hand, in multiple-sourcing, a single supplier cannot satisfy all demand so that the order need to be splitted among other selected suppliers [2]. Multiple sourcing is more preferable than single sourcing due to its order flexibility [6].

According to [7], multiple sourcing can be divided into three phase: (1) supplier evaluation: establishing a supplier base, (2) supplier selection: choosing suppliers from the base and (3) quantity allocation: determining the quantity order for each selected supplier. Supplier selection involve multiple criteria both qualitative and quantitative criteria. SS are complicated because some criteria may conflict with the other i.e. price and quality. In the past, it just focused on traditional criteria such as quality, cost, delivery not environmental criteria.

Supplier selection method can be classified as: (1) Multi criteria decision making (MCDM) technique such as Analytic Hierarchy Process (AHP), Analytic Network Process (ANP), Technique for Order Performance by Similarity to ideal solution (TOPSIS), decision making trial and evaluation laboratory (DEMATEL); (2) Mathematical programming technique such as linear programming (LP), non-linear programming (NLP), multi-objecive linear programming (MOLP), goal programming (GP) and AUGMECON; (3) Artificial intelligence technique such as Genetic Algorithm (GA), Particle Swarm Optimization (PSO), and Chaotic Bee Colony.

In some literature, researchers not only discuss about supplier selection method but also combine it with order allocation method. [4] integrated fuzzy TOPSIS that is simple for computational procedure, easy to represent human preference, allow an unlimited number of criteria and an explicit trade-off between the criteria with fuzzy multi-objective linear programming (MOLP). [2] used fuzzy TOPSIS to rank potential suppliers based on traditional and green criteria and then use AHP to get importance weight to each criteria in a multi-period model. They separated preference weight calculation between green and traditional criteria to provide flexibility for top management whether they choose to giving more or less importance weight to green criteria towards traditional criteria. [7] built news vendor single period model for quantitative sustainable sourcing to select suppliers and determine optimal order quantities restricted by uncertain end product demand and salvage value. They considered multiple suppliers, capacity constraints, fixed charges and price in their research. [8] used fuzzy TOPSIS for supplier evaluation and then usedmulti-objective integer programming with fuzzy objectives and fuzzy constraints to determine optimal order quantity for allocated supplier.

Assuming that an enterprise has already established supplier data base, this research will only focus on supplier selection and quantity allocation. For supplier selection, fuzzy set and analytical hierarchical process (AHP) are combined to get weight of criteria importance. For quantity allocation phase, AUGMECON model introduced by [9] was used. The contribution of this research is to propose new meta-model for supplier selection and order allocation by combining fuzzy AHP and AUGMECON method.

2. Research Methodology

2.1. Fuzzy AHP

Fuzzy AHP can deal with uncertainties and vagueness of human decision making. In principle, fuzzy AHP is the same as classical AHP. AHP use crisp value but fuzzy AHP use fuzzy value. There are three main steps in fuzzy AHP: (1) problem structuring, (2) evaluation of local priorities, and (3) computing the global priorities of alternatives to come to a decision [5].

In problem structuring, decision maker states the objective, defines the selection criteria and picks the alternative choices to be selected from. In evaluation of local priorities, fuzzy is applied for determining pair-wise comparisons and local priorities. In last steps, global priorities of each alternative are computed. Consistency Ratio must be computed and should be less than 0.1.

IOP Conf. Series: Materials Science and Engineering **567** (2019) 012035 doi:10.1088/1757-899X/567/1/012035

2.2. Multi-Objective Mathematical Programming (MOMP)

In MOMP, there are more than one objective functions and they are usually conflicted. So, there is no single optimal solution that can optimize all the objective functions. Decision maker look for the most preferred solution rather than optimal solution. Concept of optimality is replaced by Pareto optimality or efficiency. [9] defined Pareto optimal solutions as solutions that cannot be improved in one objective function without deteriorating their performance in the others objective functions. Decision maker look for the most preferred solution among the Pareto optimal solutions.

2.3. Augmented ε-constraint (AUGMECON)

AUGMECON is an extended ε-constraint method that was introduced by [9]. AUGMECON optimizes one objective function using the other objective function as a constraint as shown in Equation (1) where *eps* is an adequately small number (usually between10^{-3}and10^{-6}), s_p is slack or surplus variable, and $r_p$is range of i-th objective function.

$$\max\left(f_1(x) + eps \times (s_2/r_2 + s_3/r_3 + \cdots + s_p/r_p)\right) \tag{1}$$

This research was conducted through three major steps shown in Figure 1. For suppliers selection phase, data processing is done by using fuzzy AHP method and its output is the weight of criteria importance. The weights are incorporated into the supplier's assessment to obtain the qualitative value of each supplier. Order history data and qualitative values are used as input to create a multi-objective linear programming model that consisting of objective, constraint, and decision variables. After that, the calculation of order allocation is done by using AUGMECON method with the help of GAMS (General Algebraic Modeling System) software.

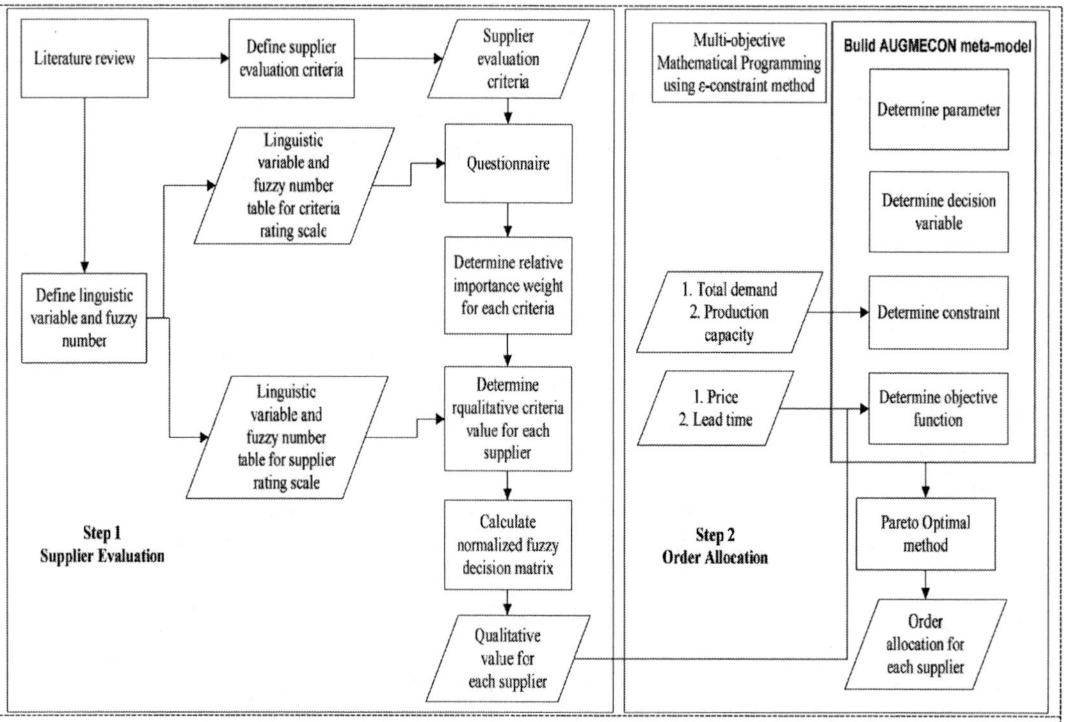

Figure 1. Research methodology.

3. Case Study

The proposed method was implemented in tire manufacture industry in Indonesia. Based on literature available, there are 41 sub-criteria for supplier selection that are grouped into 9 criteria: 8 traditional criteria and 1 green criteria. Full list of criteria and sub criteria are displayed on Table 1.

Table 1. Supplier selection criteria.

CRITERIA		SUB-CRITERIA	
A	Quality	A1	Rejected material ratio
		A2	Quality Management System/ ISO 9001 certification
		A3	Defect rate
		A4	Meeting minimum standard and requirement
B	Price/cost	B1	The average market price level of commodities
		B2	The lowest market price level of commodities
		B3	Shipping costs
		B4	Payment terms
		B5	Transaction price
		B6	Discount
C	Delivery	C1	Delivery time rate
		C2	Lead time delivery/geographical location
		C3	Delivery flexibility
		C4	Reserve capacity
		C5	Delivery quantity rate
		C6	Inventory availability
		C7	Order frequency
D	Service	D1	Quick response to customer demand
		D2	After sales support
		D3	E-transaction capability
E	Customer relations	E1	Willingness to share information
		E2	Long-term relationship development
F	Technical capability	F1	Continuous improvement ability
		F2	New product development ability
		F3	Technology sharing capability
		F4	Flexible production capability
G	Image	G1	Industry reputation
		G2	Past operation performance
H	Reliability	H1	Financial capability
		H2	Guarantee & liabilities / claim
		H3	Country's political situation
		H4	Currency exchange situation
I	Green competition / Environment Management Performance	I1	Environmental planning (program to reduce impacts, green research and development)
		I2	Selection of environmentally friendly materials (low waste: easy recycle and reuse capability)
		I3	Prohibited / toxic substances usage
		I4	Cleaner production technology (pollutant emission: CO2 equivalent, VOC, BOD and COD content and etc.)
		I5	Waste management
		I6	Environment Management System/ ISO 14001 certification
		I7	Green image
		I8	Green packaging
		I9	Distribution

Questionnaires were spread to make pair-wise comparison for each criteria in linguistic variable (represented in Table 2). The respondents for this research are manager, assistant manager and staff that are experienced in tire manufacturer procurement for several years. From 23 distributed questionnaires, only 9 questionnaires that were returned. To get preference weight for each criteria, pair-wise comparison matrix in fuzzy set was made (depicted in Table 3). Those 9 judgments were aggregated with geometric mean operation by using Equation (2) wherel_{ij} , m_{ij} and u_{ij} are respectively the lower, middle and upper values of the fuzzy membership function and K is number of decision maker.

$$l_{ij} = \left(\prod_{k=1}^{K} l_{ijk}\right)^{1/K}, m_{ij} = \left(\prod_{k=1}^{K} m_{ijk}\right)^{1/K}, u_{ij} = \left(\prod_{k=1}^{K} u_{ijk}\right)^{1/K} \quad (2)$$

Table 2. Linguistic variable for importance scale.

Linguistic variable	TFN $((l, m, u)$
Very not important	(1, 1, 1)
Not important	(1, 3, 5)
Netral	(3, 5, 7)
Important	(5, 7, 9)
Very important	(7, 9, 9)

Table 3. Linguistic variable for evaluation of qualitative criteria values for pairwise comparison.

Linguistic variable	TFN $((l, m, u)$	Reciprocal
Equally important	(1, 1, 1)	(1, 1, 1)
Moderately important	(1, 3, 5)	(1/5, 1/3, 1)
More important	(3, 5, 7)	(1/7, 1/5, 1/3)
Very important	(5, 7, 9)	(1/9, 1/7, 1/5)
Extremely important	(7, 9, 9)	(1/9, 1/9, 1/7)

Defuzzification was carried out to get a single digit of three fuzzy numbers. Center of Area (COA) is used to calculate the value of best non-fuzzy performance (BNP). If the BNP_i value is represented as the BNP value for criteria i, the BNP of the COA is calculated by Equation (3) as follows:

$$BNP_i = \frac{(l_i + m_i + u_i)}{3} \quad (3)$$

The next step is to calculate the criteria weight which is the standardization value of BNP_i (STD-BNP_i) shown as Equation (4) below:

$$Weight = STD - BNP_i = BNP_i / \sum_{i=1}^{n} BNP_i \quad (4)$$

Determining the fuzzy priority of qualitative criteria for supplier alternatives is using the linguistic variables shown in Table 4. The weights of each criteria and the value of fuzzy performance are integrated to obtain the fuzzy synthetic decision matrix. Ranking results of fuzzy synthetic decision matrix is given by defuzzification using Center of Gravity method.

Table 4. Linguistic variable for supplier evaluation.

Importance intensity	TFN $((l, m, u)$
Very good	(3, 5, 5)
Good	(1, 3, 5)
Enough	(1, 1, 1)
Not good	(1/5, 1/3, 1)
Not very good	(1/5, 1/5, 1/3)

The weight of qualitative criteria of each supplier in Table 5 will be used as input coefficients of one objective function for multi-objective optimization formulation by using AUGMECON method at the allocation order phase.

Table 5. Weight of qualitative value weight for each supplier.

Supplier	Qualitative value weight
1	0.21664
2	0.249993
3	0.249993
4	0.283375

Decision variable:
x_i = order quantity of i-th supplier for i = 1, 2, ..., n

Objective function:
There are three objective functions in determining order allocation i.e. minimizing total cost (f_1), minimizing mean delivery time (f_2) and maximizing qualitative value (f_3).
Function 1 (f_1) is obtained by multiplying the material price with order quantity then added with the shipping cost. Function 2 (f_2) is obtained by multiplying the delivery time with order quantity then divided by the total demand. Function 3 (f_3) is obtained by multiplying the weight of the qualitative value with order quantity.

Constraint:
Order quantity of each supplier should not exceed the maximum production capacity and must exceed the minimum order quantity. Total quantity of orders for all suppliers must equal with total demand. Total cost incurred should not be more than the multiplication of the average market price level of the commodity with total demand. Total cost incurred is more than equal to the multiplication of the lowest commodity market price level with total demand. All constraints are represented in Equations (8) - (12).

$$\min f_1(x) = \sum_{i=1}^{n} p_i x_i + \sum_{i=1}^{n} d_i \quad (5)$$

$$\min f_2(x) = \sum_{i=1}^{n} \frac{t_i x_i}{Q} \quad (6)$$

$$\max f_3(x) = \sum_{i=1}^{n} w_i x_i \quad (7)$$

St

$$x_i \geq M_i \quad (8)$$

$$x_i \leq C_i \quad (9)$$

$$\sum_{i=1}^{n} x_i = Q \quad (10)$$

$$f_1(x) \leq p_{AVE} \times Q \quad (11)$$

$$f_1(x) \geq p_{MIN} \times Q \quad (12)$$

where
n : number of supplier

w_i : qualitative value weight of i-th supplier that obtained from fuzzy AHP method in supplier evaluation
x_i : order quantity of i-th supplier
p_i : material price of i-th supplier
d_i : delivery cost of i-th supplier
t_i : delivery time of i-th supplier
M_i : minimum order quantity of i-th supplier
Q : total demand
C_i : maximum production capacity of i-th supplier
p_{AVE} : the average market price of commodities
p_{MIN} : the minimum market price of commodities

For case study, data on Table 6 from one of tire manufacturing company were used.

Table 6. Order data history.

Material : CCC	**Supplier 1**	**Supplier 2**	**Supplier 3**	**Supplier 4**
p_i (\$/ MT)	5250	5100	5150	5170
M_i(MT)	0	0	0	0
d_i (\$/ MT)	0	0	0	0
t_i(hari)	14	14	14	14
C_i(MT)	5000	10000	70000	15000

Total demand Q = 29000 MT
Average market price of commodities p_{AVE} = \$5167.5/MT
Minimum market price of commodities p_{MIN} = \$5100/MT

Decision variable:
x_i= order quantity for supplier *i* where *i* = 1, 2, 3, and 4

Objective function:
Minimize total cost function:

$$\min f_1(x) = p_1x_1 + p_2x_2 + p_3x_3 + p_4x_4 + d_1 + d_2 + d_3 + d_4$$
$$\min f_1(x) = 5250\, x_1 + 5100\, x_2 + 5150\, x_3 + 5170x_4$$

Minimize mean delivery time function:

$$\min f_2(x) = \frac{t_1x_1 + t_2x_2 + t_3x_3 + t_4x_4}{Q}$$
$$\min f_2(x) = \frac{14\, x_1 + 14\, x_2 + 14x_3 + 14\, x_4}{29000}$$

Maximize qualitative value function:

$$\max f_3(x) = w_1x_1 + w_2x_2 + w_3x_3 + w_4x_4$$
$$\max f_3(x) = 0.21664\, x_1 + 0.249993\, x_2 + 0.249993\, x_3 + 0.283375\, x_4$$

Constraint:
$x_1, x_2, x_3, x_4 \geq 0$
$x_1 \leq 5000$
$x_2 \leq 10000$
$x_3 \leq 70000$
$x_4 \leq 15000$
$x_1 + x_2 + x_3 + x_4 = 29000$

$f_1(x) \leq 5167.5 \times 29000$
$f_1(x) \leq 149{,}857{,}500$
$f_1(x) \geq 5100 \times 29000$
$f_1(x) \geq 147{,}900{,}000$

Calculation for order allocation model is solved by using GAMS (General Algebraic Modeling System) software.

4. Result and Discussion

Fuzzy priority estimation result for each supplier evaluation criteria were shown in Table 7. The table clearly shows that tire manufacturing industry in Indonesia are using only traditional criteria when selecting their suppliers while green competence/environmental management performance criteria are not yet used. This is evidenced by the position of quality criteria, price/cost and delivery that occupy the top three ranks. But, green competence/environmental management performance criteria has no weight value.

Table 7. Weight and ranking for supplier evaluation criteria.

Criteria	Weight	Ranking
Quality	0.923754	1
Price/cost	0.0642950	2
Delivery	0.011554	3
Service	0.0003770	4
Customer relations	0.000012	5
Technical capability	0.000008	6
Image	0	7
Reliability	0	8
Green competition/Environment Management Performance	0	9

Summary of consistency ratio (CR) for nine respondents is shown in Table 8. Table 8 indicates that all CR values are less than 0.1. The values mean that all the nine respondent ratios can be used for weighting the criteria.

Table 8. Consistency ratio for each respondent.

Respondent	Consistency Ratio
1	-0.676
2	-0.714
3	-0.680
4	-0.680
5	-0.743
6	-0.704
7	-0.647
8	-0.717
9	-0.652

For order allocation calculation, three experiments were conducted with variation of number of optimal Pareto solution to be produced (shown in Table 9). The optimal number of Pareto solutions can be set by determining the same number of intervals using the equidistant grid point in the program. Computation time is also recorded when GAMS software is run on a computer with a 3.5 GHz Quad-core processor with Microsoft Windows 10 OS.

Table 9. Trial summary

Trial	Number of equal interval	Total gridpoint	Number of total run	Number of optimal Pareto solutions	Computation time (second)
1	10	11	$11^3 = 1331$	11	6.656
2	21	22	$22^3 = 10648$	22	60.125
3	50	51	$51^3 = 132651$	51	140.453

To get the most preferred optimal Pareto solution, decision maker are involved in the interactive filtering process. Interactive filtering uses the results of Pareto's third experimental optimization experiment and performed three times as shown in Figures 2, 3, and 4. f_1, f_2 and f_3 are the minimizing total cost, minimizing mean delivery time, and maximizing qualitative value functions, respectively. Objective function results for first, second, and third run are summarized on Tables 10-12.

From the third filtering shown in Figure 4, the decision maker chose solution number 2551 as the most preferred efficient solution. It's because solution number 2251 has the greatest qualitative value compared to other solutions. The interactive filtering process is terminated until the third iteration. As the result, the order quantity variables for the chosen solution are $x_1 = 0$, $x_2 = 10{,}000$ MT, $x_3 = 4{,}000$ MT, and $x_4 = 15{,}000$ MT.

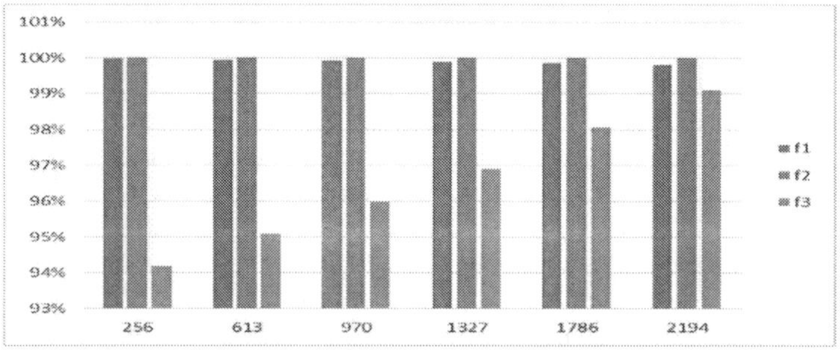

Figure 2. Subset of Pareto optimal solutions after order allocation first filtering.

Table 10. Objective function value and order quantity after order allocation first filtering

No	No. Solution	f_1	f_2	f_3	x_1	x_2	x_3	x_4
6	256	148880	406000	7299.87	0	10000	17500	1500
13	613	148922	406000	7369.97	0	10000	15400	3600
20	970	148964	406000	7440.07	0	10000	13300	5700
27	1327	149006	406000	7510.18	0	10000	11200	7800
36	1786	149060	406000	7600.31	0	10000	8500	10500
44	2194	149108	406000	7680.42	0	10000	6100	12900

IOP Conf. Series: Materials Science and Engineering **567** (2019) 012035 doi:10.1088/1757-899X/567/1/012035

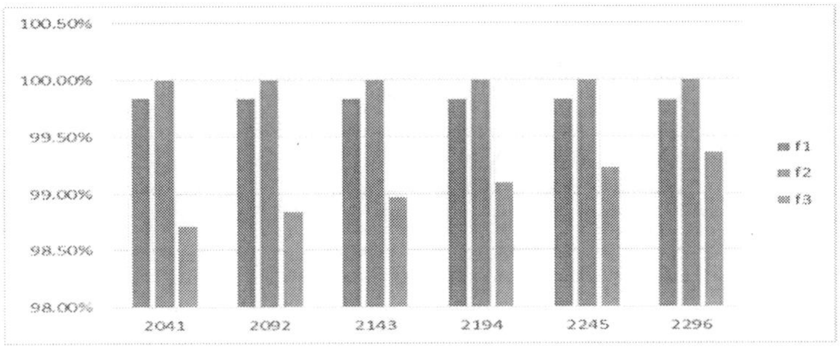

Figure 3. Subset of Pareto optimal solutions after order allocation second filtering.

Table 11. Objective function value and order quantity after order allocation second filtering

No	No. Solution	f_1	f_2	f_3	x_1	x_2	x_3	x_4
41	2041	148880	406000	7299.87	0	10000	7000	12000
42	2092	148922	406000	7369.97	0	10000	6700	12300
43	2143	148964	406000	7440.07	0	10000	6400	12600
44	2194	149006	406000	7510.18	0	10000	6100	12900
45	2245	149060	406000	7600.31	0	10000	5800	13200
46	2296	149108	406000	7680.42	0	10000	5500	13500

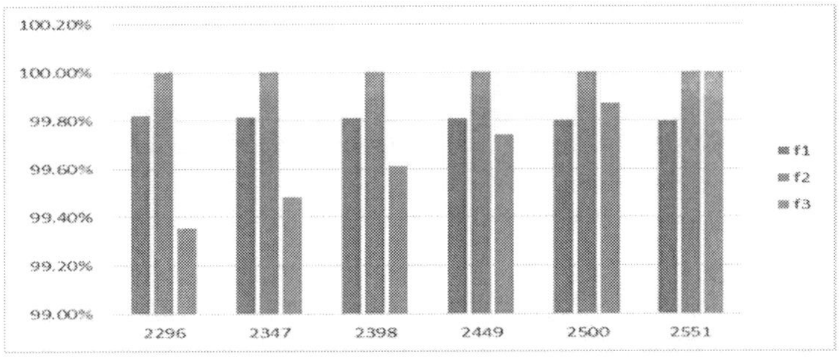

Figure 4. Subset of Pareto optimal solutions after order allocation third filtering.

Table 12. Objective function value and order quantity after order allocation third filtering.

No	No. Solution	f_1	f_2	f_3	x_1	x_2	x_3	x_4
46	2296	149120	406000	7700.45	0	10000	5500	13500
47	2347	149126	406000	7710.47	0	10000	5200	13800
48	2398	149132	406000	7720.48	0	10000	4900	14100
49	2449	149138	406000	7730.5	0	10000	4600	14400
50	2500	149144	406000	7740.51	0	10000	4300	14700
51	2551	149150	406000	7750.53	0	10000	4000	15000

Comparison of order allocation of selected solution from interactive filtering results with existing orders in terms of total cost is shown in Table 13. From these results, it appears that order allocation of the selected solution resulted $240 cheaper than the existing order in terms of total cost. Selected solution gives 0.16% of the total cost lower than the existing order. Therefore, order allocation of selected solution proves to be more advantageous rather than the existing order in terms of total cost.

Table 13. Total cost comparison for selected solutions and existing order

	Material Price ($/MT)	Selected solution(MT)	Existing order(MT)
x_1	5250	0	2000
x_2	5100	10000	8000
x_3	5150	4000	7000
x_4	5170	15000	12000
	Total cost ($)	**149150**	**149390**

5. Conclusion

This study uses the combination of fuzzy AHP and AUGMECON methods to provide order allocation in multiple-sourcing cases involving subjective human decision making. In this study, GAMS (General Algebraic Modeling System) software is used to obtain the optimal solution set in multi-objective mathematical programming (MOMP) problems. To get the most preferred optimal solution, an interactive filtering method involves decision-making preferences. As a result, the order quantity variables for the chosen solution are $x_1 = 0$, $x_2 = 10{,}000$ MT, $x_3 = 4{,}000$ MT, and $x_4 = 15{,}000$ MT. The selected solution gives 0.16% of the total cost lower than the existing order.

The advantages of this model are to provide an order allocation solution in real time and involve the decision maker's preference for selection of solutions. While the weakness is this model is only applicable to the problem of allocation of orders for one type of multi-product suppliers. However, this model can be easily developed by the mathematical model for multi-product order allocation order issues in subsequent research. In terms of novelty, this model can complete the calculation of the allocation of orders quickly and simply. The proposed model can be applied to all industrial areas because the supplier evaluation criteria used are general.

Acknowledgment
This work has been supported partially by PITTA 2018 Grant funded by DRPM Universitas Indonesia under contract No. 2436/UN2.R3.1/HKP.05.00/2018.

References

[1] Banaeian N, Mobli H, Nielsen I E, and Omid M 2015 *Production & Manufacturing Research* **3**(1) pp 149-168
[2] Hamdan S and Cheaitou A 2016 *Computer and Operations Research* **81**
[3] Bakeshlou E A and Khamseh A A 2017 *J Intell Manuf* **28** pp 913-927
[4] Govindan K and Sivakumar R 2015 *Ann Oper Res* **238** pp 243-276
[5] Özfirat P M, Taşoglu G T, and Memiş G T 2014 *J. Enterprise Information Management* **27**(3) pp 292-301
[6] Kirytopoulos K, Leopoulus V, Mavrotas G and Oulgaridou D 2010 *Supply Chain Management: An International Journal* **15/4** pp 263-276
[7] Schoell M A B and Keblis M F 2017 *Int J Development and Sustainability* **6** (7) pp 317-332
[8] Razmi J, Songhori M J, and Khakbaz M H 2009 *Int J Adv Manuf Technol* **43** pp 590-607
[9] Mavrotas G 2009 *Appl Math Comput* **213** pp 455-465

Queuing model using sojourn time distribution with single working vacation and vacation interruption

Elis Hertini[1], Chery Harisbaya[2], Julita Nahar[3], Sudradjat[4]
[1,2,3,4]Department of Mathematic Universitas Padjadjaran Bandung Indonesia
Jalan Raya Sumedang KM 21 Jatinangor-Sumedang 45363

Coresponding Author: elis.hertini@unpad.ac.id

Abstract. Queuing that occurs due to service delays caused by the server is a result of server interference. One of the server interruptions was called the vacation incident. This paper analyzes the queue caused by a lack of servers. To solve this problem, the Sojourn Time $(W^{*}(\theta))$ distribution model is used at the time the service is distributed to just one server (M / G / I), with single working vacation and vacation interruption. This model is then used to look for models of old customers waiting in queuing (W), number of customers in the system (L), and opportunities for server busyness (ρ). The values of L, W, and ρ use secondary data for one server that has an arrival pattern that has Poisson distribution and service time exponentially distributed, then the values are substituted into the model $(W^{*}(\theta))$. The results of the model analysis show that the server is not busy or does not need to add a server.

1. Introduction

Queuing due to delays in service is an event that often occurs in everyday life, this can be caused by the number of customers in the queue or can because the server is slow in serving customers. To avoid this, efforts are needed to prevent the emergence of queues in the system. If this queue continues to occur it will cause the customer's sojourn time in the system to increase, it will be detrimental to both the customer and the service owner. To avoid this, efforts are needed to prevent the appearance of queues in the system, due to server interruptions.

The service delay caused by the server is a result of server interruption that is not operating. One of the server interruptions is not operating due to maintenance or repair. Unoperated servers are referred to as vacation events, while the occurrence of taking care of once a day is called single working vacation. Many customers queue up so the server must immediately terminate the vacation called vacation interference.

According to [4], the queuing process is a process related to the arrival of customers to a queue or server system, then waiting in a queue until the waiter selects the customer according to the service discipline, and finally the customer leaves the queue system after the service is finished. The discussion in queuing theory in determining the customer sojourn time is, by determining the corresponding time distribution model. One indicator in determining the suitable model used in queuing theory is to determine the service time distribution.

In the general case, service time is always exponentially distributed, but in some cases service time does not always have an exponential distribution. If this happens, it is necessary to use a more general distribution so that all types of distribution can be applied, so that the distribution of the sojourn time used is the distribution of sojourn time for service times that are randomly distributed.

Other previous studies have been done a lot including [6], discussing the distribution of sojourn time on the M / G / 1 model where this model only discusses the determination of the sojourn time distribution model. [8], discuss the M / G / 1 / MWV queue model in this model the server does more than one vacation. [1], discuss the GI / M / 1 / SWV and GI / Geo / 1 / SWV models on this model, the server only does one vacation

Various studies on residence time distribution have been carried out. In this study, we will analyze the queue model due to the lack of a server, using the residence time distribution model $(W^*(\theta))$ with a single work holiday and vacation disruption when the service distributes one server. To obtain a numerical solution using secondary data, the model of the old customer waits in the queue (W), the number of customers in the system (L), and the chance of busyness of the server (ρ) is resolved with the help of c ++ software applications. The resulting L, W, and ρ values are included in the distribution model of Sojourn Time $(W^*(\theta))$ with single working vacation and vacation interruption to analyze the queue caused by a lack of servers.

2. Research Model

2.1. System Equations at Steady State.

According to [2], Steady state condition is a state of the system that does not depend on the initial state or time that has passed. If a system has reached a steady state condition then the chance of n customers in the system at time t is $(P_n(t))$ is not dependent on time. Assuming S_0 is the service time when the vacation is happening, S_1 is the service time when the server is available, the arrival rate of λ, and the vacation rate v, and for example λ, v, S_0, S_1 are free. Dropping [5], the events of $A_1, A_2, \ldots, A_n$ n are said to be mutually independent events if $A_i \cap A_j = \phi \;;\; i \neq j$. Given N (t) is the number of customers in the system at t, and ξ (t) is the number of servers that work or when not vacationing at time t.
When,

$$\xi(t) = \begin{cases} 0, \text{the system in the period of working vacation at time t} \\ 1, \text{the system in the normal service period at time t} \end{cases}$$

so {N (t), ξ (t), t≥0} for every i = 0.1 is Quasi Birth Death Process (QBD) in state space
$\Omega = \{(0,0)\} \cup \{(n, i) \mid n \geq 1, i = 0,1\}$.
With states (n, 1), for n≥1 shows that the system is in a normal busy state, state (n, 0), for n≥1 indicates that the system is in the working vacation period, and there are n customers in the queue. If state (1,1), states that in the system there is one customer and there is one server that is not doing vacation.
Using the probability limit is obtained

$$P_{n,i} = P(N = n, \xi = i) = \lim_{t\to\infty} P\{N(t) = n, \xi(t) = i\} \quad (n,i) \in \Omega$$

To get the probability limit it is resolved using per QGDP, from this solution the system equation is obtained at steady state, that is when it occurs:

1. Customers in the system and servers are on vacation,
$$(\lambda + v)P_{0,0} = P_{1,0}(0) + P_{1,1}(0) \tag{1}$$
2. One customer on the system when the server is on vacation,
$$\frac{d}{dx}P_{1,0}(x) = \lambda P_{0,0}S_0(x) - (\lambda + v)P_{1,0}(x) \tag{2}$$
3. There are n customers on the system when the server is on vacation,
$$\frac{d}{dx}P_{n,0}(x) = \lambda P_{n-1,0}S_0(x) - (\lambda + v)P_{n,0}(x)\,;\, n \geq 2 \tag{3}$$
4. One customer on the system when the server is in normal service,
$$\frac{d}{dx}P_{1,1}(x) = P_{2,0}(0)S_1(x) + vP_{1,0}S_1(x) + \lambda P_{0,1}S_1(x) - \lambda P_{1,1}(x) + P_{2,1}(0)S_1(x) \tag{4}$$
5. There are n customers in the system when the server is in normal service
$$\frac{d}{dx}P_{n,1}(x) = P_{n+1,0}(0)S_1(x) + vP_{n,0}S_1(x) + \lambda P_{n-1,1}(x) - \lambda P_{n,1}(x) + P_{n+1,1}(0)S_1(x)$$

$; n \geq 2$ (5)

Because the change from a server that is doing a vacation to the server under normal conditions depends on v and λ, then the chance of no customers in the system when the server is in normal service is,

$$\lambda P_{0,1} = vP_{0,0} \text{ or, } P_{0,1} = \frac{v}{\lambda} P_{0,0} \quad (6)$$

2.2. Laplace-Stieltjes Transforms (LST)

According to [7], suppose that F (t) is a function well defined from t for $t \geq 0$ and s is a complex number. If it follows the stieljes integral $F^*(s) = \int_0^\infty e^{-st} dF(t)$ converges to s_0 so $\Re(s) > R(s_0)$, then this integral is called the laplace-stieltjes transforms from F (t) which can be formulated by following the equation:

$$F(t) = \int_0^t dF(t) = \int_0^t f(x)dx$$

Then,

$$F^*(s) = \int_0^\infty e^{-st} dF(t) = \int_0^\infty e^{-st} f(t)dt \quad (7)$$

The system equation when the steady state is defined is that n is the number of customers in the system and ξ is the number of servers that work where i ϵ {0.1}, then:

a. By using the laplace stieltjes transformation (lst) the equation is obtained

$$P_{n,i}^*(\theta) = \int_0^\infty e^{-\theta x} P_{n,i}(x)\, dx\,, \quad (8)$$

b. Using the Probability generating function (pgf) the equation is obtained

$$P_i^*(z, \theta) = \sum_{n=1}^{\infty} P_{n,i}^*(\theta) z^n, \quad (9)$$

Determine the opportunity for the absence of customers in the system when the server is on vacation, namely by integrating equations (2) and (4) then substituting it to equation (1), obtained

$$P_{0,0} = \frac{\lambda v(1-\lambda E[S_1])}{\lambda^2+\lambda+v^2-\lambda^2(1+vE[S_1])S_0^*(v)} \quad (10)$$

To get the chance of no customers in the system when the server is on vacation, substitute equation (10) into equation (6), obtained

$$P_{0,1} = \frac{v^2(1-\lambda E[S_1])}{\lambda^2+\lambda+v^2-\lambda^2(1+vE[S_1])S_0^*(v)} \quad (11)$$

By integrating equation (4), it is then solved using laplace stieltjes transformation and probability generating function, then the chances of the server being vacationed are

$$P_0^*(z, \theta) = \frac{\lambda P_{0,0} z(S_0^*(v+\lambda-\lambda\ )-S_0^*(\theta)}{(\theta-v-\lambda+\lambda z)} \quad (12)$$

Using the same method, to get the laplace stieltjes transformation equation from server service opportunities under normal conditions, obtained

$$P_1^*(z, \theta) = \frac{P_{0,0} z(-z)\left(s_1^*(\lambda-\lambda z)-s_1^*(\theta)\right)[\lambda(\lambda+v)\left(1-S_0^*(v+\lambda-\lambda z)\right)+v(v+\lambda-\lambda z)]}{(\theta-\lambda+\lambda\ )(v+\lambda-\lambda\ )(S_1^*(\lambda-\lambda z)-z} \quad (13)$$

2.3. Sojourn Time Distribution Model

According to [3], a system where the customer has completed the part and the service started again for the next service in the queue, then the service time is randomly distributed. Sojourn Time from the customer that came in case i; i = 1,2,3,4 denoted by W_i and defined:

$$W_i^*(\theta) = \Pr\{case - i\}\, E[e^{-\theta W_i} | case - i\}$$

Model is used to find the customer's residence time by multiplying case-i expectations of opportunities during the i-case. Assuming that vacation time is distributed exponentially, where vacation time is represented by V which is distributed exponentially, if S_0 is the service time during the working holiday period with fkp $S_0(x)$ and lst $S_0^*(\theta)$, and if S_1 is the service time during the normal service period with fkp $S_1(x)$ and lst $S_1^*(\theta)$. To determine the model from transit time, there are a number of things that must be considered, namely by paying attention to the Laplace-Stieltjes Transformation from the transit time using the First In First Out service from customers which is divided into several cases:

Case 1.

Customers come when the system is empty and the server is on vacation, the customers who come cannot be directly served so there is a service delay. If this happens the server can serve customers at a slow rate or the server can stop vacation and serve customers with normal service rates. So the formula used for this model is the customer sojourn time when the server makes a vacation plus the customer's sojourn time when the server is available, that is

$$w_1^*(\theta) = P_{0,0}\left(\Pr\{S_0 < V\}\, E\left[e^{-\theta s_0} \middle| S_0 < V\right] + \Pr\{S_0 \geq V\}\, E\left[e^{-\theta(V+S_1)} \middle| S_0 \geq V\right]\right)$$
$$= P_{0,0}\left(S_0^*(\theta + v) + \frac{\left(v\, S_1^*(\theta)\left(1 - S_0^*(\theta + v)\right)\right)}{\theta + v}\right) \qquad (14)$$

Case 2.

Customers come when the system is empty and the server is available so the formula used for this model is the customer's sojourn time when the server is in normal condition,

$$w_2^*(\theta) = P_{0,1}\left(\Pr\{S_0 = 0\}\, E\left[e^{-\vartheta S_1} \middle| S_0 = 0\right]\right)$$
$$= P_{0,1} \times S_1^*(\theta) \qquad (15)$$

Case 3.

Customers come during the working vacation period when the server is busy. To get the appropriate sojourn time distribution, use the Total Probability Theorem. This case occurs if the server is doing a vacation but in the system as many as n customers are queuing so the server must immediately end the vacation. In this model vacation interruption occurs, so the formula used is the sum of customer service times when normal conditions are added when the server is vacationing, then

$$w_3^*(\theta) = \sum_{n=1}^{\infty} \int_{x=0}^{\infty} P_{n,0}(x) \Pr\{S_{R,0} < V \mid S_{R,0} = x\}\, x\left(S_1^*(\theta)\right)^n e^{-\theta x} dx +$$
$$\sum_{n=1}^{\infty} \int_{x=0}^{\infty} P_{n,0}(x)\, x \int_{y=0}^{x} \Pr\{V = y \mid S_{R,0} = x\} dy (S_1^*(\theta))^{n+1} e^{-\theta y} dx$$

by using (9), obtained

$$w_3^*(\theta) = \frac{\theta + v - v S_1^*(\theta)}{\theta + v} P_0^*(S_1^*(\theta), \theta + v) + \frac{v(S_1^*(\theta))^n}{\theta + v} P_0^*(S_1^*(\theta), 0) \qquad (16)$$

Case 4.

Customers come during the normal service period to see the server busy. Customers come as many as n customers queuing in the system and the server is available, so the formula used is the sum of the service times during normal conditions, then

$$w_4^*(\theta) = \sum_{n=1}^{\infty} \int_{x=0}^{\infty} P_{n,1}(x) (S_1^*(\theta))^n\, e^{-\theta x} dx$$
$$= \sum_{n=1}^{\infty} P_{n,1}^*(\theta)\, (S_1^*(\theta))^n$$
$$= P_1^*(S_1^*(\theta), \theta) \qquad (17)$$

Because in practice it cannot be estimated that the model must be used, so that a combination of models is made to facilitate decision making. To declare the unpredictable sojourn time condition (W), the sojourn time model of a customer that will be used is a combination of the

$$w^*(\theta) = \frac{w_1^*(\theta) + w_2^*(\theta) + w_3^*(\theta) + w_4^*(\theta)}{w_1^*(0) + w_2^*(0) + w_3^*(0) + w_4^*(0)}$$

By combining the equations (14), (15), (16), and (17) and equations (10), (11), (12), and (13), to obtain the sojourn time distribution model ($W^*(\theta)$), is

$$W^*(\theta) = \frac{P_{0,0}\{\lambda(\theta-\lambda)S_0^*(\theta+v)\big(\theta+v-vS_1^*(\theta)\big)+\theta S_1^*(\theta)[\lambda^2+v(\theta+\lambda+v]\}}{\lambda(v+\theta)(\theta-\lambda+\lambda S_1^*(\theta))} \quad (18)$$

3. Results and Doscussion

The vacation time is assumed to be exponential in distribution, the results of this study the resulting formula is suitable for vacation time which has an exponential distribution, the total time the server does vacation is 0.493611 so v = 0.493611.

In the analysis of the queuing model there are several assumptions that must be met, namely the arrival pattern which is poisson distribution, service time that is generally distributed either in the form of exponential distribution, erlang or other distribution, and vacation time which is exponentially distributed. Based on the results of testing the type of distribution using SPSS software, the results of the arrival pattern are poisson distribution with $\lambda = 27.4$ arrivals / hour, and service time is exponentially distributed with $S_1 = 46.97 \approx 47$ documents, Figure 1 and Figure 2.

One-Sample Kolmogorov-Smirnov Test

		VAR00001
N		4
Poisson Parameter[a,b]	Mean	27.5000
Most Extreme Differences	Absolute	.474
	Positive	.250
	Negative	-.474
Kolmogorov-Smirnov Z		.947
Asymp. Sig. (2-tailed)		.331

a. Test distribution is Poisson.

b. Calculated from data.

Figure 1. Test Distribution of the arrival pattern

One-Sample Kolmogorov-Smirnov Test

		VAR00001
N		110
Exponential parameter.[a,b]	Mean	1.2773
Most Extreme Differences	Absolute	.074
	Positive	.028
	Negative	-.074
Kolmogorov-Smirnov Z		.777
Asymp. Sig. (2-tailed)		.582

a. Test Distribution is Exponential.

b. Calculated from data.

Figure 2. Service Time Distribution Test

The vacation time is assumed to be exponential in distribution, the results of this study the resulting formula is suitable for vacation time which has an exponential distribution, the total time the server does vacation is 0.493611 so *V*= 0.493611, Table 1.

Table 1. Vacation Time

Vacation to	Start a Vacation	End of Vacation	Old Vacation
1	8:30:45	9:00:22	0:29:37
Amount			0:29:37 or 0.493611

According to [7], If in the Laplace-Stieltjes transformation the distribution function is known to have an exponential distribution, the transformation becomes:

$$F_X^*(s) = \int_0^\infty e^{-st} dF_X(t) = \int_0^\infty e^{-st}\lambda e^{-\lambda t} = \frac{\lambda}{\lambda+s} \quad (19)$$

Because the service time is exponentially distributed, then substitute equation (19) into equation (18) so that the sojourn time distribution model is obtained ($W^*(\theta)$), namely

$$W^*(\theta) = \frac{P_{0,0}\{\lambda(\theta-\lambda)\frac{\lambda}{\lambda+s_0}(\theta+v-vS_1^*(\theta))+\theta\frac{\lambda}{\lambda+s_1}[\lambda^2+v(\theta+\lambda+v]\}}{\lambda(v+\theta)(\theta-\lambda+\lambda\frac{\lambda}{\lambda+s_1})} \quad (20)$$

Then to find out the customer waiting time (W) of equation (20), so that it does not contain a variable θ, from

$$W = \lim_{\theta\to\infty} \frac{P_{0,0}\{\lambda(\theta-\lambda)\frac{\lambda}{\lambda+s_0}(\theta+v-vS_1^*(\theta))+\theta\frac{\lambda}{\lambda+s_1}[\lambda^2+v(\theta+\lambda+v]\}}{\lambda(v+\theta)(\theta-\lambda+\lambda\frac{\lambda}{\lambda+s_1})}$$

be,

$$W = \frac{\lambda v(1-\lambda E[S_1])}{\lambda^2+\lambda+v^2-\lambda^2(1+vE[S_1])\frac{\lambda}{\lambda+v}}\,\frac{2\lambda(2\lambda+s_1)+2v(2\lambda+s_0)}{(2\lambda+s_0)(2\lambda+s_1)} \quad (21)$$

In the next stage Little Law's is used to determine the number of customers in the system (L), and the server's busy schedule (ρ) is,

$$L = \frac{\lambda^2 v(1-\lambda E[S_1])}{\lambda^2+\lambda+v^2-\lambda^2(1+vE[S_1])\frac{\lambda}{\lambda+v}}\,\frac{2\lambda(2\lambda+s_1)+2v(2\lambda+s_0)}{(2\lambda+s_0)(2\lambda+s_1)} \quad (22)$$

and

$$\rho = \frac{\lambda}{s_1} \quad (23)$$

To get a numeric solution, with the help of c ++ software, substitute the parameters that have been obtained from the data into the models equation (21), (22), and (23). From the processing results, the average waiting time for documents in the system is 0.287174 hours, the average number of documents in the system is 7.897284 documents ≈8 documents, and the server busyness is 0.585106. so the server is not busy or does not need additional servers.

4. Conclusion

From the results of this study we obtained a sojourn time distribution model $(W^*(\theta))$, in service time which exponentially distributed one server with single working vacation and vacation interruption. Based on the results of data analysis by substituting parameters into the sojourn time distribution model $(W^*(\theta))$, the average waiting time of documents in the system (W), the average number of documents in the system (L), and the opportunity to get busy server (ρ) so that it can be concluded that the server is not busy or does not need to add a server.

Acknowledgment
Recognition: The main work is done with funding for that we thank the Academy Grant Leader (ALG) for all the support provided.

References

[1] Chae Kyung, Dae-Eun Lim, and Wong Yang 2009. *The M/G/1 queue and the GI/Geo/1 queue both with single working vacation.* Perform Eval, 66 (7) : 366-367.

[2] Cooper R B 1981. *Introduction to Queueing Theory.* 2nd. *Ed.* New York: Eleseveir North Holland, Inc.

[3] Gross D, and Harris C M 1998. *Fundamental of Queuing Theory* 3rd. New York: John Wiley & Sons.

[4] Hillier F S and Lieberman G J 2011. *Introduction to Operations Research.* New York: McGraw-Hill 7th edition.

[5] Hogg R V and Tanis E A 2001. *Probability and Statistical Inference*. 6th. *ed.* New Jersey: Prentice Hall International, Inc.
[6] Kim B K and Lee D H 2015. *A note on the sojourn time distribution of an M/G/1 queue with single working vacation and vacation interruption.* Operation Research perspective, 2 : 57-61.
[7] Osaki S 1992. *Applied Stochastic System Modeling.* Berlin : SpringerVerlag.
[8] Zhang M and Hou Z 2010. *Performance analysis of M/G/1 queue with working vacations and vacations interruption.* J Comput Appl Math. 234 (10) : 2977-2985.

IOP Conf. Series: Materials Science and Engineering **567** (2019) 012037 doi:10.1088/1757-899X/567/1/012037

Heuristic approach for solving capacitated location-routing problems

H N Ahmad[1], T Bakhtiar[1], P T Supriyo[1]

[1]Division of Operations Research, Department of Mathematics, Bogor Agricultural University, Kampus IPB Dramaga, Bogor 16680, INDONESIA

Corresponding author: afiesnugraha1@gmail.com

Abstract. This paper deals with the standard capacitated location-routing problem (CLRP), where two types of interdependent decisions must be made: which facilities among a number of potential ones should be operated and which vehicle routes should be built to fulfil the demand of customers from the operated facilities using a given fleet. We propose a heuristic mechanism to address the problem. The algorithm consists of four stages, namely optimally clustering customers by k-means and assessed by Dunn index, selecting the nearest depot to be opened based on Euclidean distance calculation, customers-to-depot matching and searching the optimal distribution route. This heuristic algorithm is then implemented to a dataset comprises of 20 customers and 5 potential depot facility locations.

1. Introduction

The Vehicle Routing Problems (VRP) is a terminology which refers to a problem of searching routes for a fleet of vehicles of known capacities to service a number of customers with known locations and known demands for a certain commodity, given a set of constraints, possibly integrated with production scheduling [1]. While, Facility Location Problems (FLP) concerns with selecting the placement of a facility (often from a list of integer alternatives) to minimize transportation costs and to best meet the demanded constraints [2]. If VRP and FLP are mutually considered then we have the so-called location-routing problems (LRP). In fact, classical LRP integrates the two kinds of decisions, namely opening a subset of depots, assigning customers to them and determining vehicle routes, to minimize a total cost including the cost of open depots, the fixed costs of vehicles used, and the total cost of the routes [3]. Both exact and heuristic approaches can be employed in solving LRP.

Book chapter by Marinakis [4] provides the concise reference for the basic notion of LRP that includes the model and its variants as well as the solution method. While, book by Drezner and Hamacher [2] offers more comprehensive review. Prodhon and Prins [3] analyzes the recent literatures on the standard LRP and new extensions such as several distribution echelons, multiple objectives or uncertain data, including results of state-of-the-art metaheuristics method. A more recent survey paper on standard LRP is given by Schneider and Drexl [5] as it provides concise paper excerpts that convey the central ideas of each work, discuss recent developments in the field, provide a numerical comparison of the most successful heuristic algorithms, and list promising topics for further research. Farahani et al. [6] delivers a review on recent efforts and development in multi-criteria location problems in three categories including bi-objective, multi-objective and multi-attribute problems and their solution methods.

Throughout the paper, the following main characteristics are adopted [3]:

IOP Conf. Series: Materials Science and Engineering **567** (2019) 012037 doi:10.1088/1757-899X/567/1/012037

- All relevant data are deterministic, i.e., are fully known in advance.
- There is only one planning period, i.e., a static planning situation.
- The number of potential locations for facilities is finite.
- There is only one single, scalar objective function.
- Each customer has a known demand that must be fulfilled by a delivery from exactly one of the potential facilities; there is no load transfers at intermediate locations allowed.
- Each customer must be visited exactly once by one vehicle.
- No inventory considerations apply, neither at facilities nor at customers.

2. Review on solution approach

From the perspective of solution method, a number of efficient algorithms and approaches have been introduced for addressing LRP, stimulated by the fact that it is an NP-hard combinatorial optimization problem that can be exactly solved only for limited instances of the problem. Belenguer et al. [7] is one proposing an exact approach based on branch-and-cut method to solve CLRP. While Tuzun and Burke [8] proposed a heuristic method based on two-phase tabu search. In the first phase, the best composition of opening depot was sought following by vehicle routing in the second phase. Prins et al. [9] utilized a greedy randomized adaptive search procedures (GRASP) equipped by learning process to decide depots should be opened and path relinking for routing. Prins et al. [10] suggested a solving approach by using lagrangean relaxation and granular tabu search. In the same effort, Duhamel et al. [11] also utilized GRASP and evolutionary local search (ESL) to solve the problem. Barreto et al. [12] proposed a clustering based analysis to address CLRP. Other heuristic approaches include the use of ant colony optimization [13, 14], hybrid genetic algorithm [15] and particle swarm optimization [16].

3. Method

In this work we consider the capacitated location-routing problem (CLRP), the most basic and general variant of LRP by adding capacity constraint on both depots and vehicles. We formulate the problem in term of mixed integer linear programming (MILP) and approaches heuristically by k-means clustering for grouping customers. Dunn index is measured to determine the optimal number of clusters. Our procedure consists of four stages: optimally clustering customers by k-means, selecting the best depot, allocating customers to selected depot, and searching the optimal distribution route.

3.1. Dataset

For algorithm testing, we consider a hypothetical dataset presented in [10]. The dataset consists of 20 customers with known level of demand and 5 candidates of facility with homogeneous capacity of 140 units. The scatter plot of nodes for dataset is provided in Figure 1. In this plot, black diamonds denote the customers and squares represent the potential locations of facilities.

3.2. Customers clustering

Clustering of customers is the first step of our heuristic algorithm. We employ the well-known k-means clustering for grouping customers with $2 \leq k \leq \lceil n/4 \rceil$, where n is the number of customers. The best number of clusters k is determined by means of Dunn index. Let $C = \{C_1, C_2, \dots, C_k\}$ be a set of clusters, let $\delta: C \times C \to \mathbb{R}^+$ be a cluster-to-cluster distance measure and let $\Delta: C \to \mathbb{R}^+$ be a cluster diameter measure. The Dunn index DI for the set C is defined as [18]:

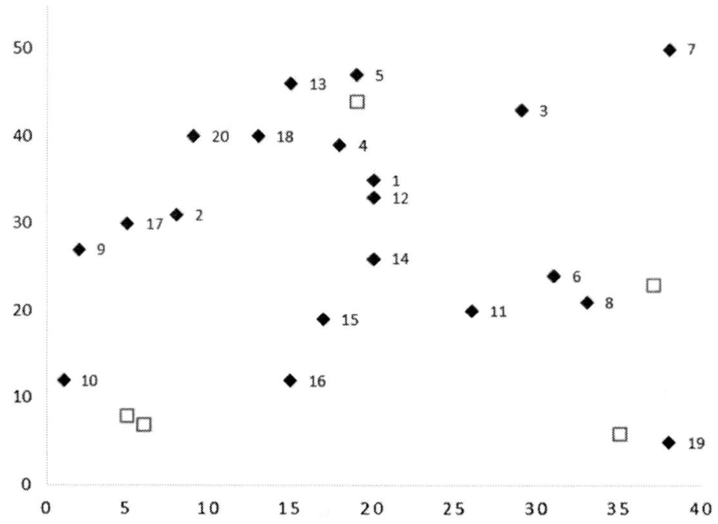

Figure 1. The location of customers and potential locations of facilities in the third dataset

$$DI(C) = \frac{\min_{i \neq j}\{\delta(C_i, C_j)\}}{\max_{1 \leq l \leq k}\{\Delta(C_l)\}}, \tag{1}$$

where $\delta(C_i, C_j) = \min_{x \in C_i, y \in C_j}\{d(x,y)\}$ and $\Delta(C_l) = \max_{x,y \in C_l}\{d(x,y)\}$ with $d: C \times C \to \mathbb{R}^+$ is object-to-object distance measure like Manhattan distance or Euclidean distance. For a given assignment of clusters, a higher Dunn index indicates better clustering. The process of customers clustering is required to decide the number of depots should be operated.

3.3. Selecting the depot facilities

The mechanism for selecting the best depot facilities relies on the calculation of Euclidean distance between centroid of cluster and depot facility. Let (a_i, b_i) denotes the centroid of cluster C_i and (p_j, q_j) denotes the location of (potential) depot facility j, then the depot facility j will be operated as long as its total distance D_j is minimum:

$$D_j = \min_{1 \leq j \leq m} \sum_{i=1}^{k} \sqrt{(p_j - a_i)^2 + (q_j - b_i)^2}, \tag{2}$$

where m is the number of (potential) depot facilities. If the total demand of all customers cannot be supplied by depot j, then the second best depot, i.e., depot facility with the next minimum total distance, should be opened. The process of opening next depots is continued until the total demand satisfied.

3.4. Customer-to-depot allocation

Customer-to-depot assignment is undertaken by measuring the Euclidean distance of each customer to each selected depot. A number of customers with minimum distance are then assigned to the nearest selected depot. The process of matching is continued until all customers assigned to depots. In this process, a customer reallocation might be needed regarding the capacity reachability of each selected depot.

3.5. Vehicle routing

The last step of the algorithm is searching the best routes from depot to customers for delivering products. A fleet of vehicles, each with known and fixed capacity, starts at a designated depot and returns to the same depot after visiting customers where service or product is demanded. The objective is either to minimize the total distance of all the routes or total cost.

Since the customers demand fulfilment by depot is guaranteed in the previous step, then in this paper we consider the most standard capacitated VRP. We here assume that a fleet of vehicles with known loading capacity is always available in each depot to deliver products. One may refer to Toth and Vigo [17] for a comprehensive account on this topic.

Suppose $\mathbb{C}$ denotes the set of all customers and $\mathbb{V}$ denotes the set of all vehicles available in the depot. If the number of customers is n, and thus $\mathbb{C} = \{1,2,\dots,n\}$, then the set of all nodes, i.e., depot and customers, is denoted by $\mathbb{N} = \{0\} \cup \mathbb{C} = \{0,1,2,\dots,n\}$, where 0 is index for the only depot. The following parameters are used: c_{ij} is the traveling cost between node i and j, W and Q are capacities of depot and vehicle, respectively, and d_j is the demand level of customer j. For decision variable, $x_{ijk} = 1$ if vehicle k travel from node i to node j, $x_{ijk} = 0$ otherwise. VRP can be formulated such that the following total cost is minimized:

$$\min_{x_{ijk}} Z := \sum_{i\in\mathbb{N}} \sum_{j\in\mathbb{N}} \sum_{k\in\mathbb{V}} c_{ij} x_{ijk}, \tag{3}$$

subject to the constraints

$$\sum_{i\in\mathbb{N}} \sum_{k\in\mathbb{V}} x_{ijk} = 1, \qquad \forall j \in \mathbb{C}, \tag{4}$$

$$\sum_{j\in\mathbb{N}} \sum_{k\in\mathbb{V}} x_{ijk} = 1, \qquad \forall i \in \mathbb{C}, \tag{5}$$

$$\sum_{j\in\mathbb{C}} \sum_{i\in\mathbb{N}} d_j x_{ijk} \le Q, \qquad \forall k \in \mathbb{V}, \tag{6}$$

$$\sum_{j\in\mathbb{C}} d_j \le W, \tag{7}$$

$$\sum_{j\in\mathbb{C}} x_{ijk} - \sum_{j\in\mathbb{C}} x_{jik} = 0, \qquad \forall i \in \mathbb{N}, \forall k \in \mathbb{V}, \tag{8}$$

$$\sum_{j\in\mathbb{C}} x_{0jk} \le 1, \qquad \forall k \in \mathbb{V}, \tag{9}$$

$$\sum_{i\in\mathbb{S}} \sum_{j\in\mathbb{S}} x_{ijk} \le |\mathbb{S}| - 1, \qquad \forall \mathbb{S} \subseteq \mathbb{C}, \forall k \in \mathbb{V}. \tag{10}$$

Constraints (4) and (5) ensure each customer is visited and exited once. Constraint (6) guarantees that the total delivered product is not exceeding the capacity of vehicle. While condition (7) requires the total demand of all customers is not exceeding the capacity of the depot. Constraint (8) enforces the route continuity, meaning that as soon as a vehicle reaches a customer to deliver products, it should leave that place at once. The requirement that all trips must be started at depot is given by (9), while (10) eliminates sub-tour.

4. Results and discussion

We implement our formulation and algorithm to a dataset that consists of 20 customers and 5 depot candidates. It is assumed that the amount of products demanded by customers vary but the capacity of depots are homogeneous. Each depot has a sufficient number of vehicles to deliver the products, however the loading capacity of vehicles are assumed to be the same, which is 70 unit/vehicle. Table 1 provides the location of all nodes as well as their demand or capacity levels, while Figure 1 depicts the scatter plot.

Based on k-means clustering analysis on customers location as well as their respecting Dunn index, it is suggested that customers can be best clustered into four clusters as indicated by its highest Dunn index (see Table 2). Under this 4-group clustering, it is known that clusters C_1 and C_3 consist of 7 customers, cluster C_2 has 4 customers and cluster C_4 has only 2 members (see Table 3).

Table 1. Nodes location, demand and capacity

Node	**Coordinate**		**Demand (Capacity)**
	x	y	
Customer 1	20	35	17
Customer 2	8	31	18
Customer 3	29	43	13
Customer 4	18	39	19
Customer 5	19	47	12
Customer 6	31	24	18
Customer 7	38	50	13
Customer 8	33	21	13
Customer 9	2	27	17
Customer 10	1	12	20
Customer 11	26	20	16
Customer 12	20	33	18
Customer 13	15	46	15
Customer 14	20	26	11
Customer 15	17	19	18
Customer 16	15	12	16
Customer 17	5	30	15
Customer 18	13	40	15
Customer 19	38	5	15
Customer 20	9	40	16
Total			**315**
Depot 1	6	7	140
Depot 2	19	44	140
Depot 3	37	23	140
Depot 4	35	6	140
Depot 5	5	8	140
Total			**700**

Table 2. Dunn index

Number of clusters	2	3	4	5
Dunn index	0.166	0.155	0.253	0.181

IOP Conf. Series: Materials Science and Engineering **567** (2019) 012037 doi:10.1088/1757-899X/567/1/012037

Table 3. Result of 4-group clustering

Cluster	Centroid		Customer
	x	y	
C_1	16.286	40	1, 4, 5, 12, 13, 18, 20
C_2	4	25	2, 9, 10, 17
C_3	25.714	18.143	6, 8, 11, 14, 15, 16, 19
C_4	33.500	46.500	3, 7

Table 4. Distance between centroid and depot

Cluster	Depot 1	Depot 2	Depot 3	Depot 4	Depot 5
C_1	34.566	4.834	26.797	38.810	33.931
C_2	18.111	24.207	33.060	36.359	17.029
C_3	22.645	26.715	12.286	15.286	23.064
C_4	48.130	14.714	23.759	40.528	47.901
Total	123.452	70.470	95.903	130.984	121.926

Table 4 provides the distance between centroid of each cluster and depot candidates. Depot 2 has the smallest total distance of 70.470. Thus, Depot 2 with capacity 140 units is selected for opening. However, according to Table 1 the total demand is 315 units. It means that in addition to Depot 2, operating two more depot is required such that the total capacity of depots is 420 units. Depots 3 and 5 which have the next smallest total distances should be opened. Table 5 gives the distance between each customer and Depot 2, 3 and 5, from which we can decide the customer-depot matching based on the smallest distance. At this stage, the total demand for Depot 2 is 171 units, which is exceeding its capacity. While those for Depots 3 and 5 are 73 and 71 units, respectively. It is imperative to reallocate a few number of customers of Depot 2 to either Depot 3 or Depot 5. The last two columns of Table 5 calculates additional distance incurred by customer reallocation from Depot 2 to Depots 3 and 5. Thus, it is suggested to reallocate Customers 17 and 2 to Depot 5, respectively. Consequently, total demand should be fulfilled by Depot 2 becomes $171 - 15 - 18 = 138$ units and total demand for Depot 5 becomes $71 + 15 + 18 = 104$ units.

Figure 2 illustrates the solution of VRP over 3 depots and 20 customers. It is known that Depot 2 should dispatch 3 vehicles to deliver products to 9 customers with total distance of 88.991, Depot 3 uses 2 vehicles to serve 5 customers with total distance of 71.930 and Depot 5 operates 2 vehicles to visit 6 customers with total distance of 85.619. In Figure 2, red squares indicate the selected depots.

5. Conclusion

A heuristic algorithm based on k-means clustering and Euclidean distance has been introduced to approach the CLRP. The algorithm comprises of customers clustering assessed by Dunn index, selecting the depot to be operated among a number of potential facilities, customer-depot matching and vehicle routing. The algorithm has successfully been implemented to a dataset consists of 20 customers and 5 potential depot facilities.

Future research direction in this area includes the application of hierarchical methods and use of route length formulae; extension to dynamic and stochastic problems; development of integrated methods in logistics, e.g., solving the location-routing-inventory and the location-routing-packing problems; and multiple objective LRP.

Acknowledgments

Authors gratefully acknowledge financial support provided by Ministry of Research, Technology and Higher Education of the Republic of Indonesia under grant No. 011/SP2H/LT/DRPM/IV/2017 through the PUPT Scheme of Bogor Agricultural University.

Table 5. Customer-depot matching

Customer	Distance			Assigned Depot	Total Demand	Additional distance due to reallocation	
	Depot 2	Depot 3	Depot 5			Depot 3	Depot 5
1	9.055	20.809	30.887	2		11.753	21.831
2	17.029	30.083	23.195	2		13.054	6.165
3	10.050	21.541	42.438	2		11.491	32.388
4	5.099	24.840	33.615	2		19.740	28.516
5	3	30	41.437	2		27	38.437
7	19.925	27.018	53.413	2	171	7.094	33.489
12	11.045	19.723	29.155	2		8.678	18.109
13	4.472	31.828	39.294	2		27.355	34.822
17	19.799	32.757	22	2		12.958	2.201
18	7.211	29.411	32.985	2		22.200	25.774
20	10.770	32.757	32.249	2		21.986	21.479
6	23.324	6.083	30.529	3			
8	26.926	4.472	30.871	3			
11	25	11.402	24.187	3	73		
14	18.028	17.262	23.431	3			
19	43.382	18.028	33.136	3			
9	24.042	35.228	19.235	5			
10	36.715	37.643	5.657	5	71		
15	25.080	20.396	16.279	5			
16	32.249	24.597	10.770	5			

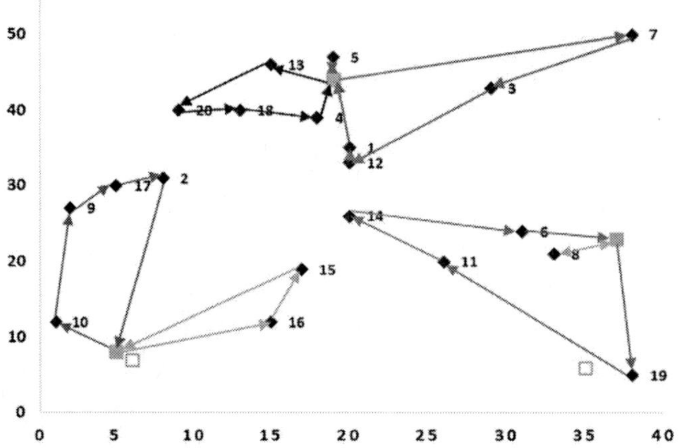

Figure 2. The optimal routes.

References

[1] Moons S, Ramaekers K, Caris A and Arda Y 2017 *Comp. & Indust. Eng.* **104** 224–245.
[2] Drezner Z and Hamacher HW 2002 *Facility Location: Applications and Theory* (Berlin: Springer-Verlag).
[3] Prodhon C and Prins C 2014 *Euro. J. Oper. Res.* **238** 1–17.
[4] Marinakis Y 2009 *Encyclopedia of Optimization Second Edition*, ed CA Floudas and PM Pardalos (New York: Springer) p 1919–1925.
[5] Schneider M and Drexl M 2017 *Ann. Oper. Res.* **259** 389–414.
[6] Farahani RZ, SteadieSeifi M and Asgaria N 2010 *App. Math. Model.* **34** 1689–1709.
[7] Belenguer JM, Benavent E, Prins C, Prodhon C and Calvo RW 2011 *Comp. Oper. Res.* **38(6)** 931–941.
[8] Tuzun D and Burke LI 1999 *Euro. J. Oper. Res.* **116(1)** 87–99
[9] Prins C, Prodhon C and Calvo RW 2006 *4OR* **4(3)** 221–238.
[10] Prins C, Prodhon C, Ruiz A, Soriano P and Calvo RW 2007 *Transport Science* **41(4)** 470–483
[11] Duhamel C, Lacomme P, Prins C and Prodhon C 2010 *Comp. Oper. Res.* **37(11)** 1912–1923.
[12] Barreto SS 2004 Análise e Modelização de Problemas de Localização-Distribuição *PhD Thesis* (Portugal: University of Aveiro)
[13] Nadizadeh A, Sahraeian R, Zadeh AS and Homayouni SM 2011 *Afric. J. Buss. Managm.*, **5(17)** 7499–7506.
[14] Ting CJ and Chen CH 2013 *Int. J. Prod. Econ.* **141(1)** 34 – 44.
[15] Lopes RB, Ferreira C, Santos BS 2016 *Comp. Scien. Res.* **70(1)** 155–162.
[16] Peng Z, Manier H and Manier A 2017 *Int. Feder. Automat. Contr.* **50(1)** 14668–14673.
[17] Toth P and Vigo D 2014 *Vehicle routing: problems, methods, and applications* (Philadelphia: SIAM-MOS).
[18] Abonyi J and Feil H 2007 *Cluster analysis for data mining and system identification* (Berlin: Birkhäuser Verlag AG).

IOP Conf. Series: Materials Science and Engineering **567** (2019) 012038 doi:10.1088/1757-899X/567/1/012038

Determination of optimal insurance company reimbursement by using exponential utility functions

Sukono[1], E. Lesmana[1], E.S. Hasbullah[1],V. Rosario[1], Y. Hidayat[2], Subiyanto[3]

[1]Department of Mathematics, Faculty of Mathematics and Natural Sciences, Universitas Padjadjaran, INDONESIA
[2]Department of Statistics, Faculty of Mathematics and Natural Sciences, Universitas Padjadjaran, INDONESIA
[3]Department of Marine Sciences, Faculty of Fishery and Marine Sciences, Universitas Padjadjaran, INDONESIA

Corresponding author: sukono@uanpad.ac.id

Abstract. In an insurance company, it is very possible that there is an imbalance between the amount of the premium and the amount of claims submitted by the insured. In managing the risk of such an imbalance, the insurance company must be able to make a decision to determine the appropriate amount of premium based on a number of claims by the insured. To make these decisions, insurance companies can do so by using certain utility functions. This paper examines the use of exponential utility functions in transferring risk between insurance companies, optimal Pareto exchanges, equilibrium, Bowley solutions, and optimal reinsurance purchases. Furthermore, using the exponential utility function is calculated the optimal amount of reimbursement that will be carried out by the insurer. The results of the discussion obtained the factors that influence the calculation and the optimal amount of reimbursement value, from a property policy or group policy that needs to be carried out by the insurer. So that based on the results of the optimal reimbursement calculation, it can be used to determine the balance between the amount of the premium and the amount of claims that may be submitted by the insured.

1. Introduction

Insurance companies are engaged in business activities that focus on risk management issues. This happens because the insurance company is willing to provide compensation or compensation to the insured or insured owner for loss, damage, or legal liability to third parties that may be suffered by insured because of an uncertain event, or give compensation for death or life someone insured [1; 2; 3].

The existence of insurance companies is because insured are willing to pay premiums on the expectation of their claims. As a result, insurance companies (insurers) collect premiums greater than the expected size of the claim [4; 5]. In the management of insurance companies, it is very possible that there is an imbalance between the amount of the premium and the amount of the claim. This can threaten the stability of insurance companies [6; 7].Therefore, insurance companies must be careful in managing the risks that may occur against insured, and risks that may occur within the insurance company itself. In managing this risk, insurance companies must be able to make decisions in determining the amount of premium for a number of claims submitted by insured [8; 9].

The decision to determine the size of the premium made by the insurance company is based on preference as a particular consequence. One thing that also needs to be considered is the determination of the premium is the occurrence of inflation [10]. Therefore, insurance companies must be able to choose an action that will lead to an uncertain event [11; 12].A theory that provides knowledge about decision making to deal with uncertain events in insurance problems, and which explains why insured is willing to pay a premium that is greater than net premium is a mathematical expectation or known as utility theory [13; 14].Therefore, this paper examines optimal reimbursement calculations in insurance companies. In the calculation of the reimbursement is done using the exponential utility function. The objective of the optimal reimbursement calculation is to find a balance between the amount of premium and the amount of claims that may be submitted by the insured. This research is useful for insurance companies in determining optimal reimbursement, especially when utility functions are exponential.

2. Mathematical models

This section discusses mathematical models including: utility functions, transfer of risk between insurance companies, transfer of optimal Pareto risk, equilibrium, Bowley solutions, and purchase of optimal reinsurance.

2.1. Utility function

A utility function is based on the choice of the decision maker for various outcome distributions, which illustrates how the utility of a particular value or circumstance is for decision makers. In general the utility scale is expressed between 0 and 1, where 1 represents the most preferred value or state and 0 represents the least preferred value or situation. An insurer who acts as a decision maker does not have to be individually, but may also be an alliance, legal entity, or government. Suppose $u(x)$ stating the utility function, the determination of the form of utility functions for insurer is probably an easy process [13; 14].

There are several types of functions, but this paper only examines the exponential utility function. In general, the exponential utility function is expressed as:

$$u(x) = \frac{1}{a}(1 - e^{-ax});\ a > 0 \tag{1}$$

with a stated the parameters, and x value or state of benefits. The properties of the utility function are $u'(x) > 0$ and $u''(x) < 0$. Based on these properties indicate that the utility function is an ascending concave. Decision makers with an elevated concave utility function are called risk avoiders. Risk avoidance coefficient of utility function $u(x)$ is given by:

$$r(x) = \frac{-u''(x)}{u'(x)} = -\frac{d}{dx}\ln u'(x), \tag{2}$$

assuming that the risk avoidance function $r(x) > 0$ [13; 14].

To determine the amount of an appropriate premium (equitable), it is assumed that insurer has a utility function u, and premiums are viewed in terms of their utility. If $u(s)$ a concave function and S random variable, then Jensen's inequality applies $E[u(S)] \geq u(E[S])$. Suppose π the amount of premium, based on Jensen's inequality can be determined that $E[u(x+\pi-S)] \leq u(E[x+\pi-S]) = u(x)$. Premium π obtained as a solution of the equation [2; 11]:

$$E[u(x+\pi-S)] = u(x). \tag{3}$$

Equation (3) generally cannot be solved analytically to search π. If distribution S have a central moment around the average, then π can be predicted using the following equation [2; 3]:

$$\pi \approx E[S] + \frac{1}{2}r(x)Var(S). \tag{4}$$

For the exponential utility function in equation (1), the amount of the premium π can be determined that:

$$\pi = \frac{1}{a} \ln E[e^{aS}]. \tag{5}$$

2.2. Risk transfer between insurance companies

Suppose n the number of companies, and X_i $(i = 1,...,n)$is a company surplus i at the end of the year. The amount of surplus at the end of the year X_i is a surplus at the beginning of the year, plus premiums received, and deducted by claims paid. Suppose x surplus early in the year ($t = 0$), c premium received per unit of time, and $S(t)$ aggregate claims paid up to time t. The surplus function can be expressed as [1; 5; 8]:

$$X(t) = x + ct - S(t),\ t \geq 0. \tag{6}$$

Because X_i random variables, it is assumed that the distribution is shared ($X_1,..., X_n$) is known. So, with the transfer of risk the company can try to improve the situation in several ways. The result of this risk transfer, for example a random vector ($Y_1,..., Y_n$) where Y_i $(i = 1,...,n)$ change in surplus from the company i. The combination of surplus after transfer of risk with a surplus before the transfer of risk must be equal, meaning:

$$Y_1 + ... + Y_n = X_1 + ... + X_n. \tag{7}$$

Furthermore, suppose $X_1 + ... + X_n = X$, one simple form of risk transfer formula is:

$$Y_i = q_i X + c_i, \tag{8}$$

where q_i proportion of company parts i ($q_1 + ... + q_n = 1$), and c_i deterministic part of payment by the company i ($c_1 + ... + c_n = 0$) [2; 5].

2.3. Pareto optimal risk transfer

Suppose the company i use the utility function u_i to evaluate risk transfer. So value ($Y_1,..., Y_n$) for companies i is $E[u_i(Y_i)], i = 1,...,n$. A company is said to achieve optimal Pareto, if there is no other risk transfer that is better for all companies. Geometrically, the Pareto optimal exchange member is identical to the solution of the selection of positive constants $k_1,..., k_n$ which maximizes the equation $k_1 E[u_1(Y_1)] + ... + k_n E[u_n(Y_n)]$. Using variations in calculus and the utility of the utility function, equivalence conditions are obtained $k_i u'_i(Y_i)$ that does not depend on each other i. Based on equation (7) can be determined Pareto optimal risk transfer [7; 8].

If it is assumed that the utility function is an exponential form like equation (1), from parameter (8) the parameters can be obtained:

$$q_i = \frac{a}{a_i},\ \text{with}\ \frac{1}{a} = \frac{1}{a_1} + ... + \frac{1}{a_n},\ \text{and} \tag{9}$$

$$c_i = \frac{1}{a}\left(\ln k_i - \sum_{j=1}^{n} \frac{a}{a_j} \ln k_j \right). \tag{10}$$

In general there will be as many ($n - 1$) Member parameters for Pareto optimal risk transfer.

2.4. Equilibrium

Suppose there is n the company formed a group. Company i $(i = 1,...,n)$ can buy a payment (reimbursement) R_i from the group for a premium π_i $(i = 1,...,n)$, where R_i is a random variable. So that the transfer of risk can be formulated with the equation $Y_i = X_i - \pi_i + R_i$, with equilibrium that:

$$R_1 + ... + R_n = \pi_1 + ... + \pi_n. \tag{11}$$

Suppose P intensity premium which is a random variable with $P > 0$ and $E[P] = 1$ unit, the appropriate premium is determined by the equation:

$$\pi_i = E[PR_i] = E[R_i] + Cov(P, R_i). \tag{12}$$

Random variable P is a function of $X_1,..., X_n$. [7; 11].

Company i want to choose R_i which maximizes $E[u_i(X_i - \pi_i + R_i)]$, where $\pi_i = E[PR_i]$. Value of R_i the optimal can be generated based on conditions:

$$\frac{u'_i(X_i - \pi_i + R_i)}{E[u'_i(X_i - \pi_i + R_i)]} = P. \tag{13}$$

It is said that ($P, R_1,..., R_n$) an equilibrium when (11) and (13) is met for $i = 1,...,n$. In the case of an exponential utility function (1), based on conditions (13) will give the equation:

$$R_i = -X_i - \frac{1}{a_i \ln P + d_i}, \tag{14}$$

where d_i an arbitrary constant. So with equation (11) and conditions that $E[P] = 1$ unit, it was found that [2; 7]:

$$P = \frac{e^{-aX}}{E[e^{-aX}]}. \tag{15}$$

If selected d_i such that $E[R_i] = 0$, using equation (15) the equation will be obtained $R_i = -(X_i - \mu_i) + q_i(X - \mu)$, where $\mu_i = E[X_i]$ and $\mu = E[X]$, then:

$$\pi_i = -Cov(P, X_i) + q_i Cov(P, X), \text{ and } Y_i = q_i X - \pi_i + \mu_i - q_i\mu. \tag{16}$$

Equation (16) shows that risk transfer is truly Pareto optimal. If it is assumed that $a_1,..., a_n$ small, then the estimation of the first order is obtained by obtaining the equation [13; 15]:

$$P \approx 1 - a(X - \mu), \text{ and } \pi_i \approx aCov(X, X_i) - aq_i Var(X). \tag{17}$$

2.5. Bowley solution

Suppose there are two insurance companies, namely insurer and reinsurer. Insurer can buy a payment $R = R_1$ for premiums $\pi = \pi_1$, which is calculated based on conditions $\pi = E[PR]$. Because given an intensity premium P, Insurer chooses R to maximize their utility expectations. As in equation (13), R generated based on conditions that:

$$\frac{u'_1(X_1 - \pi + R)}{E[u'_1(X_1 - \pi + R)]} = P. \tag{18}$$

It is assumed that the reinsurer knows the utility function and has this monopoly. So insurer will choose P to maximize utility expectations, that is $E[u_2(X_2 + \pi - R)]$. In this equation R depend on P as shown in equation (18). Results of completion of (P, R) called the Bowley solution [7; 8].

Suppose there are two utility functions known for insurer and reinsurer as follows:

$$u_1(x) = \frac{1}{\alpha}(1 - e^{-\alpha x}) \text{ and } u_2(x) = x, \tag{19}$$

where insurer has a risk avoidance coefficient $\alpha > 0$, and risk-neutral reinsurer. So the insurer request function is:

$$R = -X - \frac{1}{\alpha}\ln P + d, \tag{20}$$

where d constant changes, and reinsurer chooses P in order to maximize expectations of profits:

$$\pi - E[R] = -E(PX_1] - \frac{1}{\alpha}E[P\ln P] + E[X_1] + \frac{1}{\alpha}E[\ln P]. \tag{21}$$

Using calculus variations will be generated P determined from the equation:

$$-X_1 - \frac{1}{\alpha}\ln P + \frac{1}{\alpha P} = c, \tag{22}$$

where the constant c must be chosen so that $E[P]=1$. Equations (21) and (22) are Bowley solutions [8; 15].

2.6. Optimal Reinsurance Purchases

Suppose that a reinsurance premium is determined based on general principles $\pi = E[PR]$. Suppose also $\pi = H(R)$ is a premium that the insurer must pay for reimbursement R. Mathematically H is a function of a set of random variables. The problem is choosing R to maximize the equation:

$$E[u_1(X_1 - H(R) + R)]. \tag{23}$$

If the function H concave, then R optimal can be determined as a solution to the equation:

$$\frac{u'_1(X_1 - H(R) + R)}{E[u'_1(X_1 - H(R) + R)]} = H'(R). \tag{24}$$

Where $H'(R)$ as a gradient from H on R which is a random variable, as follows:

$$\frac{d}{dt}H(R + tQ)|_{t=0} = E[QH'(R)], \text{ for all } Q. \tag{25}$$

If $H(R) = E[PR]$, then $H'(R) = P$ [4; 15].

If for example $H(R) = \frac{1}{\beta}\ln E[Pe^{\beta R}]$, where is the parameter $\beta > 0$ and P random variable with $P > 0$ and $E[P] = 1$, then:

$$H'(R) = \frac{Pe^{\beta R}}{E[Pe^{\beta R}]}. \tag{26}$$

Furthermore, suppose u_1 exponential utility function like equation (1). Using (24) optimal reimbursement is:

$$R = -\frac{\alpha}{\alpha+\beta}X_1 - \frac{1}{\alpha+\beta}\ln P + d \tag{27}$$

where d changeable constants. Please note, that this result is only for special cases $P \equiv 1$, that is H exponentially shaped with parameters β [8; 15].

3. Numerical illustration

This section of the discussion carried out includes: data and information, as well as optimal reimbursement calculations.

3.1. Data and information

The data used in this numerical illustration is simulation data. For example, from an insurance company, property insurance data is obtained as given in Table-1. It is assumed that there are only two insurance companies, namely one insurer and one reinsurer. It is also assumed that the insurer is risk avoidance with utility functions $u_1(x)$ exponent, and risk-neutral reinsurer with utility functions $u_2(x)$ linear.

Table-1.Property insurance data

Property	Initial Surplus	Premium Received	Claims Paid
1	3,750	1,750	0
2	5,430	11,232	9,140
3	7,540	5,300	1,940
4	2,500	3,675	1,967
5	1,500	3,500	590
6	27,318	12,500	2,545
7	27,318	34,560	19,189
8	27,309	13,560	31,150
9	5,000	43,600	16,500
10	3,000	13,942	37,697
11	0	36,874	25,841
12	15,000	2,216	9,180
13	32,000	5,670	20,950
14	15,050	7,540	13,940
15	23,600	5,501	7,920
16	6,700	4,501	9,982
17	1,300	3,570	2,570
18	22,500	22,570	23,640
19	4,500	11,197	9,300
20	2,370	11,197	5,710

3.2. Optimal reimbursement calculation

Based on the data in Table-1, using equation (6) obtained surplus data at the end of the year (X_1) for each property, as in Table-2 of the Initial Surplus column. Next, it is assumed that the parameters $\alpha = 0.5$ utility function (19) become $u_1(x) = (1/0.5)(1-e^{-0.5x})$ and $u_2(x) = x$. The premium that will be paid by the insurer to the reinsurer is calculated based on the exponential principle. If assumed parameters $\beta = 0.005$, then the equation becomes $H(R) = (1/0.005)\ln E[Pe^{0.005R}]$. Because of the utility function $u_2(x) = x$ earned value $P = 1$. The amount $d_1 = E[X_1] = \sum_x x.u_1(x)$ $= \sum_x x.(1/0.5)(1-e^{-0.5x})$, by entering parameter values $\alpha = 0.5$ and $\beta = 0.005$ into equation (27), can be used to calculate the amount of reimbursement for each property. The results are given in Table-2 column *R*.

Based on the calculation results presented in Table-2 above it can be interpreted that the optimal reimbursement value for each property is less than the expected surplus amount. The greater the surplus obtained from a property insurance, the reimbursement that must be purchased will be smaller, and vice versa. In other words, if a property can generate a large enough surplus, then the reimbursement that must be paid by the insurer to the reinsurer will be smaller.

Table-2 Reimbursement of Each Property

Property	Initial Surplus X_1	$E[X_1]$	R
1	5,500	11,000	463,600.0
2	7,522	15,044	461,761.8
3	10,873	21,746	458,715.5
4	5,585	11,170	463,522.7
5	2,455	4,910	466,368.2
6	20,629	41,258	449,846.4
7	30,728	61,456	440,665.5
8	24,369	48,738	446,446.4
9	10,903	21,806	458,688.2
10	10,913	21,826	458,679.1
11	11,033	22,066	458,570.0
12	8,036	16,072	461,294.5
13	16,720	33,440	453,400.0
14	8,650	17,300	460,736.4
15	21,181	42,362	449,344.5
16	1,219	2,438	467,491.8
17	2,300	4,600	466,509.1
18	21,430	42,860	449,118.2
19	6,397	12,794	462,784.5
20	7,857	15,714	461,457.3
Total		**468,600**	

4. Conclusion

In this paper, an exponential utility function has been applied in calculating the transfer of risk between an insurance company, Pareto optimal risk transfer, equilibrium, and Bowley solutions, and an optimal reimbursement purchased by an insurer and reinsurer. The optimal reimbursement value for each property of a group of policies that is reinsured is less than the expectation of the group's surplus policy. The greater the surplus obtained, the reimbursement to be purchased is smaller, and vice versa. In other words, if a property can generate a large enough surpluses, then the reimbursement that must be paid by the insurer to the reinsurer will experience a smaller decrease. The limitation of the results of this study is that the optimal reimbursement calculation here only applies to exponential utility functions.

Acknowledgment

Acknowledgments are conveyed to the Rector, Director of Directorate of Research, Community Involvement and Innovation, and the Dean of Faculty of Mathematics and Natural Sciences, Universitas Padjadjaran, with whomthe Internal Grant Program of Universitas Padjadjaran was made possible to fund this research.The grant is a means of enhancing research and publication activities for researchers at Universitas Padjadjaran.

References

[1] Saputra, A., Sukono, and Rusyaman, E. 2018. Risk Adjustment Model of Credit life Insurance Using a Genetic Algorithm. *IOP Conf. Series: Materials Science and Engineering*, 332 (2018) 012007 doi:10.1088/1757-899X/332/1/012007.

[2] Sidi, P., M. Mamat, Sukono, and S. Supian, 2017, Insurance Premium Risk Evaluation Model

for Building Damage Caused by Flood of the River: Case Study on Citarum Watershed in South Bandung Indonesia, *Journal of Engineering and Applied Sciences*, 12 (17): 4420-4425, 2017, ISSN: 1816-949X.

[3] Delong, L. 2017. Optimal Investment for Insurance Company with Exponential Utility and Wealth-Dependent Risk Aversion coefficient. *Working Paper.* Warsaw School of Economics, Institute of Econometrics, Department of Probabilistic Methods, Niepodleg lo_sci 162, Warsaw 02-554, Poland. Email: lukasz.delong@sgh.waw.pl

[4] Simrová, S., Barták, M., Radovan Vojtísek, R., and Rogalewicz, V. 2014. The Costs and Reimbursements for Lung Cancer Treatment Among Selected Health Care Providers in the Czech Republic.*Ekonomika a Management*, 2014, XVII, 3, pp. 74-85. DOI: 10.15240/tul/001/2014-3-007.

[5] Sukono, Riaman, Lesmana, E., Wulandari, R., Napitupulu, H., and Supian, S. 2018. Model Estimation of Claim Risk and Premium for Motor Vehicle Insurance by Using Bayesian Method. *IOP Conf. Series:Materials Science and Engineering,*Vol. 300, (2018), art. no. 012027.

[6] Sidi, P., Mamat, M., Sukono, and Supian, S. 2017. Supply and demand analysis for flood insurance by using logistic regression model: case study at Citarum watershed in South Bandung, West Java, Indonesia. *IOP Conf. Series: Materials Science and Engineering* 166 (2017) 012002 doi:10.1088/1757-899X/166/1/012002.

[7] Wang, J. and Poh, K.L. 2017. Multiple Attribute Decision Making (MADM) in Proportional-Stop-Loss Reinsurance. *Journal of Risk Financial Management*, 2017, 10, 22; pp. 1-21. doi:10.3390/jrfm10040022.

[8] Quick, R., & Weglarz, D., (2012), Reinsurance Default Risk. *Working Paper,* General Reinsurance AG, Theodor Heuss-Ring 11, Cologne. www.genre.com/.../InsuranceIssues201210-en...

[9] Sidi, P., Mamat, M., Sukono, Supian, S., and Putra, A.S. 2018. Demand analysis of flood insurance by using logistic regression model and genetic algorithm. *IOP Conf. Series: Materials Science and Engineering*, 332 (2018), 012053. doi:10.1088/1757-899X/332/1/012053.

[10] Sukono, Hidayat, Y., Suhartono, Sutijo, B., Bon, A.T.B., and Supian, S. 2016. Indonesian Financial Data Modeling and Forecasting by Using Econometrics Time Series and Neural Network. *Global Journal of Pure and Applied Mathematics*. Vol. 12, No. 4, (2016), pp. 3745–3757.

[11] Garrison Jr., L.P. and Towse, A. 2017. Value-Based Pricing and Reimbursement in Personalised Healthcare: Introduction to the Basic Health Economics. *Journal of Personalized Medicine*, 2017, 7, 10; doi:10.3390/jpm7030010. www.mdpi.com/journal/jpm.

[12] Malamud, S., Rui, H., Whinston, A., Optimal Reinsurance with Multiple Tranches. *Journal of Mathematical Economics* (2016), pp. 1-45. http://dx.doi.org/10.1016/j.jmateco.2016.05.006.

[13] Sukono, Nahar, J., Putri, F.T., Subiyanto, Mamat, M., and Supian, S. 2017. Estimation of Outstanding Claims Reserving Based on Inflation Risk on Car Insurance Companies by Using the Bootstrap Method. *Far East Journal of Mathematical Sciences* (*FJMS*). Vol. 102, No. 4, (2017), pp. 687-706.

[14] Špirková, J. and Král, P. 2010. Aggregation Functions and Personal Utility Functions in General Insurance. *Acta Polytechnica Hungarica* Vol. 7, No. 4, 2010., pp. 7-23.

[15] Wei, W., & Cun Cai, (2011), Optimal Reinsurance with Positive Dependence, *Working Paper,* University of Waterloo. www.amstat.org/meetings/jsm/2010/pdfs/JSM2010ProgramBook.pdf

Mixed integer formulation for multiproduct maritime inventory routing problem

S Maulana[1], F Hanum[1] and T Bakhtiar[1]

Division of Operations Research, Department of Mathematics, Bogor Agricultural University, Kampus IPB Dramaga, Bogor 16680, INDONESIA

Corresponding author: maulsyahrulana17@gmail.com

Abstract. Companies that want to market their products to the outer islands need a large-capacity transportation mode that can distribute the product to every place. Other important consideration is the cost of the company must be efficient. The most commonly used by transportation logistics company is the mode of sea transportation. This paper presents an optimization model for determining vessel travel routes by meeting demand and paying attention to inventory levels at each ports so that the company's costs are minimum. The problem is known as the maritime inventory routing problem (MIRP) and the delivered product can be just one type (single product) or it can be multiproduct. MIRP models are formulated using integer linear programming and numerically solved by the aid of MiniZinc IDE 2.1.5.

1. Introduction

It is forecasted that world seaborne trade to increase by 2.8 per cent in 2017, with total volumes reaching 10.6 billion tons [1]. It is also projected that expansion continues with volumes growing at an estimated compound annual growth rate of 3.2 per cent between 2017 and 2022. Cargo flows are expected to expand across all segments, with containerized and major dry bulk commodities trades recording the fastest growth. These facts suggest an effective maritime transportation of utmost importance.

In maritime transportation, huge quantities of products are usually loaded and unloaded at production and consumption ports. In this situation, the loading process, inventory and transportation are time consuming and costly [2]. Thus, the routing and scheduling of the ships as well as the inventory management should be planned simultaneously. The resulting problem is called a maritime inventory routing problem (MIRP) [3, 4]. Optimization in maritime inventory routing is a well-established field of research in transportation planning with growing amount of research in the literature in various applications during the last decade [5].

This paper extends the work of Song and Furman (2013) into a multiproduct MIRP, where the model is proposed and formulated in a mixed integer programming. An optimization model for determining ships travel routes by meeting demand and paying attention to inventory levels at each place so that the company's costs are minimum is considered.

2. Related literature

Some researchers have explored the issue of MIRP, which examined the distribution of a single product on several production ports with a constant production rate and limited storage capacity taking into account travel time, the model was solved by column generation techniques [6]. Christiansen examined the problem of maritime inventory routing about the combination of inventory problems and the determination of ship routes with time constraints, and the model was solved using the Dantzig-Wolfe decomposition approach [7]. Christiansen and Fagerholt describe the usual MIRP as the transportation of one product produced in a production port and sent to a consumption port [5]. Al-Khayal and Hwang also explain the varied aspects of maritime transportation which formulate the problem of nonlinear programs into linear problems with special structures that minimize ship travel costs in loading and delivering large quantities of liquid / liquid products [8].

3. Problem description

Suppose there are several production ports J^P that will produce products sent to the port of consumption J^C in a certain period T. Each port has an initial inventory of products and has a minimum and maximum inventory limit. Travel time between ports is assumed in a day. Ships available to deliver products have different capacities and cost. A journey begins when the ship is empty and ends when it is empty again. Based on the depth level of the seawater in the port, the ships that come and go have product draft limits. Settlement of distribution and inventory problems requires costs which consist of the cost of renting a vessel, overloading which exceeds the minimum part cargo vessel as well as the cost of travel to the destination port and demurrage costs then subtracting the netback multiplied by the number of delivered products. The purpose of this problem is to find the optimal route with the minimum cost of the ship trip by taking into account certain capacities, constraints and limitations. Examples of illustrated ship trips can be seen in figure 1.

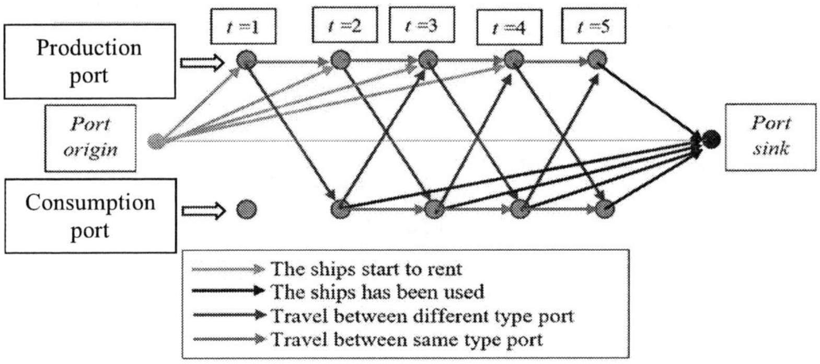

Figure 1. Ilustration of a ship trip

MIRP problems can be described by a time-space network, which consists of a set of nodes and an arc set. Nodes represent possible inter-port trips that can be written as (j, t) with j declaring a port and t declaring a specified time. The node (j, t)states the ship's visit to port j at time t. The node set is divided by each ship and each ship has its own set of arc. The set of nodes consists of the origin node $(0,0)$, sink node or end node $(0, T+1)$, and the set of regular nodes, namely nodes other than the original node and end node. Arc states the possibility of a trip by a ship that can be written as $(v, (j, t), (k, t+1)$with v denotes the ship and j is the initial port at time t and k is the destination port that reaches the time $t+1$ (Song and Furman, 2013) [4].

This problem is definedas follows. Given:

- a discrete model with units of time in days,
- horizon / planning period along T,
- the number of products / goods produced or consumed per day on each port is known,

IOP Conf. Series: Materials Science and Engineering **567** (2019) 012039 doi:10.1088/1757-899X/567/1/012039

- single shipped products for each port,
- each port has a product limit at the time of entry or exit port and has a minimum limit and maximum product inventory for each time,
- only one ship can load or discharge in one day,
- total vessel visits for loading or discharging to all ports are limited,
- travel time between ports is only one day.

This paper modifies the model by Song and Furman (2013) [4]. They study the optimal schedule for routing a heterogeneous vessels to sent a single bulk product while maintaining all capacities and inventory levels at each ports. This paper developed a multiproduct MIRP problem which involves two types of products.

4. Model

4.1 Notation

The following sets, parameters, and variables are used throughout the paper.

Sets

V	set of all vessels with index v where $V = V^C \cup V \backslash V^C$
V^C	set of vessels that hired
$V \backslash V^C$	set of vessel that unhired
J	set of all ports wit index $j$where $J = J^P \cup J^C$
J^P	set of production port
J^C	set of consumption port
N	set of all node with index n
N^R	set of regular node
A	set of all arc with index a where $\cup_{v \in V} A_v$
A_v	set of arcs belonging to vessel v
$\delta^+(n)$	set of arcs that have node n as their tail node
$\delta^-(n)$	set of arcs that have node n as their head node
B	set types of product sent with index b

Parameters

$IP_{(j,b,0)}$	the initial amount of product type b inventory level at production port
$IP_{(j,b,t)}^{min}$	minimum amount of product type b in production port at time t
$IP_{(j,b,t)}^{max}$	maximum amount of product type b in production port at time t
$IC_{(j,b,0)}$	the initial amount of product type b inventory level at consumption port
$IC_{(k,b,t)}^{min}$	minimum amount of product type b in comsumption port at time t
$IC_{(k,b,t)}^{max}$	maximum amount of product type b in consumption port at time t
b_{vb}	part cargo minimum for type b in vessel v
w_v	world scale multiplier for vessel v
r_v^o	overage multiplier scale for vessel v
DP_{vbjt}^I	draft limits which restrict the cargo capacity when entering production port
DP_{vbjt}^O	draft limits which restrict the cargo capacity when exiting production port
DC_{vbkt}^I	draft limits which restrict the cargo capacity when entering consumption port
DC_{vbkt}^O	draft limits which restrict the cargo capacity when exiting consumption port
c_{jk}	cost rate for traveling from port j to port k
C_{vb}	capacity in vessel v for product type b
c_a	the cost for using arc a

$p_{(b,j,t)}$	the number of products b produced by the production port j from time $t-1$ to time t
$d_{(b,k,t)}$	the number of products b consumed by the consumption port k from time $t-1$ to time t
fp_{bj}^{min}	minimum number of product type b that loaded on production port
fp_{bj}^{max}	maximum number of product type b that loaded on production port
fc_{bk}^{min}	minimum number of product type b that discharged on consumption port
fc_{bk}^{max}	maximum number of product type b that discharged on consumption port
m_j	maximum times to loads product
m_k	maximum times to discharge product
η	netback of each unit of product delivered by the ship.

Binary Variables

x_a	1 if vessel v use arc a, 0 otherwise
y_{vbjt}	1 if vessel v load product b, 0 otherwise
z_{vbkt}	1 if vessel v discharge product b, 0 otherwise

Continuous Variabel

fp_{vbjt}	number of product loaded
fc_{vbkt}	number of product discharged
$IP_{(b,j,t)}$	inventory levels in production port
$IC_{(b,k,t)}$	inventory levels in consumption port
IV_{vbt}	number of product in vessel v
o_{vba}	overage products in vessel v

4.2 Model formulation

The objective function of this problem is to minimize distribution costs which consist of ship usage costs and ship travel cost for overage product then reduced by netback for one ton of products delivered:

$$\min \sum_{a \in A} c_a x_a + \sum_{v \in V} \sum_{a \in A} \sum_{b \in B} w_v r_v^o c_{jk} o_{vba} - \sum_{v \in V} \sum_{k \in J^C} \sum_{t \in \{1,2,\dots,T\}} \sum_{b \in B} \eta f c_{vbkt}. \tag{1}$$

The constraints below describe a time-space network based formulation with flow conservation constraint:

$$\sum_{a \in A_v;\, a \in \delta^+(n)} x_a - \sum_{a \in A_v;\, a \in \delta^-(n)} x_a = 0, \quad \forall v \in V, \forall n \in N^R, \tag{2}$$

$$\sum_{a \in A_v;\, a \in \delta^+(0,0)} x_a = 1, \quad \forall v \in V, \tag{3}$$

$$\sum_{a \in A_v;\, a \in \delta^+(0,T+1)} x_a = 1, \quad \forall v \in V. \tag{4}$$

The nest set of constraints explain about inventory balance at each ports, namely production ports and consumption ports:

$$IP_{(b,j,t-1)} - \sum_{v \in V} fp_{vbjt} + p_{(b,j,t)} = IP_{(b,j,t)}, \quad \forall j \in J^P, \forall b \in B, \forall t \in \{1,2,\dots,T\}, \tag{5}$$

(6)

$$IC_{(b,k,t-1)} + \sum_{v \in V} fc_{vbkt} - d_{(b,k,t)} = IC_{(b,k,t)}, \quad \forall k \in J^C, \forall b \in B, \forall t \in \{1,2,\dots,T\}.$$

The next constraint is about inventory balance for products on the vessels are represented by the following expression:

$$IV_{vb(t-1)} + \sum_{j \in J^P} fp_{vbjt} - \sum_{k \in J^C} fc_{vbkt} = IV_{vbt}, \quad \forall v \in V, \forall b \in B, \forall t \in \{1,2,\dots,T\}. \tag{7}$$

The following set of constraints to ensure that a loading or discharging products by a vessel can occur only when the vessel is at that port.

$$y_{vbjt} \leq \sum_{a \in \delta^-(n)} x_a, \quad \forall n = (j,t) \in N^R, \forall j \in J^P, \forall v \in V, \forall b \in B, \forall a \in A_v, \tag{8}$$

$$z_{vbkt} \leq \sum_{a \in \delta^-(n)} x_a, \quad \forall n = (k,t) \in N^R, \forall k \in J^C, \forall v \in V, \forall b \in B, \forall a \in A_v. \tag{9}$$

If the process of loading or discharging products is happen then the loading amount or the discharging amount has to be in between the port specific minimum and maximum amounts:

$$fp_{bj}^{min} y_{vbjt} \leq fp_{vbjt} \leq fp_{bj}^{max} y_{vbjt}, \quad \forall v \in V, \ \forall b \in B, \forall j \in J^P, \ \forall t \in \{1,2,\dots,T\}, \tag{10}$$

$$fc_{bk}^{min} z_{vbkt} \leq fc_{vbkt} \leq fc_{bk}^{max} z_{vbkt}, \forall v \in V, \ \forall b \in B, \forall k \in J^C, \forall t \in \{1,2,\dots,T\}. \tag{11}$$

In many other practical cases, berthing rules specify that only one vessel can stop for loading or discharging products at a port. If you want to change the numbers of vessels desired to carry out product for load or discharge more than one, then the right side of the constraints below can be adjusted appropriately:

$$\sum_{v \in V} y_{vbjt} \leq 1, \quad \forall b \in B, \forall j \in J^P, \forall t \in \{1,2,\dots,T\}, \tag{12}$$

$$\sum_{v \in V} z_{vbkt} \leq 1, \quad \forall b \in B, \forall k \in J^C, \forall t \in \{1,2,\dots,T\}. \tag{13}$$

The next set of constraints is to uphold the assumption that total vessel visits for loading or discharging products are limited for the entire time the planning, we defined the limit total visits vessel by m_j:

$$\sum_{t \in \{1,2,\dots,T\}} y_{vbjt} \leq m_j, \quad \forall v \in V, \ \forall b \in B, \ \forall j \in J^P, \tag{14}$$

$$\sum_{t \in \{1,2,\dots,T\}} z_{vbkt} \leq m_k, \quad \forall v \in V, \ \forall b \in B, \ \forall k \in J^C. \tag{15}$$

The next set of constraints is to enforce the number of products on the vessel that will load in the production port or discharge in the consumption port so that when vessel enters or exits the port, it cannot exceed the specified limit due to water depth, but if the vessel does not load or discharge then the number of products only needs to below the vessel's capacity:

$$IV_{vb(t-1)} \leq DP_{vbjt}^{I} + (C_{vb} - DP_{vbjt}^{I})(1 - y_{vbjt}), \ \forall v \in V, \forall b \in B, \forall j \in J^P, \forall t \in \{1,2,\dots,T\}, \tag{16}$$

(17)

$$IV_{vbt} \leq DP^{O}_{vbjt} + (C_{vb} - DP^{O}_{vbjt})(1 - y_{vbjt}), \qquad \forall v \in V, \forall b \in B, \forall j \in J^P, t \in \{1,2,\dots,T\},$$

$$IV_{vbt-1} \leq DC^{I}_{vbkt} + (C_{vb} - DC^{I}_{vbkt})(1 - z_{vbkt}), \quad \forall v \in V, \forall b \in B, \forall k \in J^C, \forall t \in \{1,2,\dots,T\}, \tag{18}$$

$$IV_{vbt} \leq DC^{O}_{vbkt} + (C_{vb} - DC^{O}_{vbkt})(1 - z_{vbkt}), \qquad \forall v \in V, \forall b \in B, \ \forall k \in J^C, \forall t \in \{1,2,\dots,T\}. \tag{19}$$

The next constraint is for the overage number of products on a vessel when it exceeds the minimum part cargo of the vessel:

$$o_{vba} \geq \ (IV_{vbt} - b_{vb}) - (C_{vb} - b_{vb})(1 - x_a), \forall v \in V, \forall a \in A_v \,. \tag{20}$$

The following set of constraints are for the inventory level of products in the production portsor the consumption ports must be in between the minimum limit and the maximum limit for the product at each ports:

$$IP^{min}_{(j,b,t)} \leq \ IP_{(b,j,t)} \leq IP^{max}_{(j,b,t)}, \qquad \forall b \in B, \ \forall j \in J^{P}, \forall t \in \{1,2,\dots,T\}, \tag{21}$$

$$IC^{min}_{(k,b,t)} \leq \ IC_{(b,k,t)} \leq IC^{max}_{(k,b,t)}, \forall b \in B, \qquad \forall k \in J^C, \forall t \in \{1,2,\dots,T\}. \tag{22}$$

The next constraint is for ensure the number of products on a vessel does not exceed the capacity on the vessel:

$$0 \leq \ IV_{vbt} \leq C_{vb}, \qquad \forall v \in V, \forall b \in B, \forall t \in \{1,2,\dots,T\}. \tag{23}$$

the next is for the limitation of products over the ship when it exceeds the minimum part cargo of the vessel:

$$0 \leq \ o_{vba} \leq C_{vb} - b_{vb}, \forall v \in V, \forall b \in B, \ \forall a \in A_v. \tag{24}$$

The final set of constraints are for non-negativity and for binary constraints:

$$fp_{vbjt} \geq 0, \ fc_{vbkt} \geq 0, \quad \forall v \in V, \ \forall b \in B, \ \forall j \in J^P, \forall k \in J^C, \forall t \in \{1,2,\dots,T\}, \tag{25}$$

$$x_a \in \{0,1\}, y_{vbjt} \in \{0,1\} \ , \ z_{vbkt} \in \{0,1\}, \quad \forall v \in V, \ \forall b \in B, \forall j \in J^P, \ \forall k \in J^C, \ \forall a \in A_v, \forall t \in \{1,2,\dots,T\}. \tag{26}$$

5. Results and discussion

The optimization problem raised in this paper is solved numerically by MiniZinc, a programming language designed for optimization of problems with the decision variable in the form of real numbers. MiniZinc is designed with multiple solvers to produce optimal solutions by inputting the MiniZinc model and data files in the FlatZinc model. FlatZinc model consists of variable declarations and definitions of constraints and objective functions. Solvers contained in MiniZinc include Gecode (bundled), Gecode (Gist, bundled), Chuffed (bundled), COIN-OR CBC (bundled), Gurobi (bundled), G12 fd, G12 lazyfd, G12 MIP. MiniZinc can solve problems that are constrained easily because this software supports sets and arrays as well as types of decision variables that are diverse, namely integer, boolean, set of integer. In addition, the advantages of this software are that there are declarations of many global constraints like alldifferent, cumulative, regular tables, etc. which can be used to make the model easier. MiniZinc can also be used for different types of problems, namely satisfaction and optimization problems [9].

For an illustrative example, it is considered a simple routing problem of 4 ships connecting 5 production ports and 3 consumption ports and delivering 2 products namely Oil-1 and Oil-2 within

IOP Conf. Series: Materials Science and Engineering **567** (2019) 012039 doi:10.1088/1757-899X/567/1/012039

5 days of planning period. A number of parameters are utilized, such as the level of production and consumption in each port, including their initial levels and lower/upper bounds of inventory level, carrying capacity of ship, transportation and demurrage costs that can be seen in table 1-5. The problem was numerically solved by using software MiniZinc IDE 2.1.5 with solver COIN-OR CBC (bundled) under Windows 10 Pro operating system, processor intel/(R) core/(TM) i3-4150, CPU 3.50 GHz, RAM 4.00 GB which require 35 hours 11 minutes 43 seconds.

Table 1. Parameter values for each vessels

Notation	Vessel 1	Vessel 2	Vessel 3	Vessel 4
b_v	2	3	4	5
C_v	10	20	30	40
c_a	50	100	100	150

Table 2. Travel and demurrage cost

Port	A	B	C	D	E	F	G	H
A	3	3	4	6	4	8	7	3
B	3	4	3	3	7	3	4	3
C	4	3	2	5	2	2	3	4
D	6	3	5	3	5	2	3	7
E	4	7	2	5	1	3	2	3
F	8	3	2	2	3	2	5	5
G	7	4	3	3	2	5	4	3
H	3	3	4	7	3	5	3	2

Table 3. Parameter values for each production ports

Notation	Port A		Port B		Port C		Port D		Port E	
	Oil-1	Oil-2	Oil-1	Oil-2	Oil-1	Oil-2	Oil-1	Oil-2	Oil-1	Oil-2
$IP_{(j,0)}$	2	2	3	3	4	4	3	3	5	5
$IP^{min}_{(j,b,t)}$	1	1	1	1	1	1	1	1	1	1
$IP^{max}_{(j,b,t)}$	100	100	100	100	100	100	100	100	100	100
fp^{min}_j	2	2	3	3	2	2	3	3	2	2
fp^{max}_j	15	15	10	10	12	12	14	14	13	13
DP^I_{vjt}	10	10	15	15	12	12	20	20	17	17
DP^O_{vjt}	15	15	16	16	18	18	19	19	20	20
m_j	10	10	10	10	10	10	10	10	10	10

Table 4. Parameter values for each comsumption ports

Notation	Port F		Port G		Port H	
	Oil-1	Oil-2	Oil-1	Oil-2	Oil-1	Oil-2
$IC_{(j,0)}$	10	10	11	11	9	9
$IC^{min}_{(k,b,t)}$	1	1	1	1	1	1
$IC^{max}_{(k,b,t)}$	100	100	100	100	100	100
fc^{min}_{k}	0	0	0	0	0	0
fc^{max}_{k}	30	30	30	30	30	30
DC^{I}_{vkt}	20	20	20	20	20	20
DC^{O}_{vkt}	20	20	20	20	20	20
m_k	10	10	10	10	10	10

Table 5. Production and consumption level

Port j	*t* = 1		*t* = 2		*t* = 3		*t* = 4		*t* = 5	
	Oil-1	Oil-2	Oil-1	Oil-2	Oil-1	Oil-2	Oil-1	Oil-2	Oil-1	Oil-2
A	10	10	15	15	15	15	12	12	11	11
B	12	12	12	12	11	11	12	12	14	14
C	11	11	14	14	12	12	12	12	3	3
D	11	11	13	13	14	14	12	12	15	15
E	11	11	12	12	11	11	11	11	11	11
F	3	3	5	2	3	5	1	4	3	2
G	4	3	2	4	4	3	2	4	3	5
H	3	2	1	4	3	2	2	2	2	1

In the case of the distribution of this multiproduct, the total costs incurred were 733 and the vessels used / rented to deliver the product amounted to two vessels, Vessel 1 and Vessel 3. Both Vessel 1 and Vessel 3 transported two types of products at once. Ship 1 starts to be used when t = 1 by visiting port D to load two types of products at once with details Oil-1= 6 and Oil-2= 7. After loading the product, Vessel 1 goes to port F to send the product to when t = 2, while Vessel 3 is only used / rented on the second day by visiting port E to load the product with details Oil-1= 5 and Oil-2= 9 and immediately sent to port G the next day. The ship travel route in this case can be seen in figure 2.

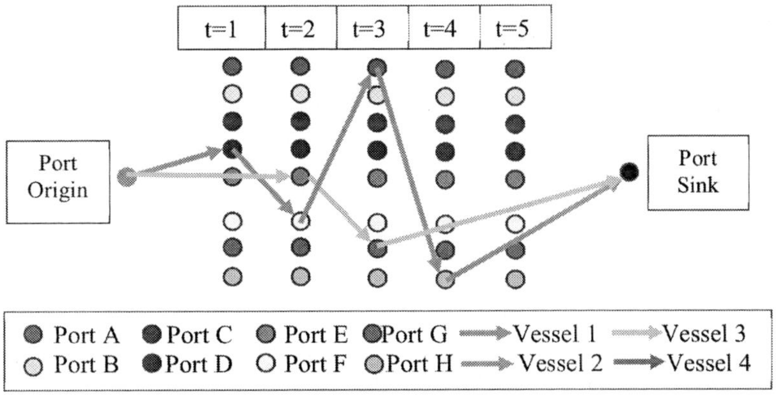

Figure 2. Vessel travel routes

6. Conclusion

In this scientific work it has been shown that the problem of determining the route of product distribution by ship, both when single and multi-product, can be seen as a mixed integer programming problem with an objective function to minimize ship operating costs. This fee consists of the cost of chartering the vessel and the overloading fee which exceeds the minimum part cargo of the vessel and is reduced by the return / netback of each delivery per product unit. Maritime Inventory Routing Model Problems in this scientific work both when multiproduct products for each optimization case were solved using the help of MiniZinc IDE 2.1.5 software with COIN-OR CBC (bundled) solver. The number of products shipped will be higher when the product return / netback value is greater than the excess charge per product unit, and the operational costs incurred by the company will be smaller. In the case of the distribution a single product can also use by this model by making set type of product sent is one.

Acknowledgements

Authors gratefully acknowledge financial support provided by Ministry of Research, Technology and Higher Education of the Republic of Indonesia under grant No. 011/SP2H/LT/DRPM/IV/2017 through the PUPT Scheme of Bogor Agricultural University.

References

[1] Song J H and Furman K C 2013 A maritime inventory routing problem: practical approach *Comp. & Ops. Res.* **40** 657–665.
[2] Christiansen M, Fagerholt K, Flatberg T, Haugen Ø, Kloster O and Lund E H 2011 Maritime inventory routing with multiple products: A case study from the cement industry *Euro. J. Oper. Res.* **208** 86–94.
[3] Agra A, Christiansen M, Hvattum L M and Rodrigues F 2018 Robust optimization for a maritime inventory routing problem *Trans. Sci.* **Articles in Advance** 1–17.
[4] Song J H and Furman K C 2013 A maritime inventory routing problem: practical approach *Comp. & Ops. Res.* **40** 657–665.Christiansen M and Fagerholt K 2009 Maritime inventory routing problem in *Encyclopedia of Optimization, Second Edition* edited by Floudasand C A and Pardalos P M (Springer, New York, 2009), pp. 1947–1955.
[5] Christiansen M and Fagerholt K 2009 Maritime inventory routing problem in *Encyclopedia of Optimization, Second Edition* edited by Floudasand C A and Pardalos P M (Springer, New York, 2009), pp. 1947–1955.
[6] Christiansen M and Nygreen B 1998 A method for solving ship routing problem with inventory constraints *Annals of Operations Research* 81:357-378.
[7] Christiansen M 1999 Decomposition of a combined inventory and time constrained ship routing problem *Transportation Science.* 33(1):3-16.
[8] Al-Khayyal F and Hwang S J 2007 Inventory constrained maritime routing and scheduling for multi-commodity liquid bulk, part I : applications and model *European Journal of Operational Research* 176:106-130.
[9] Marriott K, Stuckey P J, Koninck L D and Samulowitz 2014 *A Minizinc Tutorial.*
[10] Papageorgiou D J, Nemhauser G L, Sokol J, Cheon M S, Keha A B 2014 MIRPLib – A library of maritime inventory routing problem instances: survey, core model, and benchmark results *Euro. J. Oper. Res.* **235** 350–366.

IOP Conf. Series: Materials Science and Engineering **567** (2019) 012040 doi:10.1088/1757-899X/567/1/012040

Chemical characteristics of tuna fish bakasang

F Fatimah[1], J J Pelealu[2], S Gugule[3], A R H Korompot[1], V W Monintja[1], T E Tallei[2], J A Rorong[1]

[1]Department of Chemistry, Faculty of Mathematics and Natural Sciences, Sam Ratulangi University
[3]Department of Biology, Faculty of Mathematics and Natural Sciences, Sam Ratulangi University
[2]Department of Chemistry, Faculty of Mathematics and Natural Sciences, Manado State University

Corresponding author: fetifatimah@unsrat.ac.id

Abstract. Fish viscera are wastes produced by fish processing that have high potential as sources of biomolecules such as proteins and lipids. One of the fermented products in Indonesia that uses fish viscera as its raw material is bakasang. This study aims to determine the content the chemical characteristics of bakasang made from Tuna fish viscera under various fermentation condition. The chemical characteristics assessed included free fatty acid (FFA), peroxide value (PV), and thiobarbituric acid (TBA, which is expressed in malondialdehyde/MDA) of the bakasang product. Tuna (*Thunnus* sp.) fish viscera were fermented under the following conditions: temperature 50 and 70°C, for 10 and 15 days with the addition of 20 and 30%. The results showed that tuna fish bakasang processed at a temperature of 50-70°C with salt content of 20-30%, had the following chemical characteristics: FFA = 3.8-4.4%, PV = 2.8-10.5%, MDA = 0.35-1.15 g / 100g sample. These results indicated that bakasang tuna fish processed under the above conditions have good chemical characteristics.

Keywords: bakasang, fermented fish sauce, malondialdehyde, thiobarbituric acid, salt concentration, tuna

1. Introduction

Tuna has high protein (22.6 - 26.2 g/100 g) and low fat content (0.2 - 2.7 g / 100 g) . Furthermore, tuna is rich in calcium, phosphorus, iron, sodium, vitamin A (retinol), and vitamin B (thiamin, riboflavin thiamin, riboflavin and niacin)[1].

Bakasang, a fermented fish viscera sauce, is classified as a spontaneous fermentation, because in its production, it does not require microbial or carbohydrate starters [2] Fermented products made by adding high levels of salt cannot be used as a source of protein because. Therefor, the amount that can be consumed is limited due to its high of saltiness. Such products are usually used only as food stimulants, flavoring food or seasonings [3]. The fishes that can be used for fermentation are cakalang (*Katsuwonus pelamis*) (Fatimah et al., 2017), anchovies (*Stolephorus* sp.), sardines (*Sardinella* sp.), mackerel (*Rastrellinger* sp.), gambusia (*Affinis affinis*), Pacific whiting (*Merluccius* sp.) and other low value fish species [4].

Bakasang is made with different formulations and processing techniques, for example in the ratio of fish-water, salt content, the use of yeast / enzymes, addition of seasonings, storage time, and temperature. Fish sauce fermentation process in Japan uses ratio fish : water = 3 : 1, with 15% salt, room temperature storage and 16% *Aspergillus oryza* [5]. The results of Bahlawan's [2] study (2011) showed that there was a significant influence of salt concentration and storage time on the quality of bakasang. Sabour and Moini [6] (2009) use 20% of *Caspian kilka* in their fermentation. Desniar et al. [7] (2009) reported the increase in salt content from 30 to 50% that in fermented peda fish produced different flavor. Fermented product with a salt concentration of 30% was preferred over 50%.

One of the chemical parameters to understand the quaility of bakasang is rancidity. Rancidity can be caused by the results of hydrolysis reactions and oxidation of fish oil in bakasang. Hasan et al. [8] (2014), stated that damage to fish with salting treatment was mostly caused by fat oxidation which resulted in rancid fish, especially in high-fat fish. Oxidation reactions also run faster if the storage temperature is high enough. The purpose of this study was to determine the chemical characteristics of bakasang made from tuna viscera under several processing conditions.

2. Materials and Methods

The fermentation procedure follows the method carried out by Fatimah et al.[9] (2017). The viscera of fresh tuna (*Thunnus* sp.) were bought from a smoked tuna processing house at North Minahasa Regency. The viscera consisted of liver, intestines, heart and eggs. The viscera were washed with running tap water, dried, and cut into smaller parts (± 1 cm). Approximately 250 g of the materials were put in the fermentor, covered with various concentration of salt (20% [s1] and 30% [s2]). The fermentor lids were closed tightlty and the fermentation process was carried out at various temperature (50oC [T1] and 70oC [T2} and various fermentation time (10 days [D1] and 15 days [D2]. The fermentors were shaken once a day during the experiment. The samples were taken on the first, tenth, and fifteenth days for analysis. All samples were analyzed three times. Only two closest values of the replications which were used for statistical analysis.

The chemichal parameters analyzed were free fatty acid (FFA), peroxide value (PV), and thiobarbituric acid (MDA). Peroxide value and free fatty acid were determined base on AOCS[10] (1989). The FFA content was calculated as percentage of Eicosa-5,8,11,14,17-pentaenoic acid (EPA). Thiobarbituric acid (TBA) value was determined in accordance with the TBA distillation method described by Lynch and Frei (1993).The TBA or TBARs values were expressed in mg of malondialdehyde (MDA)/kg bakasang. Absorbance was measured using spectrophotometer (Shimadzu UV-1800 Serial No. A116351).

One-way Anova was used to analyse the results. The least significant difference (LSD) test was conducted if there were significant differences among the groups. All statistical analyses were carried out using the statistical software SAS (University Edition). The level of significance was set at α level of 0.05, $p < 0.05$. All values were presented as mean+standard deviation (SD) and statistical significance is indicated with appropriate letters on the data tables.

3. Results and Discussion

3.1. Free Fatty Acid content

Figure 1 shows that the FFA content of tuna bakasang increases along with the increasing fermentation time and temperature, but decreases at higher salt concentrations. This is in line with research conducted by Fatimah et al. [9] (2017), who stated that the FFA content decreased along with the increasing salt content.

According to Sutiah et al.[11] (2008), the storage in a certain period of time can cause the rupture of triglyceride bonds in oil and form glycerol and free fatty acids. Khoddami et al.[12] (2009) reported that the higher the temperature, the faster the formation of free radicals and free fatty acids. Free fatty acids can stimulate lipid oxidation and this condition can induce an increase in off-flavor content in the final product. Based on the standards set in IFOMA[13] (1998), the good free fatty acid content in coarse fish oil is 1-7%, but in certain industries the standard of free fatty acid content is 2-5%.

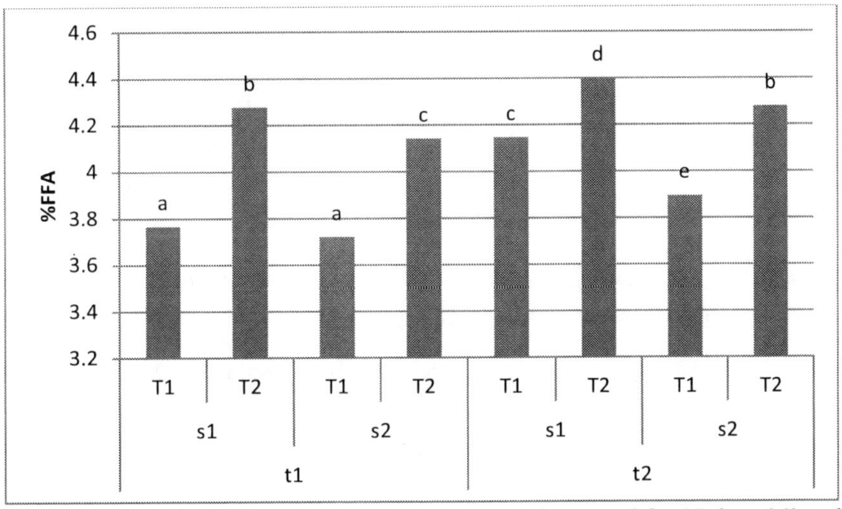

Figure 1. The free fatty acid (FFA) content of tuna bakasang fermented for 10 days (t1) and 15 days (t2), at 50oC (T1) and 70oC (T2), with 20% (s1) and 30% (s2) salts concentration Results with same superscript are not significantly different (P <0.05).

3.2. Peroxide Value

Peroxide value is a number that determines the degree of damage of oil. Oils that contain unsaturated fatty acids can bind oxygen to form peroxides. The peroxide number of tuna bakasang tuna is shown in Fig. 2.

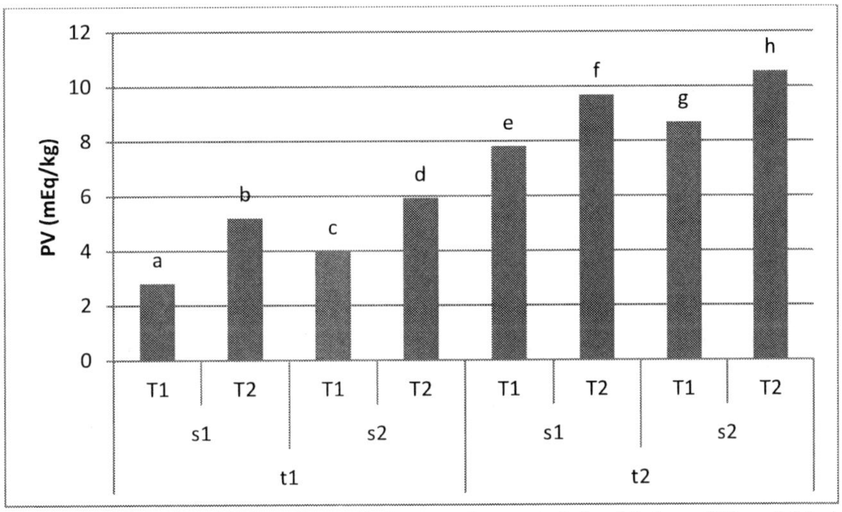

Figure 2. The peroxide value (PV) of tuna bakasang fermented for 10 days (t1) and 15 days (t2), at 50°C (T1) and 70°C (T2), with 20% (s1) and 30% (s2) salts concentration.Results with same superscript are not significantly different (P <0.05).

Peroxide value (PV) was used as measurement of oxidation. In this study, the higher the temperature and the longer duration of fermentation, the greater the peroxide value. This is in line with Husain et al.[14], (2016) that tuna fish oil peroxide (*Thunnus* sp.) increased with the increase of the storage temperature. According to Pak[15] (2005), the peroxide number is an indicator of the stability of oil against oxidation.. Lipid oxidation in tuna is easy to occur, because tuna fish oil is rich in polyunsaturated fatty acids (PUFAs). The increase of salt content can increase peroxide number. According to Raharjo[16] (2004) salt is thought to play a role in catalyzing oxidation through lipoxidase activity. According to the International Fishmeal and Oil Manufactures Association (IFOMA), the standard peroxide number is 3-25 meq / kg sample. Based on the research data above, the peroxide value of the tuna bakasang passed the predetermined standard.

3.3. Thiobarbituric Acid value

As shown in Fig. 3, the increase in fermentation time, temperature and salting caused the TBA value to increase, where the highest TBA value was in 15 days fermentation time, 70°C temperature, 30% salt content with a value of 1.1528 mg malondialdehyde/kg sample. Conversely, the lowest TBA value is 10 days fermentation time, 50°C temperature, 20% salt content with a value of 0.3261 mg malondialdehyde kg sample. Components of poly unsaturated fatty acids (PUFAs) which are high enough in fish fat are very sensitive to damage caused by the oxidation process. Based on the results of the study, the greater the salt content, the greater the value of TBA. The rancidity induced by salt can also be related to salt impurity because it is mixed with a small amount of metal. According to Boran et al[17] (2006), the good number of TBA for fish oil is 7-8 mg of malondialdehyde/kg sample. Based on this statement, it can be concluded that tuna bakasang with 10 and 15 days fermentation time, 50°C and 70°C temperature, 20% and 30% salt content are still safe because it did not exceed the established standards.

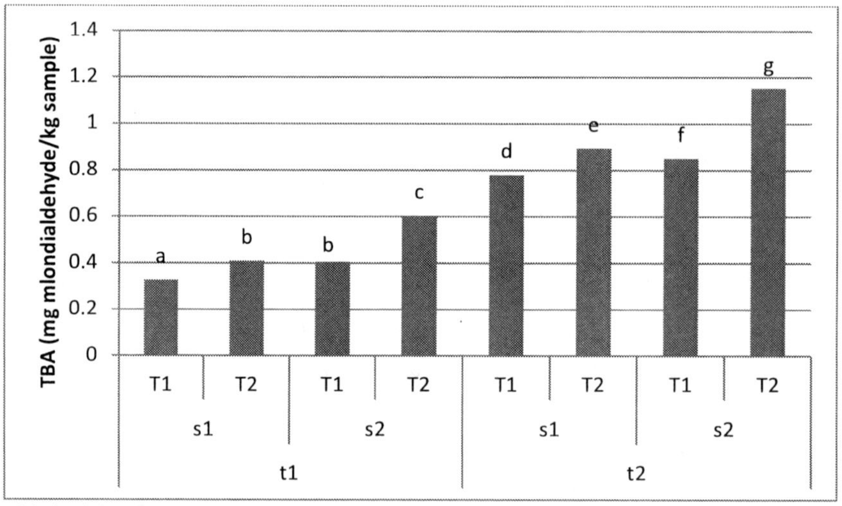

Figure 3. Thiobarbituric Acid value (TBA) value of of tuna bakasang fermented for 10 days (t1) and 15 days (t2), at 50°C (T1) and 70°C (T2), with 20% (s1) and 30% (s2) salts concentration .Results with same superscript are not significantly different ($P < 0.05$).

4. Conclusion

The results of this study showed that the chemical characteristics of bakasang processed under various conditition of fermentation time, temperature, and salt content still within the range of values allowed in accordance with existing standards for FFA, MDA, and PV values. Based of these results, the above processing conditions can be used as a reference for the fermentation process to obtain quality tuna bakasang that can be used as an additive for cooking and food.

Acknowledgement
This work was financed by Research Council and Community Service, Ministry of Research, Technology, and Higher Education, Republic of Indonesia, through excellent university research (PUPT) scheme fiscal year 2018.

References

[1] Brill R.W and P.G. Bushnell 2006 *Fish Physiol Biochem.* **32** 283294
[2] Bahalwan, F. 2011 *Bimafika* **3** 292297
[3] Hall, G. M. 1992 *Fish processing technology* (New York: VCH Publisher)
[4] Sanchez, P.C. 2008 *Philippine fermented foods: principle and technology* (Philippines: The University of the Philippines Press)
Mueda, R.T. 2015 *AACL Bioflux* **8** L4
[5] Fukui, Y.M., Yoshida, K., Shczen, Y., Funatsu, T., Takano, H., Oikawa, Y., Yano and M. Satomi 2012 *J. Gen. Appl. Microbiol.* **58** 273281
[6] Sabour A., K., S.Moini 2009 *Iranian Journal of Fisheris Sciences* **8** L2 155162
[7] Desniar, D., Poernomo and W. Wijatur 2009 *Jurnal Pengolahan Hasil Perikanan Indonesia* **12** L1 7387
[8] Hasan, B., D. Iriani., and A. Densi 2014 *Jurnal teknologi hasil perikanan* **1** L1 19
[9] Fatimah, F., Pelalu J.J., Gugule, S., Yempormase, H. V. and Tallei, T. E 2017. *Pakistan Journal of Biological Sciences* **20** L1 543551
[10] American Oil Chemists Society (AOCS) 1998 *Official methods and recommended practices of the AOCS* ed 4 (Illinois: AOCS Press)
[11] Sutiah, K. S, Firdausi, W, Setia Budi 2008. *Jurnal Fisika* **11** L2 5358
[12] Khoddami A, Arifin A. A., Bakar J. and Ghazali H., M. 2009 *Journal of World App Scs.* **7** 127131
[13] International Fishmeal and Oil Manufacturers Association (IFOMA) 1998 *International Fishmeal and Oil Manufacturers Association* (United Kingdom: Hertfordshire)
[14] Husain, R., Supramono., E, Harmayani. and Hidayat, C. 2016 *Agritech.* **36** L2 176181
[15] Pak, C.S. 2005 *Stability and Quality of Fish Oil during Typical Domestic Application Fisheries Training Progamme* (Iceland: The United Nations University)
[16] Raharjo, S. 2004 *Kerusakan oksidatif pada makanan* (Yogyakarta: Universitas Gadjah Mada)
[17] Boran, G., Karach., and M. Boran 2006 *Food Chemistry* **94** L4 693698

IOP Conf. Series: Materials Science and Engineering **567** (2019) 012041 doi:10.1088/1757-899X/567/1/012041

The potency of drug reconciliation need on elderly diabetic patient at Kandou Hospital Manado

G Citraningtyas[1], W I Wiyono[1], W A Lolo[1]

[1]Department of Pharmacy, Faculty of Mathematics and Natural Sciences, University of Sam Ratulangi, Manado, Indonesia

Corresponding author: gayatri_citra88@ymail.com

Abstract. Drug reconciliation is needed on Diabetic Mellitus patient, specially geriatry patient. North Sulawesi include in 17 provinces with high Diabetic Mellitus prevalence in Indonesia, so drug reconciliation is needed. The aim of this research is to identify the potency of drug need reconciliation on ederly Diabetic Mellitus patient. This research method was divided into two stages to identify treatment discrepancies in elderly DM patients who left the Kandou Hospital. The second stage is a qualitative descriptive study with in-depth interviews with patients including perception, choice tendency and patient willingness to pay for drug reconciliation services. Interviews were done on health service staff about perception, tendency of choice and willingness of service providers to be paid for the implementation of drug reconciliation. The results of the study included 11 intentional discrepancies from 32 patients. The most accidental nonconformity in the form of omission medication was 28.12%. Patients don't know about drug reconciliation and service providers don't know about drug reconciliation. All the parties support drug reconciliation service. Drug reconciliation has the potential to be implemented in Manado, especially in Kandow Hospital as a referral center.

1. Introduction

Drug Reconciliation is a process of evaluating a patient's medication regimen that is carried out regularly which aims to prevent medication errors that can include omission, duplication, dose errors, or drug interactions, as well as compliance and adherence of patients [1]. Drug reconciliation is needed to the patients who receive drugs that include high alert, one of them is a patient with Diabetes Mellitus (DM) who get insulin and oral antibiotics [2]. In patients with diabetes mellitus, the risk factors for incompatibility medication are greater.

The prevalence of diabetes mellitus in Indonesia has increased. The research of Basic Health (Riskesdas) in 2007 showed the prevalence of DM in Indonesia was 1.10% while in 2013 it was 2.01% [3,4]. One of the 17 provinces with a percentage of DM prevalence that is greater than the national DM prevalence percentage is North Sulawesi with the prevalence of DM diagnosed by doctors and the symptoms of 1.60% [4].

The aim of this study is to identify the potential needs of drug reconciliation in elderly people with diabetes mellitus. This aim can be achieved through identifying; the first, discrepancies that occur in the administration of diabetes mellitus patients at the time of admission to hospital and out of hospital. Second, in-depth interviews with patients and health care providers regarding their choice of perception and tendency towards drug reconciliation services. This research will be useful for the development of health services, especially drug reconciliation and as input for health services at Kandou Hospital Manado and health community in Manado.

2. Methods

2.1 Study design, Setting and Participant

The study was divided into two stages, namely: the first stage, quantitative descriptive study to identify medication discrepancies in elderly DM patients who left Kandou Hospital. This was done by comparing the treatment data of patients when they were discharged from the hospital with the best source of information related to the patient's medical history (Best Possible Medication History/BPMH) before the patients entered the hospital adjusted to the patient's condition in the medical record.

The second stage is a qualitative descriptive study with in-depth interviews with patients including perception, preference, and patient willingness to pay for drug reconciliation services. Interviews are also conducted with health care providers regarding perceptions, preferences, and willingness of service providers to be paid for implementing drug reconciliation

The study was conducted for 8 months from January to October 2018. Samples which taken in the first stage were 32 elderly patients who were diagnosed with diabetes mellitus who came out of Kandou Hospital Manado, while in the second stage there were 32 patients with Elderly Diabetes Mellitus and as many as 29 service providers included 15 pharmacists and 14 doctors.

2.2 Outcome

In qualitative research, the outcomes produced according to the nonconformity criteria based on Herrero-herrero et al [5], include:

A. There is no medication discrepancies: the medication upon discharge from the patient in the patient's medical record is the same as the medicine before entering the hospital obtained from BPMH.

B. Medication discrepancies: the difference between the patient's medication when he was discharged from the hospital and the drug before entering the hospital obtained from BPMH. The medication discrepancies in this study included:

1. Unintentional discrepancies), such as:
   a. Omission medication: drugs that patients use before entering the hospital are not re-prescribed when they leave the hospital without a clinical explanation in the patient's medical record.
   b. Giving new prescriptions at the time of hospital discharge: change in dosage, route of administration or frequency of medication when exiting the hospital in the patient's medical record compared to the dose, route of administration or frequency of the drug before entering the hospital without any changes in the patient's condition.
   c. Commission medication: patients who were prescribed at the time of discharge did not use by the patient before entering the hospital. There is no clinical explanation for the addition of the drug in the patient's therapy in the patient's medical record.
   d. The prescription is incomplete: dosage, frequency or duration of use is not recorded for one prescription drug upon discharge.
   e. Treatment duplication: prescribe more than one drug with the same class

2. Intentional discrepancies are discrepancies between drugs obtained when patients are out of hospital compared to before entering the hospital and the changes are based on the patient's condition in the medical record or carried out with the aim of correcting medication errors that occur at the time before entering the hospital. New therapies or changes in route of administration include intentional discrepancies if according to the patient's clinical status or the diagnosis recorded in the report upon discharge

C. Cannot be identified is unable to identify because there is no information related to the treatment history of BPMH patients.

2.3 Data Analysis

In the first stage, data collection was carried out in the period of 21 May 2018 to 9 July 2018. The obtained data were in the form of patient demographics, drugs before the patient was admitted to the hospital were obtained from the patient himself or care giver, while the patient's medication at the time of leaving the hospital obtained from the patient's medical record.

Medicines data before hospital admission and medication discharged from the hospital were compared to identification of the percentage of nonconformities that occurred based on the Hererro-Herrero classification [5]

In the second stage, the obtained data in the form of interview transcripts were conducted based on thematic analysis [6]. Transcripts are read repeatedly, coded and if there are the same main ideas and keywords, they will be listed and produce a theme. Analysis is done until saturation data is obtained.

3. Result

Based on research conducted on 32 inpatients, demographic data obtained as follows:

Table 1. Patient Demographic Table

Characteristics	Total	Percentage
Sex		
Male	12	37.5%
Female	20	62.5%
Age		
Early Elderly 46-55 years	10	31.25%
Late ederly 56-65 years	14	43.75%
Aged >65 years	8	25.00%
The number of diagnosis per patient		
1-2	3	9.37%
3-4	13	40.63%
≥5	16	50.00
The number of drug items obtained at the time of the patient's hospital discharge		
3-4	6	18.75%
5-6	8	25.00%
≥7	18	56.25%

[a]Age grouping based on Department of Health 2009

3.1 Medication discrepancies

There are 34 discrepancies were obtained from 32 inpatients which included the following:

Table 2. Classification table for nonconformities based on the Herrero-Herrero classification

Classification	Total	Percentage
There is no medical discrepancies	0	0.00%
Unintentional discrepancies		
Omission Medication	9	28.12%
Giving new recipes	1	3.12%
Commission Medication	0	0.00%
Not complete prescription	0	0.00%
Duplication of medication	1	3.12%
Intentional discrepancies	23	71.87%
Cannot be identified	0	0.00%

Of the 32 patients, as many as 11 patients experienced unintentional discrepancies and as many as 2 patients experienced 2 nonconforming discrepancies. Based on data obtained shows the need for drug reconciliation for patients with chronic disease patients who tend to get a lot of drugs.

3.2. Knowledge, Perception, Choice Trends and willingness to pay for Diabetes Melitus patients for drug reconciliation services

3.2.1 Knowledge of Diabetes Patients regarding drug reconciliation services

All Patients do not know about drug reconciliation.

Some statements that show patients' ignorance about drug reconciliation.

"not know. What is that?

(Patient 1, Female, 61 years old)

"I've never heard about that"

(Patient 4, female, 59 years)

3.2.2 Perception of Diabetes Patients towards drug reconciliation services.

Based on in-depth interviews conducted with DM patients, all patients did not know about drug reconciliation services. This shows a lack of patient knowledge of drug reconciliation. The patient's perception of the service of drug reconciliation is very supportive and considers good service of drug reconciliation. Some statements that support this.

"Very good to be done to be sustainable"

(Patient 4, female, 59 years old)

"Very good, because I've experienced drug interactions"

(Patient 12, female, 61 years old)

"Good to be controlled"

(Patient 30, female, 63 years old)

The patient's statements show the patient's support and desire to get drug reconciliation services. There is one patient who has experienced drug interactions so that the service of reconciliation is very necessary

3.2.3 Trends in the Choice of Diabetes Mellitus Patients for Drug Reconciliation services

3.2.3.1 Types of patient-selected drug reconciliation services

The type of service that patients want is to call the patient and visit the house. Some patients prefer the service provider to call the patient shown in the following statement:

"Just call. The problem is more practical "

(Patient 1, female, 61 years old)

"Calling saves more time, cost and distance"

(Patient 10, male, 50 years old)

"More comfortable on the phone"

(Patient 11, female, 59 years old)

Some patients who choose to call a patient care provider think it is more practical and convenient if over the telephone. Some patients who choose to call consider the distance of the house of patients who live far away in other cities or outside the island. Some patients do have a place to live far from the hospital. Some patients choose service providers to make a home visit. This is shown in the following statements:

"If you go home it's clearer to ask anything"

(Patient 6, Female, 69 years)

"A direct visit so I can ask right away the right direct information to"

(Patient 9, male, 72 years old)

'Better visit home to be more monitored"

(Patient 14, Female, 54 years old)

The patient chooses the service provider to make a home visit assuming that with a home visit, the patient can communicate directly with the service provider, have the flexibility to ask health care providers and feel more monitored. These statements show the patient's desire to interact directly with health care providers

3.2.3.2 Frequency of patient selected medication reconciliation services

Patients choose the frequency of drug reconciliation services 1x a week, 2x a week, 1x monthly and 2x monthly. Some statements that show frequency selection of drug reconciliation services are:

"2 times a week so that my communication with the pharmaceutical department is smooth"
(Patient 5, Female, 50 years old)
"It is better 1x per week, so it can be controlled"
(Patient 28, Female, 55 years old)
"1x a month only because for chronic diseases the drug is prescribed for 1 month better one month visit "
(Patient 30, Female, 63 years old)
"If I am rarely at home, it's better for 2x a month."
(Patient 3, Male, 61 years old)

The frequency of services both home visits and calling depends on each patient's situation from time availability, the patient's need for communication with the service provider and the patient's desire for the service provider to control the medication that the patient is drinking. Research on the frequency of follow-up in diabetics shows an intensive frequency of improving quality of life and clinical indications [11]

3.2.4 Willingness to pay for Diabetic Mellitus patients for drug reconciliation services

The willingness to pay patients is very diverse. There are some patients who object to paying 50,000-100,000, transportation and voluntary money.
Some patients do not have the willingness to pay. Some statements indicate the patient's objection to pay, ie

"I don't want the problem, because I entered a BPJS patient so I shouldn't pay it, right?"
(Patient 1, Female, 61 years old)
"If I just say thank you,"
(Patient 9, male, 72 years old)
"Not because I already entered the health insurance"
(Patient 26, Male, 50 years old)

Some patients who are willing to pay 50,00-100,000. Several statements that show the patient's choice are as follows:

"I was able to pay 100,000 / visit"
(Patient 3, male, 61 years old)
"Thank you, only 50,000"
(Patient 11, female, 59 years old)

Some patients show willingness to pay transportation money

"Will you pay for transportation money?"
(patient 25, female, 63 years old)

There are patients who show willingness to pay voluntarily

"Can I pay, but for the sake of gratitude, just be a participant."
(patient 28, female, 55 years old)

Patient willingness to pay for drug reconciliation services depends on the patient's economic condition and the health insurance that the patient has.

3.3. Perception, Choice Trends and willingness to be paid by Health Service providers for drug reconciliation services

3.3.1 Knowledge of health care providers regarding drug reconciliation services

Some health care providers do not know about drug reconciliation. Statement that shows this, namely:

" Not yet. Just heard first "

(Service provider 3, Female, 49 years old)

"For reconciliation ever. Yes unite, combine opinions so that it is in line. If the reconciliation of medicine has never been "

(Service provider 5, Female, 28 years old)

Some health care providers know about drug reconciliation. Some statements that show this, namely:

"Ever. Hospital reconciliation compares treatment obtained patients before entering the hospital. See the interview data for patients come to the hospital

(Service provider 26, Female, 27 years old)

Service provider knowledge is based on training and seminars followed by service providers because some service providers have conducted training on drug recommissioning usually carried out by service providers in hospitals.

3.3.2 Perceptions of Service Providers for Drug Reconciliation services

The service provider supports the service of drug reconciliation. This is indicated by the statement of the following statement:

"Nice. Very helpful on the part of the patient especially us as medical personnel"

(Service provider 7, Female, 24 years old)

"It is important, so that we can monitor the adherence to taking medication, whether the patient is regular or not"

(Service provider 14, Male, 27 years old)

There are also service providers who consider drug reconciliation services to be ideal services.

"Ideally treatment must be like that, it must be sustainable, because we are also did not close my eyes from previous treatment "

(Service provider 2, Female, 33 years old)

Statements from service providers indicate support and consider it important to carry out drug reconciliation. Service provider perceptions affect the willingness of service providers to carry out drug reconciliation

3.3.3 Trends in the choice of service providers for drug reconciliation services

3.3.3.1 Types of Reconciliation services Drug choices of service providers

Some service providers prefer to call the patient. A statement that shows the choice of calling the patient as follows:

"It's better to call if the choice is just that. The reason is saving time, power saving, time, funds more efficiently "

(Service provider 9, Male, 24 years old)

Many service providers prefer a patient's home visit. Statements that indicate this as follows:

"From myself it's better to face-toface, because in terms of direct explanation to patients is better

(Service provider 7, Female, 24 years old)

"If the phone is likely the patient can lie, it's better direct visit to home "

(Service provider 12, male, 67 years old)

Some service providers choose both depending on the patient's condition. This can be seen from the following statements.

"If you choose depending on the patient's condition, if you adhere to taking medication via the call is enough, if the patient's specific condition is certain, for example the patient is not can move or have a wound, then we also visit directly "

(Service provider 3, Female, 49 years old)

"It's better both. There is also a home visit for patients limitation of motion. Telephone to remind and home visit for see the situation "

(Service provider 10, Female, 33 Years)

Many service providers choose home visits to see the patient's condition directly and interact with patients directly, while those who choose to call are due to a lack of time available from the service provider.

3.3.3.2 Frequency of Reconciliation services Drug selected by the service provider

The frequency chosen by the service provider for drug reconciliation services are 1x per month, 1x 2-3 weeks. and 1x a week. Many service providers choose the frequency of drug delivery 1x per month. A statement that shows this as follows:

"Maybe once a month while checking fasting blood sugar"

(Service provider 12, male, 67 years old)

"Once a month, because the average DM patient takes medication a month once"

(Service provider 21, male, 25 years old)

Some service providers choose once every 2-3 weeks. Statement that shows this, that is

"2-3 weeks more efficient time"

(Service provider 14, Male, 27 years old)

Some service providers choose 1x a week. Statement that shows this, that is

"Once a week at home, because compared to once a month it can while at home, patients take other drugs (because of other illnesses), can seen from there "

(Service Provider 25, Female, 54 years old)

The frequency of drug reconciliation services chosen by the service provider depends on the time availability of the service provider, the use of drugs by the patient, the desire of the service provider to control the condition and treatment of the patient

3.3.4 Willingness to be paid by service providers for drug reconciliation services

Willingness of service providers to be paid for drug reconciliation services, ie no need to be paid, 20,000-50,000, 50,000-100,000, transportation money, voluntary patients.

"Obviously, patients don't pay. Now the era of JKN, let alone patients routine"

(Service Provider 2, Female, 33 years old)

"If patients are willing to pay around 20,00-50,000"

(Service Provider 26, Female, 27 years old)

"The applicable price range is between 50,000-100,000"

(29 service providers, male, 27 years old)

"For a visit, it takes a fee, so maybe it's just transportation money"

(Service provider 24, Female, 35 years old)

"If I am just patient,"

(Service provider 21, male, 25 years old)

The willingness of health care providers is influenced by health insurance that patients have, the type of service provided by service providers such as home visits requires more money than calling patients and a sense of service from the service provider to the patient

4. Discussion

This study shows the need for drug reconciliation services. In the first phase of the study, 11 out of 32 patients experienced unintentional discrepancies and 2 patients experienced 2 nonconforming discrepancies. The most accidental nonconformity in the form of omission medication was 28.12%.

These results are the same as those of Herrero Herrero and García-Aparicio [5] showing the highest nonconformity of omission medication was 84.6% of 52 reports of unintentional medication discrepancies. In a systematic review by Michaelsen MH et al, the general discrepancies that occur are omission medication of more than 56% and the second is a new prescription in the form of inaccuracies in dosing, route and frequency of 44% of nonconformities [7].

In the second phase of the study, it was found that there was no patient knowledge about drug reconciliation. This was due to the absence of drug reconciliation services in the community while few were in the hospital. Drug reconciliation carried out in hospitals is limited to certain diseases and DM disease is not one of them. For service providers, few know about drug reconciliation. This shows the need for socialization and drug reconciliation training to service providers. Patients and service providers support drug reconciliation. Both patients and service providers consider drug reconciliation to be an important and good service.

The tendency of patients to choose the type of drug reconciliation service is home visits and calling, while those who provide services other than home visits and calling are those who choose both depending on the patient's condition. Some studies that show both calling and home visits have a positive effect on the patient's condition and the patient's response to treatment. Research conducted on the usefulness of telephones for health services shows that the telephone is useful for motivating and reminding patients of their treatment in patients with chronic diseases. Digital communication can improve patient interpersonal relationships with service providers. [8] A meta-analysis study by Ploeg et al. Showed that home visits for elderly patients reduced mortality by around 17%. [9] Home visits make patients feel care for patients and improve patient trust in health care providers. [10]

The frequency of drug reconciliation services selected by patients is 1x per week, 2x weekly, 1x monthly and 2x monthly, while the service providers choose 1x monthly, 1x 2-3 weeks. and 1x a week. The choice of patient frequency is more frequent than the choice of the service provider due to the availability of the time of the service provider and depending on the medication that the patient gets

Willingness to pay patients for drug reconciliation services, namely objection to pay, 50,000-100,000, transportation and voluntary money. Willingness to be paid by service providers does not need to be paid, 20,000-50,000, 50,000-100,000, transportation money, voluntary patients. Willingness to pay patients and readiness is paid by the service provider depending on the economic conditions of the patient and the existence of health insurance. For Service Providers the type of service performed also affects the example of a home visit requires more costs than when calling a patient. The results obtained are in accordance with research on the willingness to pay patients for health services in Aceh by Rina S, where willingness to pay patients is influenced by the patient's ability to pay and the existence of health insurance [12]

This research limitation is only can be applied to a hospital like Kandou Hospital Manado as type A of as Type A Referral Hospital. The second stage of research can only be applied to economic conditions in countries with low to middle economies. In addition to the large number of samples that are not representative due to the length of time to get medical record data from the Hospital, for the future research there is more time span of the study.

5. Conclusion

The results of the study indicate that there is a large amount of mismatch, and the presumption of drug reconciliation is needed by health workers and patients. Drug reconciliation has the potential to be implemented in Manado, especially in Kandow Hospital as a referral center.

Acknowledgments

We would like to thank to Head of Kandou Hospital who give us a permition to conduct our research at Kandou Hospital. We also thank to Mrs Wantje Lanti, M.Farm.Klin.,Apt. as Head of Pharmacy Instalation at Kandou Hospital and all the staff who take apart in this research. We also want to thank to service staff at community which participate in this research.

IOP Conf. Series: Materials Science and Engineering **567** (2019) 012041 doi:10.1088/1757-899X/567/1/012041

References

[1] The Joint Commission, Sentinel Event Alert, 2006, available from : www.jointcommission.org/SentinelEvents/SentinelEventAlert/sea_35.htm?print=yes[9/20/2010: 35.

[2] Institute for Safe Medication Practices,2011, ISMP List of High-Alert Medications in Community/Ambulatory Healthcare, available from www.ismp.org.

[3] Badan Penelitian dan Pengembangan Kesehatan Kementerian Kesehatan Republik Indonesia, Riset Kesehatan Dasar, Jakarta, 2007.

[4] Badan Penelitian dan Pengembangan Kesehatan Kementerian Kesehatan Republik Indonesia, Riset Kesehatan Dasar, Jakarta, 2013.

[5] Herrero-Herrero JI, García-Aparicio J, 2011. Medication discrepancies at discharge from an internal medicine service, European J of Internal Medicine ; 22: 43–48.

[6] Braun V, Clarke C, Using thematic analysis in psychology, Qualitative Research in Psychology. 2006 : 3; 77-101.

[7] Michaelsen MH, McCauge P, Bradley CP, Sahm LJ. Medication Reconciliation at Discharge from Hospital: A Systematic Review of the Quantitative Literature. Wilson KA, ed. Pharmacy : Journal of Pharmacy, Education and Practice. 2015; 3(2): 53-71. Doi: 10.3390/pharmacy3020053.

[8] Watkins JOTA, Goudge J, Gomez-Olive FX, Griffiths F. Mobile pone use among patients and health workers to enhance primary healthcare : A qualitative study in rural South Africa. Science Direct Journal. 2018; 198: 139-147. Doi: 10.1016/j.socscimed.2018.01.011

[9] Ploeg J, Feightner J,Hutchison B, et al. Effectiveness of preventive primary care outreach interventions aimed at older people; meta-analysis of randomized controlled trials. Can Fam Physician. 2005;51: 1244-1245

[10] Van Kempen JAL, Zuidema SU, Rikkert MGMO, Schers HJ. Home Visit for frail older people: a qualitative study on the needs and preference of frail older people and their informal caregivers. Br J Gen Pract. 2012; 62(601); e554-e560

[11] Hu M, Zhou Z, Zeng F, Sun Z. Effect of frequency of follow-up on quality of life of type 2 diabetes patients on oral hypoglycemics. Diabetes Techno; Ther. 2012 Sep; 14(9): 777-82. Doi: 10.1089/dia.2012.0037

[12] Rina S, Rosminah M. Willingness top ay and ability to pay for health care servive at Zainoel Abidin public hospital Banda Aceh, Indonesia, Proceedings of Annual International Conference Syiah Kuala University. 2011; Nov 29-30:242-246

Grain size distribution, morphology, and elemental composition of iron sand from North Sulawesi

G H Tamuntuan[1], A Tanauma[1], G Pasau[1], H Sangian[1]

[1]Department of Physics, Faculty of Mathematics and Natural Sciences, Sam Ratulangi University, Jl. Kampus Unsrat, Manado 95114, Indonesia

Corresponding author: gtamuntuan@gmail.com

Abstract. The coast of North Sulawesi has very potential natural resources. Coastal sand with relatively high iron content or commonly known as iron sand was detected in several locations. Iron sand is a source of economical minerals, namely iron (Fe), which is easier to treat compared to sources in stone form. However, scientific studies on iron sand minerals, especially those in the coastal areas of North Sulawesi, are still poor. Therefore, a study is needed to find out how the mineralogy, morphology and characteristics of iron sand particles in North Sulawesi. The results show that distribution of sand deposit range fine to coarse in size. However, fine-sized grains were dominant in samples from Lalow, Lolan, as well as Hais regions, whereas medium-sized grains were dominant in sample from Minanga. It probably indicates that Minanga area is relatively close to rock as a source of sand formation than other sampling location. In general, SEM observations show grains with relatively rounded in their morphology as well as have a smooth surface which is indicating a long range transport of these grains. EDX results show that the major elements in representative sand grains are Fe, Ti, and O, while the minor elements are Si, Al, Mg, Na, Co, Cl, and Ca. Fe content range from 45.18 to 61.48 wt%, whereas Ti content range from 4.35 wt% to 7.1 wt%.

1. Introduction

Iron sand is one of the potential mineral resources. Various industries such as steel industry, concrete [1], as well as permanent magnets [2], ink powder and coloring agent in a number of applications [3] utilize raw materials from iron sand. Some important minerals contained in iron sand deposits are magnetite (Fe_3O_4), hematite (α-Fe_2O_3) [4], and titanium [5]. These minerals have different concentrations at each location depending on the source rock or geological conditions of an area.

North Sulawesi is an area with geological conditions that are dominated by volcanic processes [6]. Weathering of high Fe-containing rocks is then carried away by the river flow and deposited in the lower regions, including on the coast [7]. Several coastal areas in North Sulawesi were identified as containing iron sand deposits. These areas are spread from the east and west coasts of the Minahasa area as well as several areas in the Bolaang Mongondow region. However, information about the potential and characteristics of iron sand in these areas is still poor.

The characteristics of iron sand can be studied using physical or chemical methods [8]. One of these methods is a microstructure scanning technique using Scanning Electron Microscope (SEM). Several previous studies have utilized SEM which is equipped with energy dispersive X-ray (EDX) to analyze the morphology and mineralogy of materials due to various needs. Morphology and behavior of elemental composition can differ in different grain size classifications. Therefore, to enrich information about the characteristics of iron sand deposits in North Sulawesi, it is necessary to carry out granulometric studies as well as SEM-EDX observations for analysis the morphology and elemental composition of iron sand from several coastal areas in North Sulawesi.

2. Material and Methods

Sample in this study was iron sand obtained from four different locations on the coast of North Sulawesi. Two sampling sites (Lolan and Lalow beaches) are located on the west coast and two others (Hais and Minanga beaches) are located on the east coast of North Sulawesi province. The coordinates of sampling location can be seen in Table 1. Five samples were taken at different points at each location. The samples are then prepared through the drying process. The 200 mg of sand then sieved using two different sizes American Standard Test Sieve Series (ASTM), screen number 35 and 60. It was done in order to classify the grain size of sand in three categories such as coarse, medium, as well as fine sand. Weighing was done to find out the mass percentage of each grain size. Morphological and elemental composition analysis of iron sand particles then was carried out using scanning electron microscopy (SEM) and energy dispersive X-ray (EDX). SEM-EDX observation was conducted in the Central Laboratory of Malang State University using the Scanning Electron Microscope FEI Inspect S50.

Table 1. Position of the sampling location

Location	Coordinate	
	Longitude	Latitude
Hais	0°57′42″ N	124°49′49″ E
Minanga	0°58′03″ N	124°51′06″ E
Lalow	0°54′04″ N	124°02′26″ E
Lolan	0°57′06″ N	124°10′03″ E

3. Results and Discussion

Grain size distribution is obtained by separating sand grains into three fraction sizes, namely coarse grain (more than 0.5 mm), medium grain (0.25 mm until 0.5 mm), as well as fine grain (less than 0.25 mm). Figure 1 shows the mass fraction of the three classifications aforementioned. Grains in samples from Lalow, Lolan, and Hais are dominated by fine-sized grain fractions, followed by medium grain size and then coarse grain fractions. The weight percent ratio (in the form of fine: medium: coarse) for the sample from Lalow is 58.52%: 36.75%: 4.73%, the sample from Lolan is 84.33%: 8.42%: 7.25%, while the sample from Hais is 66.08%: 27.92%: 6%. In contrast, samples from Minanga were dominated by medium-sized grain fractions followed by fine-grained fractions and then coarse grains with ratio of 21.25%: 74.59%: 4.15% (fine : medium : coarse). Grain size distribution is related to wave energy in carrying and settling sand on the beach [9] as well as transport process [10]. From this, it probably means that Minanga beach is influenced by waves with higher energy than other sampling locations or the Minanga area is relatively close to rock as a source of sand formation.

Figure 2a – 2d shows the morphology of iron sand grains in samples from Lolan, Lalow, Hais, and Minanga. SEM observations show that the morphology of sand grains is irregular shape, rounded, and bitetrahedral. Grains with irregular shapes were commonly formed due to weathering or destruction of rocks, while grain morphology which tended to be rounded formed as an effect of the transportation process. Bitetrahedral grains are lithogenic titanomagnetite or magnetite minerals which are formed due to weathering of rocks. These grains generally contain high Fe and have strong magnetic properties [11]. SEM images also show the roughness of grains surface. There are grains with a rough surface as well as the smooth one. The surface roughness of the grain is estimated to be related to the length of the grains experiencing the process of transport or erosion.

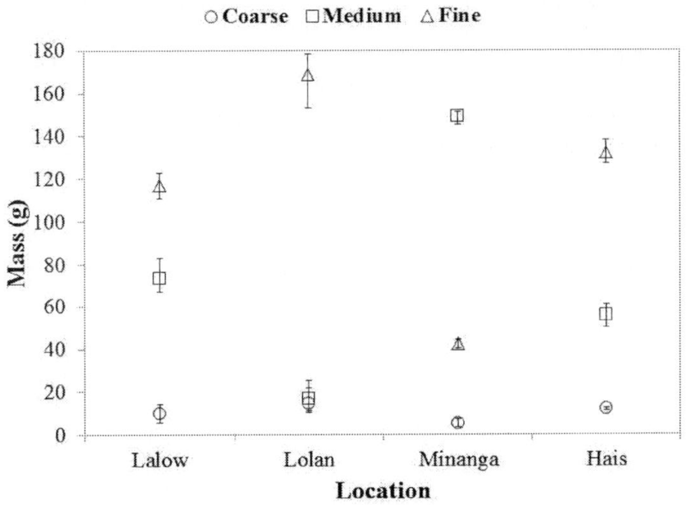

Figure 1. Grain size distribution expressed in mass

Figure 2. Morphological observations of iron sand by using SEM, (a) grains with relatively rough surface (samples from Lolan), (b) bitetrahedral grains with smooth surface (samples from Minanga), (c) irregularly shaped grains (samples from Hais) , (d) bitetrahedral and poorly rounded grains with a smooth surface (sample from Hais)

IOP Conf. Series: Materials Science and Engineering **567** (2019) 012042 doi:10.1088/1757-899X/567/1/012042

Elemental observation of sample from the four locations was done by EDX scan in the area. Until energy is around 7 keV, there were seven similar elements detected in the four samples such as O, Na, Mg, Al, Si, Ti, Fe, while elements that were not present in all samples such as Co, Cl, and Ca. Concentration ratio of these elements (expressed in percentage weight) is provided in Table 2. In general the concentrations of titanium (Ti) and iron (Fe) in all samples are more than 4.35 wt% and 45 wt% respectively. The highest Ti and Fe concentrations were found in samples from Lolan, namely 7.1 wt% and 61.48 wt%, while the lowest Ti and Fe concentrations were found in Minanga samples, which are 4.35 wt% and 45 wt%. Titanium and iron in natural samples such as sand are commonly compounded to form iron titanium oxide (Fe - Ti - O) in ferri- and canted antiferomagnetic states, namely titanomagnetite in the ulvospinel (Fe_2TiO_4) – magnetite (Fe_3O_4) series as well as titanohematite in the ilmenite ($FeTiO_3$) – hematite (Fe_2O_3) series [5].

Table 2. Elemental Comparison of samples from four locations

Elements	Weight %			
	Lolan	Lalow	Hais	Minanga
O	21.52	26.58	20.81	25.38
Na	1.23	3.61	5.83	3.25
Mg	1.80	3.13	2.76	4.03
Al	3.24	4.58	4.09	5.97
Si	3.24	5.39	5.07	7.54
Ti	7.10	5.20	5.72	4.35
Fe	61.48	49.04	51.58	45.18
Others	0.38	2.46	4.14	4.30

Figure 3 provides the results of SEM and EDX in a single grain. Observation of EDX in the whole samples shows varying of Ti and Fe content. The highest value of Ti and Fe content are 9.99 wt% and 88.27 wt%, respectively. Sampling locations if arranged based on the average Fe content in some representative grains, starting from the highest value, are Lolan, Hais, Lalow, and Minanga. There is a correlation between the size of the grain and the Fe content. In samples with finer grain sizes tend to have a higher Fe content compared to the more coarse ones. Some EDX observations in a single grain also indicate that bitetrahedral grains tend to have higher Fe content compared to another shapes. In general, mineralogy of grains in the shape of bitetrahedral is titanomagnetite derived from the results of erosion or weathering of volcanic rocks.

4. Conclusions

Iron sand deposits on the coast of North Sulawesi have grain size ranges from fine to coarse grain. The sediments on the Lolan, Lalow, and Hais beaches are dominated by sand grains on fine grades, while Minanga is dominated by sand with medium grade. SEM observations show that the morphology of sand grains is varied, but generally shaped irregularly, rounded and bitetrahedral. Based on the EDX scan area, samples from the Lolan beach have relatively higher Fe and Ti content than other samples. Fe content reaches 61.48 wt% while the Ti content is around 7.1 wt%. Sequentially, beaches with Fe content ranging from the highest are Lolan, Hais, Lalow, and Minanga. Analysis of the relationship between grain size and Fe content in the sample showed a good correlation, in this case the smaller the grain the higher the Fe content.

Acknowledgments
This research was fully funded by PNBP Sam Ratulangi University through Riset Dasar Unggulan Unsrat (RDUU) 2018 Research Grant to AT & GT.

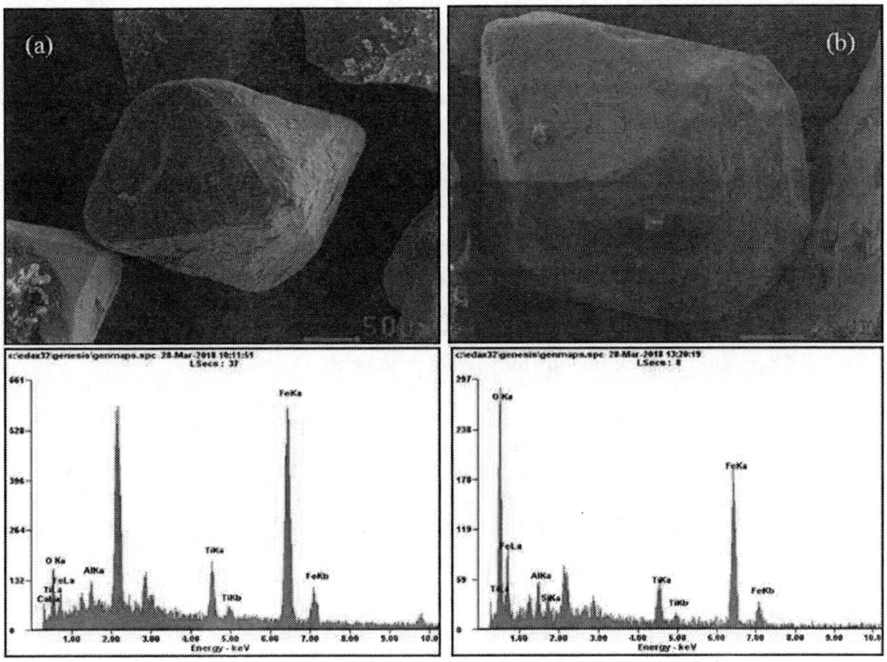

Figure 3. (a) The bitetrahedral grain was observed by SEM in fine-sized samples from Hais beach. EDX result shows that the grain has a high Fe content (Fe = 88.27 wt% and Ti = 7.03 wt%), (b) Rounded grain in sample from Lalow containing Fe = 68.3 wt% and Ti = 6.52 wt%. Note: red square is EDX spot.

References

[1] Vasudevan G 2016 *Proc. 3rd Int. Conf. on Civil, Biological, and Env. Eng.*, 10–13.
[2] Rusianto T, Wildan M W, Abraha K and Kusmono K 2012 *J. Teknol.* **5** 62–9
[3] Yulianto A, Bijaksana S and Loeksmanto W 2003 *Ind. J. of Phys.* **14(2)** 63–6
[4] Tiwow V A, Arsyad A, Palloan P and Rampe M J 2017 *IOP Conf. Series: J. of Phys.: Conf. Series* **997(2018)** 012010
[5] Hosseinzadeh M, Alizadeh M and Hosseini M R 2017 *J. of Mining and Env.* **8(2)** 191–201
[6] Siahaan E E, Soemarinda S, Fauzi A, Silitonga T, Azimudin T and Raharjo I B 2005 *Proc. World Geothermal Congress* 1–5
[7] Kurnio H 2007 *Ind. Mining J.* **10(9)** 27-38
[8] Zhang Z, Li J, Li X, Huang H, Zhou L and Xiong T 2012 *Int. J. of Mineral Processing* **114–117** 30–4
[9] Abdulkarim R, Akintoye A E, Oguwuike I D, Imhansoeleva T M, Philips I M, Ruth F B, Olubukola S O, Rashed J O and Banji A O 2011 *Nature and Sci.* **9(9)** 19–26
[10] Arens S M, Van Boxel J H and Abuodha J O Z 2002 *Earth Surf. Process. and Landforms* **27** 1163–75
[11] Tamuntuan G, Bijaksana S, Gaffar E, Russell J, Safiuddin L O and Huliselan E 2010 *ITB J. of Sci.* **42A(1)** 31–48

IOP Conf. Series: Materials Science and Engineering **567** (2019) 012043 doi:10.1088/1757-899X/567/1/012043

A simple low-cost video-based surveillance system for a flash flood warning system

H I R Mosey[1], D P Pandara[1], M D Bobanto[1], H S Sangian[1]

[1]Department of Physics, Faculty of Mathematics and Natural Sciences, Sam Ratulangi University, 95000-Manado, Indonesia

Corresponding author: hirmosey@unsrat.ac.id

Abstract. Flash flood is the most common and frequent natural disasters worldwide and can cause a loss of life and damage to property and the environment. In 2014, Manado city has experienced the most devastating flood that caused 19 people died, 10,884 homes destroyed, 2,091 people became homeless and estimated an economic loss of US$ 124 million. The huge loss was caused by the people and government did not receive information about water level rising on the watershed of the Tondano River. In conventional, decision maker needs the precipitation data and critical threshold value to issue the evacuation. There is an urgent requirement for end-to-end solution where the flash flood early warning system (FEWS) which can processed data in real-time can be integrated with Video-Based Surveillance System to take a role in saving peoples life. The aim of this study was to create a stand-alone system (Sensor Node) that could be placed on the watershed to monitor the water level rising using a microcomputer, a camera and a distance sensor. The aim of the system proposed was to measure the water level rising and when the threshold value was reached, the system would capture and display a real-time video. Future research can be done to "go-live" and posted the captured real-time video on the YouTube or Facebook in which people used them widely as a social media

1. Introduction

Flash flood is the most common and frequent natural disasters worldwide and can cause a loss of precious life and damage to property and the environment [1], [2], [3]. In 2014, Manado city has experienced the most devastating flood that caused 19 people died, 10,884 homes destroyed, 2,091 people became homeless and estimated an economic loss of US$ 124 million. The huge loss was caused by the people and government did not receive information about water level rising on the watershed of the Tondano River [4].

In conventional, decision maker needs the precipitation data and critical threshold value to issue the evacuation. There is an urgent requirement for end-to-end solution where the flash flood early warning system (FEWS) which can processed data in real-time can be integrated with Video-Based Surveillance System to take a role in saving peoples life [5].

2. Flood Early Warning System

The rapid development of information technology, natural disaster prevention are dealing with image based flood alarm system [6] and video surveillance system [7], [8]. Videos in the form of recorded videos, off-line videos and real-time video streaming are considered as the best sources of information

IOP Conf. Series: Materials Science and Engineering **567** (2019) 012043 doi:10.1088/1757-899X/567/1/012043

for various applications [9]. Application such as an advance safety vehicle (ASV)[10], home security [11], flood area detection based on UAVs (Unmanned Aerial Vehicle) [2] have been used video as source of information.

3. Hardware Design

The realization of the proposed sensor node can be seen in Figure 1. We choose to use a Raspberry Pi (RPi) minicomputer and pi camera module (PiCam) as computer and camera for the streaming hardware. We also use a ultrasonic sensor module to measure distance. All this module are cheap, simple and relatively easy to be bought on online or offline electronics store. It is also can be customised and open to be modified to meet other requirements for other purpose. We use Python programming language to program the RPi.

3.1. Raspberry Pi

The Raspberry Pi [12] is a low-cost credit card-sized single-board computers, running an open-source Linux as operating system, which is designed and developed by non-profit charitable Raspberry Pi foundation, UK. Originally, it is intent to promote the teaching of basic computer science in school and developing countries [13].

The Raspberry Pi 3 Model B specifications used in this paper has a Quad Core 1.2GHz Broadcom BCM2837 64bit CPU, 1GB RAM, onboard BCM43438 wireless LAN and Bluetooth Low Energy, 100 Base Ethernet, 40-pin GPIO, 4 USB 2 ports, 4 pole stereo output and composite video port, fullsize HDMI, CSI camera port for Raspberry Pi camera, DSI display port for connecting a Raspberry Pi touchscreen display, Micro SD port for loading operating system and storing data. It is powerfull enough to be used as a video streaming device and only cost about IDR 660,000.

3.2. Pi Camera Module

We are using the Raspberry pi camera module (PiCam) version 1.3 which can be connected directly to the Raspberry Pi board via CSI connector using a ribbon cable provided on the package. The camera module features a 5MP Omnivision 5647 CMOS sensor with 1.4 μm X 1.4 μm pixel size and focused lens able to deliver a 2592x1944 pixels resolution still image, or 1080p HD video recording at 30fps[14]. The PiCam module is cost about USD 30 (IDR 450,000)

3.3. Ultrasonic Sensor

The sensor is used for measuring the distance of the object [15]. It works by sending 40 kHz ultrasonic wave and detecting whether there is echo resulting from it. Even though its maximum distance is relatively small (5m), its accuracy is quite high (3mm) and stable. The distance itself is calculated according to the duration of the signal high level, where the distance is proportional to the duration. In addition, this sensor consists of ultrasonic transmitter, receiver and control circuit.

4. Software Design

Python programming language is used in the system to communicate with the General Purpose Input (GPIO) ports in RPi. We also use SimpleCV [16] technique to communicate with the camera to capture and stream the successive image.

The system design as follows. The ultrasonic sensor measured the distance (water level) and whenever the distance are less than the threshold value, it's value triggers the camera to start capturing image and streaming it to the monitor. If the value are more than the threshold value, the camera will be in sleep mode and display black screen.

5. Experimental Result

We tested the sensor node at a bridge by first measuring the normal height (from the bridge) to riverflow below it. The normal height was 3,402 cm. We set this "normal height" as the threshold to trigger the camera to start capturing and streaming to a monitor.

We tested the system by lowering the sensor node by 30 cm. As the sensor node becomes lower to the river, the "normal height" is less than the threshold value and the sensor node starts to streaming video to the monitor. An example of video streaming captured by the sensor node can be seen in figure 2.

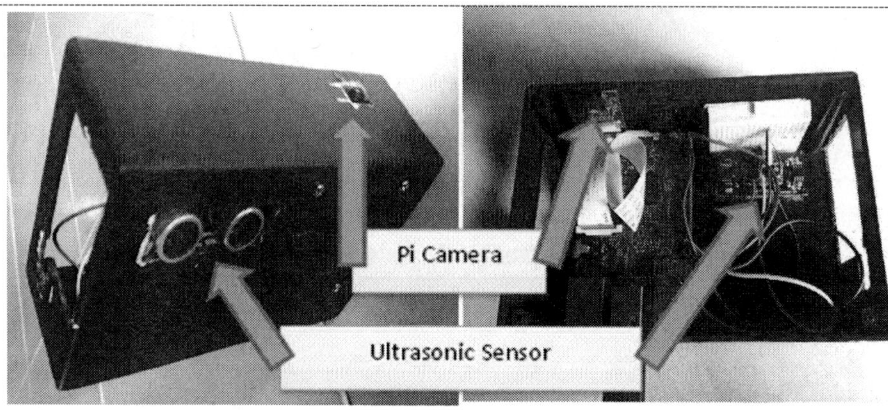

Figure 1. Realization of the sensor nodes

Figure 2. An example of the streaming video captured by the sensor nodes.

6. Conclusion and Future Work

In this paper, design of a low-cost video-based surveillance sensor node for a flash flood early warning system is implemented. It is advantageous as it low-cost, easy to use, small-size and portability. It can be place in any kind of bridges for surveillance. Future research can be done to "go-live" and posted

IOP Conf. Series: Materials Science and Engineering **567** (2019) 012043 doi:10.1088/1757-899X/567/1/012043

the captured real-time video on the YouTube or Facebook in which people used them widely as social media.

Acknowledgments
This research is financially supported by Lembaga Penelitian dan Pengabdian Masyarakat (LPPM) Sam Ratulangi University Manado in Riset Terapan Unggulan Universitas (RTUU) 2018 funding scheme.

References

[1] Basha E and Rus D 2007 *International Conference on Information and Communication Technologies and Development* p 1–10.
[2] Popescu D, Ichim L, and Caramihale T 2015 *19th International Conference on System Theory, Control and Computing (ICSTCC)* p 753–758
[3] Sunkpho J and Ootamakorn C 2011 *Songklanakarin J. Sci. Technol* **33** 227–235
[4] Karouw S and Wowor H 2016 *International Conference on Informatics and Computing (ICIC)*
[5] Filonenko A and Danilo C 2015 *IECON 2015 - 41st Annual Conference of the IEEE Industrial Electronics Society* p 4082–4085
[6] Lo S, Wu J, Lin F, Hsu C, Park H S, and Road K 2015 *Sensors* **15** 2369–2387
[7] Ko B 2012 *Optical Engineering* **51** (7), 070901
[8] Oza N and Gohil N B 2016 *International Conference on Communication and Signal Processing (ICCSP)* p 996–998
[9] Bhuvanesh D, Hegde M, and Manikandan J 2018 *Wirel. Pers. Commun* **100** (2) 477–489
[10] Chang T H and Chou C J 2009 *IEEE Intell. Veh. Symp. Proc.* 913–917
[11] Gu Y *et al.* 2013 *Int. Conf. Inf. Sci. Appl. ICISA* 1–4
[12] Raspberry Pi Foundation, http://www.raspberry.org
[13] Raspberry Pi. https://en.wikipedia.org/wiki/Raspberry_Pi
[14] Camera Module. https://www.raspberrypi.org/documentation/hardware/camera/README.md
[15] Ultrasonic Range Sensor HC-SR04. https://www.modmypi.com/blog/hc-sr04-ultrasonic-range-sensor-on-the-raspberry-pi
[16] SimpleCV. http://simplecv.org/

IOP Conf. Series: Materials Science and Engineering **567** (2019) 012044 doi:10.1088/1757-899X/567/1/012044

Anatomical characteristics of stomata, mesophyll and petiole of six varieties sweet potatoes (*Ipomoea batatas* L.) after organic fertilizer induction

H L Rampe[1] , S D Umboh[1], R Siahaan[1], P V Maabuat[1]

[1]Biology Department of Mathematics and Natural Science Faculty,
Sam Ratulangi University, Indonesia

Corresponding author : hennyrampe@unsrat.ac.id,

Abstract. The study aimed to evaluate the anatomical characteristics of six sweet potato varieties (*Ipomoea batatas* L.) after induction of organic fertilizer. Research used Completely Randomized Block Design with 3 replications. Five sweet potato varieties (J) were used i,e: Antin 3 (J1), Jago (J2), Cilembu (J3), Shi Royutaka (J4), Local Purple (J5) and Local White (J6). Petroganic Organic Fertilizers were applied 20 tons/ha, three weeks before sweet potato cuttings were planted. Observation of anatomical characteristics include: stomata density, length and width opened stomatal pores, mesophyll thickness, diameter of the petiole cortex. All characteristics were observed 60 days after planting (HST) using a microscope with a camera Optilab 2.2 and image capture with the Image Raster 3.0 program. Anatomical characters of sweet potatoes showed stomata density J1, J2, J3, J4, J5 and J6 respectively 8.56; 6.11; 5.11; 6.33; 8.56 and 9.11 in 75796.36 μm. The length of opened stomata pores were 26.99 μm, 32.08 μm, 28.96 μm, 38.72 μm, 39.52 μm and 35.86 μm. The width of opened stomata pores were 7.95 μm, 5.11 μm, 4.67 μm, 5.04 μm, 9.28 μm and 6.98 μm. The mesophyll thickness were 364,986 μm, 280,703 μm, 389,743 μm, 245,749 μm, 261,439 μm and 434,913 μm. The diameter of the petiole cortex were 454,030 μm, 373,453 μm, 456,439 μm, 373,001 μm, 275,647 μm and 318,785 μm. Five sweet potato varieties differ in anatomical characters.

1. Introduction

Sweet potato (*Ipomoea batatas*) is one of the tubers of carbohydrate sources in addition to wheat, rice, corn, potatoes and cassava. Besides containing carbohydrates, sweet potatoes also contain vitamins A and C, minerals potassium, phosphorus, calcium and sodium. Sweet potatoes are not only used as food but also as industrial raw materials and animal feed [1][2]. Sweet potatoes also contain beta carotene, anthocyanin, phenolic compounds, essential amino acids such as lysine and tryptophan [3][4].

Potential of sweet potato as a functional food continues to be developed to support food diversification programs. Various research on plant carbohydrate substitute for rice continues to be developed, including research on sweet potato plants. Research Institute for Various Nuts and Tuber of Malang City, until 2016, has released 24 superior varieties of sweet potatoes, where each variety has specific resistance properties [5]. Plant resistance to pests and pathogens can be antixenosis/non-preferences, antibiosis and tolerant. The morphoanatomy characteristics of antixenotic resistances of plants are color, shape, cell wall thickness, proliferation rate of plant tissue, hardness, characteristics of the stem, trichomes, magnitude of stomata, cuticle layer and wax [6][7].

The anatomical structure of the leaves is arranged by three tissue systems namely epidermis, mesophyll and vascular tissue. The epidermis is a layer of outer cells and covers the surface of plant organs. In leaf organs, the epidermis is on the adaxial and abaxial side [8]. Epidermal cells in later development are modified and specialized into structures in the form of stomata, trichomes, cuticles and fan cells. Stomata generally occur on the abaxial side of the leaves, but in certain plants the stomata are found on the adaxial and abaxial sides of the leaves. The shape and type of stomata are anomocytic (ranunculaceous), anisocytic (cruciferous), parasitic (rubiaceous), dichitic (caryophyllaceous) and actinositic [9][10]. Stomata together with neighboring cells are called stomata complexes.

Anatomical characters are antixenotic resistance parameters, can be in the form of color and plant shape, cell wall thickness and proliferation rate of plant tissue, hardness and characteristics of the stem, trichomes, magnitude of stomata, cuticle layer and wax [11][12][13]. The anatomical characteristics of the leaves can be seen through the length and width of the stomata, the density of the stomata, the thickness of the mesophyll and the length of the petioles [14][15]. Six varieties of sweet potatoes show specific resistance to pests and pathogens, characterized by qualitative and quantitative morphology, the ability to synthesize primary and secondary metabolites including the anatomical characteristics of leaves. Description of anatomy and morphology is a product of the response to the conditions of the environmental factors in which the plant grows. Application of organic fertilizer will provide mineral nutrients [16-17], so that plants can grow optimally and can induce the nature of antixenosis resistance, antibiosis and tolerance.

Six varieties of sweet potatoes showed specific resistance to pests and pathogens, characterized by qualitative and quantitative morphology, the ability to synthesize primary and secondary metabolites including the anatomical characteristics of leaves. Description of anatomy and morphology is a product of the response to environmental factors in which the plant grows. Application of organic fertilizer will provide mineral nutrients [16], so that plants can grow optimally and can induce the nature of antixenosis resistance, antibiosis and tolerance.

In Tomohon City, there are two local varieties of sweet potatoes with purple and white bulbs, with local names Kekeruten. These types of sweet potatoes grow wild on agricultural land. Their leaves are often used as animal feed. Their growth properties with creeping stems provide distinctive morphological structure, even showed specific properties of antibiotic resistance, antixenosis, and tolerance. Morphological hostage to find out the diversity of local cultivars will provide benefits to improve germplasm conservation.

The study of morphological diversity in sweet potatoes is the morphological diversity of local sweet potato from Muna [17], sweet potato morphology and kinship, and the correlation of anatomical characteristics of disease resistant sweet potato cultivars [18][19]. Morphological characteristic study of six sweet potato varieties was carried out to obtain morphological descriptions as scientific information that could be used to develop resistant varieties to pests and diseases. The study aimed to evaluate the anatomical characteristics of six sweet potato varieties (*Ipomoea batatas*) after the induction of organic fertilizer.

2. Research Methods

The study was conducted in the experimental garden in Lansot Village, Tomohon City. Research used Completely Randomized Block Design with three replications. Treatment of sweet potato varieties (J): Antin 3 (J1), Jago (J2), Cilembu (J3), Shi Royutaka (J4), Local Purple (J5) and Local White (J6). Petroganic organic fertilizer (20 tons/ha) were applied in three weeks before sweet potato cuttings were planted. The observed anatomy characteristics include: stomata density, length and width of opened stomata pores, thickness of mesophyll, 5) diameter of the petiole cortex. The observations were obtained at 60 days after planting (HST) using a microscope with a camera Optilab 2.2 and image capture with the Image Raster 3.0 program.

3. Results and Discussion

The results of longitudinal slices of the epidermal surface on the abaxial side of sweet potato leaves were shown in Figure 1. The stomata of sweet potato had paracytic kidney type, marked by closed cells surrounded by two neighboring cells with their longitudinal axis parallel to closed cell axis.

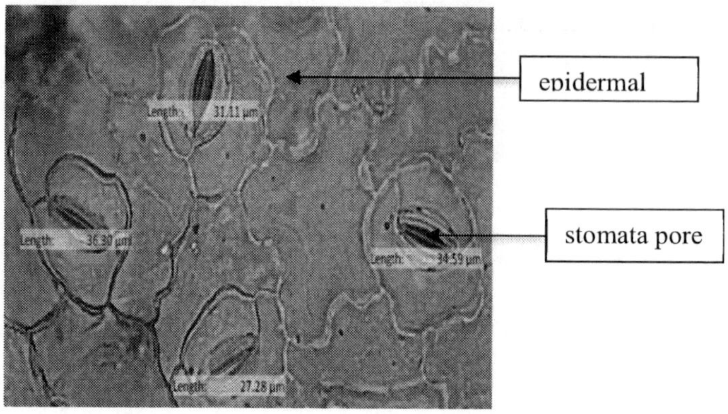

Figure 1. Anatomical structure of epidermal and stomata cells (40 times magnification)

The leaf epidermis consists of one cell layer, not plastid except in the stomata cover cell. The epidermal cell wall contains a cuticle layer. The epidermal tissue is the outermost tissue of plant organ, where the cell wall that builds it has undergone structural development according to the plant habitat. The cell wall consists of 3 layers, middle lamella, primary wall and secondary wall. The primary wall contains 25-50% hemicellulose, 10-35% pectic substances, about 10% of extensine proteins and lectin proteins[9,20]. The cell wall consists of cellulose microfibrils and nonselulose matrix. The matrix generally consists of pectin and hemicellulose. In the cell wall lignin, suberin, warts, wax, tannin, calcium carbonate and calcium oxalate, silica and other ingredients are found [20][21][22].

The results of measured anatomical characteristics of J1, J2, J3, J4, J5 and J6 were 8.56, 6.11, 5.11, 6.33, 8.56 and 9.11 respectively in 75796.36 µm. The length of opened stomata pores were 26.99 µm, 32.08 µm, 28.96 µm, 38.72 µm, 39.52 µm and 35.86 µm. The width of opened stomata pores were 7.95 µm, 5.11 µm, 4.67 µm, 5.04 µm, 9.28 µm and 6.98 µm. The highest average stomata density on J6, the highest average length of stomata pore on J5, and the highest average width of opened stomata pore on J5.

The anatomical structure of stomata is strongly associated with ontogeny stomata. Stomata are usually found in plant organs associated with air. Stomata distribution, length and width pore are closely related to plant physiological processes, namely transpiration, photosynthesis and respiration. The flexibility of the stomata pore wall allows the stomata to open and close [23]. The number and density of stomata between cultivars are influenced by the environment in which they grow. Light intensity, water availability and temperature are factors that determine stomata density [24].

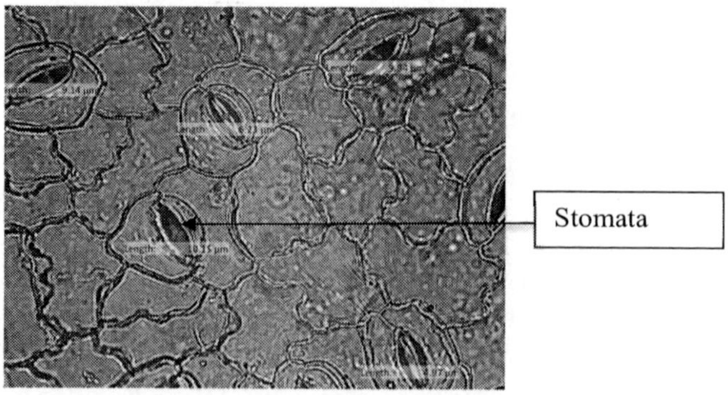

Figure 2. The width of opened stomata pore (40 times magnification)

Figures 1 and 2 show the length and width of opened stomata pore. The opening and closing the stomata is affected by closing cell turgor. Stomata open when the cell closure is high. The changing of osmotic potential value in the closing cell is caused by chemical changes that occur in the cell closure which will then change the potential of the water. Environmental factors that influence the opening of the stoma cover are the concentration of CO_2, light, water stress, temperature and wind [25].

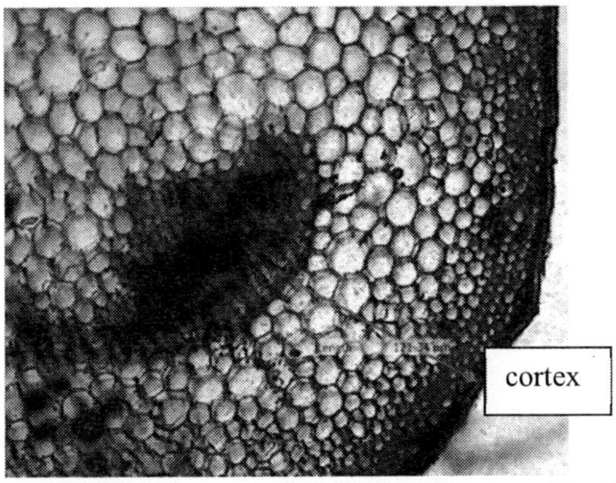

Figure 3. Petiole transverse slices (4 times magnification)

Figure 3 showed a cross section of sweet potato petiole. Petiole is composed of three regions namely epidermis, cortex and stele. The petiole epidermis is the outer layer of cells, covering the organ with a protective tissue function. The shape, size and composition of the epidermal cells are very tight to each others to form layers without space between cells. The epidermis is composed by a layer of cells, in the form of living cells, the cells have the power to divide and do not carry secondary thickening growth[22-25]. The petiole cortex is composed by rounded parenchymal tissue with real intercellular space. The results of average diameter of the petiole cortex of sweet potatoes J1, J2, J3, J4, J5 and J6 were 454,030 μm, 373,453 μm, 456,439 μm, 373,001 μm, 275,647 μm and 318,785 μm respectively. The highest petiole cortex diameter on J3.

Mesophyll leaves are built by the parenchymal tissue with a transport file. Palisade parenchyma forms elongated cells, arranged in solid bonds and containing chloroplasts. Parenchymal sponges are composed of cells that are irregular in shape, contain chloroplasts and have large interstellar space. Each cultivar has different thickness variation[26]. The average mesophyll thickness of J1, J2, J3, J4, J5 and J6 were 364,986 μm, 280,703 μm, 389,743 μm, 245,749 μm, 261,439 μm and 434,913 μm respectively. The highest mesophyll thickness was J6. The leaves that have thick mesophyll indicated the availability of chloroplast in the palisade tissue and the high parenchymal sponge caused high effectiveness of photosynthesis.

Lamina thickness is influenced by the composition of leaf mesophyll, namely parenchymal palisade and parenchymal sponge. In the growth and development of plants, the availability of essential water and nutrients is the main factor for optimal growth, including growth and development of leaves. In mesophyll leaves, there is a network of xylem and phloem transport which is a continuation of the rod transporting file. Morphological characteristic such as lamina thickness is closely related to resistance to suckling pests *Amrasca biguttula* in tobacco[26]. The thickness of the lamina is affected by the thickness of the cell wall that made up organ. The cell wall contains proteins and enzymes that actively work for the growth of thickened cell walls and strengthen cell walls during inductive resistance. When cells detect the presence of potential pathogens, enzymes catalyze and produce reactive oxygen molecules that can damage the cells of the invading organism.

The relationship with antixenotic resistance to cell wall thickness and leaf cuticle layer affect leaf hardness. The length and width of the stomata pores are related to insects that have a sucker-type mouth tool that can insert the stylet through the stomata opening. Stomata diameter does not have direct effect on movement and feeding activity but is related to the product of plant cell metabolism which is needed by insects for growth and development. Stomata diameter affects the amount of CO_2 diffusion for the synthesis of primary metabolite products, namely carbohydrates, proteins and fats, and the synthesis of metabolites secondary to plant resistance. Stomata pore size is tightly regulated by plants, and cover cells participate in resistance by closing in response to the molecular pattern of microbial association.

4. Conclusion

Six sweet potato varieties showed differences in anatomical characteristics in terms of stomata density, length and width of opened stomata pores, mesophyll thickness and petiole cortex diameter. The highest mean stomata density was local white (J6) which is 9.11 in 75796.36 μm2. The average length and width of opened stomata pore was the highest in local purple (J5) which was 39.52 μm and 9.28 μm. The highest petiole cortex diameter in Cilembu (J3) was 456,439 μm.

Acknowledgment
The writing team would like to thank to Sam Ratulangi University Institute for Research and Community Service Institute that funded this research through Applied Research Scheme 2018.

References

[1] Yang J and Gadi R L 2008 Effects of dehydration on anthocyanins, antioxidant activities, total phenols and color characteristics of purple-fleshed sweet potatoes (*Ipomoea batatas*). *American Journal of Food Technology*. http://www.academicjournals. (Accessed November 12, 2017)
[2] Logo C A 2011 The role of women in the Dani tribe in maintaining the continuity of the test of sweetness as a staple food in Jayawijaya Regency Irian Jaya. *Malang Balittan Journal* 3 (1): 353-360
[3] Mwanri A, Kogi-Wakau W and Laswai H 2011 Nutriens and antinutriens composition of raw, cooked and sun dried sweet potato leaves. *Afr.J. Food. Agric.Nutr.Dev.* 11(5) : 5142-5156
[4] Oloo B O, Shitandi A, Mahungu S, Malinga J B and Ogata B R 2014 Effects of lactic acid fermentation on the retention of beta carotene content in orange fleshed sweet potatoes. *Int. J. Food Stud.* 3 :13-33
[5] Anonymous 2017 Superior sweet potato varieties 1977-2016. Research Center for Various Nuts and Tubers. http://balitkabi.litbang.pertanian.go.id/?p=208 (Accessed November 2, 2017)
[6] Sodiq 2009 Plant resilience to pests *UPN Press*, Surabaya 80p. http://eprints.upnjatim.ac.id/47/. (Accessed July 15, 2016)
[7] Hagg J F, Zagrobelny M and Bak S 2013 Plant defense against insect herbivores *Int. J. Mol. Sci.* 14 : 10242-10297
[8] Cambell N A, Reece J B and Mitchell L G 1999 Biology (Redwood City : Benjamin Cummings Publishing Company Inc.)
[9] Fahn A 1982 Plant anatomy Third edition (England: Pergamon press Ltd.)
[10] Haryanti S 2010 Number and distribution of leaf stomata of dicot and monicotile plant species, *Bulletin of Anatomy and Physiology* XVIII (2): 21-28
[11] Hagg J F, Zagrobelny M and Bak S 2013 Plant defense against insect herbivores *Int. J. Mol. Sci.* 14 : 10242-10297
[12] Dalin P, Agren J, Bjorkman C, Huttunen P and Karkkainen K 2008 Leaf trichome formation and plant resistance to herbivory In A. Schaller (Ed) Induced plant resistance to herbivory. Springer Science. p 462
[13] Semangun S S 2001 A Colour atlas of plant structure (London : Manson Publishing Ltd.)
[14] Pradana A W, Samiyarsih S and Muljowati J S 2017 Correlation of anatomical characteristics of sweet potato leaves (*Ipomoea batatas* L.) resistant cultivars and not resistant to the intensity of leaf scabies. *Scripta Biologica* 4 (1): 21-29
[15] Bahri 2010 Chlorophyll *Capita Selekta Organic Chemistry* (Lampung : University of Lampung)
[16] Amin M E H 2010 Effect of organic fertilizer and urea on growth, yield and quality of fodder maize (*Zea mays* L.) *Int. Journal of Current Research* 8 : 35-41
[17] Warhamni, Boer D and Muzuni 2013 Morphological diversity of sweet potatoes (*Ipomoea babatas* L.) Lam.) origin of Muna Regency. *Journal of Agroteknos* 3 (2): 121-126

[18] Utari D S, Kardhinata E H and Damanik R I M Analysis of morphological character and kinship relations of sweet potato plants (*Ipomoea Batatas*) In the highlands and lowlands of North Sumatera. *Agroecotechnology Journal* 5 (4): 870- 881

[19] Frechette B, Bejan M, Lucas E, Giordanengo P and Vincent C 2009 Resistance of wild *Solanum* accessions to aphids and other potato pests in quebec field conditions. Journal of Insect Science. 10(161) : 1-16

[20] Salisbury F B and Ross C W 1985 Plant Physiology Wadsworth Publishing Company. Belmont. California. p 540

[21] Mulyana N 2006 Adaptation of morphology, anatomy and physiology of soybean plants (*Glycine max* L.) in the conditions of shade stresses (Bogor : Bogor Agricultural University)

[22] Gorgani E, Padua G and Radice S 2013 Compared anatomy of young leaves of *Prunus persica* L. Batsch with different degrees of susceptibility to *Taphrina deformans* (Berk). *Journal of Phytophatology* 161(1) : 190-196

[23] Sasmitamihardja D and Siregar A 1996 Plant physiology (Bandung : Biology Department of Bandung Institute of Technology)

[24] Woelaningsih S 2001 Structure and development of plants II (Yogyakarta : Faculty of Biology, Gajah Mada University)

[25] Indrayani I G A A 2008 Role of plant morphology to control the leaf sucker of *Amrasca biguttula* (Ishida) in Cotton Plants. Perspective 7 (1): 47-54

[26] Freeman B. C. and G. A. Beattie. 2008. An overview of plant defenses against pathogens and herbivores.http://www.apsnet.org/edcenter/intropp/topics /Pages/OverviewOfPlantDiseases.aspx. (accessed September 27, 2017)

IOP Conf. Series: Materials Science and Engineering **567** (2019) 012045 doi:10.1088/1757-899X/567/1/012045

Host plant and intensity pest attack *Paracoccus marginatus* on papaya plant in North Minahasa Regency

J M E Mamahit[1], J Manueke [1], V B Montong [1]

[1]Department of Plant Pests and Diseases, Faculty of Agriculture, Sam Ratulangi University Manado, Indonesia

Corresponding author: evamamahit@gmail.com

Abstract. The purpose of this study is to find out distribution and the host range of mealybug *Paracoccus marginatus* and their attack intensity in the field, especially in the centers of papaya cultivation in the North Minahasa region, North Sulawesi Province, Indonesia. The study uses a survey method conducted in four villages, namely: Matungkas, Laikit, Dimembe and Tetei. Sampling was done by purposive sampling. Sampling was carried out in three fields per village that had different cropping systems on the farms cultivated by farmers. Each location was sampled and observations were made on papaya plants to see whether papaya mealybug were attacked or not attacked. Samples of mealybug and host plants from the field were brought to the laboratory for identification. The results showed the attack intensity of mealybug ranged from 13.67% to 60%.Symptoms of mealybug attack on papaya plantations was variated from the symptoms of low intensity to high intensity. The results of this research found host plants of *P. marginatus* consisted of seven species of plants, besides papaya namely: *Annnona muricana, Persea gratissima, Codiaeum spp, Aglonema spp, Anthurium spp, Hibiscus spp. and Jatropha* spp.

1. Introduction

The pest insects are one of the constraints in the cultivation of papaya plants in North Sulawesi Province. Some insect pest species that attack papaya plantations are: fruit flies, armyworm, mites, aphids, shell lice, grasshoppers and mealybug. Papaya mealybug, *Paracoccus marginatus* , a serious invasive pest affecting horticultural crop especially on papaya plants[1][2]. In Indonesia this pest was first discovered in papaya plants in Bogor in May 2008 along with CRSP / Virginia Tech / Clemson University, USA IPM Team and later found it in papaya plants in the Capital City District, Manado City[3][4].These pests then spread so quickly and reportedly attacked papaya and other plants in many places throughout Indonesia. This papaya mealybug pest has distribution throughout the city of Manado so that by the end of 2009 almost all papaya that had grown in the yard of the house were attacked by these pests and eventually died. These pests have been found in the city of Manado[4]. The results of Research 2011 found that mealybug *P. marginatus* are not only spread in the Manado area and its surroundings, but have spread to almost all regions in North Sulawesi including the Minahasa, North Minahasa, to South Minahasa and Southeast Minahasa regions[5].

Mealybug *P. marginatus* attacking papaya plants also attack cassava plants, distance and other plants that release sap[6]. This pest attack on papaya plants both young and fruitful, can kill the plant less than 3 months from the beginning of pest infestation on the plant. Furthermore, research found that these pests are very dangerous attacks, because it can reduce the results of mealybug sucking liquid plants from leaves to fruit. Pests can attack on various on part of papaya plant such as the stems, leaves, petioles, flowers and fruit [7]. The spread of this mealybug can be caused by wind, seeds, people, or carried by other insects and birds. The presence of high and polyphagic mealybug has the potential to spread very quickly. Insecticides are used by farmer to kill these pests. In addition, the use of insecticides will leave toxic residues in the fruits and leaves of plants which are important food ingredients for humans as well as negative impacts on the natural enemy, environment such as pollution and pest resistance. Therefore, this pest control is the most appropriate, which is preferentially limiting the entry of these pests in crops by removing host plants that are around papaya plants. For this reason, a survey study was needed to know the species of host plants of mealybug *P. marginatus* and their level of attack in the field, especially in the centre of papaya cultivation in the North Minahasa region.

2. Methods

This study used a survey method and carried out in the center of papaya planting in North Minahasa North Sulawesi Indonesia. Sampling was done by purposive sampling. Sampling was conducted in four villages which were different from Matungkas, Laikit, Dimembe and Tetei on North Minahasa. Observations were made on the farms cultivated by farmers. Each location carried out three field and every location was taked 50 plant sampel for observation. The obsevation were made on papaya plants to see whether papaya mealybug were infested or not infested. Observations were made in the papaya farmer by seeing whether there were symptoms of papaya mealybug infested, then calculating the attack intensity of mealybug. In papaya planting, observations plant host of mealybug were carried out around the plants, namely plants that were also attacked by mealybug. In addition to near papaya planting, observations were carried out on several plants that showed symptoms of mealybug attack in the North Minahasa region. Plants that are attacked are taken as a sample by cutting the affected part, then put it in a transparent plastic bag, then take it to the laboratory. Observations in the laboratory to identify the types of plants attacked by mealybug. Observation Parameters: 1) species of plants attacked by mealybug, 2) symptoms of mealybug attack and 3) intensity of mealybug attack. The data obtained were prepared, tabulated and analyzed by variance and continued with further tests using Duncan New Multiple Range Test (DNMRT) at the level of 5%.

3. Result and Discussion

The observation area, namely North Minahasa, is a central area of papaya cultivation. The intensity of mealybug attacks in this area is summarized in Table 1. The level of attack of mealybug pests is generally low to high level attack ranging from 13.67% to 60%. Differences in attack rates are found in each village. The highest level of pest attack is found in Matungkas village which can reach 80% followed by of Laikit reached 40% and Dimembe 36,67%. The attack rate of papaya mealybug in Tetei village showed a significant difference with the level of attack of mealybug in the other three villages. The difference in the level of attack of mealybug in the papaya planting is influenced by cropping conditions, land sanitation and cropping, crop varieties, and climate. The presence of pest insects is influenced by differences in temperature, climate, geographical conditions and vegetation in the region[8]. Besides that, the determining factor is the low and high level of pest infestation in the field, namely the use of pesticides. Local farmers are very dependent on the use of pesticides. In the location of plants that have just been sprayed with insecticides especially on Tetei village, the intensity of mealybug infestation will be lower than in plants without spraying.An average yield loss of 91% on papaya crop was reported in the affected farms[9].

Table 1. Intensity of attack by papaya mealybug *Paracoccus marginatus*[a].

Village/Field	Attack intensity (%)	Notation
Matungkas	60,00	a
Laikit	40,00	a
Dimembe	36,67	a
Tetei	13,67	b

[a]The numbers in the lane followed by unequal letters differ significantly according to the DNMRT test at the 5% level

The results showed that symptoms of a distant mealybug attack appear on the part of the plant that is attacked in white flour, and from near white flour symptoms it is seen as a mass of colonies of mealybug. Mealybug are covered in a layer of wax or white flour. The wax layer is a secretion from mealybug. Infestation of mealybug appears as cluster of cotton like mass on the above ground portion of plants with long waxy filaments[2]. The survey showed that mealybug *P. marginatus* attack papaya from leaves, petioles, stems, fruit stalks and fruit. Symptoms of mealybug attack on papaya in plantations varied from symptoms of low intensity, as in Figure 1a, that were seen only one group of mealybug pests in the fruit. However, there are also plants in the plantation that are categorized as moderate intensity by mealybug attack. Where the attack has distribution to the fruit is less than half the fruit. There are also some trees are attegorized as high intensity, as in Figure 1b, that symtoms appear to be severe attacks shown from the attack of mealybug already present in the whole fruit, and the spread of attacks on plants from leaves and stems. The attack of papaya mealybug is typically seen as mealybug mass shaped like cotton on the top and bottom of plants[10]. Mealybugs cause deformity, wrinkling and rolling of the leaf edges and early leaf drop. Attack to unripe fruits causes sap running and blemishes, a source of fruit downgrading. Papaya fruit can be heavily infested with mealybugs, becoming white and essentially inedible. Under heavy infestations, the abaxial side of the lower leaves can be covered with insects that congregate near the main vein[11].

Figure 1. Symptoms of attacks by mealybug *P. marginatus* on papaya. a) low intensity b) high intensity

The results of this research found the host plants of mealybug *P. marginatus* consisted of eight species plant, namely besides papaya, Annona (*Annona muricana)*, avocado (*Persea gratissima*), croton (*Codiaeum variegatum*), Aglonema (*Aglonema* spp.*)*, Anthurium (*Anthurium* spp.) and Hibiscus (*Hibiscus rosa sinensis*) and Jatropha (*Jatropha* spp.) can be shown on Table 2 . The research showed that the mealybug *P. marginatus* was polyphagous insect which have host plants of more than one plant species from several different plant families. The host plant species of mealybug *P. marginatus* pests are various species plants. These pests have more than 60 species of host plants[12]. Furthermore, papaya mealybug had host plants including: hibiscus, advocates, oranges, cotton, tomatoes, green beans, soybeans, taro, mangoes and cherries[13[. Host plants of mealybug pests found in West Java, including: peanuts, croton, taro, waluh, hibiscus, rosella and others[14]. In Kenya reported the papaya mealybug was found to infest papaya (*Carica papaya*), cassava (*Manihot esculenta*), chili pepper (*Capsicum annuum*), guava (*Psidium guajava*), mango (*Mangifera indica*), and eggplant (*Solanum melongena*)[9]. The presence of mealybug in various host plants, accelerates the movement and spread of mealybug to healthy papaya plants that surround it. The presence of host plants in the field allows pests to be able to withstand the conditions of the main host pl *Annona muricana* ants not available. This causes the development of these pests has always been a problem for papaya farmers.

Table 2. Host Plants of mealybug *P. marginatus* on North Minahasa regency.

Number	Host Plant of mealybug	
	Common name	Species
1.	Papaya	*Carica papaya*
2.	Sirsak/Annona	*Annona muricana*
3.	Advokad/Avocado	*Persea gratissima*
4.	Puring/Croton	*Codiaeum variegatum*
5.	Aglonema	*Aglonema* spp.
6.	Anthurium	*Anthurium* spp.
7.	Kembang sepatu/Hibiscus	*Hibiscus rosa sinensis*
8.	Jarak/Jatropha	*Jatropha* spp.

4. Conclusions

The attack intensity of mealybug *Paracoccus marginatus* in North Minahasa Regency generally ranges from 13.67% to 60%. The symptoms of mealybug attack on papaya plantations was various intensity from low intensity until high attack intensity. Mealybug *P. marginatus* is polyphagous pests which have seven host plants species besides papaya (*Carica papaya*), namely: Annona (*A. muricana*), avocado (*P. gratissima*), croton (*C. variegatum)*, Aglonema (*Aglonema* spp.), Anthurium (*Anthurium*spp.), Hibiscus (*Hibiscus rosa sinensis*) and Jatropha (*Jatropha* spp.).

5. References

[1] Kunimoto SL 2006 Papaya Mealybug Department of Agriuculture. Stae of Hawaii. Biological Control Section Project.
[2] Tanwar RK, Jeyakumar P and Vennila S 2010 Papaya Mealybug its Management Strategies Natural Centre for Integrated Pest Management New Delhi
[3] Muniappan R, Meyerdirk DE, Sengebau FM, Berriger DN and Reddy GVP 2006 Florida Entomologist: 89(2): 21-217.
[4] Sembel DT 2009 New Pests of Papaya Plant, *Paracoccus marginatus*: A Threat to Papaya Farmers in North Sulawesi. Media Comments Manado
[5] Mamahit JME and Sembel DT 2011 Distribution and Natural Enemies of mealybug *Paracoccus marginatus* William and Granara de Willink (Hemiptera: Pseudococcidae) on Papaya Plants in North Sulawesi. Research Report on the collaboration of Clemsen Univ.
[6] Sembel DT and Moniaga W 2009 Survey New Pest *Paracoccus marginatus* Williams and Granara De Willink (Homoptera: Pseudococcidae) on Papaya Plants in North Sulawesi. Eugenia 15(2):108-114
[7] Mamahit JME. 2013. Attack of mealybug Paracoccus marginatus on Papaya Research Report Sam Ratulangi University Research Institute Manado
[8] Speight MR, Hunter MD and Wett AD 1999 Ecology of Insect-consept and Blackwell Science Ltd London
[9] Macharia I, Kimani E, Koome, F, Kosiom T, Heya H, Otipa M and Oronje M 2017. First Report and Distribution of the Papaya Mealybug, Paracoccus marginatus, in Kenya. J. of Agricultural and Urban Entomology 33(1):142-150. 2017
[10] William DJ, de Willink MCG. 1992. Mealybug of Central and South America. Wallingford Oxon: CAB International.
[11] Pena JE and Johnson FA1993 Insect Management in Papaya. Univerrsity of Florida. IFAS Extention
[12] Meyerdirk DE and Kauffman WC 2001 Status on the Development of a biological control program for *Paracoccus merdiatus* William, papaya mealybug. Paper presented at IV International Scientific Seminar of Plant Health. Veradera, Cuba, June 10-15

IOP Conf. Series: Materials Science and Engineering **567** (2019) 012045 doi:10.1088/1757-899X/567/1/012045

[13] Walker A, Hoy M and Meyerdick D 2003 Papaya Mealybug University of Florida Institute of Food And Agricultural Sciences
[14] Sartiami D, Dadang, Anwar R and Harahap IS 2009 Distribution of New Pests *Paracoccus marginatus* in West Java Banten and DKI Jakarta Provinces

The profile of heart attack symptoms of cardiovascular patients at XYZ hospital

R Setiadi[1], Franky[2]

[1]Department of Mathematics, Faculty of Mathematics and Science, university of Indonesia
[2]Pascasarjana Magister Management, University Prof. Dr. Moestopo (Beragama), Indonesia

Corresponding author: rianti@sci.ui.ac.id

Abstract. According to the World Heart Federation research, cardiovascular disease has the highest mortality rate in the world. Heart attacks can come suddenly without showing symptoms. Some symptoms that may appear are like chest pain and short of breath. The study will try to find the profile of attack symptoms of 1076 cardiovascular patients at XYZ Hospital based on age, gender, job, habit, weight-change, ECG, heart noise, breath noise, systolic, diastolic, pulse and sleeping problem. The method using here is the Classification Tree. The study results that the most influence factor to the heart attack symptom is the breath noise. If the breath noise is not normal, 80.5% will have symptom short of breath, for female, 93.6% will have symptom short of breath and for male 69.9% will have symptom short of breath. If the breath noise is normal, 59.3% will have symptom chest of pain. If breath noise is normal, heart voice regular, pulse is not normal, 47.1% will have symptom short of breath. If breath noise is normal, heart noise regular, pulse is normal, 72.3% will have symptom chest pain. If breath noise is normal, heart voice irregular, 47.9% will have symptom short of breath

1. Introduction

According to the World Heart Federation research, cardiovascular disease has the highest mortality rate in the world. The WHO (2017) stated that among the 56.4 million deaths in 2015 due to the top 10 causes, about 54% were attributed to cardiovascular disease. The disease accounted for 15.54% of total mortality in 2017 according to the Agency of Research and Health Development Indonesia. In general cardiovascular disease is a silent killer. Heart attacks can come suddenly without showing symptoms. Some symptoms that may appear are like chest pain (David Railton, 2018) and short of breath but they often come too late.

Sharol Ashma Menezes said that cardiovascular patients have abnormal ECG. Zhongwei Jiang at journal of biomedical engineering said that the cardiac sound measurement and analysis system for in-home use of heart abnormality monitoring heart noise. Anandeep Chaudhuri and Jayanthi Thiruvengadam (2006) wrote that cardiac abnormality can be detected using heart sound and breathe noise. Andre-Pascal Kengne and his friends compares the strengths of the associations between different baseline blood pressure variables (systolic blood pressure), diastolic blood pressure and pulse pressure. It can be concluded that abnormal cardiovascular disease patients maybe have abnormal blood pressure and pulse pressure. Naima Covassin, PhD wrote about Sleep Duration and Cardiovascular Disease Risk (2015), so it can be thought that cardiovascular disease patients can have a sleeping problem. Other factor that may be indicate cardiovascular disease is a weight change of the patients (William B. Kannel and friends 1967)

The variables above are called health condition of the cardiovascular patients. Other than that, cardiovascular disease is influenced by age, gender, job and smoking habit. These variables are called demographic status of the cardiovascular patients. If the profile of heart attack symptom based on these variables can be found, so the protection of cardiovascular disease can be increased by taking attention of the profile of heart attack symptom. The heart attack symptom consists of chest pain or short of breath.

The study will try to find the profile of the heart attack symptom of cardiovascular disease (chest pain and short of breath) based on health condition of the patient (ECG, heart noise, breath noise, blood pressure, sleeping problem and weight change) and the demographic status of the patients (age, gender, job, smoking habit). The data analyses are made separately between demographic status variables and the joint of health condition variables and demographic status variables. The study uses data 1076 cardiovascular patients at XYZ Hospital in Indonesia and the data analyses are made by using Classification Tree method.

2. Research objectives:

The research objectives in this paper is to determine the heart attack syndrome's profile based on the demographic status variables and health condition variables of the cardiovascular patients at XYZ Hospital.

3. Classification Tree

The method used here is the Classification Tree method. Classification Tree is a nonparametric statistical method that explore the data using categorical tree. The Classification tree method is used to find the variable that most influences the response variable and find the profile of the response variable using the predictor variables. By the data grouping in the tree, the relationship between response variables with one or more predictor variables can be seen. In the classification tree, response variable is a categorical variable. In this study, response variables is the symptom of heart attack including short of breath and chest pain. In the classification tree there is a series of branches and nodes. Branches are part of a classification tree that connects between nodes. The nodes contain classes of response variables consisting of root nodes, parent nodes, child nodes, and terminal nodes.

The root node is the initial node that contains all the data to construct a classification tree. The root node will generate branches and other nodes called parent nodes. Next from the parent node will split and generate the new child nodes. In binary classification, each division will generate two child nodes, namely right node and left node. Child nodes that can no longer be split are called terminal nodes.

Here is a general overview of the classification tree:

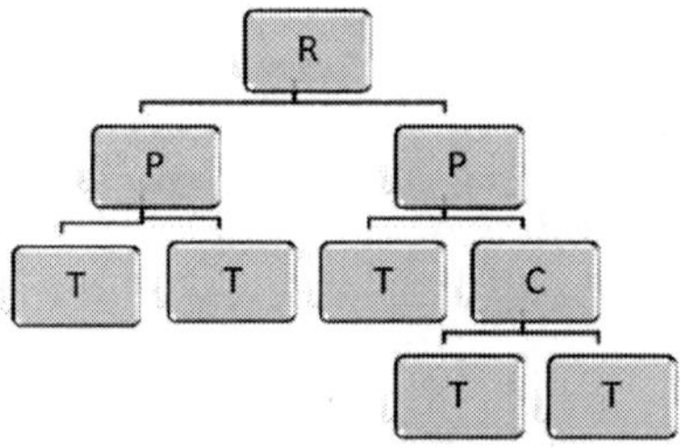

Figure 3.1 General Classification Tree

R is the root node, P is the parent node, C is the child node and T is the terminal node.

The classification tree is formed by binary recursive partitioning algorithm. The purpose of the binary is to divide the data on each parent node into two groups of child nodes that called right node and left node based on certain predictor variables. The splitting is done on each node until the terminal node is obtained (recursive).

3.1 Steps for built Tree Classification

Suppose there are N objects to be classified into S class. In this project, S = 2, short of breath and chest pain. Suppose that many nodes formed are denoted by t, where t = 1,2, ..., T.

There are three stages in the process of forming a classification tree:

3.1.1 Selection of Splitting Variable

At this stage the parent division process is done into two groups of child nodes containing all the sample data that are still heterogeneous. The samples were sorted according to the goodness of split criterion (best breaking criterion). Goodness of split criterion is a sorting evaluation by class s at node t which is defined as decreasing of heterogeneity or also called improvement. This criterion will result more homogeneous right and left node of the parent node.

Goodness of split criterion is formed based on impurity measure, denoted i (t). Impurity measure is a measure of the heterogeneity of a class of a particular node in a classification tree. The impurity function used in this research is gini index, which is formulated with $i(t) = \sum_{j \neq i} C(i|j)p_j p_i$, where pj = (Nj(t))/(N(t)). Nj (t) is the number of objects in class j at node t, N (t) is the sum of all objects in node t. Suppose C (i | j) is the weight of misclassification stating the magnitude of the given penalty if the class object j is misclassified in class i at node t. C (i | j)> 0, i ≠ j and C (i | j) = 0, i = j.
The smaller the gini index value of a node, the better the process of separating an object into a class j.

3.1.2 Goodness of split criterion

Goodness of split criterion can be formulated as $\Delta i(s,t) = i(t) - p_L i(t_L) - p_R i(t_R)$
Δi (s, t) = decrease of class heterogeneity at node t
i (t) = impurity value at node t
p_L = the proportion of the number of objects on the left node
p_R = the proportion of the number of objects on the right node
$i(t_L)$ = impurity value at left n^{th} node
$i(t_R)$ = impurity value at right-t node
s = 1,2, which denotes the number of classes in the classification tree
t = 1,2, ..., T, which denotes the number of nodes in the classification tree
The best splitting variable is one that has the greatest Δi (s, t) among all other possible divisions because greater Δi (s, t) is possible to reduce heterogeneity more significantly.

3.1.3 Terminal Determination

The splitting variable selection process continues until there is no further significant drop in heterogeneity. Generally, there is a minimum limit used in determining terminal nodes. When this condition is met, the process of tree formation stops.

3.2 Label Class Tagging

Class labeling is done based on the most number of rules, that is, the maximum proportion of an object enters into a class s in node t. Mathematically can be written as $p(s_0|t) = \max_s \frac{N_j(t)}{N(t)}$ with p (s_0 | t) is the proportion of the number of objects that enter into a class s_0 at node t.

4. Findings

4.1 The profile of the heart attack symptom of cardiovascular disease based on the demographic status

From the figure 4.1 below, the result are:

- The variable that most influences the heart attack symptom of cardiovascular disease is gender. For male 50.5% will have symptom chest pain but for female 69.8% will have heart attack symptom short of breath.
- For male with age more than 70 years old or smaller than 30 years old, 58.1% will have heart attack symptom short of breath
- For male with age 50-70 years old, not working, 49.4% will have the heart attack symptom chest pain
- For male with age 50-70 years old, working, smoking, 48.8% will have heart attack symptom short of breath
- For male with age 50-70 years old, working, not smoking, 50% will have heart attack symptom chest pain

The Classification Tree for demographic status is described in Figure 4.1 below:

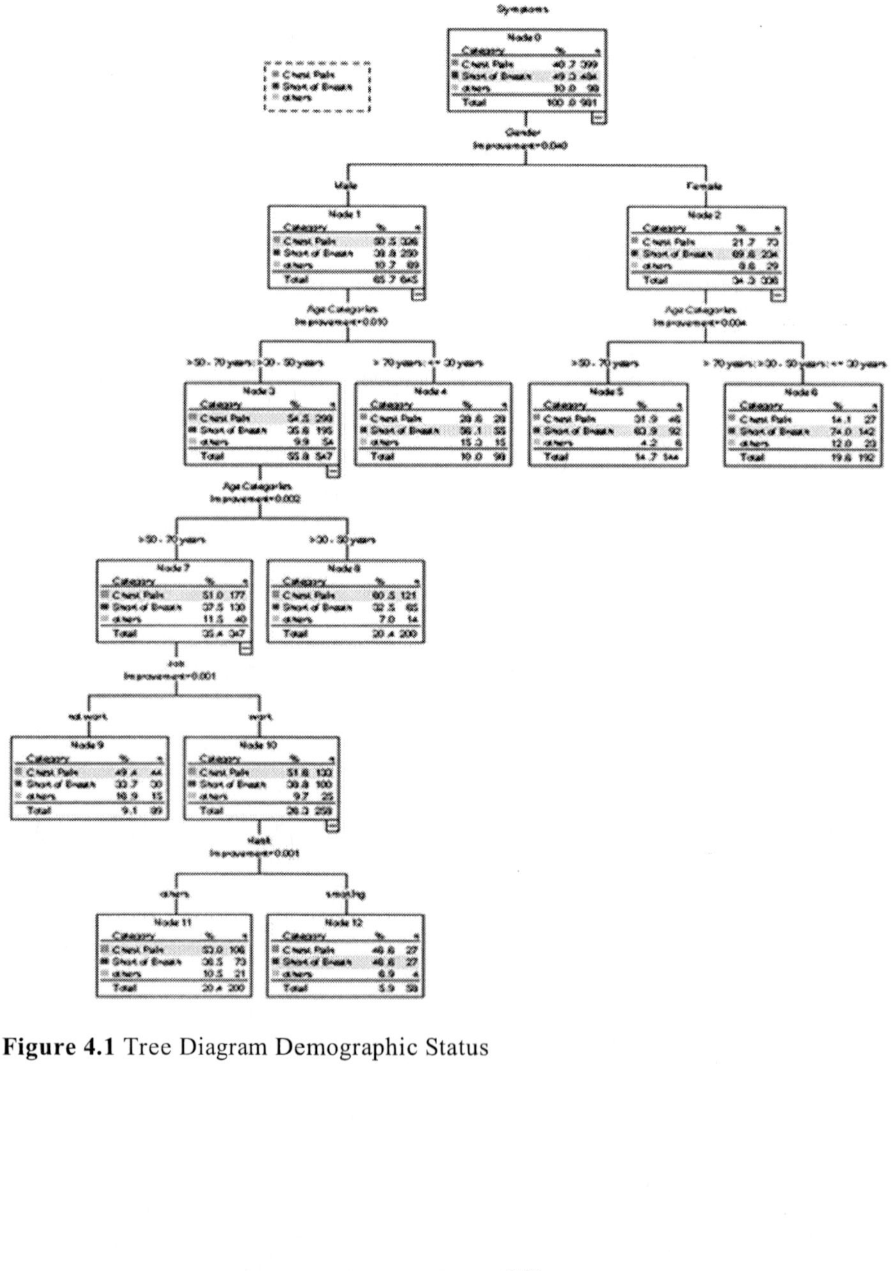

Figure 4.1 Tree Diagram Demographic Status

4.2 The profile of the cardiovascular symptom based on health condition and demographic status
The Classification Tree can be found in the Diagram Tree on Figure 4.2 below:

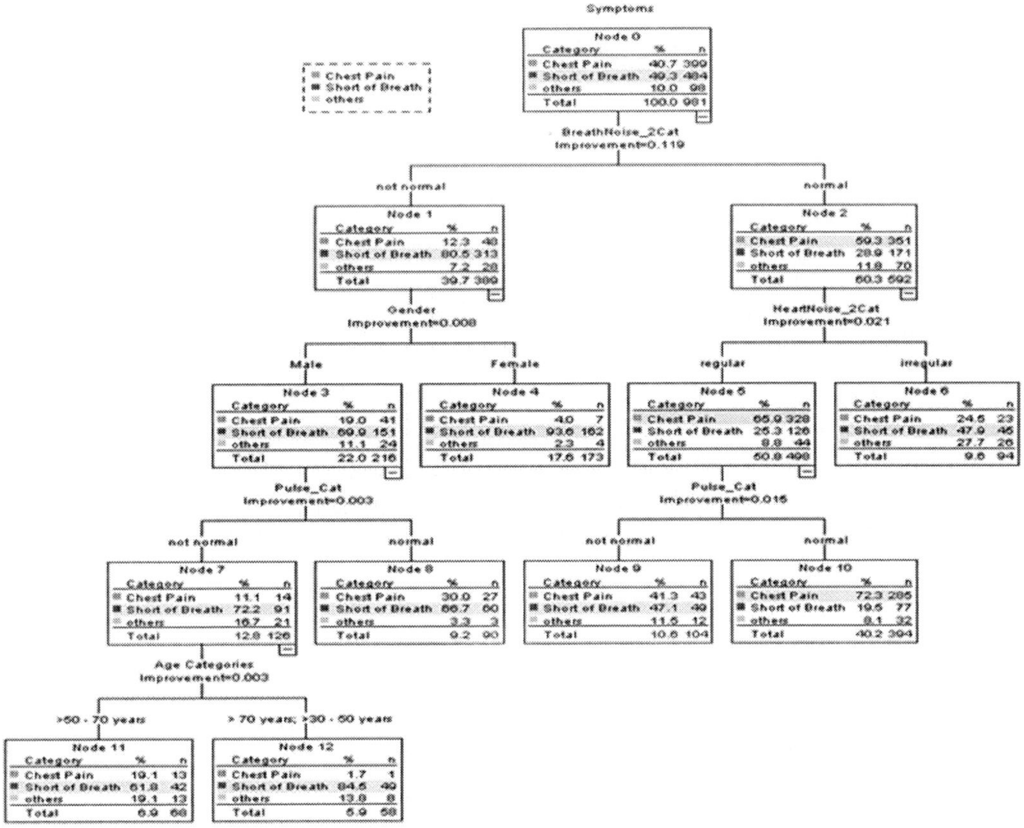

Figure 4.2 Tree Diagram for Health Condition and Demographic Status

From the Tree diagram Figure 4.2 above, the results are:

- The variables that most influences the heart attack symptom of cardiovascular disease is the breath noise. If the breath noise is not normal, 80.5% will have heart attack symptom of short of breath. If the breath noise is normal, 59.3% will have heart attack symptom of chest of pain.
- If the breath noise is not normal, for female, 93.6% will have heart attack symptom short of breath. For male 69.9% will have of the heart attack symptom of cardiovascular disease short of breath.
- If breath noise is normal, heart voice regular, pulse is nor normal, 47.1% will have heart attack symptom short of breath
- If breath noise is normal, heart noise regular, pulse is normal, 72.3% will have heart attack symptom chest pain.
- If breath noise is normal, heart voice irregular, 47.9% will have heart attack symptom short of breath.

IOP Conf. Series: Materials Science and Engineering **567** (2019) 012046 doi:10.1088/1757-899X/567/1/012046

5. Discussion

From the results above, people who do not check their health condition must be careful with their demographic status. Female must be aware with short of breath. It can be a sign of the appearance of heart attack. In general, male with age >70 years old or < 30 years old must be aware when they feel short of breath. Male who are 50-70 years old, working and smoking must also be aware with short of breath. Male must be aware with chest pain especially for men with age 50-70 years old, working and not smoking. If the sign of the profile appeared, the patients must be careful and check their health condition.

When people check their health condition and combine the result with their demographic status they must be careful with breath noise especially when their breath noise is not normal. If the breath noise is not normal, they must careful when they feel short of breath though they have regular heart noise. When they have regular heart noise, not normal pulse they must also be aware with the appearance of short of breath. When their pulse is still normal, they must be aware with the appearance of chest pain.

6. Conclusion

The conclusions of this study that describe the profile of heart attack symptom of dardiovascular disease are:

1. From demographic status, the variable that most influences the heart attack symptom of cardiovascular disease is gender. Male will have symptom chest pain but female will have symptom short of breath. Other variables that influence heart attack symptom is age, working or not and smoking habit. For male with age more than 70 years old or smaller than 30 years old will have symptom short of breath. For male with age 50-70 years old, not working will have symptom chest pain. For male with age 50-70 years old, working, smoking will have symptom short of breath. For male with age 50-70 years old, working, not smoking will have symptom chest pain
2. From demographic status and health condition of the cardiovascular disease patients, the variable that most influences the heart attack symptom is the breath-noise. If the breath noise is not normal the patients will have symptom short of breath. If the breath noise is normal the patients will have symptom chest of pain. Other variables that influence the heart attack symptom are gender, heart voice and pulse. If the breath noise is not normal, for female will have symptom short of breath. For male will have symptom short of breath. If breath noise is normal, heart voice regular, pulse is nor normal, the patients will have symptom short of breath. If breath noise is normal, heart noise regular, pulse is normal, the patients will have symptom chest pain. If breath noise is normal, heart voice irregular, the patients will have symptom short of breath.
3. Before short breath appear people must take care with their breath noise, heart noise pulse, their working and their habit to smoke. Before chest pain appear every person must take care with their breath noise, pulse, age and their working.

References:

[1] Achar, SA, Kundu, S, Norcross, W. 2005. *Diagnosis of acute coronary syndrome.* American

[2] Family Physician, Vol. 72, No. 1.

[3] Al-batah, MS. (2010). *Testing the probability of heart disease using classification and regression tree model.* Department of Computer Science and Software Engineering, Faculty

[4] of Sciences and Information Technology, Jadara University, B. O. Bo: 733, Jordan.

[5] Amborse, JA, FACC, & Barua, RS. (2004). *The pathophysiology of cigarette smoking and cardiovascular disease.* Journal of the American College of Cardiology, Vol. 43, no. 10. ISSN: 0735-1097/04/$30.00.

[6] Aminpour, S, Shahamfar, M, & Shahamfar, J. 2014. *Effects of lifestyle modification program in reduction of risk factors in patients with coronary heart disease.*

[7] Anandeep Chaudhuri, Jayanthi Thiruvengadam. 2016. *Blood Pressure Variables and Cardiovascular Risk.* Article in Biomedical Engineering Applications Basis and Communications
[8] Azen, Razia dan Cindy M Walker. 2011.*Categorical Data Analysis for the Behavioral and SocialScience*. York: Routledge Taylor and Francis Group.
[9] Breathlessness (shortness of breath) information. https://www.webmd.com/hypertension-high-blood-pressure/guide/hypertensive-heart-disease#1
[10] Boufit, C, & Lauderdale, D. 2008. *How Your Sleep Affects Your Heart. gender.* Vol.4, No.6, 309-315. doi:10.4236/health.2012.46051.
[11] *Coronary Artery Disease Can Be Silent Killer*. https://www.memorialcare.org/blog/coronary-
[12] Cohn, PF, Vokonas, PS, Williams, RA, et al. 2017. *Diastolic heart sounds and filling waves in coronary artery disease.* American Heart Association, Inc., 7272 Greenville Avenue, Dallas,
[13] Covassin N, Prachi Singh. 2016. *Sleep Duration and Cardiovascular Disease Risk: Epidemiologic and Experimental Evidence.* https://www.ncbi.nlm.nih.gov/pmc/articles/PMC4791534/
[14] Damayanti, Laila K. 2011. *Aplikasi Algoritma CART untuk Mengklasifikasi Data Nasabah Asuransi Jiwa Bersama Bumi Putera 1912 Surakarta. (Skripsi).* Fakultas Matematika dan Ilmu Pengetahuan Alam Universitas Sebelas Maret.
[15] Railton, D. 2018. *What could cause chest pain?* https://www.medicalnewstoday.com/articles/321650.php.
[16] Gupta S, Rajeev Kumar Gupta, Malini Kulshrestha, and Rajib Ratna Chaudhary. *Evaluation of ECG Abnormalities in Patients with Asymptomatic Type 2 Diabetes Mellitus*
[17] Gaziano, TA. 2017. *Lifestyle and Cardiovascular Disease*
[18] Hair, JF, Lack, WC, Babin, BJ., Anderson, RE, and Tatham, RL. 2008. *Multivariate Data Analysis* 6th Edition. New York: Pearson Prentice Hall.
[19] *High Blood Pressure and Hypertensive Heart Disease*. https://www.webmd.com/hypertension-high-blood-pressure/guide/hypertensive-heart-disease.
[20] Kannell WB, E. Joseph L, Thomas SD and Patricia MM. 1967. *Relation of Body Weight toDevelopment of Coronary Heart Disease.* The Framingham Study
[21] Kumar, A. and Cannon, CP. 2009. *Acute coronary syndromes: diagnosis and management, part* Mayo Clin Proc., 84(10):917-938.
[22] Maas, AHEM, & Appelman, YEA. 2010. *Gender differences in coronary heart disease.* Neth Heart J. 2010 Dec; 18(12): 598–602.
[23] Malik, MA, Khan, SA, et al. (2013) *Chest Pain as a presenting complaint in patients with acute myocardial infarction (AMI).*
[24] Rencher, AC, Christensen, WF. *Methods of Multivariate Analysis.* A John Wiley & Sons, Inc, Publication. Third Edition.
[25] Sharol Ashma Menezes, Anto Delasalle, Arunachalam. 2015. *A study of electrocardiographicchanges in type 2 diabetes patients*. International Journal of Research in Medical Sciences
[26] Sheldon & Sheps. 2016. *Pulse pressure: An indicator of heart health?* Mayo Clinic. https://www.mayoclinic.org/diseases-conditions/high-blood-pressure/epert-answers/pulsepressure/faq-20058189
[27] Sheng WYX. 2009. *Heart sound measurement and analysis system with a stethoscope*. Journal of biomedical engineering
[28] Sivajenani, S, Kuillini, & Madona, E. *Lifestyle Factors Influencing Coronary Heart Disease*.OUSL Journal, Vol. 8, page 51-64. Department of Health Sciences, The Open University of Sri Lanka.
[29] Souman, M, et al. 2011. *Using Decision Tree for Diagnosing Heart Disease Patients.*

Comparison between multiplicative Holt Winter and decomposition method in predicting the number of incoming international tourists to Indonesia

A Supriatna[1], E Lesmana[1], L Aridin[1], Sukono[1], H Napitupulu[1]

[1]Department of Mathematics, Faculty of Mathematics and Natural Sciences, Universitas Padjadjaran, Bandung, INDONESIA

Corresponding author: sukono@unpad.ac.id

Abstract. Indonesia has a remarkable potential in tourism sector, hence government should regulate all matters related with tourism and pay more attention to the developments happening in the tourism sector, including policy-making and planning. One way to aid the government in policy making and planning is by conducting a prediction about the number of incoming international tourists to Indonesia. This paper will elaborate the analysis about the comparison between multiplicative Holt-Winter and decomposition method in predicting the number of incoming international tourists to Indonesia. This paper aims to determine the appropriate forecasting method suitable for the available data pattern and provides more accurate forecasting result. Forecasting could be conducted using multiplicative Holt-Winter method and decomposition method. Meanwhile to determine the error level and forecast accuracy, mean absolute percentage error (MAPE) is used. Analysis result showed that prediction using Holt-Winter method provides MAPE of 4.7098%, meanwhile the MAPE for decomposition method is 3.3254%. It could be concluded that to determine the number of incoming international tourist to Indonesia, decomposition method is more appropriate.

1. Introduction

Geographically, Indonesia has shoreline length of more than 95,181 km, consisted of 13,446 islands and ocean area about 3.1 million km^2. Indonesia is known for its diversity either in its natural or human resources. In terms of human resources, Indonesia owns 1128 ethnicities, 746 languages and dialects alongside various cultural expressions and their traditional customs [1]. These properties and cultural heritages have been globally acknowledged as world cultural heritage sites (8 cultural heritage). Meanwhile, in natural resource sector, Indonesia owns 51 national parks with their own biodiversity, 35 primate species, 25% endemic species, habitat for 16% of reptiles and amphibians existing in the world, and also a habitat for 17% of all known birds species in the world. All these wealth and diversity are a potential for Indonesia to excel in tourism sector [1].

The contribution of tourism sector towards world gross domestic product (GDP) is 9%, 1 out of 11 job vacancies are created by tourism sector and its contribution towards world export value is USD 1.4 trillion or approximately 5% of all current exports in the world [2][3][4]. On national scale, the tourism sector ranked fourth in the highest foreign exchange income after oil and gas, coal and palm [1]. Though the tourism potential in Indonesia is currently up to par, this does not necessarily mean to hold back development measures. The role of the government as a facilitator and dynamist essentially means that the government has a task to create policy and planning. In terms of tourism policy and planning, Veal [5], described this framework of tourism as a system which is not to be separated from

IOP Conf. Series: Materials Science and Engineering **567** (2019) 012047 doi:10.1088/1757-899X/567/1/012047

one another. There are five interconnected elements involved in policy making and planning related to tourist destination; those elements are: man or society, organization, facility and service related to tourism, environment and process. Hence, in policy making and planning, the government should pay a better attention towards those five elements. For policy making and planning in tourism, the role of tourism arrival forecasting is highly necessary [6][7][8].

Rosy and Ponnusamy [9], applied Holt-Winter method in their study to forecast hotel room availability. This forecasting is based on a problem about the fluctuating hotel room request on day of arrival and has a tendency to be uncertain. The forecasting method on the study is a quantitative method, and its result showed that the applied forecasting method could express actual data pattern. The authors agreed that in the uncertainty of the economy variable, forecasting method could be applied in planning and policy making on hotel room availability. Similar studies which applied Holt-Winter method had also been done by Rajchakit [10], Mamula & Duvnjak [7], Blagojevic et al. [11] and Tratar [12]. Konarasinghe [3] conducted forecasting method review for tourist arrival growth incoming from Western Europe countries to Sri Lanka. The forecasting method is conducted using additive decomposition and multiplicative method. Residual plot and Anderson-Darling statistics are applied as goodness of fit test in model validation. Among those two forecasting techniques, the method which yield the least error is selected. Analysis result showed that additive decomposition method is the most suitable method to forecast incoming tourist arrival from Western Europe countries. The author also stated that tourist arrival forecasting is a very necessary discipline in planning, resources management and other decision making process in either micro or macro level. Another study which applied this decomposition method is also conducted by Yahya et al. [13], Rajchakit [10] and Çuhadar [14].

Referring to the commentary above, this paper analyzes the comparison between two methods, i.e. the multiplicative Holt-Winter method and decomposition method to predict the number of international tourist arrival to Indonesia. This study aims to obtain an appropriate forecasting method suitable for the available international tourist arrival data pattern and obtain a more accurate forecast result, i.e. one with least error level. This forecasting is necessary in aiding stakeholders to arrange tourism plan and policy in Indonesia.

2. Materials and methods

This section will elaborate the material and method involved in this study. Material refers to the data being used in this study including the source of the data. Meanwhile, method refers to the model and approach to analyze the data.

2.1. Materials

The object of this study was international tourists arriving in Indonesia through 19 main entries. The analyzed data were data comprising international tourist's arrival from July 2013 to June 2017. This tourist arrival data was a secondary data obtained from Statistics Indonesia (https://www.bps.go.id). The international tourist arrival data was shown on Table 1 and charted on Figure 1. The forecasting techniques being used in this study were Holt-Winter and decomposition. The data were analyzed using software Microsoft® Excel and Minitab® 16.

IOP Conf. Series: Materials Science and Engineering **567** (2019) 012047 doi:10.1088/1757-899X/567/1/012047

Table 1. Data of international tourist's arrival to Indonesia

Period (t)	Month	Data	Period (t)	Month	Data
1	Jul-13	681,092	25	Jul-15	775,552
2	Aug-13	735,043	26	Aug-15	813,185
3	Sep-13	734,055	27	Sep-15	822,922
4	Oct-13	682,006	28	Oct-15	790,162
5	Nov-13	765,857	29	Nov-15	723,069
6	Dec-13	802,586	30	Dec-15	862,556
7	Jan-14	716,419	31	Jan-16	698,057
8	Feb-14	662,260	32	Feb-16	768,118
9	Mar-14	724,901	33	Mar-16	806,118
10	Apr-14	689,157	34	Apr-16	811,233
11	May-14	714,353	35	May-16	841,723
12	Jun-14	815,011	36	Jun-16	784,155
13	Jul-14	742,843	37	Jul-16	931,,694
14	Aug-14	793,740	38	Aug-16	944,455
15	Sep-14	755,664	39	Sep-16	922,608
16	Oct-14	768,787	40	Oct-16	913,589
17	Nov-14	723,456	41	Nov-16	855,545
18	Dec-14	860,394	42	Dec-16	957,828
19	Jan-15	671,939	43	Jan-17	902,975
20	Feb-15	743,498	44	Feb-17	838,686
21	Mar-15	739,821	45	Mar-17	909,182
22	Apr-15	708,818	46	Apr-17	966,936
23	May-15	751,627	47	May-17	938,638
24	Jun-15	778,056	48	Jun-17	910,931

The data presented in table 1 can also be presented in graphical form as given in Figure 1.

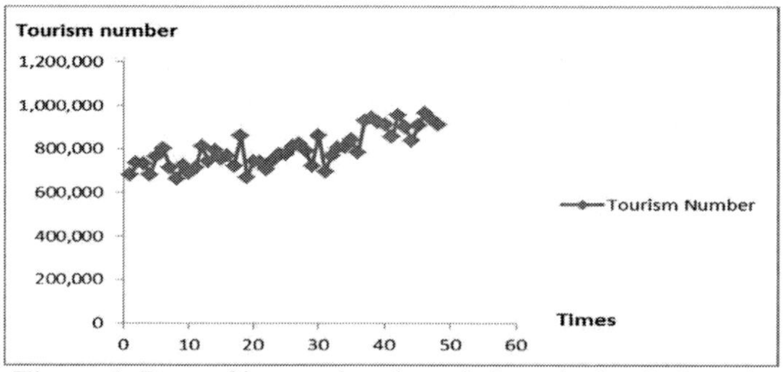

Figure 1. Data of international tourist's arrival to Indonesia

2.2. *Methods*

This section will comprise applied techniques in data analysis for international tourist arrival to Indonesia forecasting.

2.2.1. *Holt-Winter Method*

According to Sood & Jain [15], Cankurt & Subasi [16], and Padhan [17], forecasting is a branch of statistics which resembles an important element in decision making. Rosy and Ponnusamy [9] suggested that statistically, multiplicative Holt-Winter forecasting method has three equations which were slightly different with additive Holt-Winter method. The three equations in multiplicative Holt-Winter method are as follows:

- Determine the overall smoothing value using the equation below:

$$F_t = \alpha \frac{X_t}{I_{t-1}} + (1-\alpha)(F_{t-1} + T_{t-1}). \quad (1)$$

- Determine the trend smoothing value using the equation below:

$$T_t = \beta(F_t - F_{t-1}) + (1-\beta)T_{t-1}. \quad (2)$$

- Determine the seasonal smoothing value using the equation below:

$$S_t = \gamma \frac{X_t}{F_{t-1}} + (1-\gamma)S_{t-1}. \quad (3)$$

- In the end of period t, the forecasting value for the period of $t+k$ is determined using the equation below:

$$\hat{Y}_{t+m} = (F_t + T_t m)S_{t-L+m}. \quad (4)$$

With F_t as the forecasting smoothing value for the period of t; X_t as the actual value for the period of t; T_t as the trend smoothing value for the period of t; S_t as seasonality component value for the period of t; $\hat{Y}_t$ as forecasting value on period t for the m consecutive periods; m as the number of forecasted periods. The α as the smoothing parameter for trend (0< α <1); β as the smoothing parameter for trend (0< β <1); γ as the number of forecasted periods. The α as the smoothing parameter for trend (0< γ <1); and L as the number of period within a seasonal cycle.

2.2.2. Decomposition method

Konarasinghe [3] and Petrevska [18] suggested that the decomposition method is a forecasting method which requires four main components to forecast future values. Those four components are trend, seasonality, cycle and error. The decomposition method is based on the assumption that the present data are a combination of several components which could be simplified into:

$$X_t = f(T_t, S_t, C_t, E_t).$$

Simple moving data average decomposition method (classic decomposition) assumes that the multiplication could be written mathematically as follows:

$$Y_t = T_t \times S_t \times C_t \times E_t, \quad (5)$$

with X_t as periodic sequence value (actual data) during period of t; T_t as trend component value during the period of t; S_t as seasonality component value during the period of t; C_t as cycle component value during the period of t; E_t random (irregular) error component value during the period of t; $\hat{Y}_t$ as prediction value for the period of t.

To solve equation (5), several steps should be followed as shown below:

- On actual data X_t, the calculation of moving average which length was represented as p equals to the length of seasonality (e.g.: 12 months, 4 quarters or 7 days). This is to assume the association with trend and cycle. Hence, an equation is known as:

  $$M_t = T_t \times C_t. \quad (6)$$

- To determine the effect of seasonality S_t, equation (5) is divided by equation (6), thus a new equation is obtained as below:

  $$\frac{Y_t}{M_t} = S_t \times E_t. \quad (7)$$

- Identify the appropriate trend effect towards the data using the smallest regression and quadratic method, thus, an equation is obtained as follows:
  Linear trend is shown below:

  $$\hat{T}_t = a + bt,$$

  Quadratic trend is shown below:

  $$\hat{T}_t = a + bt + ct^2,$$

  Cubic trend is shown below:

  $$\hat{T}_t = a + bt + ct^2 + dt^3, \quad (8)$$

  with a as intercept constant parameter and b, c, and d as coefficient parameter.

- To determine the effect of cycle C_t is by obtaining the division result of equation (6) with trend T_t, hence, an equation obtained is as follows:

$$\frac{M_t}{T_t} = C_t . \tag{9}$$

- To obtain forecasting result, the equation below is used:

$$\hat{Y}_t = T_t \times S_t \times C_t . \tag{10}$$

The residual value obtained by subtracting actual data and forecasting result on the period of t could be used to determine the accuracy of the prediction.

2.2.3. Forecasting accuracy

Forecasting (prediction) is conducted because of the presence of complexity and uncertainty faced by the forecasting model maker. A forecasting model is said to be highly accurate if the said model has the least error level [19; 20]. Several methods exist to measure the accuracy of a forecasting model, one of which is Mean Absolute Percentage Error (MAPE). The MAPE could be determined using the equation below:

$$MAPE = \left(\frac{1}{n} \sum_{t=1}^{n} \frac{| X_t - \hat{Y}_t |}{X_t} \right) \times 100\%, \tag{11}$$

With X_t as periodic sequence value (actual data) on the period of t, F_t as the forecasting value (data goodness of fit value) during the period of t and n as the number of forecasting periods [19; 20].

3. Result and analysis

This section will be displaying results which includes: forecasting using multiplicative Holt-Winter method and forecasting using multiplicative decomposition method.

3.1. Forecasting using multiplicative Holt-Winter method

By using the data on Table 1, referring to the discussion about Holt-Winter method, it was known that the prediction of the number of international tourist arrival could be conducting by following these steps: to determine overall smoothing value, equation (1) was used; to determine trend smoothing value, equation (2) was used; and to determine seasonal smoothing value, equation (3) was used. Hence, on the end of period t, the forecasting value for period $t+k$ was determined by using equation (4).

Meanwhile, the accuracy of the forecast was measured by using equation (10). Using Microsoft Excel software, parameter value which minimized the MAPE was obtained $\alpha = 0.143$, $\beta = 0.636$, and $\gamma = 0.94$. Forecasting result using multiplicative Holt-Winter method was presented on Table 2.

Table 2. Forecasting result using multiplicative Holt-Winter method

Month	$\hat{Y}_t$	Month	$\hat{Y}_t$	Month	$\hat{Y}_t$
Jul-14	683,156	Jul-15	737,535	Jul-16	892,039
Aug-14	754,606	Aug-15	774,094	Aug-16	952,440
Sep-14	770,908	Sep-15	732,836	Sep-16	955,799
Oct-14	723,681	Oct-15	744,308	Oct-16	906,907
Nov-14	835,308	Nov-15	723,936	Nov-16	828,869
Dec-14	863,882	Dec-15	859,799	Dec-16	990,174
Jan-15	775,225	Jan-16	694,630	Jan-17	792,520
Feb-15	698,363	Feb-16	768,337	Feb-17	895,362
Mar-15	771,060	Mar-16	783,189	Mar-17	926,257
Apr-15	725,596	Apr-16	769,945	Apr-17	918,149
May-15	744,715	May-16	841,663	May-17	958,946
Jun-15	845,977	Jun-16	907,931	Jun-17	910,851

The forecasting result presented on Table 2 could also be presented in forms of graph or chart as presented on Figure 2. This forecast using multiplicative Holt-Winter method yields MAPE error of 4.7098%, and thus, gave forecast accuracy of 95.2902%.

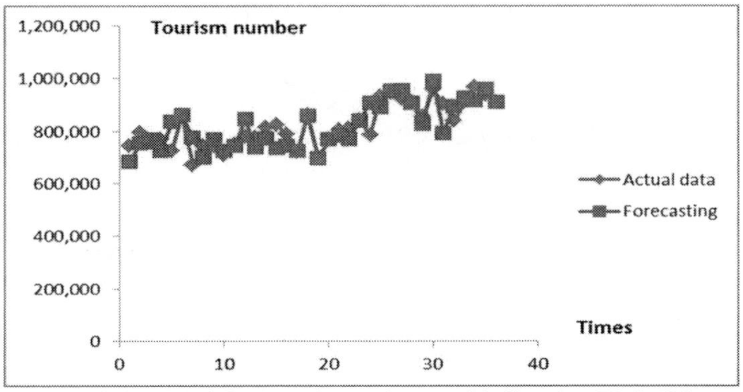

Figure 2. Actual data and forecasting using multiplicative Holt-Winter method

3.2. Forecasting using multiplicative decomposition method

By using data on Table 1 and referring to the discussion about multiplicative decomposition method, it was known that the prediction of the number of international tourist arrival could be conducted by following these steps: assumed effect of trend and cycle was obtained by using equation (6); seasonality effect S_t was determined using equation (7); the appropriate trend effect was identified by using equation (8), where a new equation was obtained as follows:

$$\hat{T}_t = 722031.313 + 385.303t + 50.082t^2 + 0.894t^3; \qquad (12)$$

Hence, cycle effect C_t was obtained by using equation (9). While forecast result could be obtained by using equation (10). The final forecasting result using multiplicative decomposition method was presented on Table 3.

Table 3. Forecasting result using multiplicative decomposition method

Month	$\hat{Y}_t$	Month	$\hat{Y}_t$	Month	$\hat{Y}_t$	Month	$\hat{Y}_t$
Jul-13	740,226	Jul-14	758,673	Jul-15	779,898	Jul-16	876,827
Aug-13	779,559	Aug-14	798,182	Aug-15	821,035	Aug-16	932,859
Sep-13	771,317	Sep-14	791,513	Sep-15	813,797	Sep-16	927,783
Oct-13	754,185	Oct-14	773,046	Oct-15	800,554	Oct-16	915,483
Nov-13	702,439	Nov-14	720,300	Nov-15	751,212	Nov-16	860,307
Dec-13	809,210	Dec-14	827,468	Dec-15	867,376	Dec-16	998,569
Jan-14	653,182	Jan-15	675,648	Jan-16	714,472	Jan-17	821,753
Feb-14	700,466	Feb-15	721,743	Feb-16	772,400	Feb-17	885,189
Mar-14	728,910	Mar-15	751,173	Mar-16	809,647	Mar-17	927,172
Apr-14	692,974	Apr-15	713,244	Apr-16	773,725	Apr-17	885,713
May-14	735,283	May-15	755,840	May-16	829,448	May-17	947,628
Jun-14	761,748	Jun-15	782,373	Jun-16	868,210	Jun-17	991,319

The forecasting result presented on Table 3 could also be presented in forms of graph or chart as presented on Figure 3. This forecast using multiplicative Holt-Winter method yields MAPE error of 3.3254%, and thus, gave forecast accuracy of 96.6746%.

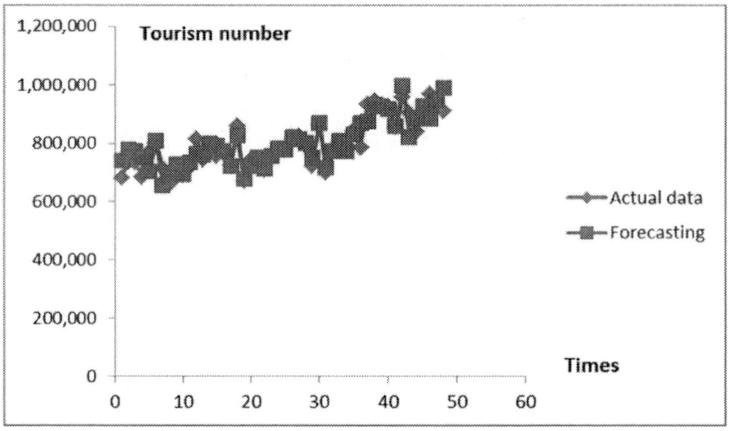

Figure 3. Actual data and forecasting result using multiplicative decomposition method

3.3. Discussion

Referring to the data presented on Table 1 and the graph on Figure 1, it was known that international tourist arrival to Indonesia during July 2013 to June 2017 was fluctuating each month; occasionally increased on certain month and decreased on the other. However, after paying a closer attention, Figure 1 showed that during the period of July 2013-June 2017, the tourist arrival trend had a tendency to increase. This showed that the tourist arrival to Indonesia was gradually increasing. Since tourism sector was a prime sector in economic growth in Indonesia, the tourism sector should be managed carefully. To be managed well, the tourism sector should be supported with proper planning and policies as well. Forecasting the international tourist arrival was one of the basic ways to plan and make certain policies in the tourism sector. The forecasting on this study was conducted using two approaches, i.e. the multiplicative Holt-Winter method and multiplicative decomposition method.

The forecasting using multiplicative Holt-Winter method was based on in sample data, with the result being presented on Table 2 and graph being presented on Figure 2. Referring to Figure 2, it was known that the forecast data graph plot using multiplicative Holt-Winter method almost overlaps the actual data. This forecast result yields MAPE error of 4.7098%, hence the forecast accuracy using multiplicative Holt-Winter method was 95.2902%. This simply meant that the multiplicative Holt-Winter method yields a high accuracy result given that the result was able to express actual data with an accuracy of 95.2902%.

The forecasting using multiplicative decomposition method was based on in sample data, with the result being presented on Table 3 and graph being presented on Figure 3. Referring to Figure 3, it was known that the forecast data graph plot using multiplicative decomposition method also almost overlaps the actual data. This forecast result yields MAPE error of 3.3254%, hence the forecast accuracy using multiplicative Holt-Winter method was 96.6746%. This simply meant that the multiplicative Holt-Winter method yields a high accuracy result given that the result was able to express actual data with an accuracy of 96.6746%.

Comparison of these two forecasting method as applied for forecasting international tourist arrival within the period of July 2013-June 2017 showed that the method with higher accuracy level was multiplicative decomposition method. Thus, in the future, this multiplicative decomposition method was proposed to be applied for forecasting as a basis in planning and policy making in tourism sector of Indonesia. The forecasting using out sample data for the period of July 2017-June 2018 was presented on Table 4.

Table 4. Forecast result for out sample data using multiplicative decomposition method

Month	S_t	$\hat{T}_t$	C_t	Forecasting ($\hat{Y}_t$)
Jul-17	101.7854	966,342.25	1.0136	996,972
Aug-17	106.8588	978,257.46	1.0117	1,057,585
Sep-17	105.4039	990,541.04	1.0095	1,053,988
Oct-17	102.7501	1,003,198.35	1.0071	1,038,106
Nov-17	95.4320	1,016,234.75	1.0046	974,274
Dec-17	109.6324	1,029,655.60	1.0021	1,131,207
Jan-18	89.5433	1,043,466.28	0.9996	933,980
Feb-18	95.3706	1,057,672.14	0.9972	1,005,884
Mar-18	98,7941	1,072,278.54	0.9950	1,054,051
Apr-18	93.3512	1,087,290.86	0.9930	1,007,894
Mey-18	98.8069	1,102,714.46	0.9914	1,080,188
Jun-18	102.2713	1,118,554.69	0.9900	1,132,521
TOTAL				12,466,650

Where S_t were calculated using equation (6), $\hat{T}_t$ equation (11) and C_t equation (8)? Meanwhile, forecasting $\hat{Y}_t$ was calculated using equation (9). Based on the forecasting result on Table 4, it could be concluded that the international tourist arrival during the period of July 2017-June 2018 was approximately 12,466,650 tourists.

4. Conclusion

This paper had analyzed the comparison between multiplicative Holt-Winter method and multiplicative decomposition method and their application in forecasting the number of international tourist arrival to Indonesia. The analyzed data was the international tourist arrival data during the period of July 2013-June 2017. Analysis result showed that the MAPE error using multiplicative Holt-Winter method was 4.7098%, hence forecasting accuracy of 95.2902%. Meanwhile, forecasting using multiplicative decomposition method yielded MAPE error of 3.3254%, hence forecasting accuracy of 96.6746%. Thus, forecasting using multiplicative decomposition method was better and recommended to analyze the current international tourist arrival data. Out sample data forecast result using multiplicative decomposition method revealed that the international tourist arrival during the period of July 2017-June 2018 was approximately 12,466,650 tourists.

Acknowledgment

Acknowledgments are conveyed to the Rector, Director of Directorate of Research, Community Involvement and Innovation, and the Dean of Faculty of Mathematics and Natural Sciences, Universitas Padjadjaran, with whomthe Internal Grant Program of Universitas Padjadjaran was made possible to fund this research.The grant is a means of enhancing research and publication activities for researchers at Universitas Padjadjaran.

References

[1] Ratman, D.R. 2015. *Laporan Akuntabilitas Kinerja Deputi Bidang Pengembangan Destinasi dan Industri Pariwisata (LAKIP) Tahun 2015*. Kementerian Pariwisata Republik Indonesia. http://www.kemenpar.go.id/userfiles/LAKIP%20DEPUTI% 20BPDIP%202015.pdf

[2] Karimi, A., Faroughi, P., and Rahim, K.A. 2015. Modeling and Forecasting of International Tourism Demand in ASEAN Countries. *American Journal of Applied Sciences*, 2015, 12 (7):pp. 479-486. DOI: 10.3844/ajassp.2015.479.486.

[3] Konarasinghe, K.M.U.B. 2016. Decomposition Techniques on Forecasting Tourist Arrivals from Western European Countries to Sri Lanka. *Working Paper*, 13th International Conference on Business Management 2016. Institute of Mathematics and Management, Sri Lanka.

[4] Konarasinghe, K.M.U.B. 2017. Model Comparison on Forecasting Tourist Arrivals to Sri Lanka from Western Europe. *International Journal of Novel Research in Physics Chemistry & Mathematics*, Vol. 4, Issue 2, pp: (28-35), Month: May- August 2017, Available at: www.noveltyjournals.com

[5] Veal, A. J. 2002. *Leisure and tourism policy and planning*, 2nd Edition. Wallingford, UK: CABI Publishing.

[6] Akuno, A.O., Otieno, M.O., Mwangi, C.W., and Bichanga, L.A. 2015. Statistical Models for Forecasting Tourists Arrival in Kenya. *Open Journal of Statistics*, 2015, 5, pp. 60-65. Published Online February 2015 in SciRes. http://www.scirp.org/journal/ojs. http://dx.doi.org/10.4236/ojs.2015.51008

[7] Mamula, M. and Duvnjak, K. 2017. Modelling the Employment in Tourism–Case Study of Croatia. *International Journal of Business Administration*, Vol. 8, No. 3; 2017; pp. 37-444. oi:10.5430/ijba.v8n3p37 ; URL: https://doi.org/10.5430/ijba.v8n3p37.

[8] Xiaoxuan, L., Qi, W., Geng, P., and Benfu, L. 2016. Tourism Forecasting by search Engine Data withNoise-Processing. *African Journal of Business Management*, Vol. 10(6), pp. 114-130, 28 March, 2016, pp. 114-130. DOI: 10.5897/AJBM2015.7945. Article Number: 9217D8757660. ISSN 1993-8233. Copyright © 2016Author(s) retain the copyright of this article. http://www.academicjournals.org/AJBM

[9] Rosy, C.P. and Ponnusamy, P. 2017. Evaluating and Forecasting Room Demand in Tourist Spot using Holt-Winters Method. *International Journal of Computer Applications* (0975 – 8887), Volume 172 – No.2, August 2017, pp. 22-25. IJCA™: www.ijcaonline.org.

[10] Rajchakit, M. 2017. A New Method for Forecasting via Feedback Control Theory. *Proceedings of the International MultiConference of Engineers and Computer Scientists*, 2017, Vol I, IMECS 2017, March 15 - 17, 2017, Hong Kong, pp. 1-6. ISSN: 2078-0958 (Print); ISSN: 2078-0966 (Online).

[11] Blagojevic, N.P., Vujko, A., and Gajic, T. 2016. Comparative Analysis of Exponential Smoothing Models to Tourists Arrivals in Serbia. *Economics of Agriculture*, 2016 (63) 3, pp. 835-845.

[12] Tratar, L.F. 2013. Improved Holt-Winters Method: A Case of Overnight Stays ofTourists in Republic of Slovenia. *Economic And Business Review*, VOL. 16, No. 1, 2013, pp. 5–17.

[13] Yahya, N.A., Samsudin, R., and Shabri, A. 2017. Tourism Forecasting using Hybrid Modified Empirical Mode Decomposition and Neural Network. *Int. J. Advance Soft Compu. Appl,* Vol. 9, No. 1, March 2017, pp. 14-31. ISSN 2074-8523.

[14] Cuhadar, M. 2014. Modelling and Forecasting Inbound Tourism Demand to Istanbul: A Comparative Analysis. *European Journal of Business and Social Sciences*, Vol. 2, No.12, pp. 101-119, March 2014, pp. 101 – 119. URL: *http://www.ejbss.com/recent.aspx*. ISSN: 2235-767X.

[15] Sood S, Jain K. 2017. Comparative Analysis of Techniques for Forecasting Tourists. Arrival. *Journal of Tourism Hospit* 6: 285, pp. 1-4. doi: 10.4172/21670269.1000285

[16] Cankurt, S. and Subasi, A. 2015. Developing tourism demand forecasting models using machine learning techniques with trend, seasonal, and cyclic components. *Balkan Journal of Electrical & Computer Engineering*, 2015, Vol.3, No.1, pp. 42-49. http://www.bajece.com.

[17] Padhan, P.C. 2011. Forecasting International Tourists Footfalls in India: An Assortment of Competing Models. *International Journal of Business and Management*, Vol. 6, No. 5; May 2011, pp. 190-202. www.ccsenet.org/ijbm.

[18] Petrevska, B. 2013. Empirical Analysis of Seasonality Patterns in Tourism. *(JPMNT) Journal of Process Management – New Technologies, International,* Vol. 1, No.2, 2013, pp. 87-95.

IOP Conf. Series: Materials Science and Engineering **567** (2019) 012047 doi:10.1088/1757-899X/567/1/012047

[19] Sukono, Sholahuddin, A., Mamat, M., Prafidya, K. 2014.Credit Scoring for Cooperative of Financial Services Using Logistic Regression Estimated by Genetic Algorithm. *Applied Mathematical Sciences*, Vol. 8, 2014, no. 1, 45 – 57. doi: 10.12988/ams.2014.310600.
[20] Sukono, Nahar, J., Mamat, M., Putri, F.T., and Supian, S. 2016. Indonesian Financial Data Modeling and Forecasting by Using Econometrics Time Series and Neural Network. *Global Journal of Pure and Applied Mathematics*. Vol.12, No.4, pp. 3745-3757.

IOP Conf. Series: Materials Science and Engineering **567** (2019) 012048 doi:10.1088/1757-899X/567/1/012048

Comparison of hierarchical clustering methods (case study: data on poverty influence in North Sulawesi)

C E Mongi[1], Y A R Langi[1], C E J C Montolalu[1], N Nainggolan

[1]Department of Mathematics, Faculty of Mathematics and Natural Science, University of Sam Ratulangi, Manado, Indonesia

Corresponding author: charlesmongi@unsrat.ac.id

Abstract. Grouping of Large data has been carried out in various fields. One method for grouping is cluster analysis where this method consists of hierarchy and non-hierarchy method. The aim of this study was to compare the use of cluster analysis on aspects of the causes of poverty data. The method used is agglomerative hierarchical clustering, that is, the average linkage, centroid methods and ward methods. The results obtained are compared with the RMSSTD value and the smallest value is the ward method with a value of 2.0937. So the ward method is good for this case.
Keywords : Hierarchical clustering, average linkage, centroid method, ward method.

1. Introduction

The problem of grouping large amounts of data has become important at this time. Grouping of Large data has been carried out in various fields, for example in the fields of soil and spatial [1][2], business [3], medicine [4], health [5] and other fields. One method for grouping is cluster analysis where this method consists of hierarchy and non-hierarchy method.

In cluster analysis we search for patterns in a data set by grouping the multivariate observations into clusters. The goal is to find an optimal grouping for which the observations or objects within each cluster are similar but the clusters are dissimilar to each other[6]. Cluster analysis is a more primitive technique in that no assumptions are made concerning the number of groups or the group structure. Grouping is done on the basis of similarities or distances (dissimilarities)[7].

Hierarchical methods are among the traditional techniques of cluster analysis. They consist in successive aggregation or division of the observations and their subsets. Resulting from this kind of procedure there is a tree-like structure, which is referred to as dendrogram. The agglomerative techniques start from the set of observations, each of which is treated as a separate cluster. Clusters are aggregated in accordance with the decreasing degree of similarity (or the increasing degree of dissimilarity) until one, single cluster is established [8].

The application of the agglomerative hierarchy method will be used in data that affects poverty with aspects of population, economy, education and health. Population aspects are used data on population and unemployment, economic aspects are used gross regional domestic product data, educational aspects used net enrollment rate of elementary school, junior high school and senior high school, while the health aspect is used health insurance member data, health personal, and number of hospital.

In this article used 3 hierarchical method that is average linkage, centroid method, and ward method. In section 2 explained the formula of each method. In section 3 explained the use of the hierarchical method on the data that causes poverty in the North Sulawesi province. The application of the hierarchical method will be compared to each method using the RMSSTD value. The smallest RMSSTD value is a good method used for poverty influence data.

2. Hierarchical Methods

2.1. Average Linkage

The distance between two cluster is define by (1)

$$D_{KL} = \frac{1}{N_K N_L} \sum_{i \in C_K} \sum_{i \in C_L} d(x_i, x_j) \quad (1)$$

D_{KL} is any distance or dissimilarity measure between cluster C_K and C_L. In average linkage the distance between two clusters is the average distance between pairs of obsevations, one in each cluster. Average linkage tends to join clusters with small variances, and it is slightly biased toward producing clusters with the same variance.

2.2. Centroid Method

The distance between two clusters is define by (2)

$$D_{KL} = \|\bar{x}_K - \bar{x}_L\|^2 \quad (2)$$

In the centroid method, the distance between two clusters is defined as the Euclidean distance between their centroid or means.

2.3. Ward Method

The distance between two clusters is define by (3)

$$D_{KL} = \frac{\|\bar{x}_K - \bar{x}_L\|^2}{\frac{1}{N_K} - \frac{1}{N_L}} \quad (3)$$

The distance between two cluster is the ANOVA sum of squares between the clusters added up over all the variables.

2.4. Evaluating the cluster

The root mean squared standard deviation of a cluster C_K is define on (4)

$$\text{RMSSTD} = \sqrt{\frac{W_K}{v(N_K - 1)}}, \quad (4)$$

where $W_K = \sum_{i \in C_K} \|\boldsymbol{x}_i - \bar{\boldsymbol{x}}_K\|^2$, N_K = number of observation in C_K [1]

$\boldsymbol{R^2 = 1 - \frac{P_G}{T}}$, P_G is $\sum \boldsymbol{W_J}$, where summation is over the *G* clusters at the *G*th level of the hierarchy

3. Aplication use influence poverty data in North Sulawesi Province

The data used consists of 15 objects which are 15 districts / cities in the province of North Sulawesi with variables namely the factors that cause poverty consist of population, unemployment, gross regional domestic product, elementary school net enrollment rate, junior high school net enrollment rate, senior high school net enrollment, health insurance member, personal health, number of hospital. From the data will be analyzed with three hierarchical methods. Data in the form of 9 variables and 15 objects, namely districts / cities in North Sulawesi province, were analyzed using the SAS University Edition program. Cluster formation results can be seen in section 3.1 for the average linkage, section 3.2 for the centroid method and section 3.3 for the ward method.

3.1. Result with Average linkage

In table 1. that can see the cluster formation process where column 1 is the number of clusters. The second column is the object that joins into a new cluster. The third column is the frequency of each cluster and the fourth column is the root mean square standard deviation value of each cluster. The fifth column is the norm value for root mean square distance. Figure 1 shows the dendogram using average linkage.

Table 1. Cluster history with average linkage

Number of Clusters	Clusters Joined		Freq	New Cluster RMS Std Dev	Norm RMS Distance
14	Talaud	Bolmut	2	0.3939	0.1313
13	Sitaro	Boltim	2	0.5114	0.1705
12	CL14	CL13	4	0.7350	0.2801
11	Minut	Bitung	2	1.0321	0.344
10	Sangihe	CL12	5	1.0511	0.4657
9	Minsel	CL11	3	1.3007	0.472
8	Bolmong	CL9	4	1.4852	0.5497
7	CL10	Mitra	6	1.4506	0.6752
6	CL7	Bolsel	7	1.6695	0.7067
5	Tomohon	Kotamo	2	2.2676	0.7559
4	CL8	CL6	11	1.9961	0.7647
3	CL4	CL5	13	2.1830	0.8629
2	CL3	Minahasa	14	2.3543	1.0649
1	CL2	Manado	15	3.0000	1.87

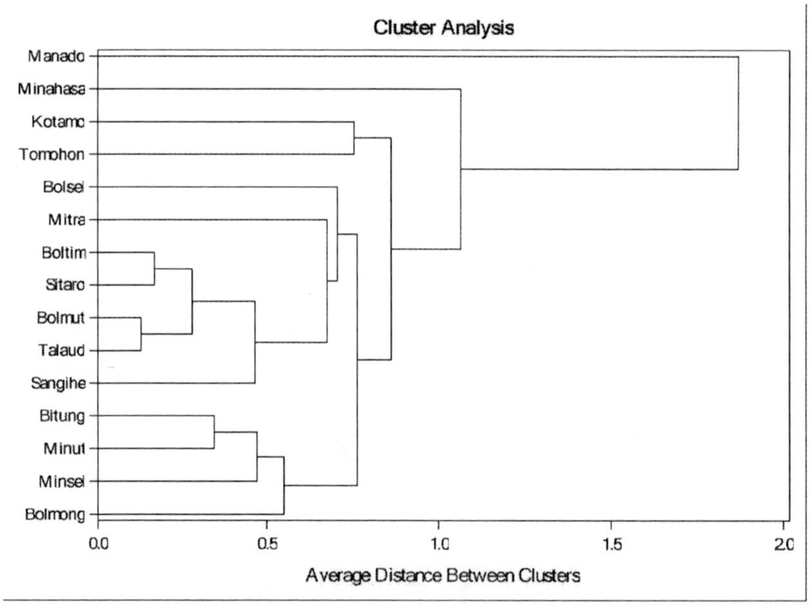

Figure 1. Dendogram used average linkage

IOP Conf. Series: Materials Science and Engineering **567** (2019) 012048 doi:10.1088/1757-899X/567/1/012048

3.2. Result with centroid method

Table 2. Cluster history using centroid method

Number of Clusters	Clusters Joined		Freq	New Cluster RMS Std Dev	Norm Centroid Distance
14	Talaud	Bolmut	2	0.3939	0.1313
13	Sitaro	Boltim	2	0.5114	0.1705
12	CL14	CL13	4	0.7350	0.2586
11	Minut	Bitung	2	1.0321	0.344
10	Minsel	CL11	3	1.3007	0.4396
9	Sangihe	CL12	5	1.0511	0.4409
8	Bolmong	CL10	4	1.4852	0.4894
7	CL8	CL9	9	1.7560	0.5837
6	CL7	Bolsel	10	1.8939	0.6856
5	CL6	Mitra	11	1.9961	0.6798
4	CL5	Tomohon	12	2.0753	0.6756
3	CL4	Kotamo	13	2.1830	0.7685
2	CL3	Minahasa	14	2.3543	0.9432
1	CL2	Manado	15	3.0000	1.7919

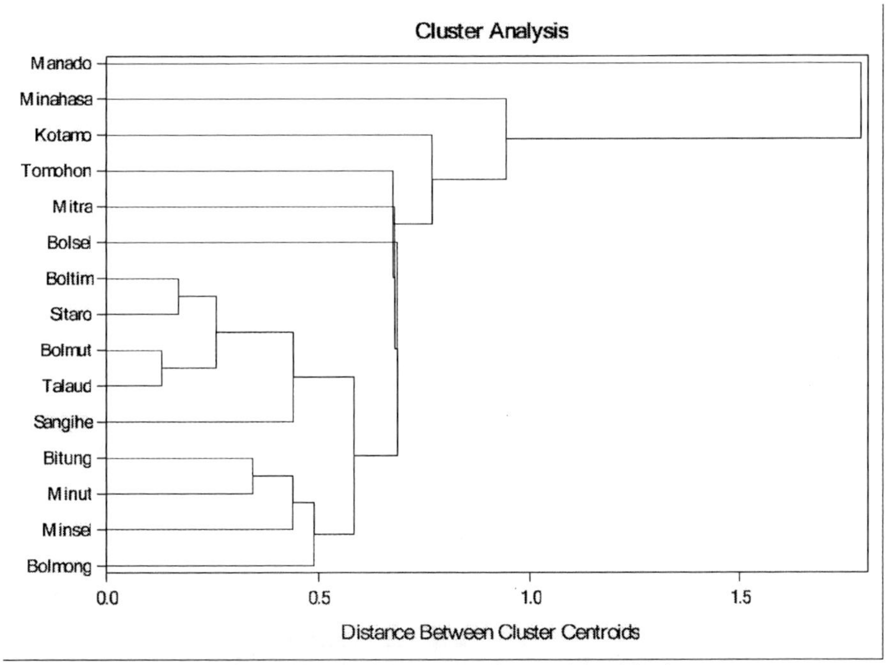

Figure 2. Dendogram used centroid method

3.3. Result with ward method

Table 3. Cluster history using ward method

Number of Clusters	Clusters Joined		Freq	New Cluster RMS Std Dev	Semipartial R-Square	R-Square
14	Talaud	Bolmut	2	0.3939	0.0012	.999
13	Sitaro	Boltim	2	0.5114	0.0021	.997
12	Minut	Bitung	2	1.0321	0.0085	.988
11	CL14	CL13	4	0.7350	0.0096	.979
10	Sangihe	Tomohon	2	1.5075	0.0180	.961
9	Minsel	CL12	3	1.3007	0.0184	.942
8	Bolmong	CL9	4	1.4852	0.0257	.917
7	CL11	Bolsel	5	1.3194	0.0424	.874
6	CL10	Kotamo	3	1.9714	0.0437	.831
5	CL7	Mitra	6	1.6218	0.0491	.781
4	CL8	Minahasa	5	2.0185	0.0768	.705
3	CL6	CL5	9	2.0937	0.1123	.592
2	CL4	CL3	14	2.3543	0.1642	.428
1	CL2	Manado	15	3.0000	0.4281	.000

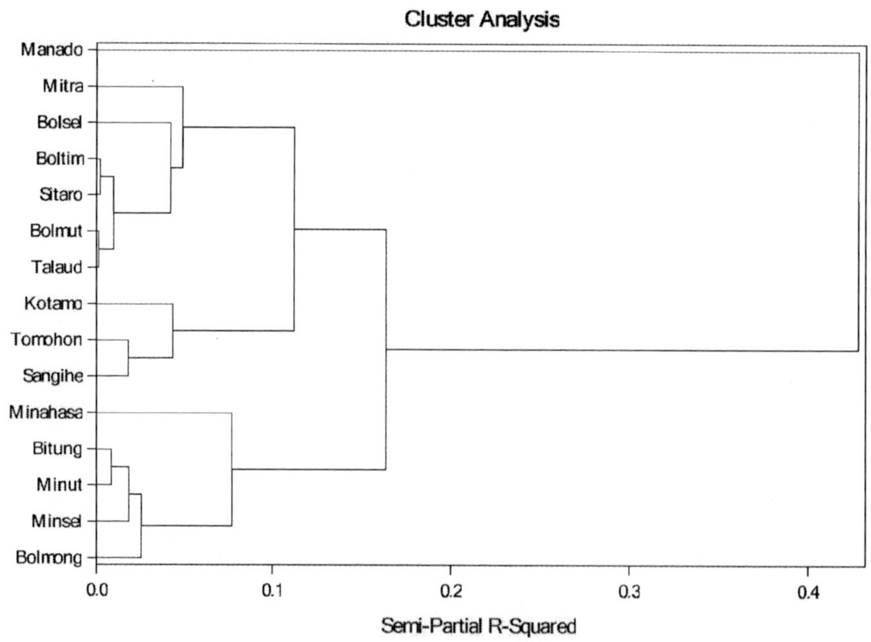

Figure 3. Dendogram used ward method

Based on the RMSSTD value obtained in Table 1,2,3 for the number of clusters 4 the smallest value at the average linkage is 1.9961. For the number of clusters 3 the smallest RMSSTD value is the ward method of 2.0937. Figure 2 and Figure 3 show the dendogram grouping using the centroid method and the ward method. In grouping using the ward method for 3 clusters, cluster 1 namely bolmong district, minsel, minut, bitung city, minahasa district. Cluster 2 is Sangihe District, Tomohon City, Kotamobagu City, Talaud District, Bolmut, Sitaro, Boltim, Bolsel, Mitra, and Cluster 3 only in Manado City.

4. Conclusion

A good method to use in the case of poverty data factor is the ward method based on the small RMSSTD value for the number of clusters selected by 3 clusters compared to other methods.

5. Acknowledgments

Thank you to the Sam Ratulangi University Manado ("Lembaga Penelitian dan Pengabdian Masyarakat"), for funding this research.

References

[1] A. M. Astel, L. Chepanova, and V. Simeonov. 2011. Soil contamination interpretation by the use of monitoring data analysis, *Water. Air. Soil Pollut.*, 216, 1–4, 375–390.
[2] L. Ke, L. Fan, and X. Kunqing. 2007. An efficient high dimensional cluster method and its application in global climate sets, *Data Sci. J.*, 6, SUPPL., S690–S697.
[3] H. Sun, M. Zhu, and F. He. 2013 . Cluster Analysis on PEAD Using SUE Model with Quarterly Data, *iBusiness*, 5, 31–34.
[4] M. Zhang, J. Zhang, Q. Gao, Y. Liu, and F. Lu. 2015. Evaluation Procedure for Quality Consistency of Generic Nifedipine Extended-Release Tablets Based on the Impurity Profile, *Am. J. Anal. Chem.*, 6, 776–785.
[5] C. Suhaeni, A. Kurnia, and Ristiyanti, 2018. Perbandingan Hasil Pengelompokan menggunakan Analisis Cluster Berhirarki , K-Means Cluster , dan Cluster Ensemble (Studi Kasus Data Indikator Pelayanan Kesehatan Ibu Hamil), *J. media infotama*, 14, 1, 31–38.
[6] A. C. Rencher and W. F. Christensen, 2012. *Methods of Multivariate Analysis 3rd edition*. New Jersey: John Wiley & Sons, Inc.
[7] R. a. Johnson and D. W. Wichern, 2007. *Applied multivariate statistical analysis sixth edition*. New Jersey: Pearson Education, Inc.
[8] S. Wierzchon and M. Klopotek. 2018. *Modern Algorithms of Cluster Analysis*. Cham: Springer.
[9] SAS Institute Inc. 2017. *SAS/STAT® 14.3 User's Guide*. Cary, NC: SAS Institute Inc.

IOP Conf. Series: Materials Science and Engineering **567** (2019) 012049 doi:10.1088/1757-899X/567/1/012049

Prediction of foreign tourist visits to Indonesia using moving average and exponential smoothing methods and acceptance of admission analysis

E Lesmana[1], A Supriatna[1] , Riaman[1]

[1]Padjadjaran University, Bandung, Indonesia

Corresponding author : eman.lesmana@unpad.ac.id

Abstract. This paper discusses the foreign exchange analysis of foreign tourists based on the daily expenditure and the country of origin concerned. Furthermore, there will be presented statistical data analysis on the length of stay and prediction of foreign tourist arrivals to Indonesia based on the country of origin of tourists using the method of moving average and exponential smoothing. The calculation results show that the country's foreign exchange (revenue) from the tourism sector compared with the 11 largest export commodities ranked the highest in the period of 2011 to 2015 with an average of 10,224.24 in millions of USD. Based on the calculation of foreign tourist arrivals to Indonesia for the period of 2002 to 2016 an average of 7,041,315 people per year and an average growth rate of 8.55% per year for the last five years.

1. Introduction

The national tourism sector is now a new belle for national development. Foreign exchange contributions and employment in this sector are very significant for foreign exchange. In fact, it is estimated that in 2019 it has outperformed foreign exchange earnings from the palm oil industry (CPO). Foreign exchange from the tourism sector in 2016 amounting to US $ 13.568 billion was in second place after CPO of US $ 15.965 billion. In 2015, foreign exchange from the tourism sector amounted to US $ 12.225 billion or was in the fourth position under Oil and Gas at US $ 18.574 billion, CPO at US $ 16.427 billion, and coal at US $ 14.717 billion. The acquisition of state foreign exchange from the tourism sector since 2016 has outperformed revenues from oil and gas and is below the income from CPO. Based on the results of World Bank research, the tourism sector is the easiest contributor to foreign exchange and gross domestic income (GDP) of a country. The reason is that the derivative impact of investment in the tourism sector on GDP is indeed very large [1].

2. Visit of Foreign Tourists

The definition of foreign tourists in accordance with the recommendations of the UNWTO (United Nations World Tourism Organization) is every person who travels to a country outside the country of residence less than one year for business, vacation or other personal purposes. International tourists are divided into two categories, namely tourists (tourist) and travelers (excursionist). Tourists are visitors who live longer than 24 hours but less than 12 months. While travelers are visitors who live less than 24 hours.

3. Methods

The research methodology used in this paper uses descriptive quantitative methods. This paper describes systematically the facts and characteristics of foreign tourist visits data from country of origin, length of stay and foreign exchange receipts. To predict foreign tourist arrivals, the moving avrerage method and exponential smoothing are used.

3.1. Moving average method

This method uses all the data that has been observed as a basis for determining the average value which is then used as a prediction of future conditions. The mathematical models used are as follows: [2]

$$F_{t+1} = \frac{\sum_{t=1}^{T} X_t}{T} \tag{1}$$

With :
T = observation period
F_{t+1} = forecasting value for (t + 1) period
X_t = observation value for t period

3.2 Exponential smoothing method

In general, the data obtained in observations have different importance values when viewed from the time of getting them. In the exponential smoothing method the data obtained is now given greater weight than the data obtained in the past. The weighting of the data is given exponentially. The mathematical model used to predict future conditions with the Single Smoothing exponential method is as follows: [2]

$$F_{t+1} = \alpha X_t + (1-\alpha) F_t \tag{2}$$

with :
X_t = data at t periode
F_t = prediction for t periode
F_{t+1} = prediksi for t+1 periode
α = exsponential constant (1/N)
N = periode of observation

3.3 Data Collection Method

Data collection methods in this paper use secondary data obtained from the Central Statistics Agency (BPS). Data on international tourists includes all visitors arriving in Indonesia through airports and ports. There are 122 doors for the arrival of foreign tourists spread throughout the country of Indonesia. The four main doors for foreign tourist arrivals are through Jakarta's Soekarno-Hatta Airport, Bali's Ngurah Rai Airport, Kuala Namu North Sumatra, and Batam Riau Islands .

3.4 Data Analysis Method

After the data is collected then plotting the data is done to find out past data patterns. Furthermore, it is also calculated the size of descriptive statistics to analyze the data of the number of foreign tourists' visits such as the mean (mean), standard deviation, average absolute deviation as follows: [Sudjana, 2005]

a. Average Value

$$\overline{X} = \sum_{i=1}^{n} \frac{X_i}{n} \tag{3}$$

b. Standard deviation

$$SD = \sqrt{\frac{1}{n-1}\sum_{i=1}^{n}(X_i - \overline{X})^2} \quad (4)$$

c. MAD (*mean absolut deviation*)

$$MAD = \frac{1}{n}\sum_{i=1}^{n}\left|X_i - \overline{X}\right| \quad (5)$$

3.5 Size of Accuracy of Forecasting

To analyze errors in the prediction of the number of foreign tourists used the following statistical measures: [2]

1. *Error* (e_i)

$$e_i = X_i - F_i \quad (6)$$

2. *Mean error* (ME)

$$ME = \frac{1}{n}\sum_{i=1}^{n} e_i \quad (7)$$

3. *Mean Absolute Error* (MAE)

$$MAE = \frac{1}{n}\sum_{i=1}^{n}\left|e_i\right| \quad (8)$$

4. *Mean Squared Error* (MSE)

$$MSE = \frac{1}{n}\sum_{i=1}^{n}(e_i)^2 \quad (9)$$

with :
e_i = error in periode i
X_i = actual data in periode i
F_i = forecast value in periode i
n = number of data

4. Results and Discussion

Data on foreign tourist visits to Indonesia from 2002 to 2016 are presented in table 1.

Table 1: Data of Foreign Tourist Visits

Year	Tourist	Year	Tourist
2002	5 033 400	2010	7 002 944
2003	4 467 021	2011	7 649 731
2004	5 321 165	2012	8 044 462
2005	5 002 101	2013	8 802 129
2006	4 871 351	2014	9 435 411
2007	5 505 759	2015	10 406 759
2008	6 234 497	2016	11 519 275
2009	6 323 730		

Source: Badan Pusat Statistik [2017]

The results of plotting data on foreign tourists visiting from 2002 to 2016 are presented in Figure 1.

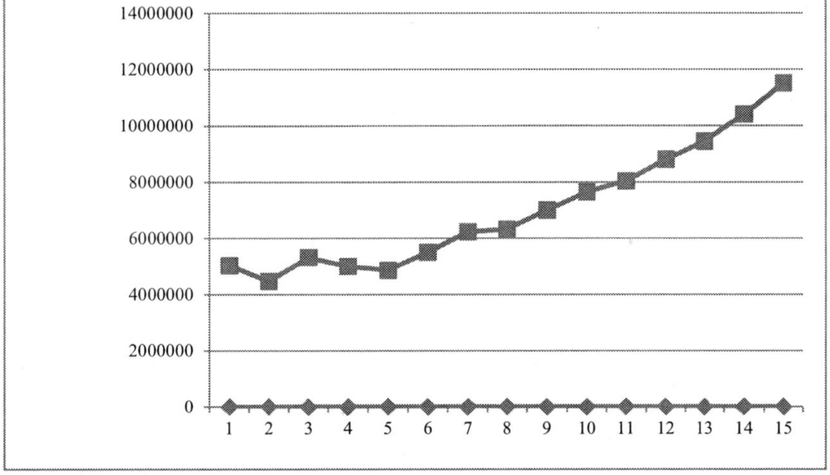

Figure 1 Results of data plotting

Based on Table 1 it can be seen that foreign tourist arrivals to Indonesia have increased with an average visit of 7,041,316 people per year. Increased foreign tourist arrivals to Indonesia in the last five years can be seen in table 2 below

Table 2 Increase in foreign tourists visiting Indonesia

Year	Tourist	Increase	% increase
2012	8 044 462	394 731	0,05
2013	8 802 129	757 667	0,09
2014	9 435 411	633 282	0,07
2015	10 406 759	971 348	0,10
2016	11 519 275	1112 516	0,11

From table 2 it can be seen that the increase in foreign tourists visiting Indonesia in the last five years continues to increase by an average of 8.55% per year. Monthly foreign tourist visits data from January 2015 to August 2017 are presented in the following table 3 :

Tabel 3 Data of Tourist Visits per Mounth

Mounth	Tourist	Mounth	Tourist
Jan 2015	785973	Mei 2016	915206
Feb 2015	843928	Jun 2016	857651
Mar 2015	841071	Jul 2016	1032741
Apr 2015	801873	Ags 2016	1031986
May 2015	852388	Sep 2016	1006653
Jun 2015	872385	Okt 2016	1040651
Jul 2015	877584	Nov 2016	1002333
Ags 2015	911704	Des 2016	1113328
Sep 2015	920128	Jan 2017	1032930
Okt 2015	877798	Feb 2017	957583
Nov 2015	835408	Mar 2017	1066588
Des 2015	986519	Apr 2017	1142180
Jan 2016	814303	May 2017	1150067

Feb 2016	888309	Jun 2017	1111616
Mar 2016	915019	Jul 2017	1379661
Apr 2016	901095	Ags 2017	1404664

Source : Badan Pusat Statistik 2017

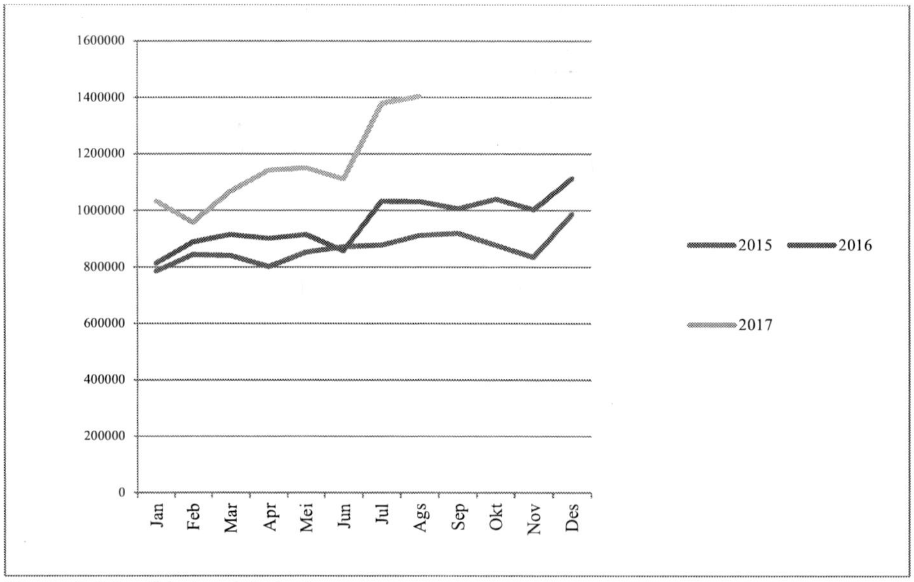

Figure 2 Monthly tourist chart

4.1 Results of prediction of foreign tourists visit

The results of the prediction of foreign tourist arrivals to Indonesia using the 4-monthly moving average method and the exponential smoothing method can be seen in Table 3 and Table 4 as follows: [Makrridakis, 1983] [POM-QM 3.0]

Table 4 Results of predictions of foreign tourist visits by using Moving Average (4)

Mounth	Tourist	Mounth	Tourist
Jan 2015		Mei 2016	879682
Feb 2015		Jun 2016	904907
Mar 2015		Jul 2016	897243
Apr 2015		Ags 2016	926673
Mei 2015	818211	Sep 2016	959396
Jun 2015	834815	Okt 2016	982258
Jul 2015	841929	Nov 2016	1028008
Ags 2015	851058	Des 2016	1020406
Sep 2015	878515	Jan 2017	1040741
Okt 2015	895450	Feb 2017	1047311
Nov 2015	896804	Mar 2017	1026544
Des 2015	886260	Apr 2017	1042607
Jan 2016	904963	Mei 2017	1049820
Feb 2016	878507	Jun 2017	1079105
Mar 2016	881135	Jul 2017	1117613
Apr 2016	901038	Ags 2017	1195881
		Sep 2017	1261502

The results of the prediction of tourist visits to Indonesia in September 2017 using the four monthly moving average method is 1,261,502 people. With the help of the 3.0 version of POM (Production and operation Management) software, ME (mean error) was obtained 43,985, MAE (mean absolute error) 59,749, and MAPE (mean absolute percentage error) 0,06. [POM-QM 3.0]
Complete results of predictions of foreign tourist arrivals using exponential smoothing method with the help of POM (Production and operation Management) software version 3.0 can be seen in Table 5. Predictions of foreign tourists for September 2017 were 1,220,247 obtained by ME (mean error) 56,035, MAE (mean absolute error) 71,277, and MAPE (mean absolute percentage error) 0,07. [POM-QM 3.0] [Untung, 1991]

Table 5 Results of predictions of foreign tourists visiting the exponential smoothing method ($\alpha = 0.25$). [2] [POM-QM 3.0]

Mounth	Tourist	Mounth	Tourist
Jan 2015		Mei 2016	891183
Feb 2015	785973	Jun 2016	897189
Mar 2015	800462	Jul 2016	887304
Apr 2015	810614	Ags 2016	923664
Mei 2015	808429	Sep 2016	950744
Jun 2015	819418	Okt 2016	964721
Jul 2015	832660	Nov 2016	983704
Ags 2015	843891	Des 2016	988361
Sep 2015	860844	Jan 2017	1019603
Okt 2015	875665	Feb 2017	1022935
Nov 2015	876198	Mar 2017	1006597
Des 2015	866001	Apr 2017	1021595
Jan 2016	896130	Mei 2017	1051741
Feb 2016	875674	Jun 2017	1076322
Mar 2016	878832	Jul 2017	1085146
Apr 2016	887879	Ags 2017	1158775
		Sep 2017	1220247

Table 6 Error Analysis Results Prediction of foreign tourists visiting Indonesia

Statistik Descritif	*Moving Average*	*Eksponentias Smoothing*
Mean Error	43.985	56.053
MAE	59.749	71.277
MAPE	0.06	0.07

From table 6 shows that the moving average method has a smaller error analysis than the exponential smoothing method. Thus the results of the prediction of foreign tourist arrivals to Indonesia with the moving average method have a better level of accuracy than the exponential smoothing method.

4.2 Foreign Exchange Analysis of foreign tourists visiting
Foreign exchange of state revenue from the sector of foreign tourist visits to Indonesia from 2010 to 2014 can be seen in table 7 below.

Table 7 Foreign Exchange Revenue (in millions of USD)

COUNTRY OF ORIGIN	2010	2011	2012	2013	2014
Brunei	34,83	36,43	26,53	17,31	18,29
Malaysia	864,34	930,85	972,1	1.002,53	1.053,89
Philipine	161,97	175,90	195,5	206,31	212,90
Singepore	927,97	1.054,21	1.000,	1.049,41	1.145,83
Thailand	97,46	127,92	144,3	139,10	127,13
Hong Kong	75,39	91,53	91,19	119,47	125,24
India	147,29	175,13	171,4	221,62	254,95
Japan	409,87	419,80	477,8	558,85	597,71
South of Korea	251,05	295,82	290,3	381,83	420,89
Pakistan	6,72	5,06	12,97	6,15	7,42
Bangladesh	10,49	13,71	13,49	9,53	8,74
Sri Lanka	7,65	6,73	11,75	9,33	11,04
Taiwan	184,76	188,15	204,5	231,09	254,66
China	433,38	520,61	714,5	810,79	981,46
Arab Saudi	170,03	211,06	135,6	178,73	198,20
Austria	24,26	24,01	28,88	31,74	35,66
Belgia	40,38	32,73	40,20	52,12	48,08
Denmark	26,99	29,48	38,45	39,48	37,27
Perancis	244,25	229,33	273,63	316,26	345,96
Germany	217,38	229,41	245,07	251,54	312,27
Italia	49,63	68,66	70,76	105,19	101,17
Belanda	269,20	263,02	243,16	280,64	320,03
Spanyol	47,44	35,99	46,34	61,07	75,20
Swedia	40,94	48,51	37,24	51,76	49,78
Swiss	55,46	61,11	64,75	69,97	93,50
England	277,14	269,61	321,92	349,20	401,48
Finlandia	23,07	21,80	21,89	34,07	28,79
Norwegia	21,22	27,78	33,20	29,32	26,60
Rusia	136,80	178,64	170,00	191,97	176,91
U S A	252,23	317,28	312,55	363,91	406,57
Canada	67,70	80,97	80,95	96,35	122,12
Central of America	3,20	4,28	2,22	1,82	5,80
South of America	42,64	41,00	46,31	49,27	54,05
Australia	1.171,87	1.502,10	1.452,31	1.470,89	1.802,8
New Zealand	43,25	60,27	93,56	82,80	120,73
Egypt	4,02	5,18	5,17	6,35	10,80
Others	742,39	754,56	1.005,85	1.144,32	1.135,5
T O T A L	**7.603,45**	**8.554,39**	**9.120,85**	**10.054,15**	**11.166,13**

Sumber : Badan Pusat Statistik 2017

IOP Conf. Series: Materials Science and Engineering **567** (2019) 012049 doi:10.1088/1757-899X/567/1/012049

From table 7 it can be seen that tourists from Australia are the biggest foreign exchange earners during the period 2010 to 2014.

4.3 Contribution of tourism to foreign exchange revenue of the Republic of Indonesia
Indonesia's foreign exchange rating from 2011 to 2015 can be seen in table 8

Table 8 Foreign Exchange Acceptance Ranking

Rank	2011		2012		2013		2014		2015	
	Commodities	Value (million USD)	Commodities	Value (million USD)	Commodities	Value (million USD)	Commodities	Nilai (million USD)	Commodities	Nilai (million USD)
1	Minyak & gas bumi	41,477. 10	Minyak & gas bumi	32,633.20	Minyak & gas bumi	18,552.10	Minyak & gas bumi	30,318.80	Minyak & gas bumi	36,977.00
2	Batu bara	22,221.80	Batu bara	24,501.40	Batu bara	15,943.00	Batu bara	20,819.30	Batu bara	26,166.30
3	Minyak sawit	17,261.30	Minyak kelapa sawit	15,839.10	Minyak kelapa sawit	15,385.20	Minyak kelapa sawit	17,464.90	Minyak kelapa sawit	18,845.00
4	Karet	14,258.20	***Pariwisata***	***9,120,85***	***Pariwisata***	***10,054,15***	***Pariwisata***	***11,166.13***	Pariwisata	12,225,89
5	Pariwisata	8,554.39	Karet olahan	9,316.60	Pakaian jadi	7,371.90	Pakaian jadi	7,450.90	Pakaian jadi	***9,120.85***
6	Pakaian jadi	7,801.50	Pakaian jadi	7,501.00	Makanan olahan	6,456.30	Karet olahan	7,021.70	Makanan olahan	7,304.70
7	Alat listrik	7,364.30	Alat listrik	6,418.60	Karet olahan	5,842.00	Makanan olahan	6,486.80	Karet olahan	6,481.90
8	Tekstil	5,563.30	Makanan olahan	5,434.80	Alat listrik	5,644.80	Alat listrik	6,259.10	Alat listrik	5,278.10
9	Makanan olahan	4,802.10	Tekstil	5,293.60	Tekstil	4,996.00	Tekstil	5,379.70	Tekstil	5,135.60
10	Bahan kimia	4,630.00	Kertas dan barang dr kertas	3,802.20	Kayu olahan	3,815.80	Kayu olahan	3,914.10	Kayu olahan	3,972.00
11	Kertas	4,214.40	Kayu olahan	3,514.50	Kertas dan barang dr kertas	3,605.50	Bahan kimia	3,853.70	Kertas dan barang dr kertas	3,636.30
12	Kayu olahan	3,288,90	Bahan kimia	3,501.60	Bahan kimia	2,807.60	Kertas dan barang dr kertas	3,780.00	Bahan kimia	3,337.70

Source : Badan Pusat Statistika [2017]

From table 8, it can be seen that the tourism sector contributed to Indonesia's increasing foreign exchange revenue and ranked fourth in 2015 with an average of .10,224.24 million USD per year during the period 2011 to 2015.

4.4 Analysis of the length of stay of tourists
The average length of stay of foreign tourists in Indonesia by country of origin from 2009 to 2014 is presented in table 9 below.

Table 9 Average length of stay of tourists

Negara Asal	Tahun 2009	Tahun 2010	Tahun 2011	Tahun 2012	Tahun 2013	Tahun 2014
Brunei	6,38	5,23	5,15	5,43	4,98	7,29
Malaysia	5,28	5,03	5,23	5,26	4,82	5,30
Filipina	7,92	7,09	6,11	6,38	5,45	5,46
Singapura	4,51	4,67	4,44	4,27	4,17	4,16
Thailand	5,77	5,78	6,25	6,32	6,33	5,88
Hong Kong	5,90	5,85	5,92	5,59	5,81	5,96
India	8,89	8,68	7,90	7,18	7,39	7,75
Jepang	6,50	6,14	6,37	6,43	6,86	6,75
Korea Selatan	5,49	5,93	5,93	5,57	6,40	6,35
Pakistan	7,50	11,69	4,68	7,94	7,80	7,41
Bangladesh	11,05	11,42	9,08	6,82	8,80	5,71
Sri Lanka	6,44	11,20	5,63	7,52	6,00	6,88
Taiwan	5,79	5,84	5,63	6,12	5,78	5,89
Cina	6,21	5,98	6,01	6,31	6,21	6,28
Arab Saudi	10,31	9,37	9,14	8,69	7,69	7,42
Austria	12,58	12,16	13,03	11,36	11,93	13,31
Belgia	12,60	13,68	10,74	12,33	12,15	14,90
Denmark	12,51	13,24	15,20	13,49	16,31	9,89
Perancis	11,85	12,59	13,30	12,06	13,06	12,92
Jerman	12,93	12,58	13,55	12,59	12,61	12,47
Italia	10,95	10,69	11,62	10,18	11,34	10,04
Belanda	14,10	15,52	13,97	14,59	13,39	15,14
Spanyol	12,65	12,18	10,32	11,30	11,89	15,26
Portugal	9,69	12,63	10,77	10,25	11,13	9,45
Swedia	10,21	18,68	14,33	15,14	12,83	14,53
Swiss	13,78	14,91	13,83	12,68	13,97	15,85
Inggris	10,75	11,76	10,26	11,37	10,47	10,14
Finlandia	19,21	12,20	13,15	13,17	12,21	11,41
Norwegia	11,80	11,06	13,38	11,61	12,89	9,20
Rusia	9,84	11,38	13,26	12,00	12,67	13,03
Amerika Serikat	11,11	10,55	10,98	10,41	10,49	11,07
Kanada	12,13	11,89	11,08	11,36	11,94	14,02
Amerika Tengah	7,63	11,50	10,00	6,00	9,75	12,58
Amerika Selatan	8,95	15,88	11,59	11,68	10,89	
Australia	10,22	10,56	9,47	9,26	9,62	9,08
Selandia Baru	10,06	10,16	10,97	11,20	10,10	10,24
Mesir	8,27	6,38	9,64	12,75	5,75	5,00
Lainnya	9,29	9,37	10,77	9,50	8,26	8,79
RATA - RATA	**7,69**	**8,04**	**7,84**	**7,70**	**7,65**	**7,66**

Source : Badan Pusat Statistik & pessenger exit survey

From table 9 it can be seen that the average length of stay of foreign tourists in Indonesia is 8 days

4.5 Visit of Tourists in 2016-2017
Foreign tourist visits during 2016 to August 2017 are presented in table 10 below.

Table 10 Foreign Tourists Visit in 2016 - 2017

Month	Year 2016	Year 2017	Growth
January	814.303	1.032.930	26.8 %
February	888.309	957.583	7.8 %
March	915.019	1.066.588	16.6 %
April	901.095	1.142.180	26.7 %
May	915.206	1.150.067	25.7 %
June	857.651	1.111.616	29.6 %
July	1.032.741	1.379.661	33.6 %
August	1.031.986	1.404.664	36.1 %
September	1.006.653		
October	1.040.651		
November	1.002.333		
Desember	1.113.328		

Source : Badan Pusat Statistika [2017]

4.6 Choice of Tourist Destinations in Wisman
Indonesia has a tourist attraction that has the attraction of beauty and uniqueness of each. Of the many destinations there are eight locations that are a favorite place for foreign tourists. The eight tourist destinations are as follows:

Table 11 Favorite tourist destinations for foreign tourists

NO	TOURISM DESTINATION	DESCRIPTION
1	Bali Island	Bali is the most sought after tourist destination by foreign tourists due to the beauty of its beaches such as Kute Beach, Pandawa Beach, Jimbaran, Nusa Penida, Nusa Lembongan, Siring, Tanah Lot, and unique cultural customs and various dances and pure etc.
2	Komodo National Park East Nusa Tenggara	The national park located in West Manggarai district, NTT has the beauty of Kanawa beach, Komodo animals, trekking to Gili Lawa and Padar Island
3	Bromo Tengger National Park, East Java	It has the beauty of the sunrise on the penanjakan, the crater of Mount Bromo, the desert savanna, and the mountain Semeru
4	Kelimutu Flores Mountain National Park	The beauty of the sunrise and the three-color lake and the hospitality of the villagers of Moni
5	Raja Ampat	Unique natural landscapes and beautiful underwater nature and the hospitality of the Wayag, Waigeo, Waisai and Misool residents
6	Lombok	Lombok Island is surrounded by beaches that have beautiful panoramas such as Sengigi, Gili Trawangan, Tanjung Aan, Mawun Beach, Mawi beach and Selong Belanak
7	Jogyakarta	Has many beautiful beaches, majestic temples, Borobudur temple, Prambanan temple, Ratu Boko, Yogyakarta palace and Malioboro
8	Manado Bunaken National Park North Sulawesi	Located in Manado, North Sulawesi is famous for the beauty of the underwater park. Other tourist destinations are Manado Tua Island, Siladen Island, Nain Island and Mantehage Island.

Source : Government Tourism Office

5. Conclusion

Tourism is a potential commodity in its role in increasing Indonesia's foreign exchange income. This is evident in the period of 2011 to 2015, the tourism sector ranks continuously increasing. The tourism sector contributes to Indonesia's increasing foreign exchange revenue and ranks fourth in 2015 with an average of 10.224.24 million USD per year. Based on the results of the calculation of foreign tourist visits to Indonesia for the period 2002 to 2016 an average of 7,041,315 people per year and an average growth rate of 8.55% per year for the last five years.

Acknowledgments

This research was funded by Competency-Based Research or Competency-Based Research from the Indonesian Ministry of Research and Technology for 2017 contract number 718 / UN6.3.1 / PL / 2017.

References

[1] Badan Pusat Statistik, 2017
https : // www.bps.go.id/subjek/view/id/16# subjekviewTab2
[2] Makridakis, S.1983. *Forecasting Methods and Applications*. Jhon Wiley & Sons
[3] *POM-QM for Windows (version 3). Software.informer.com/3.0*
[4] S.A. Untung. 1991. *Metode dan Aplikasi Peramalan.* Erlangga
[5] Sudjana, 2005. *Metoda Statistika*. Tarsito Bandung.

Institute of Physics
Dirac House, Temple Back
Bristol BS1 6BE UK

ISSN: 1757-8981
ISBN 978-1-5108-9393-1

CURRAN ASSOCIATES INC.
proceedings.com